The Comparative Ethology and Evolution of the Sand Wasps

HOWARD E. EVANS

Harvard University Press · Cambridge, Massachusetts · 1966

© Copyright 1966 by the President and Fellows of Harvard College

All rights reserved

Distributed in Great Britain by Oxford University Press, London

Library of Congress Catalog Card Number 66-18245

Printed in the United States of America

Acknowledgments

These studies were supported by successive grants from the National Science Foundation, extending from 1952 through 1964. A portion of the work was conducted at the Archbold Biological Station, Lake Placid, Florida, and a portion at the Jackson Hole Biological Research Station, Moran, Wyoming. I wish to express my thanks to the directors and staffs of these stations for the many kindnesses rendered.

A number of persons have assisted me from time to time in various aspects of the field work. These include particularly C. S. Lin, C. M. Yoshimoto, and J. E. Gillaspy. Frank E. Kurczewski has generously turned over to me all his field notes on members of the subfamily Nyssoninae, and K. V. Krombein and M. A. Cazier have sent me copies of their field data prior to publication.

In the identification of the wasps and of their hymenopterous parasites I have been greatly assisted by K. V. Krombein and by R. M. Bohart. A great many persons have been involved in the identification of prey and of the dipterous parasites. The majority of the Hemiptera were determined by J. P. Kramer, H. Ruckes, L. M. Russell, and D. A. Young; the Lepidoptera by H. W. Capps, J. F. G. Clarke, and W. T. M. Forbes; the Diptera by R. H. Foote, C. H. Martin, R. H. Painter, L. L. Pechuman, E. G. Reinhard, C. W. Sabrosky, G. Steyskal, H. W. Weems, and W. W. Wirth. I am deeply indebted to all of these specialists for their time and consideration.

I particularly wish to thank Mrs. Morna MacLeod for typing the tables and parts of the manuscript, in some cases through several drafts. My wife, Mary Alice Evans, assisted with the bibliography and provided encouragement at times when it was much needed.

Acknowledgment is made to the William Morton Wheeler Fund of the Museum of Comparative Zoology, contributed by Dr. Caryl P. Haskins, for assuming part of the cost of publication.

H. E. E.

Contents

Chapter I. An introduction to the nyssonine wasps — 1
- A. Classification of the Nyssoninae   3
- B. Structure of the nyssonine wasps   5
- C. Ethological considerations   12

Chapter II. *Alysson* and *Didineis*, two genera of hygrophiles — 16
- A. *Alysson melleus* Say   17
- B. Ethology of other species of *Alysson*   28
- C. Summary of ethology of the species of *Alysson*   29
- D. Ethology of the genus *Didineis*   30

Chapter III. *Gorytes* and some closely related genera — 31
- A. Genus *Hoplisoides* Gribodo   33
- B. *Hoplisoides nebulosus* (Packard)   34
- C. *Hoplisoides spilographus* (Handlirsch)   46
- D. *Hoplisoides costalis* (Cresson)   49
- E. *Hoplisoides tricolor* (Cresson)   53
- F. Ethology of other species of *Hoplisoides*   55
- G. Genus *Gorytes* Latreille   57
- H. *Gorytes canaliculatus* (Packard)   58
- I. *Gorytes simillimus* Smith   66
- J. Nesting behavior of other species of *Gorytes*   67
- K. Ethology of some genera closely related to *Gorytes*   68
- L. Genus *Clitemnestra* Spinola   73
- M. Genus *Ochleroptera* Holmberg   74
- N. *Ochleroptera bipunctata* (Say)   75
- O. Genus *Ammatomus* Costa   78
- P. *Ammatomus moneduloides* (Packard)   78
- Q. Summary of the ethology of *Gorytes* and related genera   80

Chapter IV. The Nyssonini, a complex of cleptoparasites   83

    A. Ethology of the species of *Nysson*   85
    B. Ethology of other genera of Nyssonini   89

Chapter V. The cicada killers, *Sphecius* and *Exeirus*   91

    A. Genus *Sphecius* Dahlbom   92
    B. *Sphecius speciosus* (Drury)   92
    C. Ethology of other species of *Sphecius*   110
    D. Genus *Exeirus* Shuckard   111
    E. *Exeirus lateritius* (Shuckard)   112

Chapter VI. The stizine wasps: *Stizus, Stizoides,* and *Bembecinus*   116

    A. Genus *Stizus* Latreille   117
    B. *Stizus pulcherrimus* Smith   118
    C. *Stizus fasciatus* (Fabricius)   124
    D. Ethology of other species of *Stizus*   125
    E. Genus *Stizoides* Guérin   127
    F. *Stizoides unicinctus* (Say)   127
    G. General remarks on the ethology of *Stizoides*   131
    H. Genus *Bembecinus* Costa   132

Chapter VII. *Bicyrtes,* a genus of stinkbug-hunters   144

    A. *Bicyrtes quadrifasciata* (Say)   146
    B. *Bicyrtes ventralis* (Say)   160
    C. *Bicyrtes fodiens* (Handlirsch)   167
    D. Ethology of other species of *Bicyrtes*   170
    E. Summary of the ethology of the species of *Bicyrtes*   174

Chapter VIII. Three genera of Bembicini with recessed ocelli: *Stictiella, Glenostictia,* and *Steniolia*   176

    A. Genus *Stictiella* Parker   177
    B. Ethology of the species of *Stictiella*   179
    C. Genus *Glenostictia* Gillaspy   185
    D. Ethology of the species of *Glenostictia*   186
    E. *Glenostictia pulla* (Handlirsch)   186
    F. *Glenostictia gilva* Gillaspy   188
    G. *Glenostictia scitula* (Fox)   189
    H. Genus *Steniolia* Say   205

Contents

    I. Ethology of the species of *Steniolia*   207
    J. Summary of the ethology of the genera with recessed ocelli   219

Chapter IX. *Stictia* and some South American genera   222

    A. Genus *Stictia* Illiger   223
    B. *Stictia carolina* (Fabricius)   223
    C. *Stictia signata* (Linnaeus)   243
    D. *Stictia vivida* (Handlirsch)   247
    E. *Stictia heros* (Fabricius)   252
    F. Ethology of other species of *Stictia*   253
    G. Genus *Zyzzyx* Pate   253
    H. *Zyzzyx chilensis* (Eschscholz)   254
    I. Genus *Rubrica* Parker   257
    J. *Rubrica surinamensis* (DeGeer)   258
    K. *Rubrica denticornis* (Handlirsch)   261
    L. *Rubrica gravida* (Handlirsch)   262
    M. Genus *Editha* Parker   264
    N. *Editha adonis* (Handlirsch)   265
    O. Summary of behavior of *Stictia*, *Zyzzyx*, *Rubrica*, and *Editha*   265

Chapter X. Some generalized species of *Bembix*   267

    A. *Bembix amoena* Handlirsch   269
    B. *Bembix sayi* Cresson   288
    C. *Bembix nubilipennis* Cresson   298
    D. *Bembix truncata* Handlirsch   308
    E. *Bembix cameroni* Rohwer   310
    F. *Bembix spinolae* Lepeletier   311
    G. *Bembix belfragei* Cresson   315
    H. *Bembix u-scripta* Fox   317

Chapter XI. Some specialized species of *Bembix*   322

    A. *Bembix texana* Cresson   323
    B. *Bembix troglodytes* Handlirsch   335
    C. *Bembix multipicta* Smith   337
    D. *Bembix pruinosa* Fox   345
    E. General remarks on the ethology of *Bembix*   349

Chapter XII. *Microbembex:* a genus of scavengers   360

    A. *Microbembex monodonta* (Say)   361
    B. *Microbembex ciliata* (Fabricius)   388
    C. Ethology of other species of *Microbembex*   390
    D. Summary of the ethology of the species of *Microbembex*   390

Chapter XIII. Fossil history, distribution, and comparative morphology of the Nyssoninae — 392

    A. Fossil record of the Nyssoninae   392
    B. Distribution of the genera of Nyssoninae   398
    C. Comparative morphology of the Nyssoninae   401
    D. Further discussion of important structural features   407
    E. Major trends in structural modification   415

Chapter XIV. Comparative ethology of the Nyssoninae — 421

    A. Ecology and general features of adult behavior   421
    B. Nesting behavior   433
    C. Hunting and provisioning   442
    D. Cleptoparasitism   453
    E. Cocoon spinning   455
    F. Natural enemies   467

Chapter XV. The evolution of the behavior of sand wasps — 473

    A. Behavior and the classification of the Nyssoninae   473
    B. Phyletic trends in the behavior of sand wasps   477
    C. Examination of specific behavior patterns   480
    D. Some general considerations regarding the evolution of behavior   495
    E. A final look at the sand wasps   500

Bibliography — 507

Index — 521

Tables

1. Nest data for *Alysson melleus*	22
2. Prey records for *Alysson melleus*	26
3. Nest data for *Hoplisoides nebulosus*	37
4. Prey records for *Hoplisoides nebulosus*	40
5. Prey records for *Hoplisoides costalis*	51
6. Nest data for *Gorytes canaliculatus*	60
7. Prey records for *Gorytes canaliculatus*	63
8. Prey records for *Ochleroptera bipunctata*	77
9. Host relationships of the species of *Nysson*	86
10. Prey records for *Sphecius speciosus*	104
11. Nest data for several species of *Bembecinus*	138
12. Nest data for *Bicyrtes quadrifasciata*	150
13. Prey records for *Bicyrtes quadrifasciata*	155
14. Nest data for *Bicyrtes ventralis*	162
15. Prey records for *Bicyrtes ventralis*	166
16. Nest data for species of *Stictiella*	181
17. Nest data for *Glenostictia scitula*	198
18. Prey records for *Glenostictia scitula*	201
19. Nest data for species of *Steniolia*	217
20. Families of flies used as prey by species of *Steniolia*	217
21. Nest data for *Stictia carolina*	232
22. Selected records of outer closure during provisioning in *Stictia carolina*	235
23. Prey records for *Stictia carolina*	237
24. Data on final closure in *Stictia carolina*	241
25. Nest data for *Bembix amoena*	277
26. Prey records for *Bembix amoena*	282
27. Nest data for *Bembix sayi*	291
28. Prey records for *Bembix sayi*	293
29. Duration of several nests of *Bembix sayi*	295

30. Nest data for *Bembix nubilipennis*	302
31. Prey records for *Bembix nubilipennis*	305
32. Recent data on nest depth in *Bembix spinolae*	314
33. Nest data for *Bembix u-scripta*	320
34. Length of false burrows in nests of *Bembix texana*	328
35. Nest data for *Bembix texana*	329
36. Prey records for *Bembix texana*	330
37. Comparison of females of *multipicta-troglodytes* complex	338
38. Prey records for *Bembix pruinosa*	347
39. Nest data for *Microbembex monodonta*	372
40. Distribution of the genera of Nyssoninae	399
41A. Thirty characters of adult Alyssonini, Gorytini, and Nyssonini	402
41B. Thirty characters of adult Stizini and Bembicini	403
42. Ten characters of larval Nyssoninae	404
43. Comparison of some characters of the genera of Bembicini	405
44. Number of nest cells in various genera of Nyssoninae	443
45. Prey of genera of Nyssoninae	444
46A. Data on cocoons of Alyssonini, Nyssonini, Gorytini, and Stizini	462
46B. Data on cocoons of Bembicini (other than *Bembix*)	463
46C. Data on cocoons of 15 species of the genus *Bembix*	464
47. Natural enemies of the Nyssoninae	469

Figures

1. Frontal view of head of female cicada killer, *Sphecius speciosus* — 6
2. Dorsal aspect of mesosoma of *Sphecius speciosus* — 7
3. Lateral aspect of mesosoma of *Sphecius speciosus* — 7
4. Wings of *Sphecius speciosus* — 9
5. Apical portion of metasoma of male *Sphecius speciosus* — 10
6. Sternite VIII of male *Sphecius speciosus* — 11
7. Male genitalia of *Sphecius speciosus* — 11
8. Generalized unicellular nest of a sand wasp — 14
9. Head of female *Alysson melleus* — 17
10. Wings of *Alysson melleus* — 17
11. Mesosoma of *Alysson melleus* — 17
12. Male genitalia of *Alysson melleus* — 17
13. Sternite VIII of male *Alysson melleus* — 17
14. Apex of hind femur and base of tibia of female *Alysson melleus* — 17
15. Front tarsus of female *Alysson melleus* — 17
16. Middle tibia and basitarsus of female *Alysson melleus* — 17
17. View of *Alysson melleus* nesting site along Fisheating Creek, Florida — 18
18. Three typical nests of *Alysson melleus* — 21
19. Female *Alysson melleus* entering nest with prey — 24
20. Adult leafhopper bearing the egg of *Alysson melleus* — 27
21. Head of female *Gorytes simillimus* — 32
22. Wings of *Gorytes simillimus* — 32
23. Mesosoma of *Gorytes simillimus* — 32
24. Male genitalia of *Gorytes simillimus* — 32
25. Sternite VIII of male *Gorytes simillimus* — 32
26. Front tarsus of female *Gorytes simillimus* — 32
27. Middle tibia and basitarsus of female *Gorytes simillimus* — 32
28. Head of female *Hoplisoides nebulosus* — 33
29. Mesosoma of *Hoplisoides nebulosus* — 33

30.	Typical habitat of *Gorytes canaliculatus* and *Hoplisoides nebulosus*	34
31.	Three typical nests of *Hoplisoides nebulosus*	37
32.	Female *Hoplisoides nebulosus* entering nest with prey	38
33.	Adult treehopper bearing egg of *Hoplisoides nebulosus*	41
34.	*Nysson tuberculatus* entering a covered nest of *Hoplisoides nebulosus*	43
35.	Cell of *Hoplisoides nebulosus* opened to show contents	45
36.	Immature treehopper bearing egg of *Nysson daeckei*	46
37.	Nest of *Hoplisoides costalis*	50
38.	Three typical nests of *Gorytes canaliculatus*	61
39.	Immature leafhopper bearing egg of *Gorytes canaliculatus*	63
40.	Mesopleuron of *Argogorytes campestris*	69
41.	Mesopleuron of *Ammatomus moneduloides*	69
42.	Two typical nests of *Ochleroptera bipunctata*	76
43.	Head of female *Ammatomus moneduloides*	79
44.	Male genitalia of *Ammatomus moneduloides*	79
45.	Sternite VIII of male *Ammatomus moneduloides*	79
46.	Head of female *Nysson lateralis*	84
47.	Wings of *Nysson lateralis*	84
48.	Mesosoma of female *Nysson lateralis*	84
49.	Male genitalia of *Nysson lateralis*	84
50.	Sternite VIII of male *Nysson lateralis*	84
51.	Dorsal aspect of propodeum of female *Nysson lateralis*	84
52.	Front tarsus of female *Sphecius speciosus*	93
53.	Apex of titia and basal segments of tarsus of middle leg of female *Sphecius speciosus*	93
54.	Apex of tibia and basal segments of tarsus of hind leg of female *Sphecius speciosus*	93
55.	Nest entrances of *Sphecius speciosus*	99
56.	Three nests of *Sphecius speciosus*	102
57.	Three frames from motion picture film of *Sphecius speciosus* dragging prey to nest	106
58.	Adult female cicada bearing egg of *Sphecius speciosus*	108
59.	Head of female *Exeirus lateritius*	113
60.	Sternite VIII of male metasoma of *Exeirus lateritius*	113
61.	Mesosoma of female *Exeirus lateritius*	113
62.	Front tarsus of female *Exeirus lateritius*	113
63.	Head of female *Stizus brevipennis*	117
64.	Wings of *Stizus brevipennis*	117
65.	Apex of metasoma of male *Stizus brevipennis*	117
66.	Mesosoma of female *Stizus brevipennis*	117
67.	Front tarsus of female *Stizus brevipennis*	117
68.	Male genitalia of *Stizus brevipennis*	117
69.	Two nests of *Stizus pulcherrimus*	120

Figures xiii

70.	Grasshopper bearing the egg of *Stizus pulcherrimus*	122
71.	Grasshopper bearing the egg of *Stizus fasciatus*	122
72.	Head of female *Stizoides unicinctus*	127
73.	Mesosoma of female *Bembecinus neglectus*	133
74.	Male genitalia of *Bembecinus neglectus*	133
75.	Apical sternite of male *Bembecinus neglectus*	133
76.	Typical nests of three species of *Bembecinus*	138
77.	Typical position of egg in cell of *Bembecinus hungaricus*	139
78.	Typical position of egg in cell of *Bembecinus neglectus*	139
79.	*Bembecinus neglectus* female arriving at nest entrance with prey	140
80.	Head of female *Bicyrtes quadrifasciata*	145
81.	Wings of *Bicyrtes quadrifasciata*	145
82.	Mesosoma of female *Bicyrtes quadrifasciata*	145
83.	Male genitalia of *Bicyrtes quadrifasciata*	145
84.	Apical sternite of male *Bicyrtes quadrifasciata*	145
85.	Front tarsus of female *Bicyrtes quadrifasciata*	145
86.	*Bicyrtes quadrifasciata* clearing the nest entrance	149
87.	Four typical nests of *Bicyrtes quadrifasciata*	152
88.	Egg of *Bicyrtes quadrifasciata* on an immature coreid bug	158
89.	Egg of *Bicyrtes ventralis* on an immature stinkbug	158
90.	Egg of *Bicyrtes fodiens* on an adult stinkbug	158
91.	Typical nests of *Bicyrtes ventralis* and *B. fodiens*	163
92.	Female *Bicyrtes ventralis* closing nest entrance	165
93.	Female *Bicyrtes ventralis* arriving at nest entrance with prey	165
94.	Sleeping posture of *Bicyrtes capnoptera* on *Melilotus alba*	172
95.	Head of female *Stictiella formosa*	178
96.	Wings of *Stictiella formosa*	178
97.	Mesosoma of female *Stictiella formosa*	178
98.	Tibia and tarsus of middle leg of male *Stictiella formosa*	178
99.	Male genitalia of *Stictiella formosa*	178
100.	Apex of metasoma of male *Stictiella formosa*, lateral aspect	178
101.	Apex of metasoma of male *Stictiella formosa*, ventral aspect	178
102.	Sternite VIII of male *Stictiella formosa*	178
103.	Head of female *Glenostictia scitula*	185
104.	Sternite VIII of male *Glenostictia scitula*	185
105.	*Glenostictia pulla* leveling soil at nest entrance	187
106.	Nesting area of *Glenostictia scitula* near Lajitas, Texas	190
107.	Sleeping aggregation of male *Glenostictia scitula*	192
108.	Female *Glenostictia scitula* leveling soil at nest entrance	195
109.	Profile of nest of *Glenostictia scitula*	198
110.	Head of female *Steniolia obliqua*, anterior aspect	206
111.	Head of female *Steniolia obliqua*, lateral aspect	206
112.	Male genitalia of *Steniolia obliqua*	206

113.	Female *Steniolia obliqua* taking nectar from *Erigeron*	208
114.	Nesting area of *Steniolia obliqua* in Jackson Hole, Wyoming	208
115.	Sleeping cluster of *Steniolia obliqua* on lodgepole pine	210
116.	Sleeping cluster of *Steniolia obliqua* on *Helianthus*	211
117.	Female *Steniolia obliqua* entering nest without prey	215
118.	Typical nest of *Steniolia obliqua* with contents removed	215
119.	Head of female *Stictia carolina*	224
120.	Wings of *Stictia carolina*	224
121.	Apex of metasoma of male *Stictia carolina*, lateral aspect	224
122.	Apex of metasoma of male *Stictia carolina*, dorsal aspect	224
123.	Base of middle legs of male *Stictia carolina*	224
124.	Male genitalia of *Stictia carolina*	224
125.	Apical segments of metasoma of male *Stictia carolina*, ventral aspect	224
126.	Sternite VIII of male *Stictia carolina*	224
127.	Nesting area of *Stictia carolina* in Pottawatomie County, Kansas	226
128.	Nesting area of *Stictia carolina* at Highlands Hammock State Park, Florida	227
129.	Female *Stictia carolina* digging nest	230
130.	Newly completed nest of *Stictia carolina*	231
131.	Typical nests of three species of *Stictia*	233
132.	Female *Stictia carolina* taking prey into nest	239
133.	Nest of *Stictia carolina* that has received final closure	242
134.	Nesting area of *Stictia vivida* near Progreso, Yucatan, Mexico	248
135.	Male *Stictia vivida* in flight over nesting area	249
136.	Nest of *Stictia vivida* with contents removed	250
137.	Egg of *Stictia vivida* on a fly	251
138.	Head of female *Zyzzyx chilensis*	255
139.	Head of female *Editha magnifica*	255
140.	Head of female *Bembix spinolae*	268
141.	Wings of *Bembix spinolae*	268
142.	Apex of metasoma of male *Bembix spinolae*, lateral aspect	268
143.	Sternite VIII of male *Bembix spinolae*	268
144.	Front tarsus of female *Bembix spinolae*	268
145.	Front tarsus of male *Bembix spinolae*	268
146.	Male genitalia of *Bembix spinolae*	268
147.	Apex of metasoma of male *Bembix spinolae*, dorsal aspect	268
148.	Apex of metasoma of male *Bembix spinolae*, ventral aspect	268
149.	Female *Bembix amoena* digging nest	275
150.	Female *Bembix amoena* digging nest	275
151.	Five typical nests of *Bembix amoena*	276
152.	Several burrows of *Bembix amoena* showing various types of curvature observed	276

Figures

153.	Nest entrances of four *Bembix amoena* showing false burrows, back burrows, and back furrows	280
154.	*Bembix amoena* female making final closure of nest	286
155.	Mounds of several successive nests of *Bembix sayi*	290
156.	Typical nest of *Bembix sayi*	291
157.	Final closure and back burrow in *Bembix sayi*	297
158.	Female *Bembix sayi* working at nest entrance	298
159.	A three-celled nest of *Bembix nubilipennis*	303
160.	A male *Bembix spinolae similans* in typical position on sand	313
161.	Typical nest of *Bembix belfragei*	316
162.	Cell of *Bembix u-scripta* showing egg on fly	318
163.	Three stages in mound leveling in *Bembix u-scripta*	319
164.	Typical nest of *Bembix u-scripta*	321
165.	Nesting area of *Bembix texana* at Highlands Hammock State Park, Florida	323
166.	*Bembix texana* female entering nest with prey	326
167.	*Bembix texana* female opening nest entrance while carrying prey	327
168.	*Bembix texana* female sweeping away debris cleaned from cell	332
169.	*Bembix texana* female making final closure of nest	334
170.	Female mutillid, *Dasymutilla pyrrhus*, entering false burrow of *Bembix texana*	335
171.	Nesting site of *Bembix multipicta* near Progreso, Yucatan, Mexico	339
172.	Female *Bembix multipicta* digging at nest entrance	340
173.	Nest of *Bembix multipicta* showing inner and outer closures	342
174.	Resting burrow of male and brood nest of female *Bembix multipicta*	343
175.	Cells of three *Bembix multipicta* nests showing variation in position of egg	343
176.	Head of *Bembix spinolae*, lateral aspect	361
177.	Head of *Microbembex monodonta*, lateral aspect	361
178.	Head of female *Microbembex monodonta*, frontal aspect	362
179.	Wings of *Microbembex monodonta*	362
180.	Apical segments of metasoma of male *Microbembex monodonta*	362
181.	Front tarsus of female *Microbembex monodonta*	362
182.	Front tarsus of male *Microbembex monodonta*	362
183.	Male genitalia of *Microbembex monodonta*	362
184.	Apical segments of metasoma of male *Microbembex monodonta*, ventral aspect	362
185.	Sternite VIII of male *Microbembex monodonta*	362
186.	Open sleeping burrows of *Microbembex monodonta*	366
187.	Female *Microbembex monodonta* digging	369
188.	Typical burrows of *Microbembex monodonta*	371
189.	Female *Microbembex monodonta* entering nest with a dead caddis fly	381

190.	Nest cell of *Microbembex monodonta*, showing egg in typical position	384
191.	Type specimen of *Hoplisus archoryctes*	395
192.	Wings of *Entomosericus kaufmanni*	410
193.	Wings of *Kohlia cephalotes*	410
194.	Head of female *Clitemnestra gayi*	411
195.	Wings of *Clitemnestra gayi*	411
196.	Mesosoma of female *Clitemnestra gayi*	411
197.	Male genitalia of *Clitemnestra gayi*	411
198.	Sternite VI of male metasoma of *Clitemnestra gayi*	411
199.	Sternites VII and VIII of male metasoma of *Clitemnestra gayi*	411
200.	Sternite VIII of male metasoma of *Entomosericus kaufmanni*	412
201.	Sternite VIII of male metasoma of *Harpactostigma gracilis*	412
202.	Ocelli of selected examples of Bembicini	414
203.	Tentative phylogenetic arrangement of major genera of Nyssoninae, based on morphological evidence	416
204.	Cross section of a typical "pore" in the wall of the cocoon of a *Bembix*	457
205.	Cocoons of *Alysson melleus, Ochleroptera bipunctata, Gorytes canaliculatus, Hoplisoides costalis,* and *Sphecius speciosus*	458
206.	Cocoons of *Bembecinus neglectus, Glenostictia pulla, Microbembex monodonta, Bicyrtes quadrifasciata, Bembix cinerea,* and *Stictia vivida*	460
207.	Cocoons of *Stizus pulcherrimus* and *Bembix texana*	461
208.	Cocoon of *Steniolia nigripes* surrounded by outer sheath bearing fragments of prey	461
209.	Relationship of size of cocoons to the number of pores	466
210.	Miltogrammine fly perching over nesting site of a sand wasp	470
211.	Phylogeny of Nyssoninae with reference to five major ethological characters	481
212.	Prey selection in the Nyssoninae shown on an adaptive grid	487
213.	Ovary of a species of *Gorytes* as compared with that of a species of *Bembecinus*	490
214.	Diagram showing probable evolution of major types of oviposition behavior in the Nyssoninae	493
215.	Model suggesting two possible courses of behavioral change	498

The Comparative Ethology
and Evolution of the Sand Wasps

It is true, of course, that the various structural categories from the phylum down to the species, subspecies, variety, sex and individual—all show what may be regarded as correlated or corresponding ethological characters, although this correspondence is often very loose, vague and irregular, for it is evident that slight morphological may be correlated with complex ethological characters, and conversely . . . We are certainly justified in regarding ethological characters as very important, as belonging to the organism and as being at least complementary to the morphological characters. If this is true, our existing taxonomy and phylogeny are deplorably defective and one-sided. To classify organisms or to seek to determine their phylogenetic affinities on purely structural grounds can only lead, as it has led in the past, to the trivialities of the species monger and synonym peddler . . .

The fact that the morphologist has so consistently either neglected or opposed the use of ethological characters in classification shows very clearly that in his heart of hearts he has never very earnestly concerned himself with the parallelism of structure and function. He is inclined to regard function, especially psychical function, as something utterly intangible and capricious . . . Are not our museums largely mausoleums of animal and plant structures which we can forever describe and redescribe, tabulate and retabulate, arrange and rearrange, without troubling ourselves in the least about anything so volatile as function?

It is, indeed, not only conceivable, but very desirable, that a taxonomy should be developed in which the ethological will receive ample consideration, if they do not actually take precedence of the morphological characters . . . One great desideratum in ethology at the present time is a satisfactory and sufficiently elastic working classification of the instincts and reactions, like that of the organ systems of the morphologist. Such a classification can be developed only by a comprehensive, comparative study of behavior in a number of genera and families and not by any amount of intensive study of a few reactions in a few species.

William Morton Wheeler, writing in *Science*, April 7, 1905

Chapter I. An Introduction to the Nyssonine Wasps

Wasps of the genus *Bembix* occur throughout the world, and wherever they occur they attract attention because of their relatively large size and bright colors, their large nesting aggregations, and various aspects of their digging, predatory, and homing behavior. Their behavior often appears simple at first, but to the serious student it proves to be rich in detail and quite strikingly different among the various species. The genus has had many students, chief among whom are E. T. Nielsen, who in his "Moeurs des *Bembex*" (1945) reviewed the ethology of the European *B. rostrata* and presented comparative data with other species; K. Tsuneki, who in a series of three papers on *B. niponica* (1956–1958) monographed the behavior of this species in depth, including the activities inside the nest; and the present writer, who presented a comparative study of thirteen North American species of the genus in 1957. These three authors have largely reviewed all the other published work on this genus, except of course papers published within the last few years.

Just as the proper study of mankind is not only man, but his primate relatives, and in fact all the higher vertebrates, so one may say that a full understanding of the behavior of *Bembix* must rest upon an understanding of other digger wasps, especially those regarded as belonging to the same phyletic stock. The species of *Bembix* are in many ways remarkable wasps, surely close to the apogee of sphecid evolution by any reckoning. There is fairly good agreement on the more general aspects of the phylogeny and evolution of the Sphecidae, and in the present study I propose to survey all that is known of the ethology of members of the subfamily to which *Bembix* belongs, the Nyssoninae.

There are some formidable difficulties in undertaking such a task. There are nearly a thousand known species of Nyssoninae, and the proper arrangement of the species into genera and tribes is by no means finally settled. No more than perhaps 5 percent of the species have been studied in the field in any detail, and for only another 10 percent, roughly, are there even fragmentary records (such as prey records, excavations of single nests, and so forth). Also, there are often uncertainties as to whether published reports are based on correctly identified material, as well as observations that are obviously or probably erroneous in some details. Then there are the difficulties inherent in all comparative studies (for example, cases of convergence, differing rates of evolution for different features, lack of knowledge of the genetic basis of characters); as well as difficulties special to comparative ethology (learning and other modifications of innate behavior, variations in motivation, thresholds, and diverse unsuspected factors, and so forth); and finally, difficulties special to solitary wasps (the fact that much behavior is underground, that they will behave poorly if at all under artificial conditions, and so on).

On the credit side, it should be stated that under ideal field conditions these wasps behave normally even in the close presence of a human observer, that much behavior does occur outside the nest, and that in a large colony one can often gather much data rather quickly and record most of it readily on motion picture film. Many of the difficulties listed above, of course, also present challenges: the challenge to correct erroneous facts and impressions; the challenge to gather data on more species, and in as much detail as possible; the challenge to ferret out the evolutionary background of specific behavior patterns and to try to understand the variation that is observed and the adaptive significance of behavioral modifications; and finally the challenge to improve our understanding of the phylogeny and classification of the Nyssoninae by the use of ethological data.

It goes without saying that no comparative study is likely to be entirely sound if it is based upon data of only one type. Some aspects of these wasps have been studied little or not at all (for instance, internal anatomy, chromosomes). It would, however, be unwise of me to ignore the external structure of these wasps and their larvae. Indeed, one cannot intelligently discuss behavior and structure separately. Behavior is what an animal does with its structure; structure is what an animal uses to behave. Discussions of structure will seem a digression in the earlier chapters and can be skimmed over by persons primarily interested in behavior. In the

A. Classification of the Nyssoninae

final chapters, comparisons of selected structural features will serve to strengthen (or weaken) hypotheses based on the comparison of behavior patterns.

A. Classification of the Nyssoninae

The use of the term "sand wasps" to apply strictly to the Nyssoninae is not universally established, but I find it convenient to have a vernacular term for this taxon. All Nyssoninae are associated with the soil, and nearly all nest in soil that could be described as "sand" at least in a broad sense. It is, of course, true that other wasps nest in sand, just as it is true that wasps other than the digger wasps (family Sphecidae) dig in the soil (while some Sphecidae do not). It may also be wondered why these wasps are called Nyssoninae, when logic would seem to suggest the use of the best-known genus, *Bembix*, as the basis of the subfamilial name. *Nysson*, as a matter of fact, is a rather aberrant genus both in structure and behavior. However, zoological nomenclature, like other languages, is not dictated by logic. In the contemporary literature, the name Nyssoninae is used for the complex I wish to discuss here. The term Bembicinae, recently more often used as the tribal name Bembicini, is generally considered to include only *Bembix* and its immediate relatives.

The modern classification of these wasps may be said to have begun with Handlirsch (1887–1895) and with Kohl (1896). Handlirsch treated the species of the world under the title "Monographie der mit Nysson und Bembex verwandten Grabwespen." Kohl treated them in "Die Gattungen der Sphegiden" as generic groups VI and VII. Both of these authors had a clear concept of the limits of the group as presently defined, except for the inclusion of a few genera which now seem likely to belong elsewhere (*Mellinus, Bothynostethus*).

Unfortunately many subsequent authors did not recognize this complex as a natural group. Within a very short time Ashmead (1899) had split the Sphecidae into several families, separating the Bembicidae widely from the three other families formed of this complex because these wasps lack one of the mid-tibial spurs. Rohwer (1916) went even further; impressed by the lack of an epicnemium in *Bembix* and its allies, he placed these wasps in the Bembicidae, all the rest of the complex in the Sphecidae as the subfamily Nyssoninae, coordinate in rank with the Sphecinae, Larrinae, and others. In Brues and Melander's *Classification of Insects* (1932) the digger wasps are classified in no less than 17 families, 5 of which make up the Nyssoninae as here understood.

It was Pate (1938) who first returned to the saner days of Handlirsch and Kohl. The Nyssoninae were conceived by Pate as a subfamily of Sphecidae with six tribes. He considered the Gorytini the most generalized, the Stizini probably derived from them, the Bembicini probably derived from the Stizini; the Alyssonini and Nyssonini were considered somewhat divergent, the latter somewhat closer to the Gorytini than the former. I find it hard to disagree with Pate's arrangement except for one point. He also included the Ampulicini as a tribe of Nyssoninae, albeit tentatively. Subsequent authors have not followed him on this, and I have found that the ampulicids have larvae much like those of the Sphecinae; they certainly do not belong in the Nyssoninae.

Pate did not include *Mellinus* in the Nyssoninae, but later workers (Krombein, 1951; Beaumont, 1954) have done so, following Handlirsch and others. However, Beaumont indicates that his placement is provisional, and more recent studies of the larvae have suggested quite a different placement for *Mellinus* (Evans, 1959a). The ethology of *Mellinus* also tends to disqualify it as a nyssonine wasp (for summary, see Hamm and Richards, 1930; Olberg, 1959).

I will return to the classification of this complex in the final chapter. For the present I merely wish to indicate my faith in the group as a natural and probably monophyletic complex of wasps, best considered as a subfamily divisible into several tribes, those of Pate (listed above, minus the Ampulicini) being acceptable at least as a point of departure. For the sake of defining this complex as a natural taxon, the following summary of the more important structural features is provided.

Adults: eyes subparallel or convergent below (rarely weakly convergent above); mandibles often dentate on the upper margin, never with a notch or tooth below; mesoscutum with a pair of laminae which overlie the tegulae; propodeum with a dorsal enclosure, usually subtriangular; middle tibia with two apical spurs (except Bembicini, where there is only one); metasoma sessile; fore wing with the marginal cell not appendiculate; three submarginal cells present (except third one absent in some Nyssonini), the second one receiving both recurrent veins (or at least nearly so); hind wing with the anal lobe shorter than the submedian cell.

Larvae: more or less fusiform, with well-developed pleural lobes and a terminal anus; integument not spinulose except sometimes on parts of the prothorax; head with the parietal bands absent or relatively short and weak; antennal papillae present; labrum with setigerous punctures, usually with an apical series of sensory cones; epipharynx in large part spinulose (rarely papillose); mandibles slender, roughly twice as long as their basal width, with from one to four teeth or lobes in a series along the inner margin; maxillae projecting as free lobes, spinulose or papillose on the inner margin, palpi subequal to or longer than the galeae; hypopharynx prominent, spinulose; spinnerets paired, not much if any exceeding the labial palpi.

B. Structure of the nyssonine wasps

A preliminary discussion of some of the major features of external structure is necessary at this point in order to establish the terminology to be employed. For the most part I have followed Richards (1956), to whose paper the reader is referred for comparative data with other Hymenoptera and for his bibliography. Snodgrass' *Anatomy of the Honeybee* (1956) is also useful as a detailed study of a not unrelated insect. The following discussion and series of figures are based on the cicada killer, *Sphecius speciosus* (Drury), a large and reasonably typical member of the Nyssoninae, tribe Gorytini.

(1) *Head.* The major features of the head and of the mouthparts are labeled in Fig. 1, and no detailed discussion is necessary. In the specimen figured, the mouthparts have been fully extended, as they are when the wasp is feeding on nectar; in their resting position, the mouthparts are drawn back into the proboscidial fossa, so that only the palpi are visible in frontal view. That portion of the head immediately behind the eyes is often called the *temples*, while the extreme posterior surface of the head, inside the occipital carina, is the *occiput*.

(2) *Mesosoma.* The major body region behind the head consists of the three thoracic segments and the first abdominal segment (propodeum), all closely consolidated and providing a strongly sclerotized box containing the muscles of locomotion and the articulations of the legs and wings. Michener (1944) proposed the term mesosoma for this body region, and his term is tending to replace the term alitrunk, which has also been applied to this same region. The major features of the mesosoma are labeled in Figs. 2 and 3.

Some of the more significant features of the thorax concern the mesopleural region. *Sphecius* possesses a strong *epicnemial ridge*, followed closely by a groove, marking off the *epicnemium* (=prepectus); the ridge is discontinuous below, so that the epicnemium is not separated below from the sternal region; however, in some other

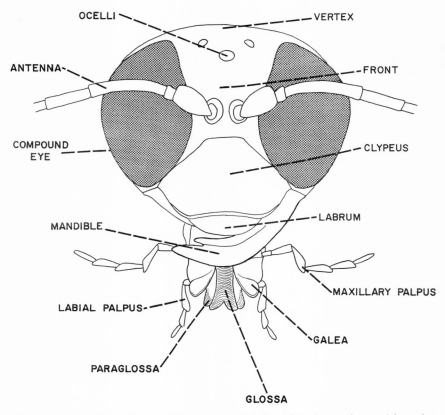

Fig. 1. Frontal view of head of female cicada killer, *Sphecius speciosus,* with major structures labeled.

genera the epicnemial ridge is continuous across the sternum, separating a small anterior portion from the main part of the sternum (see Fig. 29). Other important landmarks in *Sphecius* are three elongate pits: (1) the *subalar pit*, just beneath the base of the fore wing, (2) the *scrobe*, near the posterior margin of the mesopleurum, and (3) the *signum*, in front of the middle coxae. The area in front of the subalar pit, which is convex and set off by a weak groove, is the *subalar prominence*. In some sand wasps there is a strong ridge above the signum, which may connect the base of the middle coxa to the epicnemial ridge; this is the *precoxal ridge* (= sternaulus) (Fig. 23).

In *Sphecius* and in many other sand wasps one finds an arching groove running from the subalar pit to the scrobe and beyond the scrobe to the posterior margin of the mesopleurum. This I term the

§B. Structure of the nyssonine wasps

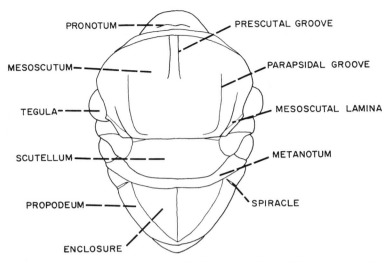

Fig. 2. Dorsal aspect of mesosoma of *Sphecius speciosus* (wings removed), with major structures labeled.

scrobal groove, following Richards (1956), although other workers have called it the epimeral suture (for instance, Beaumont, 1954) or the pleural sulcus (for example, Snodgrass, 1956). Snodgrass believes that since this groove marks the position of an internal ridge and

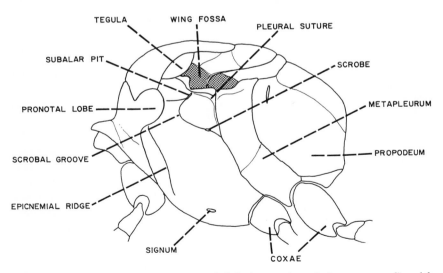

Fig. 3. Lateral aspect of mesosoma of *Sphecius speciosus* (wings removed), with major structures labeled.

apodeme, it must be homologous with the pleural sulcus of other insects; in this case the sclerite behind it represents the true epimeron. Beaumont believes this to be a secondary development, but uses the terms epimeral suture and epimeron since these are in common usage among wasp specialists. I prefer Richards' more noncommittal term.

Comparison with other wasps seems to support the view that the scrobal groove is, in fact, a secondary development, along with its internal ridge and apodeme. As Snodgrass himself pointed out in 1910, the pleural suture is reduced or greatly modified throughout the lower Hymenoptera, most of which are relatively weak fliers. It is only in the higher, stronger-flying wasps and bees that one finds an arching groove and ridge simulating a pleural sulcus; this development is lacking in the Scolioidea, Bethyloidea, Pompilidae, and many Sphecidae. In the smaller, relatively weak-flying Nyssoninae, one finds an arrangement of grooves quite different from that in *Sphecius*. There may be an *oblique groove* (= pre-episternal suture of Michener, 1944, episternal suture of Beaumont, 1954) passing from the epicnemium to the subalar pit (Fig. 40); this is typically connected to the scrobe by a short, horizontal groove. In some forms the upper part of the oblique groove disappears, and the lower part comes to form part of a horizontal groove crossing the pleurum, as in the genus *Gorytes* (Fig. 23). In others it is the lower part of the oblique groove that disappears, so that the scrobal groove forms with the upper part of the oblique groove an angulation or a broad arc (Fig. 29), later smoothing off to a gentle arc as in *Sphecius*. There is good reason to believe that the true pleural suture in all Nyssoninae (and in all wasps, bees, and ants) is represented by an impression close to the posterior margin of the mesopleurum and evident only near the top of this sclerite (Fig. 3; see Michener, 1944:173–176, for a fuller discussion).

In the sand wasps, as in most other Hymenoptera, the thoracic pleura are not separated from the sterna by grooves or ridges. On the ventral surface, one observes a median groove, which represents the line of invagination of a portion of the sternum to form the endosternum, which, along with the phragmata formed by invaginations of portions of the terga, form the major part of the internal skeleton of the thorax. In these wasps the coxae of the front and hind legs are contiguous, while the middle coxae are separated from each other only very slightly. Two pairs of small sternal flaps underlie the median bases of the middle and hind coxae.

The propodeum bears a pair of large spiracles which in some

§B. Structure of the nyssonine wasps

Nyssoninae gives rise to a strong groove passing downward and backward, the *stigmatal groove* (Fig. 23).

The legs require no special discussion except to mention that the term *pecten* (= tarsal comb) refers to a series of more or less evenly spaced strong spines lining the outer side of the front tarsus; the two combs are used for raking soil, and are brought into play by bending the front tarsi strongly toward the midline of the body so that the spines are directed downward. The front tibia possess a single apical *spur* (= calcar), which forms the antennal cleaner, while the middle and hind tibiae have two simple spurs (middle tibia sometimes only one, the shorter spur being lost). The tarsus terminates in the usual two *claws* between which is the padlike *arolium*.

The wings are fully developed in all digger wasps, and in this complex they differ only in details of the venation from those of *Sphecius* (Fig. 4). I find it convenient to use the terminology in common use among wasp specialists, as labeled in Fig. 4. The wings of

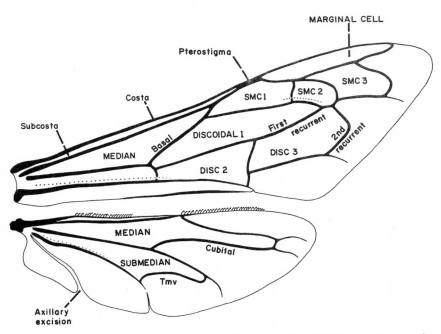

Fig. 4. Wings of *Sphecius speciosus*, with veins and cells labeled according to the arbitrary system in common use among hymenopterists. SMC = submarginal cell; DISC = discoidal cell; Tmv = transverse median vein.

a bee, labeled according to the Comstock-Needham system as modified by Ross, are illustrated by Michener (1944, Fig. 35); these differ in no important way from the wings of sand wasps. A relatively few simple terms suffice for discussions of differences in wing venation within the Nyssoninae.

(3) *Metasoma.* The metasoma of all higher Hymenoptera consists of the true abdomen minus the first segment, which is fused with the thorax and is called the *propodeum.* In the female *Sphecius* there are six externally visible metasomal segments, each with a dorsal plate or *tergite* and a ventral plate or *sternite.* The six tergites form a more or less smooth dorsal surface, but there is a distinct constriction between sternites I and II. The apical tergite has a pair of carinae marking off a more or less flat, densely punctate, dorsal plate, the *pygidium.* Beyond the pygidium may be seen the slender *sting* embraced at the base by rather stout *sting-sheaths.*

The abdomen of the male *Sphecius* has been figured and discussed by Snodgrass (1941:50–51, Pl. 18, Figs. E–O). Snodgrass numbers the segments of the true abdomen and thus, when he speaks of tergite VIII, he refers to what I call tergite VII. In Fig. 5 I have presented a sketch of the apical segments as they look when treated in KOH and drawn apart manually. In their normal position, the tergites overlap the sternites on the sides; dorsally, tergite VI overlaps the base of VII, and tergite VII completely covers tergite VIII, which is weakly sclerotized. Sternite VI is prolonged apically and

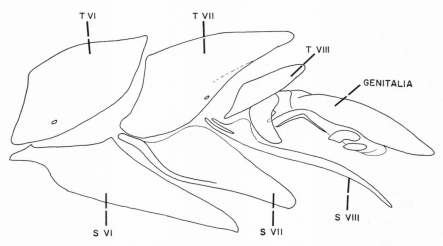

Fig. 5. Lateral aspect of apical portion of metasoma of male *Sphecius speciosus,* the segments drawn apart. T = tergite; S = sternite.

§B. Structure of the nyssonine wasps

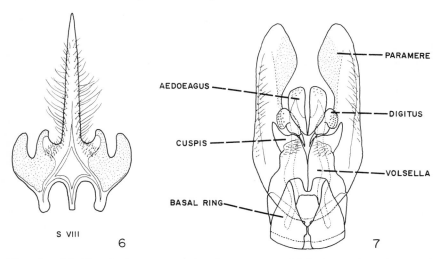

Figs. 6 and 7. Terminal structures of male metasoma of *Sphecius speciosus:* 6, ventral aspect of sternite VIII; 7, ventral aspect of genitalia. In both figures membranous parts are stippled.

in resting position covers sternite VII completely; sternite VIII projects well beyond the tip of the abdomen as a slender spine or *pseudosting*. Thus the tip of the abdomen, in resting position, consists of tergite VII opposed to sternite VI, with sternite VIII protruding well beyond the tip. Sternite VII is actually sclerotized only ventrally and on the apical half, elsewhere hyaline. Sternite VIII is of complex structure basally and in part membranous; it is shown in ventral view in Fig. 6.

The male genitalia (Fig. 7) consist of a large *basal ring* which is somewhat produced ventrally, overlying the base of the *volsellae*, which terminate in paired structures, the median, movable *digiti* and the more lateral *cuspides*. The apex of the *aedoeagus* is embraced by the digiti. The very large lateral structures, which extend far beyond the volsellae and aedoeagus, are the *parameres*.

(4) *Larval characters*. The larvae of digger wasps have recently been treated in a series of papers by Evans and Lin (1956a,b); a general discussion of terminology will be found in 1956a, a review of the Nyssoninae, in 1956b. Further notes on the Nyssoninae were presented by Evans in 1959a and in 1964a. The larvae are simple in structure, and no complex terminology is required. The under surface of the *labrum* is spoken of as the *epipharynx*. A pair of *sensory areas*, each containing several pores with rather thick rims, occurs

near the center of the epipharynx. The *maxillae* terminate in two elongate structures, the lateral *palpi* and the more median *galeae*. The apical part of the *labium*, the *prementum*, bears the *labial palpi* and the paired *spinnerets*.

C. Ethological considerations

All species of Nyssoninae nest in the soil, usually in soil with at least some sand content. Most of them are somewhat restricted ecologically and geographically, and I have tried to summarize briefly the range and habitat of each species considered in the text. Most of the species do not fly far from their nesting sites (not more than a few meters up to perhaps 1 km in some of the larger and more strong-flying forms). Nearly all species are also strongly restricted as to their season of activity. In the subtropics as well as in the temperate zone, most species are active for only a few months of the year, spending the cooler months as diapausing larva in cocoons in the soil. Species with only one generation of adults a year are said to be *univoltine*, those with two or more generations, *bivoltine* or *multivoltine*.

(1) *General features of behavior*. As in most wasps and bees, the males typically emerge a few days before the females. The males commonly fly about the nesting site and often have characteristic prenuptial flights; in some of the larger species the males are territorial. The females are generally fecundated immediately upon emergence from the soil, and often do not mate again once they have begun to nest. The reproductive behavior of solitary wasps has been little studied. I have tried to summarize what is known of courtship and mating, but for the most part the information is fragmentary. The males play no role in the nesting process, although they may live for several weeks, flying about the nesting area and, like the females, visiting flowers for nectar (at least in most species).

Some nyssonine wasps spend the nights and periods of inclement weather in burrows in the soil, while others rest on vegetation, in a few cases in dense clusters. I speak of such behavior as *sleeping*, fully realizing that the sleep of insects may not be quite comparable to that of the higher vertebrates. Actually, the sleeping behavior of sand wasps is one of the most interesting aspects of their ethology and perhaps of considerable phylogenetic significance. However, the amount of information on this subject is meager.

(2) *Nesting behavior*. It is convenient to consider digging behavior and nest structure separately even though the latter is obviously

§C. Ethological considerations

the result of the former. *Digging* proper involves the actual loosening and pushing away of soil with the mandibles and forelegs. Nyssoninae use the two forelegs simultaneously (rather than alternately as in Pompilidae), and tend to move the abdomen up and down synchronously with the legs as the soil shoots out beneath it. *Clearing* involves backing up the burrow and out the entrance while scraping soil with the forelegs, thus freeing the burrow and entrance of accumulated loose soil. *Leveling* involves scraping movements outside the entrance which result in the accumulated soil being scattered away from the front of the entrance. *Closing* refers to the scraping of soil into the entrance or into a portion of the burrow. Closure may be temporary or final (that is, permanent); a temporary closure may be made from outside or inside the entrance, in the latter case the wasp normally remaining inside the nest for some time. *Concealing* refers to the scraping of soil or moving of objects over the closed and leveled nest entrance with the evident function of erasing traces of the nest.

The terms employed in describing the nest are for the most part self-evident. The accumulation of soil outside the entrance I call simply the *mound*, as always prefering a simple Anglo-Saxon word to a longer Latin word (*tumulus*) when there is no possibility of confusion. A *cell burrow* is a branch burrow leading to a cell, used especially when several cells are constructed from the main burrow or when there is a short, blind, terminal burrow or *spur* which is used for sleeping and for soil for the inner closure. An *inner closure* is one separating a cell from the main burrow, an *outer closure* one at the nest entrance. As pointed out above, the latter may be made either from the outside or inside of the nest; thus one may have an "outer closure from the inside," which is quite different from an "inner closure" (Fig. 8).

Some wasps dig a short, blind burrow from the soil surface close to the true burrow. When such a burrow is dug shortly after completion of a new nest, it is termed an *initial false burrow*; when it is dug at the time of final closure, it is called a *terminal false burrow*. Most false burrows are lateral, that is, off to one side of the nest entrance and directed at an angle to the true burrow (45–135°); occasionally they are directed away from the true burrow (that is, at about 180°), in which case it is convenient to refer to the false burrow as a *back burrow*. False burrows are not normally closed by the wasp, but they may be filled in by weathering or in the course of subsequent clearing of the true burrow.

No North American Nyssoninae are known to maintain more than one brood nest at a time. I have worked with marked individ-

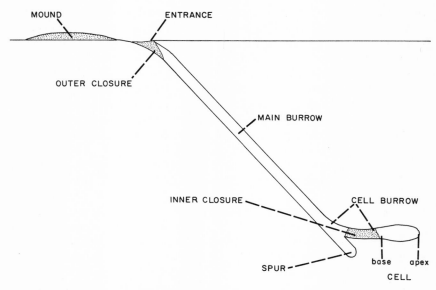

Fig. 8. Generalized unicellular nest of a sand wasp, with major features labeled.

uals of several genera (*Gorytes, Bembecinus, Glenostictia, Bembix, Microbembex*) and have found no exceptions. In every case a new nest (or at least a new cell) is dug only after final closure of a fully provisioned nest (or cell). (Possible exceptions from other regions are the Chilean *Zyzzyx chilensis* and *Bembix brullei*, as described by Janvier, 1928; also, Tsuneki, 1956, found that some populations of *Bembix niponica* exhibit simultaneous care of more than one larva.)

(3) *Provisioning the nest*. Provisioning may occur rapidly and be completed before the egg hatches, in which case it is called *mass provisioning*. Or it may extend over 2 or 3 days, so that the last prey is added shortly after the egg hatches; this is usually of irregular occurrence, often associated with delays due to inclement weather, and is called *delayed provisioning*. Wasps that do not start provisioning until just before the egg hatches and then bring in prey over a period of days as the larva grows are said to exhibit *progressive provisioning*. One may distinguish *fully progressive provisioning*, which implies day to day provisioning until the larva is ready to spin up, from *truncated progressive provisioning*, which means simply that if prey is available in abundance and weather conditions are favorable, the cell may be filled and closed off when the larva is no more than half grown. These four types of provisioning form a more or less continuous

§C. Ethological considerations 15

spectrum; the simple dichotomy of mass versus progressive provisioning of much of the older literature is no longer fully tenable.

Host specificity is as characteristic of sand wasps as of other wasps, although none of the Nyssoninae can be described as "highly" host specific. I have presented all available prey records for North American species and have tried to summarize those for extralimital species, in order to define as fully as possible the actual limits of predation of the various species and genera. Wasps studied in only one locality often tend to show a high degree of host specificity, but in other areas quite different prey (generally of the same major group) may be employed. Some aspects of prey selection by digger wasps I have recently reviewed elsewhere (Evans, 1963b).

(4) *Outline of ecology and behavior*. Because of the differences in the behavior of the various genera and species, and particularly because of the variable amount of information available, it is impossible to follow a hard-and-fast outline for every species considered. In general, I have tried to follow approximately this outline:

1. Range of species
2. Location and author of studies made
3. Habitat and season of activity
4. General features of adult behavior (including sleeping, feeding, relative degree of gregariousness)
5. Reproductive behavior
6. Digging the nest
7. Nest structure
8. Hunting and provisioning
9. Immature stages and development
10. Natural enemies

The species discussed at greatest length are North American species which I have studied myself; other species of a given genus, both North American and exotic, are treated more briefly at the end. Summaries are provided following major genera or groups of genera.

In my original observations, I have often cited my field notes by number. These notes, along with associated specimens of wasps, prey, and parasites, have been placed on permanent file at the Museum of Comparative Zoology, Harvard University. Future workers may examine these notes and specimens as desired, and in this way avoid the many difficulties I have encountered in trying to tie some of the published observations with certainty to a particular species.

Chapter II. *Alysson* and *Didineis:* Two Genera of Hygrophiles

The genera *Alysson* and *Didineis,* both mainly Holarctic in distribution, make up the tribe Alyssonini of most authors, the family or subfamily Alyssonidae or Alyssoninae of some. Two other genera sometimes included in this tribe (*Entomosericus* and *Bothynostethus*) are of unknown ethology. The latter surely belongs in the Larrinae, as indicated by Krombein (1951); the position of the former is discussed in Chapter XIII.

These are small, slender, delicately constructed wasps which most commonly occur in damp or shady situations. The larvae of *Alysson* are remarkable in possessing six prominent, nipple-like protuberances on the front and unusually long antennal papillae (Yasumatsu and Masuda, 1932; Evans and Lin, 1956b; the larva of *Didineis* has not been described). The important structural features of the adults may be summarized as follows:

Mandibles stout, dentate; labrum small, exserted only slightly beyond the margin of the clypeus; labium very short, blunt; clypeus transverse, mainly below the bottoms of the eyes; front broad, eyes subparallel; mesosoma relatively elongate and with well-marked constrictions between the tergites, the pronotum and propodeum especially elongate; mesopleurum with an epicnemial ridge and a shallow oblique groove, but without a signum or precoxal ridge and without a scrobal groove; pecten very weakly developed; middle and hind tibiae with two apical spurs, but those of the middle tibia rather small; hind femora with a strong apical tooth; metasoma slender; female with a pygidial area set off by carinae; male with metasomal tergite VII simple, sternite VIII terminating in two slender processes; male genitalia with the digiti very much longer than the simple, reduced cuspides. Fore wing with the basal vein reaching subcosta close to the base of the stigma, which is quite large; second and third submarginal cells small, the second petiolate; recurrent veins both received by

§A. *Alysson melleus* Say 17

the second submarginal cell, or nearly interstitial with the intercubital veins, the second recurrent sometimes received by the third submarginal at its basal corner (Figs. 9–16).

The ethology of the species of *Alysson* is summarized following that genus; a brief review of what is known regarding *Didineis* concludes the chapter.

A. *Alysson melleus* Say

This small, brightly colored wasp occurs throughout the eastern half of the United States. It has been studied briefly by Hartman

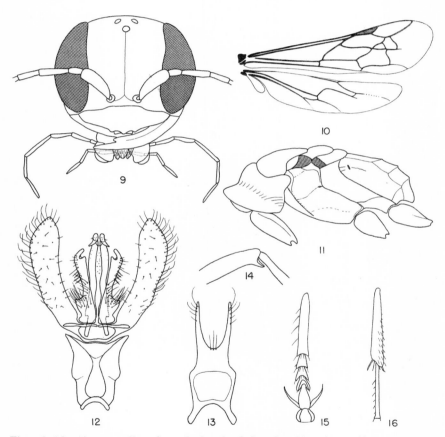

Figs. 9–16. *Alysson melleus* Say: 9, head of female; 10, wings; 11, mesosoma, lateral aspect; 12, male genitalia, ventral aspect; 13, sternite VIII of male, ventral aspect; 14, apex of hind femur and base of tibia of female; 15, front tarsus of female, showing pecten; 16, middle tibia and basitarsus of female.

(1905) in Texas and by the Raus (1918) in Missouri and Kansas. The present report is based upon 24 field notes made in Kansas, 10 in Florida, and 6 in central New York, several of the latter having been contributed by Frank E. Kurczewski.

(1) *Ecology.* The Raus remark that they never found these wasps "elsewhere than in a cool, damp bank of mud or sandy clay, near to a body of water." According to Hartman the female "always selects the sloping sides of a pit as a location for her nest." I have found *melleus* nesting in either flat or slightly sloping soil, rarely in the sides of relatively steep banks. The type of soil selected varies from firm sand to clay-sand or sandy loam. The Kansas observations were all made along Blackjack Creek, Pottawatomie County, either within 1 or 2 m of flowing water or actually in the slightly moist soil of portions of the stream bed, places that must actually be flooded at times. The Florida observations were made at two localities along Fisheating Creek, Glades and Highlands counties, where the wasps nested in the sloping, sandy banks of the creek only 2 or 3 m from the water, again in situations that must occasionally be flooded (Fig. 17). In central New York, *melleus* nests in sand and gravel pits

Fig. 17. View along Fisheating Creek, near Palmdale, Glades County, Florida. *Alysson melleus* nested in some numbers in the white sand along the stream, along with a few *Bicyrtes quadrifasciata* somewhat farther from the water (immediate foreground).

§A. *Alysson melleus* Say

often at some distance from water, but the soil in the places selected is always at least slightly moist. In Kansas and Florida the nesting sites are often in such situations that they are shaded much of the day.

The Florida observations were made in April and May, those in Kansas in June. Evidence from museum specimens and from rearing of larvae indicates that the species has more than one generation during the summer in these states. In New York the species is definitely univoltine; although I have taken the species as early as late June, the peak of its nesting appears to be in August. The Raus' observations were made during June and July. The authors remark that these wasps "are seldom to be seen in the heat of the day but become most numerous and active at about 5 p.m." I have often found *melleus* very active on cloudy days or even between light rain showers, and there is little question that on clear, warm days the wasps are active principally during the morning and late afternoon. I have never taken *Alysson melleus* on flowers, but males are often taken in considerable numbers on foliage covered with honey-dew.

(2) *Reproductive behavior.* Males are often observed about the nesting areas of the females. They fly about singly, close to the ground, and frequently land on the ground facing a female, their wings slightly raised and their antennae extended rigidly. I observed mating only once, at Ithaca, New York, on 15 July 1956 (no. 1186). On a cool, partly cloudy morning (0930) a female was seen walking over the soil of a small sandpit in the usual antlike manner of this wasp. A male landed 3 cm behind her, facing her, then after a moment took wing, made several circles about her, then landed again about 3 cm behind her. This was repeated several times, the female continuing to walk over the sand and showing no obvious reaction to the male's activities. Then the male landed directly on top of the female, straddling her body and appearing to embrace her with all three pairs of legs. The wings of the male were thrust way forward and held in this position, while the tip of the abdomen was twisted sideways so as to make contact with the genital orifice of the female. The pair remained in this position for 15 sec, the female being completely motionless during this time. The male then flew off and the female resumed her walking over the sand.

(3) *Nesting behavior.* The concentration of nests in any one area seems to vary greatly, perhaps dependent upon the availability of suitable sites and the success of the species in that area. In central New York one finds only a nest or two here and there, and obtains the impres-

sion that this is a highly solitary species. Along Fisheating Creek, in Florida, I counted nine nests in an area of about 2 m² in what appeared to be the place of greatest concentration of nests; some of the nest entrances were only 3-10 cm apart. Along Blackjack Creek, Kansas, the concentration of nests was not greatly different from this, with one exception. On 16 June 1952 I found a nesting site in damp sand beneath a willow tree in a draw leading into the main stream bed. In an area about 0.75 by 1.5 m there were an estimated 300 nests. A few seemed to be closed, completed nests, but the majority had the entrances open, and many females were bringing in prey. Many of the nests were so close together that the mounds at the entrances were more or less confluent. The wasps in the aggregation appeared to have no difficulty finding their own nests in the honeycombed earth, and in no case was any aggression between females noted. Unfortunately the pressure of other work prevented my spending more than about 1 hr with this remarkable aggregation.

Although Hartman remarks that the excavation of the nest "is carried out after the usual manner of wasps," this relatively meaningless statement most certainly does not apply to *Alysson melleus*. The small, vertical bore of the burrow is dug out with the mandibles and front legs. After a bit of soil is loosened by the mandibles, it is formed into a small lump which is pushed backward beneath the body by the front legs, then passed out behind the body either from the initial thrust of the front legs or with assistance from the middle and hind legs. When the burrow is a few millimeters deep, the wasp assumes a vertical position in it, head down, rotating in a clockwise direction while drilling into the earth and pushing the soil upward. As the wasp digs deeper, the small lumps of earth clog the entrance and hide the wasp; from time to time the wasp backs up and pushes them out of the entrance. The earth is not cleared away from the hole, and forms a pile of gradually increasing height and diameter, like that at the entrance to the nests of many ants. The entrance at first passes vertically down through the mound, but later, during provisioning, assumes a more lateral position, so that the entrance is actually oblique for a few millimeters before reaching the vertical tunnel.

I found most digging to occur in the late afternoon (1600-1800). Hartman remarks that "the work of digging the nest is all done at once," but it is doubtful if he actually observed this, since the wasp is not easily observed once the nest is a few centimeters deep. He also remarks that "some dirt is brought up from time to time after the provisioning has begun." This is true; some earth is removed from time to time during provisioning when the burrow is blocked

§A. *Alysson melleus* Say

by cave-ins, and it is probable that some soil is brought to the surface at the time that additional cells are constructed.

(4) *Nest structure.* As already noted, the burrow is vertical or nearly so when dug in flat soil. Burrows dug into sloping surfaces tend to start out perpendicularly with the soil surface and then become nearly vertical (noted by the Raus, and once by myself). Burrow diameter is only about 2 mm. The terminal cell is small, slightly longer than high and wide or occasionally nearly circular, measuring 4-6 mm in height and width by 6-10 mm in length. The initial cell is normally the deepest, the additional cells being constructed gradually upward toward the entrance, though the cells are separated by no more than 0.5-2.0 cm. In some nests the cells assume the form of a small cluster, while in others they are strung out in more or less of a single series back toward the entrance (Fig. 18). In all three areas in which studies were made, the maximum number of cells per nest was five. In all areas the majority of nests were unicellular, but probably all of these were incomplete when dug out. One nest in Florida (no. 1973) was dug out following final closure, but had only two cells.

The length of the burrow (that is, the distance of the initial cell from the surface) showed considerable variation in all three areas, the total range being from 5 to 17.5 cm (see Table 1). The maximum spread in depth in the cells of one nest was 6 cm (no. 47, Kansas); in this nest the initial cell was at 17.5 mm, the last cell (of five) at

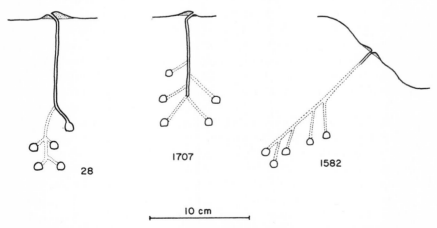

Fig. 18. Three typical nests of *Alysson melleus:* no. 28, Kansas; no. 1707, Florida; no. 1582, New York (see text for specific localities). Burrows indicated by broken lines were filled and could not be traced exactly, hence are somewhat hypothetical.

TABLE 1. NEST DATA FOR ALYSSON MELLEUS*

Locality	Number of nests					Depth initial cell (cm)	Depth final cell (cm)
	1-cell	2-cell	3-cell	4-cell	5-cell		
Kansas	5	3	3	1	2	12.7 (7.5-17.5)	11.2 (7.5-15.0)
Florida	5	3	1		1	8.0 (5.0-14.0)	7.5 (6.0-11.0)
New York	2				1	10.8 (8.0-13.5)	10.0

* In this table, as in others of a similar nature, measurements are presented in the form of a mean and range of variation; locality data are presented as name of the state only, necessitating reference to the text for specific localities and dates.

11.5 mm. It will be noted that the Kansas nests averaged slightly deeper than those in Florida and New York (12.7 cm as compared with 8.0 and 10.8 cm). Hartman states that the nest is 12 in. (30 cm) deep, which is considerably greater than any depth found in the present studies. The Raus state that the burrow is 1.5 to 4 in. long (about 4 to 10 cm); it is possible that some of the very shallow burrows were incomplete, as the Raus report finding few completed nests. Neither Hartman nor the Raus report multicellular nests, but the Raus state that upon finding a cell they would often find in the earth nearby "little cells whose connection could not be traced." They assumed these to be the cells of other nests, but it seems much more probable that they were, at least in part, earlier cells of the same nest. In my own studies, I tried to use somewhat isolated nests that had been under observation for more than 1 day, so there is little question that the data on cell number are substantially correct.

In the Florida site I measured several mounds at nest entrances and found the average height to be 0.8 cm, the average diameter, 3.5 cm. As already noted, active nests typically have the entrance on the side of the mound so that it is somewhat sloping at first. The Raus describe some of the variation they found in the form of the tops of the nests, some having "a little chimney or canopy of mud." In another place they speak of the wasps "making" canopies. I suspect that any such canopies are simply fortuitous results of the adhesion of moist bits of earth; there is no evidence that these wasps construct earthen canopies comparable to those of some of the Vespidae.

§A. *Alysson melleus* Say

(5) *Closure*. Hartman reports that the nest entrance is left open at all times, day and night. I observed no temporary closures whatever in any of the nests studied.

Because of the nature of the nest, returning wasps occasionally find the entrance or burrow partially or completely blocked by the collapse of part of the mound. When this occurs, the wasp deposits her prey beside the entrance and attempts to force her way through the barrier; if successful, she comes out, grasps her prey, and drags it in. If she is not able to force her way into the burrow, she digs it out in much the way that a new nest would be dug. Typically, females proceed directly into the open nest entrance carrying their prey. Below the entrance, the burrow is completely open to the cell, that is, there is no inner closure and no storage place for prey. When a cell is fully provisioned it is closed off solidly with soil, the soil apparently often being taken from the excavation of the next cell. It is usually impossible to trace cell burrows that have been closed off, even though they may be quite short.

(6) *Hunting and provisioning*. Females hunt for their leafhopper prey on herbs and low trees by running over the branches and leaves much like ants. I sometimes saw leafhoppers escape capture by moving around to the opposite side of a leaf. Hartman speaks of "the easy grace with which she flits from place to place when on the hunt . . . She runs swiftly up and down the stems and over the leaves, both the upper and under sides, often darting like a flash to another branch or to another plant." Along Fisheating Creek, Florida, I watched a female flush an immature leafhopper from a low herb on to the ground. She bent her abdomen beneath and forward and stung the leafhopper once, very quickly, then seized it with her mandibles and proceeded over the ground a few centimeters. She then paused in the shade briefly and took flight with her prey.

The manner of prey carriage in this wasp is highly characteristic. The leafhopper is held venter up, the wasp grasping it with her mandibles near the base of the head (presumably by the base of the beak). The body of the leafhopper extends backward beneath that of the wasp, but the wasp does not grasp it with her legs, except possibly in flight (Fig. 19). Hartman reports carriage by the mandibles, and all my notes and photographs indicate that the wasp walks on all three pairs of legs and employs the mandibles alone for holding the prey (as in Olberg's excellent photographs of *A. fuscatus*). The wasp flies into the nesting area but lands at some distance (10 cm to over 1 m) from the nest entrance and walks without pause to the nest entrance, and typically directly into it, carrying her prey.

Fig. 19. *Alysson melleus* female entering her nest with a leafhopper (Palmdale, Florida). Note that the open entrance is on the side of a somewhat conical mound of sand.

If the nest entrance is partially or wholly blocked, as mentioned above, the hopper is deposited on the ground while the entrance is cleared; in this event the wasp eventually comes out and grasps the prey by its head and backs into the burrow with it. Commonly the wasp remains in the nest only a very short time, then walks out a few centimeters beyond the entrance and takes flight. However, it is not unusual to see a female leave a nest, walk about, clean herself, even fly off briefly, then re-enter the nest. This may actually be repeated several times before she finally leaves to take another leafhopper.

When walking along with prey, *Alysson* typically does not move the wings, but moves the metasoma up and down rhythmically. Hartman is in agreement that the wasps generally fly from the site of prey capture into the nesting area, then walk to the nest. However, certain variations in the manner of locomotion have been noted. I saw two females at some distance from the nest walk up

§A. *Alysson melleus* Say

stems and take flight from them (nos. 21 and 22). Occasionally females walk along for several centimeters, then take a short flying hop of several more centimeters, then walk for a distance, fly again, and so on (nos. K101–K103). These variations may be related to the nature of the terrain, the weight of the prey, or possibly the individuality of the wasps.

Some impression of the speed of provisioning may be obtained from the following observations made at Blackjack Creek, Kansas, 7 June 1952 (no. 28). This wasp was using small cicadellid nymphs and was carrying them in her mandibles alone. The nest entrance was open at all times:

0929: in with prey, out at 0930
0937: in with prey, out at 0939
0943: in with prey, out at 0944
0946: in with prey, out at 0947
0949: in with prey, out at 0950
0952: in with prey, out at 0953
0959: in with prey, out at 1000
1001: in with prey, out at 1002
1006: in with prey, out at 1008
1013: in with prey, out at 1014
1020: in with prey, did not reappear although nest was watched until 1110. May be digging new cell.

As already indicated, a wide variety of leafhoppers, both nymphs and adults, are employed as prey. The number used per cell is very variable. In several fully provisioned cells I found as few as 3 adult leafhoppers (nos. 26A, 26B, 1582A, and 1973B), while the maximum was 23 very small adult leafhoppers (no. 1706). As might be expected, cells containing small nymphs or adults contain more leafhoppers than those containing larger ones. About two thirds of the cells I examined contained between 6 and 11 leafhoppers. The Raus reported "about a dozen" leafhoppers in a cell, and Hartman found 7 leafhoppers in a cell that contained no egg and was therefore probably not fully provisioned. The leafhoppers are packed in the cell head in, venter up.

Although the total list of known prey is long (Table 2), any one cell tends to contain only one or a few species, indicating that individual females tend to do most of their hunting in a circumscribed area. Adults and nymphs are used in approximately equal numbers; some cells contain only one or the other, but most cells contain some adults and some immatures. Immature leafhoppers can be

TABLE 2. PREY RECORDS FOR ALYSSON MELLEUS

Species of prey	No. of specimens*				
	Kan.	Fla.	N.Y.	Mo.	Tex.
CICADELLIDAE					
Aceratagallia sp.	1				
Agalliopsis novella Say				x	
Balclutha sp.		1			
Chlorotettix sp.		2			
Ciminius hartii Ball	1	1			
Colladonus clitellarius Say				x	
Deltocephalus flavicosta Stal	1	3			
Draeculacephala antica Walker			6		
Draeculacephala mollipes Say			1		
Draeculacephala paludosa B. and C.	6				
Draeculacephala portola Ball			1		
Draeculacephala spp.		1	22		
Empoasca fabae Harris				x	
Exitianus exitiosus Uhler	2			x	
Graminella pallidula Osb.	1				
Graminella sp.		9			
Hortensia similis Walker		5			
Keonolla dolobrata Ball	15				
Macrosteles fascifrons Stal	1				
Paraphlepsius irroratus Say	3				
Sanctanus sp.		1			
Tylozygus bifidus Say		3			x
DELPHACIDAE					
Genus and species ?		4			

* Missouri records from Rau and Rau, 1918, Texas record from Hartman, 1905. These authors did not indicate the number of specimens found.

identified only with difficulty (if at all), so the table includes chiefly adult records.

Frequently the leafhoppers found in the cells of *Alysson melleus* appear dead, and after a day or two in a rearing tin they often appear somewhat desiccated. The Raus also found this to be true. They suggest that these wasps nest in damp, cool situations because "only in such surroundings could the food be kept fresh and moist." I found much variation in the condition of the prey; sometimes the leafhoppers were fresh and responded to stimuli up to 48 hr after being stung.

(7) *Immature stages and development.* The egg is laid longitudinally on the ventral side of the thorax, close to the coxae (Fig. 20). With

§A. *Alysson melleus* Say

rare exceptions, the egg is laid on an adult rather than a nymph. The newly emerged larva begins its feeding at the cervical membrane or the front coxal cavity. In every one of the many instances in which the egg or small larva was found *in situ*, it was found on one of the topmost leafhoppers in the cell. There is much evidence that the egg is laid after the cell is fully provisioned and just before it is sealed off.

After the leafhopper on which the egg was laid has been consumed, the larva rapidly consumes the other leafhoppers in the cell. Legs, wings, and most of the exoskeleton are not eaten, but remain in the bottom of the cell and eventually more or less surround the cocoon. The larva reaches full size after only 4 or 5 days of feeding. The greater part of 1 day is required to spin the cocoon. One cocoon spun on 16 July 1962, in the Kansas locality, gave rise to an adult female only 18 days later.

(8) *Final closure.* I watched one female making a final closure at 1030 on 26 April 1964 (no, 1973, Fisheating Creek, Venus, Florida). She came out of the entrance again and again, scraping soil with her forelegs, then backing in. Each time she came out at a slightly different point, so that the top of the mound was reduced to a small crater. When the burrow was filled, she walked over the top of the mound in an irregular pattern, scraping soil here and there and tapping the soil lightly with the tip of her abdomen. Then she walked off, leaving a small, flat-topped mound about 0.3 cm high and 3 cm in diameter. When this nest was dug out the burrow was found to be solidly filled and impossible to trace except in small part. The two cells both contained eggs.

(9) *Natural enemies.* Only 1 of over 30 cells excavated in Kansas contained any evidence of parasites. This cell (no. 47A) contained

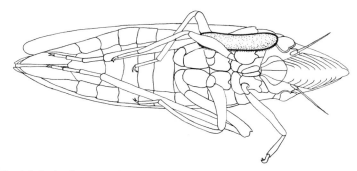

Fig. 20. Adult leafhopper (*Draeculacephala*) bearing the egg of *Alysson melleus.*

an *Alysson* larva when first dug out, but 2 days later the larva had disappeared and a dipterous puparium occupied the cell. A miltogrammine fly emerged from this puparium 13 days later, but unfortunately it escaped. Three of 19 cells excavated in Florida contained maggots (all 3 cells in nest no. 1707). In each case (as in the Kansas example) only one maggot occurred in each cell, and in each case the *Alysson* larva had apparently been killed. These maggots formed their puparia in the rearing tin, and two flies emerged only 13 days later. The flies were in poor condition when preserved; they were determined doubtfully as *Phrosinella fulvicornis* (Coq.).

At Groton, New York, Frank E. Kurczewski noticed a *Nysson* sp. landing on a grass blade overhanging the entrance of an *Alysson* nest. After a moment it flew off, without entering the *Alysson* nest.

B. Ethology of other species of *Alysson*

Several species of *Alysson* besides *melleus* occur commonly in the United States, but nothing has been recorded concerning their behavior. None of the species appear to take nectar at flowers, but males are often encountered on honeydew, and to a lesser extent females. *A. oppositus* Say is sometimes common around sandbanks and along watercourses in the northeastern United States, but I have never found it nesting. There is one female *oppositus* in the Museum of Comparative Zoology, Harvard University, taken at Chilmark, Massachusetts, pinned with a female *Draeculacephala mollipes* (Say). This and other species may nest amid vegetation where they are not readily observed. Frank E. Kurczewski has sent me a female *A. guignardi* Provancher taken as prey of the asilid fly *Dioctria baumhaueri* Meigen.

Several Palaearctic species of this genus have been studied in some measure. Kohl (1880) reported that *A. fuscatus* Panzer (= *bimaculatus* Panzer) nests in sand and provisions its nest with various species of leafhoppers. Olberg (1959:336–338) found this species carrying leafhoppers to the nest on foot, holding them in the mandibles; his brief account is illustrated with several excellent photographs of wasps carrying prey.

Ferton (1901) found *A. ratzeburgii* Dahlbom nesting in a slope of compact, cool clay; the nests were from 10–15 cm deep, each cell provisioned with two or three leafhoppers. The egg is laid in the manner already described and figured for *melleus*. Ferton provides no information on the number of cells per nest or on whether the egg is laid on the first or the last leafhopper in the cell. He states that the wasp carries its prey by holding a leg with its mandibles and also supporting the prey with the middle legs. Adlerz (1910) studied this same species and reported that the wasps carried their prey merely by seizing the beak of the leafhopper in their mandibles, a report more in accord with what is known of *fuscatus* and *melleus*. Ferton (1909) also reported *A. tricolor* Lepeletier and Serville preying upon leafhoppers, but provided no further details regarding this species.

In 1932 Yasumatsu and Masuda described a new species of *Alysson* from Japan, *cameroni*, and presented some notes on its nesting behavior. Their paper is in Japanese, and I have read only the English summary. This species apparently also nests in damp, cool situations. They found as many as 50 nests per square meter in one place. The nests are reported to be unicellular, with 10 to 20 adult or immature leafhoppers being provided per cell. They found that provisioning females fly with the prey but land on the ground at some distance from the nest and proceed to it on foot; the nest entrance is left open day and night. These authors compiled the host data for all species of *Alysson* as of 1932; the list includes 5 species of *Alysson* that collectively prey upon 18 species of Homoptera, chiefly Cicadellidae but also including records of Cercopidae and Issidae.

C. Summary of ethology of the species of *Alysson*

Although the available information on the species of *Alysson* is rather fragmentary, the various reports are reasonably consistent with one another and permit several tentative generalizations:

(a) The species nest in relatively cool, moist situations, either in sandy or argillaceous soil. Usually several females nest in close proximity, and at times the concentration of nests may be very great.

(b) The burrow of the nest is in large part vertical or nearly so, and is dug by "corkscrewing" down through the soil and pushing the soil upward, where it forms a pile of small pellets, which is left in place. The entrance comes to occupy a lateral position in this mound, and is not ordinarily closed by the wasp at any time.

(c) The nest may have as many as five cells, which are constructed progressively back toward the entrance (but none of the Palaearctic species are known to make more than one cell per nest).

(d) The prey consists of immature or adult leafhoppers (Cicadellidae, less commonly Cercopidae or Fulgoroidea). Most species take a considerable variety of leafhoppers, which they capture on herbs and bushes in the vicinity of the nesting site. The prey does not recover from the paralysis and usually dies at most within a few days.

(e) The prey is carried to the nest partly in flight, partly on the ground, the wasp always proceeding the last several centimeters (up to 2 or 3 m) to the nest on foot and entering the nest directly, without depositing the prey (unless the entrance has accidentally been clogged). The wasp seizes the base of the leafhopper's beak in its mandibles and holds the hopper venter up beneath the anterior part of its body; the legs of the wasp do not usually assist in holding the hopper while on the ground, but may do so in flight.

(f) The egg is laid longitudinally on the side of the venter of the thorax of one of the top leafhoppers in the cell. It is laid after completion of provisioning of the cell, and the cell is then closed off.

D. Ethology of the genus *Didineis*

Most of the several North American species of this genus are poorly known. I have often taken *D. texana* (Cresson) in damp, grassy or weedy situations, even in marshes, where the females presumably hunt for their prey. The only report on the biology of a North American *Didineis* is Strandtmann's (1945) brief note on *texana*. Strandtmann found a female of this species running along the edge of a field of turnips near Carrizo Springs, Texas, carrying a *Cixius stigmatus* (Say) (Fulgoroidea).

Ferton (1911) has provided a brief account of the nesting behavior of the Palaearctic *D. lunicornis* Fabricius. He found this species nesting in the sides of hoofprints, the burrows being horizontal for 2–3 cm, then extending almost vertically to a depth of 15 to 20 cm, then becoming horizontal again and, after 3 to 5 cm, terminating in an ovoid cell. At this depth the soil was compact and argillaceous, much like that in which he found *Alysson ratzeburgii* nesting, although the topsoil here was sandy. The prey of *Didineis lunicornis* was found to consist of adult and immature Homoptera of the genera *Thamnotettix, Eupelix, Chiasmus, Agallia,* and *Delphax* (Cicadellidae and Delphacidae). The prey is carried by short flights and by walking and is carried directly into the nest, the entrance to which is left open at all times. The egg was not found.

Ferton remarks that the nidification of *Didineis lunicornis* differs hardly at all from that of *Alysson ratzeburgii*, and we may extend that generalization so far as to say that the genera *Alysson* and *Didineis* exhibit no known differences in their ecology and nesting behavior. Pending evidence to the contrary, we may assume that the summary provided for the behavior of *Alysson* probably applies equally well to the closely related genus *Didineis*.

Chapter III. *Gorytes* and Some Closely Related Genera

The wasps considered in this chapter were all included in *Gorytes* by Handlirsch in his "Monographie" (1888). These wasps are, in fact, all quite similar in general appearance and in gross behavior, although most contemporary workers place them in several genera. The most important recent systematic work is that of Pate (1938) and of Beaumont (1954).

As compared to *Alysson* and *Didineis*, *Gorytes* and its relatives have a more robust body and a more compact mesosoma, the pronotum and propodeum being quite short. The majority of species are black with limited pale yellow markings, but some are extensively yellowish or rufous. The following summary of the more important structural features is provided.

Adults: mandibles stout, dentate; labrum small, rounded, exserted only slightly beyond the clypeus; tongue short; clypeus transverse and mostly below the bottoms of the eyes, or in some cases higher and more narrow, extending upward between the eyes such that the antennal orbits are pushed upward toward the center of the front; pronotum short, transverse; pterothorax short, robust; propodeum short, unarmed; mesopleura with an epicnemial ridge and with a scrobal groove, the latter often extending forward to join an oblique groove arising from the epicnemial ridge; pecten usually well developed in female; middle and hind tibiae with two apical spurs; hind femora simple; female with a pygidial area set off by a carina; male with the eighth sternite produced to form a slender process apically, or sometimes tongue-like, the apex bifid; fore wing with the basal vein reaching subcosta close to the base of the stigma, the latter well developed; three submarginal cells present, the second and third four sided (Figs. 21–27).

Larvae: frontal protuberances absent; labrum more strongly setose than in *Alysson*, and with a series of barrel-shaped sensilla along the mar-

Chapter III. *Gorytes* and Related Genera

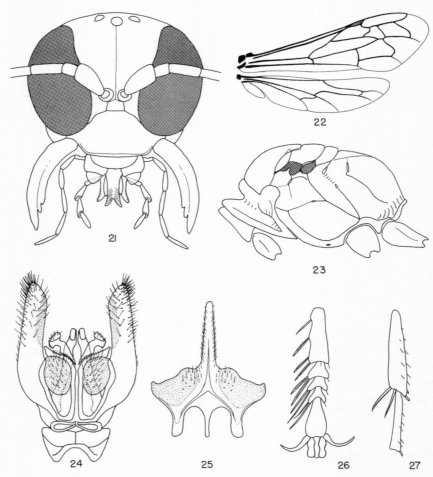

Figs. 21–27. *Gorytes simillimus* Smith: 21, head of female, mouthparts extended; 22, wings; 23, mesosoma, lateral aspect; 24, male genitalia, ventral aspect; 25, sternite VIII of male, ventral aspect; 26, front tarsus of female; 27, middle tibia and basitarsus of female.

gin; mandibles with a single tooth on the inner margin beside the apical tooth and with from one to several setae on the sides toward the base; maxillary palpi exceeding the galeae; body smooth and without setae. (See Evans and Lin, 1956b; Evans 1959a.)

Members of this complex occur throughout the world, although little is known of the ethology of species occurring in the tropics or the south temperate regions. Nor are the comparative morphology and systematics of this group well understood. I have for the most

§A. Genus *Hoplisoides* Gribodo

part followed Beaumont's classification and that of Krombein (1951) in the recent catalogue of North American Hymenoptera; various groups considered as subgenera by Beaumont are ranked as genera by Krombein and in the present work. I discuss the genus *Hoplisoides* first because it is best known. This is followed by brief discussions of the following genera, in turn: *Gorytes, Psammaecius, Argogorytes, Lestiphorus, Psammaletes, Harpactostigma, Dienoplus, Hapalomellinus, Clitemnestra, Ochleroptera,* and *Ammatomus*. A short summary of the ethology of members of this complex follows.

A. Genus *Hoplisoides* Gribodo

These are among the more familiar of gorytine wasps in the north temperate regions. Some species are not uncommon in areas of bare, sandy soil, and although small in size are likely to attract attention when they are seen carrying the often bizarre treehoppers which comprise the prey of many species. Reinhard has written a popular account of one species in his *Witchery of Wasps* (1929).

As compared with *Gorytes*, wasps of this genus are rather more robust and covered with large, well-spaced punctures; the epicnemial ridge is strong, and separates below into two ridges, one of which passes back to the middle coxa (the precoxal ridge), the other of which crosses the anterior part of the mesosternum and joins the one from the opposite side; there is no oblique groove, and the scrobal groove is indistinctly prolonged upward toward the subalar pit; the front is broad, the eyes subparallel, the propodeum lacks a groove passing backward from the spiracles; and the metasoma is robust, the first segment not at all nodose or petiolate (Figs. 28 and 29).

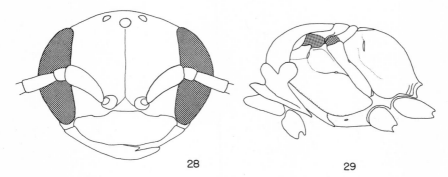

Figs. 28 and 29. *Hoplisoides nebulosus* (Packard): 28, head of female; 29, mesosoma, lateral aspect.

B. *Hoplisoides nebulosus* (Packard)

This is one of the more commonly encountered Gorytini in eastern North America. Typical habitats include small sand hills, blowouts, man-made sand or gravel pits, and sandy roads. The type of soil selected for nesting varies from pure sand to coarse, sandy gravel with numerous pebbles. The nesting area is frequently surrounded by woods or scrub, such areas presumably being rich in the treehoppers on which this wasp preys. This species does not appear to fly far from its hunting and nesting sites, and specimens are only rarely collected on flowers.

The present report is based on 21 field notes, all but 4 made in two sand pits at Ithaca, New York, in the summers of 1954–1960. One of these sand pits is shown in Fig. 30. Two were made in sandy areas just west of Boston, Massachusetts, June–August 1962–63, the remaining two at "Little Gobi Desert," a hilltop blowout in prairie dissected by ravines, Pottawatomie County, Kansas, in the summers of 1952 and 1953. Published accounts of the biology of this species

Fig. 30. Small sandy area on South Hill, Ithaca, New York, typical habitat of *Gorytes canaliculatus* and *Hoplisoides nebulosus*.

§B. *Hoplisoides nebulosus* (Packard)

consist of a prey record from Wyoming (Pate, 1946) and two brief accounts of females nesting in coastal North Carolina, both along a sandy road in the woods (Krombein, 1953, 1959). It is possible that Barth's (1907) observations purported to deal with *Gorytes canaliculatus* were wholly or partly based on this species; a specimen in the Barth material at Milwaukee proved, however, to be *canaliculatus*, and his observations are therefore included under that species, with some reservations.

Most records of this species nesting are for the month of July, but I have found it nesting at Ithaca as early as 7 June and as late as 28 August, so there may be two generations in the course of one summer. This is a decidedly solitary wasp, a single female tending to select a small section of sand or gravel for a series of nests continuing over a period of days, but two or more females (in my experience) only occasionally being encountered at one time and place. The movements of *Hoplisoides nebulosus*, when searching for a place to dig, when digging, closing, or bringing prey, are completely silent and seem very slow as compared with wasps such as *Anacrabro ocellatus* and *Bembix spinolae*, to mention two wasps which often nest in close proximity to *nebulosus*. When walking about on the sand, working outside the nest entrance, or arriving at the entrance with prey, the wings are elevated at about a 35–45° angle with the abdomen. As the female enters the burrow, either when digging or provisioning, she depresses her wings flat against her body. This characteristic elevation and depression of the wings has been noted in many Gorytini.

(1) *Digging the nest.* The female frequently starts to dig in several places before finally remaining at one site and completing the nest. About half the nests I observed were started from the bare surface of flat or slightly sloping sand or gravel, the other half started in such a place that the entrance was overhung by a stone, a leaf, or a lump of earth. One individual (no. 1483) was followed through a series of four unicellular nests over a period of as many days, digging each nest beneath a partially decayed poplar leaf (each beneath a different leaf, but all within an area of about 2 m^2). One individual in Kansas started her nest in the side of a hoofprint of a horse, and Krombein found a nest in a wheel rut.

The nest is dug with the front legs, which work synchronously, the abdomen moving up and down rhythmically and allowing the soil to shoot out behind the body. The mandibles are used for breaking hard soil and for dragging pebbles from the burrow. In the early stages of excavation, the wasp backs out frequently and sweeps the soil away from the entrance, piling it up in a small, semicircular mound 2–4 cm away. After backing out of the nest a short distance, the wasp works forward raking soil, either in a straight line or with weak, irregular zigzagging, and re-enters the nest. As the digging progresses, the wasp is seen less frequently, but she reappears from time to time to sweep the entrance clear. One individual watched by Krombein often turned around inside the burrow, then came out head first pushing sand ahead of her.

During the later phases of nest excavation, the female may from time to time rise obliquely into the air, hover facing the entrance, then descend rather slowly to resume her digging. When the nest is complete, the wasp closes the entrance by scraping in a small amount of soil from in front of the entrance. She then scrapes additional soil over it from the sides, thoroughly concealing the hole. There are no true leveling movements such as occur in some species of *Bembix*, but the mound of earth is inconspicuous and partially dispersed by other movements. The movements of concealment are normally interrupted several times by additional short, oblique, hovering flights, 20–40 cm high. These flights tend to become longer, higher, and with more side-to-side movements, and finally the wasp disappears instead of returning to her nest. During these flights the wasp undoubtedly learns the surroundings of her nest, as on the return flight she invariably descends obliquely to the entrance from a height of half a meter or more, retracing the flight pattern she established during the orientation flights.

Nests are commonly begun in the morning and finished before noon. One female (no. 64) was seen starting a nest at 0905; this nest was completed at approximately 1130. Krombein saw a female at 0835 with a nest which had apparently just been started. This nest was completed by 0850. Where the females spend the night is not known.

(2) *Nature and dimensions of the nest*. The burrow is oblique and forms an angle of from 45° to 70° with the soil surface. The burrow diameter is about 5 mm, the horizontal terminal cell 7–9 mm in diameter and 9–11 mm long. Of the 17 nests which I dug out (most of them after the final closure), 11 had only one cell, 5 had two cells, and 1 had three (Fig. 31). As mentioned earlier, I followed one individual (no. 1483) for 4 days and found that she dug and provisioned a new one-celled nest each day. Another wasp (no. 1072) was seen provisioning, then digging for about 15 min, then provisioning again the same afternoon, requiring as little as 45–50 sec

§B. *Hoplisoides nebulosus* (Packard)

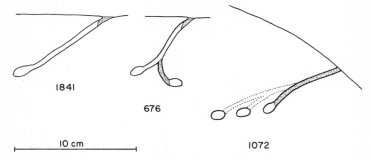

Fig. 31. Three typical nests of *Hoplisoides nebulosus:* no. 1841, Lexington, Massachusetts; nos. 676 and 1072, Ithaca, New York. Filled burrows are indicated by stippling, burrows that could not be traced, by broken lines.

to obtain a treehopper and spending only 20 sec in the nest each time. When this nest was excavated the next day it was found to contain three cells. Apparently this female had found such a rich source of prey that she was able to complete three cells in 1 day. In no case did I observe a wasp to remain with a given nest for more than 1 day. In bicellular or tricellular nests the cells are commonly arranged in series, separated by 1–2 cm of earth fill, the oldest cell in each series being the one farthest from the entrance. In some two-celled nests it appeared that the second cell had been constructed at the end of a short branch from the side of the main burrow. Data on burrow length and cell depth are presented in Table 3. The two Kansas nests were in notably drier soil than the others, and this may account for their greater depth.

TABLE 3. NEST DATA FOR HOPLISOIDES NEBULOSUS

| Locality | Number of nests | | | Burrow | Cell |
	1-cell	2-cell	3-cell	length (cm)	depth (cm)
Kansas	1	1		16.5 (15–18)	11.0 (8–15)
No. Carolina (Krombein)	1			6.5	3.0
New York	9	3	1	9.0 (5–12)	7.0 (4–10)
Massachusetts	1	1		9.5 (9–10)	6.0 (5–7)

38   Chapter III. *Gorytes* and Related Genera

(3) *Provisioning the nest*. These wasps prey upon immature and adult treehoppers (Membracidae) of several genera. Even though the different kinds of prey vary much in size and in shape, they are always carried in the same manner: venter up, head forward, held well back beneath the wasp by her middle legs (Fig. 32). When the wasp lands at the nest entrance she scrapes it open with her forelegs while standing on her hind legs and continuing to hold the prey with the middle legs. The prey is apparently released by the middle legs and grasped by the hind legs as the wasp enters, so that the prey actually follows the wasp down the burrow, as occurs in the Bembicini.

The approach to the nest is characteristic. The wasp glides in slowly, obliquely, from a considerable height (0.5–1 m), usually landing directly in front of the closed nest entrance, occasionally landing on a leaf overhanging the nest and then descending to the nest. During pauses before entering the nest, the wings are elevated obliquely and the abdomen moves up and down rhythmically.

Fig. 32. Female *Hoplisoides nebulosus* entering nest with a treehopper (Ithaca, New York). Note that the middle legs grasp the prey while the front legs are thrust forward in digging position. (From Evans, 1963d.)

§B. *Hoplisoides nebulosus* (Packard)

When leaving the nest, the wasp flies up rather slowly, retracing the course of her arrival. Reorientation occurs if there has been some disturbance to the area, also after the construction of a new cell. I found it easy to misguide provisioning wasps. Wasps nesting under dried leaves glided down directly to the side of the leaf where they normally went under it, even when I had displaced the leaf by several centimeters. Such wasps invariably took an orientation flight after they left their nest the next time.

Females usually spend only 20–60 sec in the nest each time they bring in prey. They leave the burrow with their wings flat against their back, then raise them obliquely as they close. Closure is effected by scraping a small amount of soil into the entrance from directly in front, sometimes also from the sides. Not uncommonly the wasp pounds the fill slightly with the tip of the abdomen. I observed many individuals closing, and saw only three which omitted the closure. One wasp (no. 1072) brought in two treehoppers over a period of 5 min while the entrance was open, then began a second cell; after completion of the second cell she began provisioning again, this time making a closure each time. Thus it appears that the closure is occasionally omitted in the very last stages of filling a cell, as sometimes occurs in some *Bembix*. A second wasp (no. 1483) failed to close consistently for two successive nests. Both of these nests were well concealed under dried poplar leaves.

Wasps may take a very short time, actually less than 1 min, to capture and return with another treehopper. On the other hand, some individuals are away from the nest for much longer periods, even as much as 1 hr. Number 1215, watched for 2 hours at Ithaca, on 14 August 1956, brought in nine treehoppers during this time, each time remaining inside the nest 30–90 sec. The actual times of arrival were: 1422, 1438, 1453, 1506, 1520, 1531, 1546, 1557, and 1612.

As shown in Table 4, this wasp preys upon a wide variety of treehoppers. However, individual nests normally contain only one or a few species, indicating that females locate aggregations of these insects and return again and again to the same place for their prey. The treehoppers are piled in the cell head in and mostly venter up, though it is not uncommon to find some of them on their side or even dorsum up. The number of treehoppers in fully provisioned cells depends in part upon the size of the treehoppers utilized. The largest number found in one cell was 20 adult *Entylia sinuata* in one of the two cells of nest no. 676. The smallest number was 4 large adult membracids, *Palonica virida* and *Telamona decorata,* in nest no. 1587. Most cells had between 10 and 15 treehoppers. Many cells

TABLE 4. PREY RECORDS FOR HOPLISOIDES NEBULOSUS

Species of prey	No. of cells	No. of adults	No. of nymphs	Locality*
MEMBRACIDAE				
Campylenchia latipes Say	3	9	10	N.Y., Kan.
Ceresini spp.	5		36	N.Y., N.C.
Enchenopa binotata Say	2	6	8	N.Y.
Entylia sinuata Fabricius	7	50	7	N.Y.
Microcentrus spp.	2		6	N.Y., N.C.
Palonica virida Ball	1	1		N.Y.
Palonica sp.	1		1	Wyo.
Publilia concava Say	8	54		N.Y.
Spissistilus festinus Say	1		4	Kan.
Telamona decorata Ball	1	3		N.Y.
Vanduzea arcuata Say	2	6		N.Y.
Vanduzea triguttata Burm.	1	1		Kan.

* North Carolina records are from Krombein (1953, 1959), the Wyoming record from Pate (1946), others original.

contained a mixture of nymphs and adults, either of the same species or of different species. None of the treehoppers found in the nests of this wasp ever showed any recovery from paralysis, although they seemed to remain in fresh condition for several days.

It should be noted that many treehoppers exhibit protective resemblance to the branches on which they live. For example, *Enchenopa binotata* is a form closely resembling a rose thorn and assumed to gain some protection from this resemblance. However, this species was used in some numbers by *nebulosus*, as were nymphs and adults of several other bizarre forms.

As already noted, *Hoplisoides nebulosus* appears to visit flowers only rarely. There is evidence that the females feed from time to time on the treehoppers they capture. Number 1485 (Ithaca, 24 August 1957) was seen to land on a blade of grass and hang there by one hind leg. She held her treehopper beneath her by her front and middle legs and inserted her mouthparts into the cervical membrane of the bug and appeared to suck up the body fluids. The other hind leg of the wasp extended upward loosely. This wasp remained in this position for three minutes and then flew off with the bug. Whether this treehopper was used for provisioning a cell

§B. *Hoplisoides nebulosus* (Packard)

I do not know; at least it was not discarded immediately after feeding. Very similar behavior has been described for certain bembicine wasps.

(4) *Oviposition and development of the larva.* I found eggs in 13 cells, and in each case the egg had been laid on one of the top treehoppers in the cell. Cells dug out before completion of provisioning invariably did not contain an egg. Thus it seems clear that the egg is laid after the cell has been fully provisioned. I measured several eggs of this species and found them to range from 2.0 to 2.2 mm long by about 0.6 mm wide. The egg is invariably laid on the side of the venter of the thorax, beside the coxae (Figs. 33 and 35); when the prey is small the egg may extend along part of the abdomen and part or all of the head.

The egg of this wasp hatches in about 2 days, and the small larva begins feeding through one of the anterior or middle coxal cavities on which the egg was laid. In about 2 days it begins to feed on other treehoppers on top of the pile, and in another 4-6 days it has consumed all the treehoppers and begins to spin its cocoon. The cocoon is elliptical and the outer wall contains sand grains as in the Bembicini, but there are no pores in the wall as in that tribe.

(5) *Final closure.* This was observed several times. Soil is scraped from the walls of the burrow and later from the outside to use as fill for the burrow. The entire final closure takes only 20-30 min. For the first few minutes the wasp is not seen outside the nest, but the entrance is left open. Then the wasp comes out of the nest (elevating the wings obliquely in the usual manner), scraping sand to about 2 cm from the entrance. She then turns around and enters the

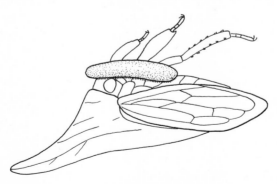

Fig. 33. Adult treehopper (*Campylenchia*) bearing egg of *Hoplisoides nebulosus*.

nest head first, then turns around inside and repeats the performance (no. 1188); or she may simply back into the burrow each time (no. 1194). As the burrow becomes nearly full, it can be observed that the wasp pounds the earth in place with strong blows of the tip of the abdomen. A large number of trips to the periphery of the hole may be made before the burrow is completely full. When it is full, the wasp goes out in several directions a distance of 2–4 cm scraping sand toward the entrance. One individual (no. 1188) dug mostly at one point, leaving a shallow pit or rudimentary "false burrow." Most individuals leave the soil surface completely smooth.

(6) *Natural enemies*. Miltogrammine flies were observed following provisioning female *Hoplisoides nebulosus* on two occasions at Ithaca. In both cases the flies were *Senotainia trilineata* (Wulp). Only one nest (no. 1587, at Ithaca) was found infested with maggots. This nest was dug out 1 day after it had been provisioned, and at this time the maggots had already destroyed the wasp's egg. Within a few days the maggots completely consumed the contents of the cell. No adult flies were successfully reared from these maggots.

Krombein observed the mutillid wasp *Dasymutilla v. vesta* (Cresson) exploring the vicinity of a *H. nebulosus* nest in North Carolina. The mutillid was eventually driven off by the *Hoplisoides* and did not actually enter the nest. Another subspecies of *Dasymutilla vesta* is known to parasitize *Bembix cinerea* (Evans, 1957).

Wasps of the genus *Nysson* are important parasites of these wasps. Krombein reported *Nysson opulentus* Gerst. investigating a *nebulosus* nest in North Carolina, but the parasite was not actually seen to enter the nest. It is probable that Krombein's wasp was *Nysson tuberculatus* Handlirsch, which has until recently been misidentified as *opulentus* (R. M. Bohart, personal correspondence). *N. tuberculatus* was seen entering *Hoplisoides nebulosus* nests at Ithaca on three occasions (nos. 676, 1194, and 1653), and in one instance (no. 1194) an adult was reared from a *nebulosus* cell. The following is a synopsis of my note no. 1653, taken on 2 July 1960.

At 1300 in a small patch of bare, sandy gravel, a female *Hoplisoides nebulosus* was seen starting a nest beneath a leaf. I left the area and returned at 1430. At this time the nest entrance was closed and well concealed. I assumed the female was away hunting treehoppers and decided to move the leaf several centimeters to one side to see whether she returned to the leaf or to the exact spot where the nest was located. After I had done this, a female *Nysson tuberculatus* appeared, walking over the ground with her antennae continually tapping the soil. She came to the *Hoplisoides* nest and

§B. *Hoplisoides nebulosus* (Packard) 43

seemed to detect the entrance immediately; she paused, her antennae in rapid motion, then dug into the entrance (Fig. 34). After only about 10 sec she came out, closing the entrance behind her exactly as a *Hoplisoides* female would do. The *Nysson* (or perhaps another one) was seen in the area later, but was not seen to re-enter the nest.

At 1445 the *Hoplisoides* female arrived with a treehopper, as usual gliding down obliquely from a height of more than 0.5 m. She homed in correctly, but turned slightly to one side before landing, so as to land beside the leaf rather than at the true nest entrance. After digging a bit, she flew off and returned again, again being misled by the fact that I had moved the leaf. After 10 min and several unsuccessful attempts by the wasp to find the nest, I replaced the leaf to its former position. This time she had no difficulty in finding the nest and entered in the usual manner, coming out in a few seconds and taking a reorientation flight. She returned with another treehopper at 1505. She had not returned again by 1530. I left the area at that time. It appeared that the *Nysson* had found

Fig. 34. *Nysson tuberculatus* entering a covered nest of *Hoplisoides nebulosus* (Ithaca, New York; in area shown in Fig. 30).

the nest by odor, using her antennae as the main organs of detection. The *Hoplisoides*, however, obviously relied upon sight, as she was easily misled by a slight change in landmarks.

I dug out this nest at 1630 on the following day. The single cell contained ten adult treehoppers of two species. The egg of the *Hoplisoides* was on the top treehopper (an *Entylia sinuata*) in the usual manner. No other egg was visible, but the bottom treehopper, a *Publilia concava*, appeared slightly "disjointed," that is, the pronotum appeared to have been pulled away slightly from the remainder of the body. I pulled it away slightly farther and saw the *Nysson* egg beneath the pronotum (Fig. 35). It was smaller than the *Hoplisoides* egg (only about 1 mm long as compared to 2.2 mm for the *Hoplisoides*) and the surface was wholly covered with microscopic projections (smooth in *Hoplisoides*). Unfortunately neither of these eggs was reared successfully, although on another occasion I did rear this same species of *Nysson* from *H. nebulosus*. (See *Hoplisoides costalis*, Chapter III: D, for a summary of Reinhard's account of a similar relationship between *costalis* and *Nysson hoplisivora*, in which Reinhard was able to follow the entire story.)

Nysson daeckei Viereck was also seen entering a *Hoplisoides nebulosus* nest at Ithaca, New York (no. 1557, 21 July 1958). This species attacks *Gorytes canaliculatus* perhaps more commonly and is discussed further under that species. In the present instance, the *Nysson* appeared on the scene while the *Hoplisoides* was making a final closure. After standing about at various points not far from the nest for several minutes, the *Nysson* entered the nest while the *Hoplisoides* was still closing. Several minutes later I caught the *Hoplisoides* and dug out the nest. The *Nysson* was still in the cell and was captured for identification. The cell contained nothing but nymphs of one species of Membracidae. The *Hoplisoides* egg was found on one of them in the usual position. The *Nysson* egg was not detected in the field, and I supposed that the parasite had failed to lay her egg, since there were no adult treehoppers and therefore no pronota under which to hide the egg. However, when I examined the cell contents under a microscope I found the *Nysson* egg (only about 0.8 mm long) laid transversely behind a hind coxa of a nymph, so that it was partially concealed (Fig. 36). The contents of this cell were eventually destroyed by mold, but the *Nysson* larva reached about half its normal full size before dying, and I was able to determine that it had the process on the front and other features of the larva of *Nysson*. The *Hoplisoides* egg was apparently destroyed by the *Nysson*, but I did not observe the act of destruction.

At Billerica, Massachusetts, in June 1963, *Nysson daeckei* was

§B. *Hoplisoides nebulosus* (Packard)

Fig. 35. Cell of *Hoplisoides nebulosus* opened to show contents (no. 1653, Ithaca, New York). The topmost treehopper bears a *Hoplisoides* egg, one of the bottom ones, an egg of *Nysson tuberculatus;* in the latter case the pronotal shield of the treehopper was pulled away from the body so that the egg could be photographed. (From Evans, 1963d.)

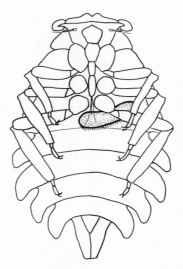

Fig. 36. Immature treehopper (Ceresini) bearing the egg of *Nysson daeckei*, from a nest of *Hoplisoides nebulosus* (no. 1557, Ithaca, New York).

common in a small nesting aggregation of *Hoplisoides nebulosus*. One *Nysson* was seen to remain motionless for a considerable period near a *Hoplisoides* that was digging a nest, and another entered a partially completed nest that had been abandoned. No nests were dug out in this locality.

C. *Hoplisoides spilographus* (Handlirsch)

This species resembles the preceding very closely, but is confined to the western United States. It has usually gone under the name *adornatus* (Bradley), but Dr. R. M. Bohart, who has studied the type specimens of *spilographus* as well as of *adornatus*, informs me that the two are synonyms. Powell and Chemsak (1959) published an account of a nesting aggregation of this species found in Contra Costa County, California, in May 1958. I made five field notes on this species in a sandy area along the Snake River at Moran, Wyoming, 18-27 July, 1964. My notes are in close agreement with those of Powell and Chemsak.

At Jackson Hole, Wyoming, some 20-30 females nested in flat, moderately firm, sandy clay only 3-6 m from the banks of the river. Nest entrances were separated by from 30 to over 100 cm but were interspersed with those of various other digger wasps, especially *Bembix spinolae, Ammophila azteca,* and *Oxybelus uniglumis quadrino-*

§C. *Hoplisoides spilographus* (Handlirsch)

tatum. Powell and Chemsak found 60 to 80 females nesting along the sandy border of a level, grass-covered parking area at 1500 ft on Mt. Diablo in California. Over half the wasps were concentrated in a zone measuring about 1.5 by 3 m. These authors were able to show that the females spend the night away from the burrows, but they were unable to locate the sleeping sites. They observed males at the nesting site in the early afternoon and also on oak foliage in late afternoon.

Powell and Chemsak found that this species maintains a fairly strict daily regimen, much as reported for *nebulosus*. In the morning, the females would arrive at about 0845–0915, shortly thereafter starting to scratch the sand here and there before finally selecting a nesting site and completing a nest. Digging activities continued for about 2 hr, and these were followed by a period of active provisioning that lasted until late afternoon, when the females made final closures of their nests. A very similar daily regimen was observed at Jackson Hole.

(1) *Nesting behavior.* Powell and Chemsak reported that most nests were on open sand, often in slight depressions, but some were close to pebbles or other debris, while others were under leaves. The five nests studied at Jackson Hole were all on smooth soil, not close to or covered by debris. Powell and Chemsak described the digging behavior as follows:

> The digging was commenced with the fore legs, the wasp always being directed head first. As the burrow deepened, sand was kicked back almost continuously with the hind legs . . . As the burrow progressed, the length of time spent in the burrow increased with the wasp periodically reappearing. Upon each reappearance the wasp would back up four or five centimeters from the entrance, pause a few seconds, then gradually progress forward to the entrance kicking the loose sand behind . . . In the final stages the wasp reappeared head first forcing the sand out ahead of her. After this type of emergence the female would repeat the previously described procedure of removing the loose sand. Construction of the burrow took two to two and one-half hours in the few cases in which the entire process was observed.

After completing the burrow, according to these authors, the female closes the entrance by kicking sand backward from all sides, "periods of kicking being alternated with short hovering flights one to two cm above the burrow." Shortly, the hovering flights are increased to 6–8 cm high, then the female makes "several longer hovering arcs 1 m or more above the burrow site before flying away." When returning with prey, the wasps fly to a position 2 or

3 m above the nest, then descend obliquely, with a hovering flight and gradually diminished speed almost directly to the burrow entrance. "After a few seconds pause the female quickly opened the burrow entrance with her fore legs and entered, without changing the position of the prey. Usually about 20 seconds (8 to 30 seconds) were spent in the deposition of the prey. Usually the burrow entrance was covered upon each departure."

These observations were confirmed in almost every detail at Jackson Hole. I noted no females that did not close the entrance when they left. In every instance females arriving with prey descended to the nest entrance from a considerable height. Motion pictures of digging females show that the front legs move synchronously, as in *Bembix* and nearly all other nyssonine wasps. It was noted that females landing on the sand without prey (for example, prior to starting a new nest), always elevate and lower the body rhythmically by extending and then flexing the legs. The significance of this behavior is unknown.

Powell and Chemsak state that the prey is carried with the hind legs. My own observations indicate that prey carriage is similar to that of other gorytine wasps, that is, the middle legs provide the major grasp on the prey.

At Mt. Diablo, California, the prey was found to consist entirely of nymphs of Ceresini (Membracidae), possibly *Stictocephala*, except for two females taken with nymphs of a larger species, possibly some species of Telamonini. Powell and Chemsak also reported a colony of this species in Monterey County, California, employing "an undetermined membracid nymph." At Moran, Wyoming, all prey found in nests or taken from females were nymphs of a single species of Ceresini, probably *Stictocephala* sp. (19 records). All records indicate that the egg is laid in the manner described and illustrated for *H. nebulosus* (Figs. 33 and 35). I found a maximum of 7 treehoppers per cell, while Powell and Chemsak report 12 to 14 (presumably smaller) treehoppers.

Powell and Chemsak dug out seven nests and found that the burrow varied in length from 6.5 to 17 cm (mean about 7 cm), while the cell was at a depth of from 5 to 6 cm. My own figures, based on five nests, are almost identical: burrow length 6 to 17 cm (mean 10.5 cm), cell depth 4 to 6.5 cm (mean 5.3 cm). Powell and Chemsak reported that most burrows had a bend part way down. I found some burrows to be straight, while others had several lateral bends, apparently around the numerous small stones in the soil. The size of the cell is about 6 × 15 mm.

Powell and Chemsak reported that the female constructs a new

§D. *Hoplisoides costalis* (Cresson)

burrow each morning. They found that females often brought in enough membracids each day to provision two cells, but they were not able to definitely associate two cells with any burrow. At Jackson Hole, females appeared to provision very slowly and probably did not usually stock more than one cell per day. One female (no. 1998) was observed provisioning a nest that had been dug the day before, and another (no. 2006) completed a two-celled nest closely resembling the nest of *H. costalis* shown in Fig. 37. The others contained only one cell when dug out.

(2) *Natural enemies*. Powell and Chemsak found *Nysson moestus* Cresson to be common within the *spilographus* colony: "Individuals were frequently seen investigating or entering open burrows or other holes. One was seen to enter a burrow with the digging occupant inside . . . Another was observed to dig its way into a temporarily closed burrow and remain within for almost eight seconds." *Nysson pumilis* Cresson and *N. rusticus* Cresson were also collected in some numbers in the nest area. Powell and Chemsak list certain species of chrysidid and mutillid wasps and bombyliid and miltogrammine flies which occurred in the same area, but none were definitely associated with *Hoplisoides spilographus*.

Nysson rusticus Cresson was also common in the nesting area at Jackson Hole, but I did not observe it to be closely associated with *Hoplisoides* nests. Miltogrammine flies were also common, and on several occasions they were seen hovering behind female *H. spilographus* laden with prey; however, none of the nests excavated contained maggots. Several flies captured as they followed provisioning females were identified as *Senotainia* sp. (*trilineata* complex).

D. *Hoplisoides costalis* (Cresson)

This wasp is widely distributed in the eastern United States but is more common toward the south. Reinhard (1925a,b 1929) has studied it in some detail in Maryland, Rau (1922) presented a brief note on a female nesting in Missouri, and Krombein (1953, 1959) has studied it twice in coastal North Carolina. Krombein found the species nesting in loose sand in open woods, while Reinhard found it in a sandy path loosely paved with bricks, a place occupied also by a large colony of *Philanthus gibbosus* (Fabricius). Rau's wasp had her nest in the ground under a small piece of loose bark, but Reinhard's and Krombein's wasps apparently nested on bare sand. I found one female nesting in bare, fine-grained sand only 15 cm from the base of a raspberry bush and well overhung by the branches; my single observation (no. 1952) was made at Andover, Massachusetts, 6 August 1963. Apparently this is a relatively solitary species, although Reinhard found some 20 nests scattered among the *Philanthus* nests. Reinhard reports two generations a year in Maryland.

(1) *Nesting behavior.* This species may prepare a new nest each day, as Reinhard speaks of females digging their nests in the morning before other activities have begun. However, Krombein found that the egg in one cell of a two-celled nest hatched soon after the nest had received the final closure, suggesting that it had been laid the previous day. Reinhard remarks that each completed nest is marked by a small mound of sand. He found that the burrow entered the soil at about at 30° angle with the horizontal; the burrow is said to be about 13–15 cm long, terminating in "a scattered group of cells" at a depth of about 5 cm. The nest found by Krombein was in a 45° slope; the burrow was at an angle of 30° and about 8.5 cm long. This nest, which was dug out during the final closure, contained two cells. Rau and Reinhard both indicated that the burrow entrance is kept closed when the female is away.

The nest I studied was found to contain two cells (Fig. 37), but it was dug while the female was still provisioning so might eventually have contained more. There was a small mound at the entrance of this nest; the entrance was closed while the female was away. The burrow was at a 45° angle with the surface and was 9 cm long, reaching a cell at a depth of 6 cm. Only 1 cm deeper there was a second cell, fully provisioned and closed off.

Hoplisoides costalis apparently preys upon a considerable variety of adult treehoppers (Table 5), but individual nests may contain all or mostly one species. For example, Krombein took 14 treehoppers from two cells of one nest, and all were alike. I found 6 adult *Stictocephala,* all one species, in the fully provisioned cell of the nest I

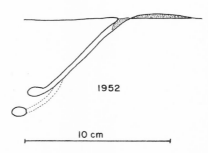

Fig. 37. Nest of *Hoplisoides costalis* (Andover, Massachusetts).

§D. *Hoplisoides costalis* (Cresson)

TABLE 5. PREY RECORDS FOR HOPLISOIDES COSTALIS

Species of prey	Locality	Reference
MEMBRACIDAE (ADULTS)		
Archasia galeata Fabr.	Md.	Reinhard, 1925
Campylenchia latipes Say	Md.	Reinhard, 1925
Ceresa bubalis Fabricius	Md.	Reinhard, 1925
Ceresa palmeri VanDuzee	N.C.	Krombein, 1953
Cyrtolobus arcuatus Emmons	Md.	Reinhard, 1925
Glossonotus crataegi Fitch	Md.	Reinhard, 1925
Platycotis vittata Fabr.	Md.	Reinhard, 1925
Stictocephala borealis Fairmaire	Md.	Reinhard, 1925
	N.C.	Krombein, 1959
	Mass.	Evans, no. 1952
Telamona monticola Fabricius	Md.	Reinhard, 1925
Telamona tristis Fitch	Md.	Reinhard, 1925
Telamona unicolor Fitch	Md.	Reinhard, 1925
Thelia bimaculata Fabricius	Md.	Reinhard, 1925
Vanduzea arcuata Say	Md.	Reinhard, 1925

dug out. Krombein found 5 to 9 treehoppers per cell. Reinhard found 3 to 6, the majority of cells having 4. He found that the membracids are stung to complete immobility but remain in good condition for about 1 week, when they die. However, he found in one nest a treehopper that was able to move its legs and wings vigorously, suggesting that the wasp "had bungled." Reinhard notes that the prey is held venter up with the middle legs. The wasps usually descend "squarely in front of their burrow" and scrape the entrance open with the front legs. The wasp I watched landed on a raspberry leaf, pumped her abdomen up and down a few times, then descended quietly to the entrance, scraped it open, and entered.

The egg is about 0.75 × 3.0 mm and is laid in the manner described for *nebulosus*. Krombein found that in both of the cells he excavated the wasp egg was on the membracid that had been brought in first. However, Reinhard reported that the egg is laid on the last treehopper placed in the cell, which is more in accord with what is known of other members of this group. The egg apparently hatches in 1 or 2 days, and larval development is rapid, requiring only 4 or 5 days. Reinhard (1929) described the spinning of the cocoon in considerable detail (see Fig. 205 D, a photograph of a cocoon collected by Reinhard).

(2) *Natural enemies.* Reinhard found two cells provisioned with three treehoppers each, each cell containing a dipterous maggot that 2 days later formed its puparium. He also found two empty puparia in an old cell containing the remains of four treehoppers. The puparia were identified as those of the miltogrammine fly *Pachyophthalmus signatus* Meigen (currently called *Amobia aurifrons* Townsend).

Reinhard (1925b, 1929) found *Nysson hoplisivora* Rohwer to be an important parasite of *Hoplisoides costalis* at Woodstock, Maryland. Since his observations are among the most detailed and important regarding the relationships of *Nysson* and the gorytine wasps, I shall quote from his paper (1925b) at some length.

The *Nysson,* says Reinhard, could always be found flying about or resting near a *Hoplisoides* burrow; the *Philanthus* burrows "did not interest them."

> They would fly very low, just skimming over the surface, to pause at every sand heap, inspect it, circle about inquisitively, and dart along to the next mound . . . They delayed longest at the doors of [*Hoplisoides*] to gather information with trembling taps of their antennae. If the owner was about, the visitor would sometimes settle herself on the ground close by, and, with watchful gaze directed at the doorway, wait quietly until [*Hoplisoides*] had departed. At times she would enter an open gallery but come out hastily again, and warily take up a post of observation to sit awaiting. But when the coast was clear and the nest vacated, the *Nysson* would boldly break through its barricaded portal to enter the nurse-chambers below, where her depredations could be committed without detection.

Reinhard notes that the *Nysson* opens the *Hoplisoides* nest with her front legs and remains inside only a few seconds. Then she comes out, he says,

> and with wonderful nonchalance, as it were, carefully rearranges the sand over the doorway. To see her you would think she was the dwelling's rightful owner and housewife. When the tunnel is once more blocked and the entrance obliterated, the *Nysson* calmly departs. Was the *Nysson's* action in closing the burrow really an attempt to eliminate the traces of her forced entrance, or was it merely a tropism and remnant of the days when her more industrious ancestors were wont to throw a curtain of sand over the doorway at every departure?

The *Hoplisoides* female continues her provisioning and eventually lays her egg on one of the top treehoppers and closes off the cell. The egg of the *Nysson* is well concealed, having been laid dorsally beneath the wings of one of the lower treehoppers in the cell. Reinhard reports that the *Nysson* egg is only 0.44 × 1.34 mm, that is,

§E. *Hoplisoides tricolor* (Cresson) 53

less than half the length of the egg of its host, and is "dull white, tough, with a 'goose-flesh' texture that shows under the microscope as numerous small excrescences placed with regularity." Reinhard found that the *Nysson* egg hatches half a day in advance of that of the *Hoplisoides*. "The intruding larva takes a little nourishment, then advances through the cell to seek out and devour the egg of its host. When its rival is destroyed, *hoplisivora* feasts without danger of molestation upon the store of provisions which its parasitism has appropriated." According to Reinhard, the first-instar *Nysson* larva has a pair of long, thin mandibles and a series of ventral "blisters," the latter presumably assisting the larva in moving about in the cell. Reinhard states that the cocoon of the parasite is indistinguishable from that of its host except that it is slightly smaller.

Reinhard's account of the relationship of these two wasps is well worth reading in its entirety, especially as retold in chap. xiii of his *Witchery of Wasps* (1929). Reinhard is flagrantly anthropomorphic and teleological, but this should not be allowed to conceal the fact that his account represents the most complete study made of the behavior of a *Nysson* at any time or place.

Krombein (1958d) has recorded the chrysidid wasp *Elampus viridicyaneus* Norton as a parasite of *H. costalis*.

E. *Hoplisoides tricolor* (Cresson)

This western species has been the subject of a brief study by Evans, Lin, and Yoshimoto (1954). This study was made in August 1952, in a sandy strip along a field near Ulysses, in western Kansas. About 40 or 50 females nested here, all of the nests being located near the bases of plants, none of the nests being closer together than about 10 cm, and most of them being separated by at least 30 cm.

(1) *Nesting behavior*. All observations were made between 1100 and 1900 on 1 day, so it is difficult to be certain of the exact daily regimen. Several females were seen digging new nests in the late afternoon (1700-1900). In digging, the female backs out of the burrow as much as 10 cm scraping sand, then works forward toward the burrow, producing a small trough in front on the entrance. About 45 min are required to complete the nest, and the sand at the entrance is then leveled and the entrance closed. One nest dug out at 1850 contained an adult wasp in the empty cell behind a closure of sand, where she would perhaps have spent the night (no. CMY90). Since other gorytine wasps are reported to spend the night away from the nest, this point should be checked further before being

accepted as fact. We obtained the impression that one nest may serve a female for several days, as some of the nests contained cells with eggs and others with small larvae (no. CMY91). However, it is possible that some of these cells actually belonged to other, completed nests.

The nest is a simple oblique burrow forming about a 35° angle with the flat surface; burrow diameter is about 3 mm, cell diameter about 7 mm. In the 11 nests excavated, burrow length varied from 7 to 14 cm (mean 10.5 cm), cell depth from 4 to 8 cm (mean 6.0 cm). Most of the burrows were straight or nearly so. Four of these nests had received the final closure; of these one had two cells, two had three, and one had four. The deepest cell is invariably the first cell, others being added a short distance off from the main tunnel progressively back toward the entrance.

Only one species of leafhopper was used as prey in this nesting aggregation: *Parabolocratus brunneus* Ball (Cicadellidae). Of 194 leafhoppers taken from 19 cells of the wasp, 191 were adult females of this leafhopper, the other three last instar nymphs. According to Evans, Lin, and Yoshimoto (1954):

> The flight of this species is rather slow, and the wasp characteristically lands several inches away from the nest entrance (sometimes on a plant) rather than flying directly to it. On landing, the wasp stands on its front and hind legs and holds the leafhopper with the middle legs, venter-up, tightly against the under side of its thorax and abdomen. After landing and before entering the burrow the wasp invariably moves its abdomen up and down rather slowly in a broad arc, while holding the wings motionless at a slight angle from the horizontal.

After a moment, the wasp walks or takes a short flight to the burrow, depresses the wings and enters, of course digging through the closure with the front legs. Usually only 30 sec or so are spent inside the burrow, and the entrance is invariably closed upon leaving. From time to time wasps are seen to stay inside the nest for some time (about 15 min) and to throw up some additional sand. These intervals are undoubtedly used for oviposition and for closing off the cell and preparing a new one.

The number of leafhoppers in fully provisioned cells varied from 7 to 19 (mean 11). All were venter up and seemed thoroughly paralyzed. The egg of the wasp was invariably found on the top leafhopper in the pile; it was found to be laid longitudinally alongside the coxae in the manner already described for several species.

Final closure of the nest was observed several times. Sand is scraped from the walls of the burrow, later scraped in from the

§F. Ethology of other species of *Hoplisoides*

outside. The soil is packed into the burrow with vigorous blows of the pygidium, which describes small circles as the abdomen moves up and down fairly rapidly. Finally, the wasp scrapes sand over the nest from a distance of up to 7 cm, producing a small mound which is then flattened out by kicking sand in various directions. Most final closures were in the late afternoon, the wasps then beginning a new nest not far from the old one.

(2) *Natural enemies*. Several specimens of the small, brightly colored wasp *Nysson bellus* Cresson were seen in the nesting area, behaving in the usual manner of wasps of that genus. One specimen (no. HE102) was seen digging near a *Hoplisoides* nest. When a provisioning *Hoplisoides* landed nearby, the *Nysson* remained motionless. When the *Hoplisoides* entered the nest, the *Nysson* followed her in but came out almost immediately. The *Hoplisoides* remained inside the nest for 15 min, which was unusual since she was not preparing a new cell. During this time the *Nysson* entered the nest three times and came out twice. Finally the *Hoplisoides* came out, leaving the *Nysson* inside, and made a closure in the usual manner. The nest was dug out and found to have two cells, but unfortunately we were unable to locate the eggs of either wasp. One cell of another nest (no. CY91) was found to contain two larvae, one slightly larger than the other, but these larvae were not reared successfully.

F. Ethology of other species of *Hoplisoides*

A small amount of information is available on three other American species of this genus. Pate (1946) recorded a female of the western North American species *H. spilopterus* (Handlirsch) from Wenatchee, Washington, with *Stictocephala wickhami* VanDuzee (Membracidae). I found this species nesting along the San Diego River, San Diego County, California, on 3 August 1954 (no. 782). A female was seen descending obliquely to her nest in open, rather moist sand in a manner much resembling *nebulosus*. No movements of the abdomen were noted as described for *nebulosus* and *tricolor*. She scuffed open the entrance and went in holding the prey in the usual manner. After 1.5 min she left and closed the entrance. This nest was dug out at this point and found to contain a single cell with seven treehoppers, two adults and five nymphs, all *Spissistilus festinus* (Say) (Membracidae). The egg had not yet been laid. The burrow was straight, 17 cm long, the cell 13 cm beneath the surface. On 18 August 1964, I encountered a female of this same species at Cornish, Utah, carrying an adult membracid, *Campylenchia latipes*

(Say). She descended slowly and obliquely to her nest in a sandbank near an irrigation ditch. I did not examine the nest.

Krombein (1959) took a female *Hoplisoides denticulatus* (Packard) in North Carolina with a deltocephaline nymph 5.5 mm long (Cicadellidae). The Neotropical *H. umbonicida* Pate is reported to prey upon the membracid *Umbonis spinosa* (Houttyn) in Trinidad (Pate, 1941).

The nesting behavior of two Eurasian species of this genus is fairly well known. Ferton (1901, 1908, 1910, 1911) studied *Hoplisoides punctatus* (Kirschbaum) in both Provence and in Corsica. Several species of *Tettigometra* (Tettigometridae, Fulgoroidea), both nymphs and adults, provided the sole prey in both areas. The nest is dug in sandy soil and attains a depth of up to 15 cm. Ferton remarks that in this species and several other Gorytini he studied, the burrow is at first nearly horizontal and close to the surface for several centimeters, then bends down sharply before reaching the horizontal cells. The female *punctatus* brings in prey very rapidly, and may supply as many as 60 *Tettigometra* per cell. The nest is always closed between trips for prey. The egg is laid longitudinally on the side of the venter of the thorax as in other Gorytini.

Iwata (1964) studied *H. punctatus manjikuli* Tsuneki in Thailand. The burrow was 7 cm long and was dug into a sloping sandbank along a stream; the entrance was, as usual, kept closed between trips for prey. Iwata found only one cell, 8 × 10 mm in size, containing 13 prey. Twelve of these were nymphs of a species of Membracidae, one of them bearing an egg in the usual position for wasps of this genus. The remaining prey was a fairly large adult chalcid wasp (4.5 mm long) of an unknown genus of Encyrtidae. This is a most remarkable prey record for a gorytine wasp, and one supposes it must represent an unusual behavioral aberration. The use of Membracidae by this form leads one to wonder if it may not be specifically distinct from *punctatus*.

The nesting behavior of *H. latifrons* (Spinola) is much like that of *punctatus*. Maneval (1936, 1937) studied the species in France, and Grandi studied it over a period of years in Italy (summarized in Grandi, 1961). In both countries *latifrons* preys on several species of *Tettigometra* (Tettigometridae). The burrow is dug in loose, sandy soil and reaches a depth of 6–7 cm (Maneval) or 8–10 cm (Grandi). Most nests found were unicellular, but Maneval found one nest in which two burrows, each terminating in a single cell, diverged from a common orifice. The nest entrance is kept closed between trips for prey. Some 14 to 19 *Tettigometra* are provided per cell; oviposition is similar to that of other members of the genus. Grandi has described the digging behavior, Maneval the larva. According to Maneval, the species is quite solitary, but a given female tends to build successive nests in the same small area.

Maneval (1939) found *Nysson dimidiatus* Latreille attacking *Hoplisoides latifrons* quite commonly. Maneval watched a small group of *Hoplisoides* nests and saw two or three *Nysson* in the area, from time to time entering the *Hoplisoides* nests while the females were away, digging into the entrances and closing them upon leaving "just like the owners." He was able to dig out one cell carefully while the *Nysson* was still in it. He found the *Nysson* motionless in the middle of the pile of leafhoppers, "apparently holding on with its claws and feeling about with the tip of its abdomen." Finally the parasite left the cell, and Maneval examined the cell contents under a microscope. One of the leafhoppers had the wings slightly elevated, and beneath the wings he found the egg of the *Nysson*. This cell had about two thirds the full complement of prey. In another, fully provisioned cell he found an egg of the *Hoplisoides* and a newly hatched *Nysson* egg, the larva from which was presumably in the cell and would have destroyed the egg or small larva of the host. Later Maneval dug out several cocoons from the nesting area and found one *Nysson* cocoon, from which he extracted a larva. The *Nysson* cocoon differs scarcely at all from that of its host. Maneval has provided an excellent series of figures of the egg of the parasite on a leafhopper, of the larva and various details of its structure, and of the cocoon.

G. Genus *Gorytes* Latreille

The genus *Gorytes* is here used as in the catalogue of North American Hymenoptera (Krombein, 1951) and is the equivalent of the subgenus *Gorytes* of Beaumont (1954). These wasps occur in much the same situations as do the species of *Hoplisoides;* in fact, it is not uncommon to find species of the two genera nesting close together. The species of *Gorytes* tend to be a little more slender and more shining than *Hoplisoides,* and in general they lack the large integumental punctures of that genus. As in *Hoplisoides,* the epicnemial ridge is strong and continuous with a strong precoxal ridge; however, there is no ridge extending across the mesosternum as in that genus. The upper part of the mesopleurum is crossed by a strong oblique groove. The clypeus is narrower than in *Hoplisoides,* and the eyes are strongly convergent below; the propodeum has a groove extending back from the spiracles; the metasoma tends to be fusiform and more slender than in *Hoplisoides* (Figs. 21-27).

H. *Gorytes canaliculatus* (Packard)

This small wasp occurs throughout the northern half of the United States and the southern part of Canada. It is characteristic of restricted areas of fine-grained sand, especially small sandpits or sandy strips along streams. Of the 44 field notes summarized here, 3 were made in a small sandy area at Lexington, Massachusetts, 2 in a similar area at Andover, Massachusetts, 13 in two sandpits near Ithaca, 8 in a sandy area near Groton, and 1 in a small dune near Granby Center, Oswego County, New York, 2 at Presque Isle, Pennsylvania, 11 at two localities in Pottawatomie County, Kansas, 3 along the Snake River at Jackson Hole, Wyoming, and 1 in a sandy patch along a mountain stream near Aspen, Colorado. Several of the Ithaca notes were made by C. S. Lin, and all of those from Groton and Presque Isle were contributed by Frank E. Kurczewski. The observations made in Kansas and Wyoming pertain to the form described as *asperatus* by Fox, the others to typical *canaliculatus*, which is somewhat darker and lacks the strong yellow tints on the wings characteristic of *asperatus*. There seems no reason to treat the two separately here, as they are at most subspecifically distinct, and no behavioral differences were discovered.

Published reports on *canaliculatus* include prey records and nest data from Plummers Island, Maryland, by Krombein (1964a), and a brief ecological note from Presque Isle, Pennsylvania, by Kurczewski and Kurczewski (1963). The observations of Barth (1907), purported to deal with *canaliculatus*, may have been based partly or wholly on this species. Barth's wasps preyed upon Membracidae rather than Cicadellidae, and they also tended to nest under stones or tufts of grass rather than in bare sand: both features suggesting *nebulosus* rather than *canaliculatus*. However, I borrowed from the Milwaukee Museum one of the specimens on which Barth based his studies, and found it to be *canaliculatus* (actually the form known as *asperatus*). Possibly he observed both species and failed to distinguish them. In any case, Barth's observations are considered briefly in their proper place below.

This species typically occurs in small nesting aggregations of from 1 or 2 to 10–20 individuals. Either flat soil or gentle slopes and the sides of depressions are utilized. In several localities I have found the species nesting along streams, but it is not restricted to streamside localities and may, in fact, occupy very dry, powdery soil, as in a large hilltop blowout in Pottawatomie County, Kansas. Nests are usually widely spaced (0.5 m or more), but in relatively populous aggregations one often finds nests only 6–10 cm apart.

§H. *Gorytes canaliculatus* (Packard)

The nesting period of this species appears to be relatively brief. On South Hill, Ithaca, I found an aggregation of about 12 individuals on 10 June 1955, but 2 weeks later nesting activity had completely ceased in this place. At Lexington, Massachusetts, I found females nesting from 20–27 June 1963, and none thereafter, but at Andover, Massachusetts, the nesting dates were 31 July–6 August. Nearly all other observations are from the month of July. I would judge that the species is univoltine throughout its range, but that local populations emerge on slightly different dates. *G. canaliculatus* has occasionally been taken at flowers (for instance, *Pastinaca sativa*), but in general adults do not seem to fly very far from their nesting sites. Adults do not spend the nights in their nests, but I have no information on where they do spend them.

(1) *Digging and orientation.* At Ithaca, several females were seen digging new nests in the afternoon (1430–1700). Apparently no more than about 1 hr is required for completion of a nest. Most nests are dug on open sand, but I have found a few close to the base of tufts of grass, under sticks, or under dried leaves. The oblique burrow is dug with the forelegs working synchronously and throwing the sand beneath and behind the body in small spurts as the abdomen moves up and down rhythmically. During digging, the wasp backs out of the entrance about 5 cm, raises her wings at an angle for a moment, then flattens them against her back as she moves forward kicking sand. When the nest is complete, the wasp comes out, turns about, and scrapes sand into the entrance to make a temporary closure. She then backs across the mound 5–7 cm and works toward the entrance in almost a straight line, kicking sand. This is repeated several times, over slightly different paths, so that the mound is partially leveled. Afterwards sand is scraped in various directions over the entrance, concealing it completely.

Following the initial closure and concealment of the new nest, the wasp makes an orientation flight. This was observed several times and appears to follow a pattern quite different from that of *Hoplisoides nebulosus*. The wasp flies in a series of loops and figure eights only about 5 cm above the nest entrance, then, after five or more such loops, increases her height slightly and takes several more loops at a gradually increasing height from the nest entrance. Then she swiftly flies off to take her first leafhopper, which may require only a few minutes. Females occasionally interrupt their orientation flight by landing briefly at the nest or landing on a leaf or other object in the vicinity.

One individual in Kansas (no. 131) abandoned her burrow when it was about 8 mm long, filling it up as if making a closure before starting another burrow 0.5 m away. I found that many females in Kansas dug their nests in the late afternoon (1500-1830); presumably nests dug in the afternoon are provisioned the next day, as reported by Barth in Wisconsin.

(2) *Nature and dimensions of nest.* The burrow is oblique and forms roughly a 45° angle with a horizontal surface; in a sloping surface the burrow may be nearly perpendicular to the surface, although approximating 45° with the horizontal. Burrow diameter is about 3 mm; the cells are horizontal or nearly so, and measure about 6-8 mm in diameter and 9-12 mm in length. The maximum number of cells found in any nest was four. As in other gorytine wasps, additional cells are constructed progressively back toward the entrance as the burrow is shortened. Cells of any one nest are separated by from 2 to 4 cm (rarely only 1 cm). Data on burrow length, cell depth, and number of cells are presented in Table 6. It should be remembered that most nests were dug out before they had received final closure; doubtless most completed nests are multicellular. However, several nests that had received the final closure were found to have only a single cell (nos. 1073, 1085, and 1951). One

TABLE 6. NEST DATA FOR GORYTES CANALICULATUS

Locality	Number of nests				Burrow length (cm)	Cell depth (cm)
	1-cell	2-cell	3-cell	4-cell		
Lexington, Mass.	2	2			11.5 (9-16)	8.0 (6-10)
Andover, Mass.	1	1			13.5 (9-16)	9.5 (6-11)
Ithaca, N.Y.	5	4	2		8.5 (4-15)	7.0 (4-10)
Groton, N.Y.	5	1	3		9.5 (8-13)	8.0 (5-11)
Presque Isle, Pa.	1				14.0	11.5
Plummers Isl., Md. (Krombein)	1				14.0	13.0
Pottawatomie Co., Kan.	1	3	1	1	10.0 (8-12)	8.0 (6-10)
Wisconsin (Barth)		2	1		11.0	8.0 (4-10)
Aspen, Colo.	1				7.5	4.5
Jackson Hole, Wyo.	1	1			8.5 (7-10)	7.5 (7-8)

§H. *Gorytes canaliculatus* (Packard)

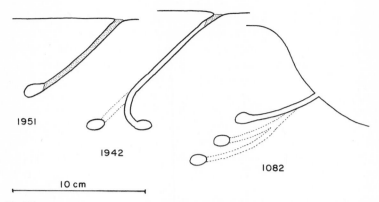

Fig. 38. Three typical nests of *Gorytes canaliculatus:* nos. 1951 and 1942, Andover, Massachusetts; no. 1082, Ithaca, New York. Burrows indicated by broken lines were filled and could not be traced exactly.

two-celled nest (no. 1942) is known to have been provisioned over 2 days, while another two-celled nest (no. 353) was completely provisioned in 1 day. Barth reported that the three-celled nest he excavated represented the work of 3 days (Fig. 38).

The nest entrance is kept closed at all times except when the wasp is actually inside the nest taking a leafhopper to the cell or digging a new cell. Upon leaving the nest, the wasp faces away from the hole and scrapes sand over it with the forelegs. Only three to five scuffs of sand are made and only 1–2 sec required for the closure. The wings are held flat until the closure is finished, then elevated obliquely for a moment before the wasp takes flight.

(3) *Provisioning the nest.* The prey is carried in flight, supported by the middle and usually the hind legs. The wasp usually flies directly to the nest entrance, but may occasionally land on a leaf overlooking the nesting area. When the wasp lands, she holds the prey with the middle legs only. Upon landing, the wings are held obliquely high above the abdomen for a moment, then flattened against the abdomen as the entrance is opened with the forelegs. Typically the wasp remains inside the nest only a very short time (10–20 sec). There is much variation in the amount of time required to take a leafhopper. For example, no. 597 brought in only eight leafhoppers between 1500 and 1730, requiring a minimum of 10 min for each trip. Number 1074, on the other hand, brought in eight leafhoppers in only 42 min (1458–1540), requiring from 3 to 10 min for each trip. Most provisioning occurs on sunny days between 1030 and

1700. As usual, the prey is placed in the cell head in, venter up.

The prey consists of either nymphal or adult leafhoppers (Cicadellidae and Fulgoroidea); many cells contain both nymphs and adults, while some contain all adults or all nymphs. The number of leafhoppers in fully provisioned cells varies from 6 to 19 (most commonly from 10 to 15). As might be expected, when nymphs are employed in whole or large part, more are required per cell. It is interesting to note that in most localities leafhoppers of the genus *Idiocerus* were employed, in some cases exclusively. It is probable that these leafhoppers merely happen to be common at the season when *canaliculatus* nests and on the type of trees (willows, poplars) that tend to grow in sandy areas. In Kansas, *canaliculatus* preyed extensively on genera other than *Idiocerus*. In all areas, individual wasps tended to take all or mostly leafhoppers of the same species. For example, 22 prey taken from the two cells of nest no. 366 were all adult *Haplaxius pictifrons*, while 23 taken from another nest in the same locality (no. 111) were all adult female *Macropsis* sp. However, some nests contained mixtures of two or more species. Prey records are summarized in Table 7.

As noted earlier, I believe that Barth's records of this species preying upon Membracidae may possibly be the result of his confusing this species with *Hoplisoides nebulosus*.

The leafhoppers found in the cells of this wasp may show movements of the legs for several hours, but soon they become immobile and appear dead. In no case were any movements of body parts noted after the first day, and in no case did any of the prey placed in rearing tins recover from paralysis. However, the leafhoppers appear to remain in fresh condition for several days.

(4) *Immature stages and development.* The egg is laid on one of the topmost leafhoppers after the cell is fully provisioned. It is laid longitudinally on the side of the venter of the thorax as in other gorytine wasps; in most cases the middle and hind femora of the leafhopper lie across the egg (Fig. 39). The egg is about 2.0 mm long and 0.70 mm in maximum width. The egg hatches in about 2 days, and the larva reaches maturity after about 4 days of feeding. One larva spun its cocoon in a rearing tin on 29 June 1952 and produced an adult female on 19 May 1953 (no. 111, Pottawatomie County, Kansas).

(5) *Final closure.* Final closure of the nest was observed several times. The wasp comes out and walks up to 1–2 cm from the entrance, then backs in kicking sand. When the burrow is nearly full, she

§H. *Gorytes canaliculatus* (Packard)

TABLE 7. PREY RECORDS FOR GORYTES CANALICULATUS

Species of prey	No. of specimens*					
	Mass.	N.Y.	Pa.	Md.	Kan.	Colo.
CICADELLIDAE						
Idiocerus lachrymalis Fitch						2
I. populi pallidus Fitch		1				
I. populi suturalis Fitch		30				
I. stigmaticalis Lewis		19				
I. spp. (mainly immatures)	7	85	6	12		
Macropsis viridis Fitch				1		
Macropsis sp.	1				23	
Norvellina helenae Ball					2	
Oncopsis sp.	5					
Orientus ishidae Mats.	1					
Paraphlepsius sp.					1	
Stragania alabamensis Bak.					24	
FULGOROIDEA						
Haplaxius pictifrons Stal					22	

* Maryland records from Krombein (1964); Pennsylvania records supplied by F. E. Kurczewski. See introductory remarks under this species for exact localities. Barth (1907) reported this species preying upon two species of Membracidae in Wisconsin (Cyrtolobus fenestratus Fitch and Atymna inornata Say).

may bite soil from around the entrance and scrape it into the hole. Much use is made of the tip of the abdomen for pounding soil into the burrow. Number 1074 began her final closure at 1540, and at 1548 had filled the burrow to the surface. She remained in the area

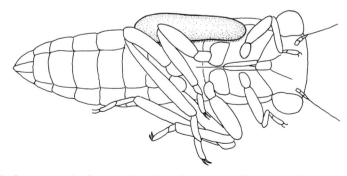

Fig. 39. Immature leafhopper bearing the egg of *Gorytes canaliculatus*. (From a field sketch made at Groton, New York, by F. E. Kurczewski.)

until 1555 scraping sand in various directions over the well-concealed nest entrance.

One wasp, after final closure and concealment, began a new nest only a few centimeters away.

(6) *Natural enemies.* This wasp is rather commonly attacked by miltogrammine flies and by wasps of the genus *Nysson*. In one instance, at Ithaca, a provisioning female was followed several times by the fly *Phrosinella fulvicornis* (Coq.), but the fly was not actually seen to larviposit on the prey or enter the nest; however, two maggots were later found in the cell. A puparium was found in one other nest at Ithaca, but no fly was successfully reared from it. At Lexington, Massachusetts, two of six cells contained one maggot each, and at Andover, Massachusetts, one of three cells contained a single maggot. In each case the maggot had destroyed the egg or larva of the host. The maggot from Andover was reared successfully and found to be *Phrosinella* sp. (?*fumosa* Allen).

Both cells of nest no. 1802, at Jackson Hole, Wyoming, contained maggots when the nest that the wasp was seen provisioning was dug out late on the same day. One cell contained one maggot, the other two; the eggs of the wasp had apparently been destroyed by the maggots. These three maggots were placed together in an artificial cell in a rearing tin. Three days later only one large maggot was present. This maggot formed its puparium 8 days after the nest was excavated, and an adult miltogrammine fly emerged in June of the following year. It was determined as *Metopia argyrocephala* (Mg.).

Female *Nysson daeckei* Viereck were seen in the nesting area of *Gorytes canaliculatus* on numerous occasions both in New York and Massachusetts. They flew close to the sand, now and then landing and walking along tapping the sand with their antennae; often they would remain motionless on the ground or on a stick for some time. I did not actually see any enter *Gorytes* nests at Ithaca, but one of the larvae reared (no. 677B) proved to be that of a *Nysson* (easily told from that of *Gorytes* by the mamilliform process on the head). At Lexington, the parasites seemed much more common than their hosts; one was seen digging into a known *Gorytes* nest, but no *Nysson* eggs or larvae were found in the nests dug out. Frank Kurczewski found female *Nysson daeckei* entering *canaliculatus* nests at Presque Isle, Pennsylvania, on two occasions (reported briefly by Kurczewski and Kurczewski, 1963). The following is an extract from Kurczewski's notes made at that locality:

§H. *Gorytes canaliculatus* (Packard) 65

[The first *Nysson daeckei*] sat on a flat rock four centimeters away from the closed entrance of a *G. canaliculatus* nest. A *Gorytes* with its leafhopper held underneath flew in, landed on the sand, and opened its entrance. It came out in about 10 seconds, made a temporary closure, and flew off. The *Nysson,* which had been motionless on the rock during all this, then made a short flight to approximately the general area of the closed entrance and began tapping the sand with its antennae while walking erratically in small circles. Within five seconds, the *Nysson* had located the *Gorytes* entrance and opened it, its fore legs throwing the loose sand beneath the abdomen and behind it. The *Nysson* then entered head first, reappeared at the entrance in 27 seconds, and made a very brief closure, standing below the entrance while throwing sand backwards into it using the fore legs; the tip of the abdomen was at all times pointed toward the entrance while closing. The *Nysson* then walked slowly away following a zig-zag course. Another female *daeckei* showed similar behavior only without having watched a *Gorytes* bring prey to her nest. At least six other female *N. daeckei* in this same area were observed making short, hopping flights of not more than a few centimeters, and then walking slowly over the sand surface. These walks described erratic circles; the antennae were held somewhat outstretched and constantly tapped the sand. Only rarely did females dig into the sand and when they did they had usually located a *Gorytes* nest.

Nysson daeckei is apparently a univoltine species with a short flight period, usually in early summer, like its host. At Groton, New York, Frank Kurczewski noted *Nysson lateralis* females behaving in exactly the same way as *daeckei,* and one *lateralis* was seen entering and leaving a *canaliculatus* nest.

Barth (1907) reported *Nysson fidelis* Cresson entering two nests of "*Gorytes canaliculatus*" in Wisconsin; in each case the *Nysson* remained in the nest a short while and closed the entrance upon leaving. In one cell Barth found the egg of the *Gorytes* on a treehopper plus another egg lying unattached on the floor of the cell. The second egg was reported to be smaller and less shiny than that of the *Gorytes,* so undoubtedly it was a *Nysson* egg. Possibly the egg was knocked loose in the course of digging out the burrow; or it may have been deposited loosely abnormally. All other records indicate that *Nysson* oviposits on one of the prey and in a concealed location.

A mutillid wasp was found associated with *G. canaliculatus* on one occasion. Nest no. 26, in Pottawatomie County, Kansas, was dug in the afternoon of 17 July 1952 and at 1700 had been closed from the outside. At this time some movement was seen in the earth covering the entrance, and a female *Timulla leona* Blake (Mutillidae) emerged from the nest. Unfortunately this nest could not be followed further, but it seems doubtful that the mutillid would have laid her egg in a freshly prepared nest. However, it is possible that the wasp was actually preparing a new cell in an old nest and that the mutillid was attacking the contents of an older cell.

I. *Gorytes simillimus* Smith

This is not an uncommon *Gorytes* in the eastern United States, occurring in small, sandy areas similar to those occupied by *G. canaliculatus*. The peak of the nesting appears later than in *canaliculatus*, records from New York state covering the period 6 July–31 August. On one occasion I found the two species nesting at the same time and place near Ithaca, but in general *simillimus* makes its appearance after *canaliculatus* has disappeared. I have taken several female *simillimus* with prey, but found only one nest (at Ithaca, 31 August 1958, no. 1588). Krombein has published notes on this species nesting at Buffalo, New York (1936) and at Westmoreland State Park, Virginia (1952).

Prey carriage and manner of entry into the nest are similar to those of the preceding two species. The prey consists of nymphs and (more commonly) adults of relatively large leafhoppers. Rohwer (1911) described *gyponacinus* as a predator on *Gyponana flavilineata* (Fitch); Krombein (1958d) has recently placed *gyponacinus* in the synonymy of *simillimus*. Krombein records two different wasps at Buffalo using adult *Gyponana octolineata* (Say), and I found adults and nymphs of this same species to be used at Ithaca. I also took a female at Ithaca with a female *Gyponana* of undetermined species, as did Krombein at Westmoreland State Park, Virginia. This wasp apparently does not specialize on this genus of leafhoppers, however, as I took one at Ithaca with *Scaphoideus productus* Osborn.

Krombein found the wasp nesting at Buffalo in the vertical surface of a sand-covered stump, although in Virginia the nest was apparently in flat soil. I found one female at Pittsford, New York, entering a nest in the side of a vertical sandbank, and the one nest successfully excavated at Ithaca entered a sloping sandbank (about a 45° slope). This nest was at about a 90° angle with the surface. The burrow was 8 cm long and terminated in a cell which was not yet fully provisioned. Krombein dug one nest in Virginia and found the burrow to be at a 45° angle with the horizontal surface; the single cell was only about 4 cm beneath the surface. Neither Krombein's nest nor mine yet contained an egg, indicating that the egg is laid after completion of provisioning, as usual in this tribe of wasps. No parasites have been found attacking *simillimus*.

J. Nesting behavior of other species of *Gorytes*

Krombein (1958a) has recently described *G. deceptor,* a species closely resembling *simillimus* and previously confused with it. However, *deceptor* apparently preys upon Membracidae; at least there is a single record of a female taken by K. W. Cooper at Princeton, New Jersey, with an adult *Spissistilus constans* (Wlk.).

The widely distributed and relatively common North American species *G. atricornis* Packard has never been found nesting, and may nest not in bare places but amid vegetation. Dr. Ellis G. MacLeod took several specimens with prey at Belmont, Massachusetts, during July 1963. These wasps were in an open meadow of tall grass and herbs. They may have been flying to bare, sandy places in the vicinity, but no such places were located. The prey consisted of adult Cercopidae, *Aphrophora parallela* (Say). This species is also known to prey upon the membracid *Cyrtolobus tuberosus* (Fairmaire) at Ithaca (Pate, 1946).

Apparently nothing has been recorded regarding the behavior of the remaining North American species of this genus, including all members of the subgenus *Pseudoplisus.* One species of *Pseudoplisus, phaleratus* Say, is very widely distributed and not uncommon. It is curious that it has not been found nesting. Frank Kurczewski watched one female at Groton, New York, and found her flying slowly about various crevices in sandy soil, digging in pre-existing holes and in pits under stones. She dug with the forelegs in the usual manner of these wasps, holding the wings flat over her back while digging. It is possible that members of this subgenus characteristically nest in situations where they cannot be readily observed.

Several European species of *Gorytes* (subgenus *Gorytes*) have been studied in some measure. The most detailed studies are those of Maneval (1939) on *planifrons* (Wesmael). Maneval found a small colony of four individuals nesting in sloping, fine-grained sand in southern France. The nest is a simple, straight, nearly horizontal burrow. Although he studied the wasp over several days, Maneval found no nests with more than one cell. The nest was as usual closed between trips for prey. The prey consisted of adult *Issus coleoptratus* Fabr. (Issidae, Fulgoroidea), four being used per cell. The prey is well paralyzed and remains immobile but fresh for several days. The egg is laid on the side of thorax in the manner described above for *canaliculatus*. Maneval figured the egg on the prey, the nest, the cocoon, and the full-grown larva and various structural details thereof.

Maillard (1847) long ago reported *Gorytes laticinctus* Lepeletier nesting in a flowerpot in his garden in France. The burrow was found to be about 10 cm long, the prey to consist of nymphs of the cercopid *Philaenus spumarius* L. Olberg (1959) has recently presented a series of photographs of this wasp digging its nest. Ferton (1905) has also found *G. sulcifrons* Costa preying upon *Philaenus spumarius*, as has Dubois (1921). In Italy, Grandi (1961) found *G. pleuripunctatus* Costa nesting in soil and provisioning with *Dictyophara* (Dictyopharidae, Fulgoroidea). Maidl and Klima (1939) list additional references on the biology of Palaearctic species, but none of these references contain detailed information on nests or prey.

The Neotropical *Gorytes brasiliensis* Shuckard has been studied briefly by Williams (1928). Williams found females nesting "in banks of rich soil along the margin of the jungle" near Belem, Brazil. One nest later dug out was situated in masses of earth adhering to the roots of overturned trees near Japaty, Brazil. The cells of this nest were found to be provisioned with adult and immature Fulgoroidea of several genera, including *Dictyophara* (Dictyopharidae), *Thionia* (Issidae), and a species of Flatidae. Six prey were provided in each of two cells, the prey being "limp and sometimes capable of very slight movement." The egg was found to be laid "alongside the thorax so that it margined the basal portion of the forewings." The cocoon is described as "a soil cask that was gently rounded at the fore end and more narrowed and drawn out a little, nipple-like at the base."

K. Ethology of some genera closely related to *Gorytes*

In this section I shall discuss seven genera I have had no opportunity to study in the field myself. Two of the genera, *Psammaecius* and *Dienoplus*, differ from *Gorytes* only in minor details and are considered subgenera of that genus by Beaumont (1954). Four others (*Psammaletes, Harpactostigma, Hapalomellinus,* and *Lestiphorus*) are also very similar to *Gorytes*, but they differ in having the first metasomal segment small and often constricted apically, so that the base of the metasoma is subpetiolate or nodose (a condition also found in *Ochleroptera*, a genus considered in a later section of this chapter). The remaining genus, *Argogorytes*, requires special comment; it is superficially similar to *Gorytes*, but there are some interesting differences.

The genus *Argogorytes* contains rather hairy species possessing a broad front and slightly emarginate eyes; the clypeus is transverse and located mostly below the bottoms of the eyes. The mesosoma is somewhat less compact and smooth contoured than in many Gorytini; the mesoscutal laminae are very narrow and have no trans-

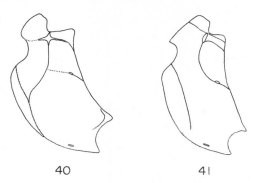

Figs. 40 and 41. Lateral aspect of mesopleuron: 40, of *Argogorytes campestris;* 41, of *Ammatomus moneduloides.*

verse ridge; the epicnemial ridges are strong and continuous across the sternum, but precoxal ridges are absent; on the upper part of the mesopleurum there is a strong oblique groove and a weak scrobal groove (Fig. 40); the pecten is essentially absent. The structure of the metasoma differs in no important way from that of *Gorytes.*

The genus *Argogorytes* is widely distributed, the North American and European species being placed in the subgenus *Archarpactus.* Nothing is known regarding the nesting behavior of the single North American species, but two European species, *campestris* (Mueller) and *mystaceus* (Linnaeus) have been studied (see summary and references in Hamm and Richards, 1930). Rather numerous prey records for both species indicate that they are specialists on spittle insects of the genus *Philaenus* (Cercopidae). The nymphs alone are employed, and these are extracted from the spittle masses on vegetation in a manner best described by Adlerz (1906) for *campestris.* The *Argogorytes* is said to land on the stem and walk to the spittle, then plunge her legs and sting into it. The prey is carried in flight with the middle legs. The burrow is said to have from six to nine cells and be left open while the wasp is away. The prey are placed in the cell head in, from 19 to 27 being supplied per cell. *Nysson spinosus* Foerster is recorded as a parasite of both *campestris* and *mystaceus.*

Another European species of a related genus, *Lestiphorus bicinctus* (Rossi), is also reported to attack *Philaenus spumarius* Linnaeus (Bernard, 1934). The North American species *Psammaletes pechumani* Pate is recorded by Pate (1946) as preying upon the fulgoroid *Ormenoides venusta* (Mel.). Gittins (1958) has recently published a note on *Harpactostigma laminiferum* (Fox), a species occurring in the northwestern United States. He found the prey to consist of immature *Scolops* sp.

(Fulgoroidea). This nest was unicellular (but still being provisioned when dug out); the burrow was about 17.5 cm long and was constructed from the bottom of a crack in a vertical clay bank.

Psammaecius is a small Palaearctic genus closely related to both *Gorytes* and *Hoplisoides;* in fact, the most recent catalogue of North American Hymenoptera (Krombein, 1958) lists *Hoplisoides* as a subgenus of *Psammaecius. P. punctulatus* (Van der Linden) has been studied briefly by Ferton (1901, 1905) in France, and there appears to be nothing strongly individualistic about its nesting behavior. Ferton reports that the prey is always adult *Selenocephalus obsoletus* Germ. (Cicadellidae), four or five being used per cell. The nest is in rather compact sand and reaches a depth of about 12 cm. The prey is carried with the middle legs, as in other Gorytini, and the egg is laid longitudinally on the side of the venter of the thorax.

A fairly large genus of small gorytines is *Dienoplus,* which in much of the older literature is called *Harpactus. D. gyponae* (Williams) was studied by Williams (1914) in western Kansas. This species was found nesting in sandy soil with clumps of weeds and grass; the prey, *Gyponana cinerea* Uhler (Cicadellidae), was apparently being captured in the grass. Williams states that the bugs are malaxated after stinging, then carried to the nest in flight supported with the middle legs. In digging, the wasp throws out soil with "her well-fringed fore feet," but carries out "larger pieces of dirt in her jaws." When the nest is finished, the wasp comes out head first and makes a short locality study before taking off to hunt for leafhoppers. The one nest excavated was about 7.5 cm long and had two cells, each with four bugs (only one immature); the wasp was apparently preparing a third cell in the same nest. The egg is laid on the side of the venter as usual in this tribe.

Of the Palaearctic species, *D. laevis* (Latreille) is perhaps best known (Ferton, 1901; Maneval, 1928; Iwata, 1937).[1] The nest is a simple burrow which may contain two successive cells. Maneval reports the depth as 15 cm, Iwata as only 2.5-4 cm. The entrance is closed upon leaving, the egg laid in the usual manner of these wasps. The prey consists of several genera of Cicadellidae (Ferton, Maneval) as well as Cercopidae (Iwata), mainly immatures. The nesting behavior of *D. tumidus* (Panzer) is also fairly well known (Adlerz, 1903, 1906; Hamm and Richards, 1930; Maneval, 1932; Bristowe, 1948), and very similar to that of *laevis* so far as is known. Adlerz and Maneval both found that the nest entrance is closed

---

[1] According to Tsuneki (1963b), Iwata was probably working with *D. tumidus japonensis* Tsuneki. *D. laevis* is not definitely reported from Japan.

while the wasp is away, but Bristowe found it to be left open. Maneval states that there are several cells per nest. The prey consists of adults and nymphs of various Cicadellidae and Cercopidae. Recorded parasites are *Hedychridium roseum* Rossi (Chrysididae), *Nysson maculatus* Fabricius, and *N. dimidiatus* Jurine. Adlerz found an egg of *N. maculatus* in four cells of three nests he saw the *Nysson* entering; in each case the egg was laid longitudinally beneath the wings of a leafhopper. In another nest he found a small *Nysson* larva in the same position.

References on the biology of several other Palaearctic species of *Dienoplus* are listed by Maidl and Klima (1939). So far as these various reports go, all of the species appear strikingly similar to the three species considered above and to the species of *Gorytes*. All of the known species make simple burrows in sandy soil. *D. affinis* (Spinola), *D. fertoni* (Handlirsch), *D. leucrurus* (Costa), and *D. lunatus* (Dahlbom) all take Cicadellidae, while *D. concinnus* (Rossi) and *D. elegans* (Lepeletier) take both Cicadellidae and Issidae. *D. concinnus* is known to make multicellular nests with an average of 15 cells per nest; it is attacked by *Nysson trimaculatus* Rossi (Deleurance, 1946). *D. elegans* is the host of *Nysson dimidiatus* Jurine (Ferton, 1901). The *Nysson* is said to enter the nests of its hosts, stay in a short while, then close them upon leaving, but Ferton failed to find the egg. *D. lunatus* is said to be the host of the chrysidid wasp *Hedychridium integrum* Dahlbom (Olberg, 1959, after Haupt).

There remain to be considered the recent observations of Cazier and Mortenson (1965) on *Hapalomellinus albitomentosus* (Bradley). This wasp is restricted to semidesert areas of the southwestern United States. It is a slender wasp with a nodose first metasomal segment, richly ornamented with whitish pubescence, although otherwise very similar to Gorytes. The observations of Cazier and Mortenson were made in Cochise County, Arizona, at about 4700 ft. The nests were located in bare, dry, sandy soil, mostly on level ground. Adults were observed taking nectar from *Salsola*. I have taken this species in numbers on *Croton* and *Chilopsis* in western Texas, and Krombein (1961) recorded *Hapalomellinus* sp. from *Euphorbia* in Arizona.

According to Cazier and Mortenson, these wasps either back out or come out head first when bringing soil from the burrow. "When they come out head first, pushing a little dirt in front of them, they turn around immediately in front of the entrance and start throwing dirt out behind them with their front legs. During most of the excavating activities outside the nest entrance the body is held up at about a 45° angle from the ground surface by the long hind legs. This gives them the appearance of almost standing on their heads . . ."

Cazier and Mortenson present considerable data on digging behavior. The mound is at first allowed to accumulate outside the entrance, but later the females interrupt their digging and spend "from 5 to 15 minutes spreading the dirt out in front of the entrance to a distance of from 5 to 10 cm . . . These movements are very erratic and the dirt is thrown in all directions, occasionally even back over the burrow entrance, in a thin layer that leaves no telltale evidence at the entrance site except for the entrance hole." The latter is closed from the outside at all times when the female is away from the burrow, from the inside "when the diurnal activities cease or after periods of prey collecting and excavation." One assumes that the females spend the night in the nest.

After the initial closure, the female makes an orientation flight described by Cazier and Mortenson as follows: "She flies slowly back and forth in a short arc several times [facing the entrance], gradually moving further away and upward, widening the arc into a half circle and finally into a complete circle around the area and between 30 and 40 cm. away and above the entrance site. After completing 3 to 5 circles she flies away."

The burrow of this species is only 4 to 5 mm in diameter and after numerous changes in angle and direction reaches a depth of from 7 to 10 cm. Two of the nests excavated contained two cells, but Cazier and Mortenson believe that the upper cell in each case was merely a storage cell. This point should be studied further, since storage in special cells or in the burrow is otherwise unknown in the Nyssoninae, although it is common enough in certain other subfamilies (Philanthinae, Astatinae). In any event, it seems certain that a second cell would eventually have been prepared in these nests, possibly more than two.

The prey in this area consisted entirely of adult and (less commonly) immature Cicadellids, *Stragania robusta* (Uhler), 14 to 15 being supplied per cell. Details of prey carriage and oviposition are essentially as described for *Gorytes;* the egg is said to be laid on the last leafhopper placed in the cell. Most of the *Hapalomellinus* females observed were being followed by "from one to three small sarcophagid flies, *Senotainia* (*trilineata* Wulp complex) sp. as they came in with prey." They observed several of these flies actually landing on the prey while the wasp was in flight. One of the two fully provisioned cells studied contained maggots.

Much more detail on most aspects of behavior will be found in Cazier and Mortenson's paper.

L. Genus *Clitemnestra* Spinola

This is a poorly known genus which is confined to Chile and to Australia. The species familiar to me are superficially *Gorytes*-like, but the abdomen is spotted rather than banded. There is a small constriction between the pro- and mesonota, the mesoscutal laminae are simple, and the oblique groove of the mesopleurum is well developed. The wing venation is unusual in that the recurrent veins are received close to the corners of the second submarginal cell. The terminal abdominal segments and genitalia are also of unusual form. The structure of this genus is discussed further in Chapter XIII: D, 2, and some of the major structural features are shown in Figs. 194–199.

We are indebted to Janvier (1928) for a description of the nesting behavior of two Chilean species of this genus, *chilensis* (Saussure) and *gayi* (Spinola). *C. chilensis* was found nesting in hillocks of clay soil, along with various species of *Cerceris, Philanthus,* and *Sphex*. The colonies are said to be dense, a hundred or more nests occupying a small area. The wasps make much use of their mandibles in digging, and from time to time clear away the soil with their front legs, which have a weak pecten. The burrows are from 15 to 30 cm in length and eventually have a cluster of cells (five to ten), the first cells being closest to the entrance. The nest entrance is left open during provisioning.

In two different colonies, Janvier found the prey of *C. chilensis* to consist entirely of nymphs and adults of the genus *Dictyophara* (Fulgoroidea, Dictyopharidae). These insects were captured on vegetation near the colony. The female wasps fly slowly about the foliage, descending quickly upon their prey and seizing it with their mandibles and front legs; the sting is inserted between the middle legs of the prey. The wasps then fly quickly to their nests, holding the prey between their mandibles and legs. The hoppers are piled in the cell with their heads toward the apex of the cell. The egg is laid on the first prey in the cell, "fastened over one of the middle legs, so that the cephalic pole is directed toward the thorax."

The wasp is said to bring in about ten more *Dictyophara* "in two or three days"; however, eclosion of the egg requires 5 or 6 days, so there is apparently no contact between mother and larva. Janvier's description and sketches of the larva are not sufficiently detailed to be meaningful. Particles of earth are incorporated into the walls of the cocoon, which bears a general resemblance to that of *Bembix*.

Clitemnestra gayi was found to form populous colonies in the walls of clay banks in forested areas, the nest entrances often being more or less concealed among mosses and lichens. As in the preceding species, several successive cells are prepared from the same burrow, the later cells being deeper in the soil than the first ones. Both males and females are said to spend the night in the nests.

The prey of this species consists of nymphs of a small homopterous insect (not identified) plus a few adult Membracidae. The prey is captured and stung in much the same manner as in *chilensis*. It is carried to the nest in the mandibles. The egg is laid on the first prey in the cell "across the second pair of legs." Smaller prey are said to be thoroughly paralyzed, but larger ones only incompletely so. In both species the colonies are said to remain in the same area for several successive years.

There appear to be several unusual features in the behavior of these two species of *Clitemnestra* as compared with that of *Gorytes* and *Hoplisoides*. The manner of carrying the prey is different, the egg is laid on the first prey in the cell and in a manner different from that of other Gorytini, and the nest is provisioned over a period of several days. It is also unusual that the first cell in the nest is the one closest to the entrance and that the egg requires 5 or 6 days to hatch. Despite the undoubted excellence of much of Janvier's work, one notes with surprise that eclosion of the egg in many Chilean wasps requires much longer than is reported for related wasps in other parts of the world, and one notes various other discrepancies which suggest either that many Chilean wasps have certain common peculiarities or that Janvier may have at times drawn certain assumptions which more detailed studies might have proved false. Several (or even all) of the unusual behavioral characteristics of *Clitemnestra* may prove to be real, but I find myself unable to accept them without the proverbial grain of salt until they have been confirmed by other workers.

M. Genus *Ochleroptera* Holmberg

This is a small genus of small wasps which is confined to the New World. The basal segment of the metasoma is strongly nodose, and the eyes are unusual in possessing much larger facets below than above. The front is narrow, the eyes converging strongly below; on the mesopleura the epicnemial ridge merely comes to an end below, there being neither a precoxal ridge nor a transverse ridge across the sternum; the legs are rather spinose, but there is no well-developed pecten on the front tarsus. Pate (1947) regards *Ochleroptera* as a derivative of the closely related, archaic genus *Clitemnestra*. Presumably *Ochleroptera* evolved in South America during the Tertiary, as most of the species are still restricted to that continent.

N. *Ochleroptera bipunctata* (Say)

This very small gorytine wasp occurs throughout much of temperate and subtropical North America. It does not appear to be strongly restricted ecologically, but occurs in many types of open country. The females apparently hunt their prey in grass, weeds, and bushes, and the nests are dug in small patches of bare soil, often in the sides of vertical banks. These wasps do not often visit flowers, but they sometimes occur on honeydew in some numbers. Strandtmann (1945) has published a note on a few individuals found nesting in flower boxes at Dallas, Texas, and Pate (1946) published a prey record from Ithaca, New York. My seven field notes were made at two sandbanks near Ithaca, and are not nearly as detailed as might be desired. New York nesting records cover the period 26 July–30 September; the species is especially characteristic of late summer and may be univoltine in the northeastern states.

Of the seven nests located at Ithaca, five were in vertical banks, in every case within a few centimeters of the top of the bank, where the soil was in some measure held together by the roots of grass and herbs growing in the flat soil above. The other two were in the side of sloping banks. In every case the soil was a rather coarse sand of medium moisture content. Strandtmann's wasps nested in the flat soil of flower boxes, presumably not especially sandy. He states that the flowers were watered each day "but the wasps were undismayed. As soon as the surface dried enough to become workable, they would dig open their holes and go back to work provisioning them."

Digging of the nest was not observed. In no case was a mound of soil observed at the nest entrance; in the case of the nests in vertical banks, of course, the soil from the burrow must have merely rolled down the bank. In vertical banks the burrow is straight and horizontal or it may curve downward somewhat; in sloping banks it is oblique, at roughly a 90° angle with the bank. Burrow diameter is 3–4 mm, often slightly more than this at the entrance. The length of the burrow is rather variable, in the four nests studied in detail varying from 9 to 20 cm (to the deepest cell). The cells are ovoid and measure about 6 × 10 mm. In the six active nests dug out, the number of cells varied from one to three. On one occasion (no. 1584), while digging out the nest of another wasp, I accidentally uncovered seven cells of an *Ochleroptera*. These cells were not all close together, and they may have represented cells of more than one nest; however, all cells were provisioned with adults of a single species of leafhopper, suggesting that they were the work of one female. In nests that were studied more carefully, the cells were found to be separated by about 2–3 cm. The first cell prepared is the deepest; additional cells are constructed from short side burrows (1–2 cm long) back toward the entrance. At least one of the nests studied (no. 1490) represented the work of 2 days, one cell being provisioned each day (Fig. 42).

As mentioned above, one group of seven cells contained prey of only one species. However, other nests contained several species. For example, the one cell of no. 1380 contained four species of Cicadellidae and one of Psyllidae. Both adults and nymphs are employed, adults much more commonly than nymphs. In all, this wasp is known to utilize as prey members of five families of Homoptera, including both Sternorhyncha and Auchenorhyncha (Table 8). The prey is carried venter up and held with the middle legs, as usual in this tribe; the prey often extends out behind the body of the wasp during transport. Provisioning may be quite rapid; one individual (no. 1385) brought in five leafhoppers in 15 min.

The number of prey in fully provisioned cells varies from 6 to 18, most commonly 9 to 12. The prey are stacked in the cells head in, venter up. The egg is laid longitudinally on the side of the venter of the thorax, as in other Gorytini. When the cell is completely provisioned and the egg laid, it is closed off with a short plug of sand and another cell begun (or more probably the soil from the new cell burrow is used for plugging the old cell). During provision-

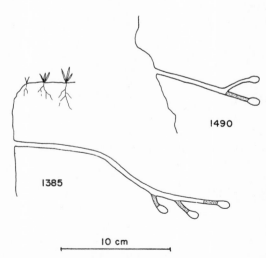

Fig. 42. Two typical nests of *Ochleroptera bipunctata* (Ithaca, New York).

§N. *Ochleroptera bipunctata* (Say)

TABLE 8. PREY RECORDS FOR OCHLEROPTERA BIPUNCTATA

Species of prey	Locality	Note no. or reference
CICADELLIDAE		
Aceratagallia sanguinolenta Prov.	N.Y.	1386
Coelidia olitoria Say	N.Y.	1490
Japananus hyalinus Osb.	N.Y.	1380
Macrosteles fascifrons Stal	N.Y.	1380
Macrosteles sp.	N.Y.	1386
Orientus ishidae Mats.	N.Y.	1380
Paraphlepsius irroratus Say	N.Y.	1386
Scaphytopius sp.	N.Y.	1380
Strangania apicalis Osb. and Ball	N.Y.	1584
CERCOPIDAE		
Clastoptera obtusa Say	N.Y.	1490
Philaenus leucophthalmus L.	N.Y.	1490
Philaenus lineatus L.	N.Y.	Pate, 1946
MEMBRACIDAE		
Cyrtolobus acutus VanDuzee	Texas	Strandtmann, 1945
FULGOROIDEA		
Haplaxius radicis Osb.	N.Y.	1562
PSYLLIDAE		
Psylla annulata Fabr.	N.Y.	1380

ing, the nest entrance is closed when the female leaves the nest and reopened with a few scuffs of the front feet when she returns. However, one individual nesting in a slightly overhanging bank (no. 1490) appeared to omit the closure; it would appear to be difficult to make a closure under these conditions. Final closure of the nest was not observed. The egg hatches in about 2 days, and only about 3 days are required for the larva to reach maturity. The cocoon is hard, containing sand grains in the walls, and is typically surrounded with the remains of the homopterous prey.

Miltogrammine flies were occasionally seen trailing females bringing in prey, and maggots were found in one cell of nest no. 1562 and one of no. 1490; in both cases they had destroyed the wasp egg or larva. Unfortunately the flies were not identified and the maggots not reared successfully.

O. Genus *Ammatomus* Costa

This genus is rather strongly differentiated in structure from other typical Gorytini. The head is striking because of the clavate antennae and the very large eyes, which extend to the flattened posterior surface of the head; the clypeus is high and narrow and is located entirely between the bottoms of the eyes, with which it is in broad contact on each side, thus causing the antennal sockets to be situated much higher than usual, near the middle of the front of the head; the mandibles are edentate. The mesosoma is unusually compact and boxlike, sometimes not more than about one and one-third times as long as wide, and the wings are characterized by an extremely long marginal cell and a rather slender stigma. Beaumont states that the epicnemial suture is absent, but in the North American species (subgenus *Tanyoprymnus*) this suture is present, fading out below as in the preceding genus. There is no oblique suture on the mesopleurum, but the scrobal suture is present and similar to that in *Sphecius, Stizus,* and *Bembix*. In our species the abdomen is relatively robust, although the first segment is constricted in some exotic species. The female possesses a fairly strong pecten. In the male, sternite VIII is parallel-sided and notched apically, the genitalia also of striking and unusual form. The larvae have not been described (Figs. 41 and 43–45).

Unfortunately very little is known of the ethology of these insects. So far as I know, nothing has been published on any of the Old World species, and our one species has been studied only briefly.

P. *Ammatomus moneduloides* (Packard)

This species occurs transcontinentally in the southern United States, south into Mexico. I have seen females flying around the sandy banks of streams in Kansas, but I have not been able to locate the nests. Krombein (1959) has found the species nesting at Kill Devil Hills, North Carolina, and since his notes represent nearly all that is known of the ethology of this genus, I shall quote them almost in full:

> This species begins its nests in the vertical surface of sand banks. One female . . . 10 mm. long was caught while she hovered in front of her burrow entrance at 1545 on August 7. She was carrying a paralyzed adult fulgorid, *Rhynchomitra microrhina* (Wlkr.) [Dictyopharidae], 13 mm. long.
>
> A second female . . . 11 mm. long was captured at 1325 on the following day as she flew toward her burrow entrance with a paralyzed fulgorid

§P. *Ammatomus moneduloides* (Packard)

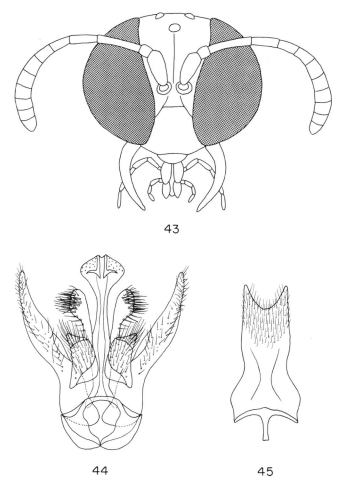

Figs. 43–45. *Ammatomus moneduloides* (Packard): 43, head of female, mandibles opened and mouthparts extended; 44, male genitalia; 45, sternite VIII of male.

nymph 8.5 mm. long . . . The wasp was released so that she could continue storing her nest. At 1351 she returned with an adult fulgorid, carried it into the burrow, and left a minute and a half later. I excavated the burrow at 1330 on August 9. It had a diameter of about 8 mm., went into the bank at a slight downward angle for 6.5 mm. and then curved slightly downward for another 7.5 cm. I found no cell at the bottom of this burrow, but there was one 5 cm. to the left at the same level containing four fifth (?) instar nymphs, *Rhynchomitra microrhina*, 8–10 mm. long. Presumably this cell was not fully stored, for there was no wasp egg. The wasp was captured at 1410, when she returned with another *Rhynchomitra* nymph.

Frank E. Kurczewski has sent me a female *Ammatomus moneduloides* collected by him at Arcadia, Florida, 4 July 1962, with a nymphal dictyopharid, probably of the genus *Scolops*.

Q. Summary of the ethology of *Gorytes* and related genera

This chapter has been highly repetitious because of the close similarity of the wasps considered. Nevertheless, the gorytine wasps are commonly considered to provide the ancestral stock for the higher nyssonine wasps, and for this reason it has seemed desirable to try to depict their general behavioral features in some detail. The following summary covers the major behavioral similarities of these wasps and attempts to point out the important exceptions. It must be born in mind that these are relatively uncommon, little-studied wasps, and any conclusions regarding their behavioral repertory must be considered tentative.

(a) The habitat generally consists of limited areas of bare, sandy, gravelly, or light clay soil; there is no indication of species being specialized for digging in hard or loamy soil or in open sand dunes. Flat or gently sloping soil is preferred by most species, but several species have more commonly been reported from vertical or nearly vertical banks (*Gorytes simillimus*, and the known species of *Clitemnestra, Ochleroptera,* and *Ammatomus*).

(b) Females tend to build successive nests in the same area, and colonies may persist in the same place for more than 1 year. There is much variation in the size and density of nesting aggregations; many species nest in very small aggregations, but relatively populous colonies are reported for *Hoplisoides spilographus* and *H. tricolor* and for the species of *Clitemnestra.*

(c) Adults of many species are reported to take nectar from flowers, and others (for example, *Ochleroptera bipunctata*) commonly visit honeydew. In other instances the adults rarely if ever visit sources of nectar or honeydew, and it is probable that the females feed on the blood of their prey (*vide Hoplisoides nebulosus*).

(d) Virtually nothing has been reported on the behavior of the males, suggesting that there is no conspicuous precopulatory flight or territoriality.

(e) In some species (*Hoplisoides spilographus* for instance) the adults apparently spend the night away from the nesting area, presumably on vegetation. *H. tricolor* and *Hapalomellinus albitomentosus* females may sleep in the burrows. In *Clitemnestra,* Janvier reports that both sexes spend the night in the burrows.

(f) Nests are dug in friable soil with the mandibles and front legs;

§Q. Summary of *Gorytes* and related genera 81

the nests are shallow and only 30 min to about 2 hr are normally required for completion. Relatively little soil accumulates at the nest entrance and this is usually fairly well dispersed; simple leveling movements have been described for *Gorytes canaliculatus* and some other species.

(g) Many Gorytini hold the wings at a strong angle from the body when outside the nest, either during digging or provisioning, then depress them against the abdomen when entering. Some species fly in to the nest from a considerable height, and their orientation flight is relatively high (for example, at least two species of *Hoplisoides*), while in others this is not the case.

(h) The nest entrance is kept closed and well concealed at all times when the female is not actually in the nest (exceptions: *Argogorytes campestris, Clitemnestra chilensis*). In closing, and especially at the final closure, the pygidium may be used for pounding soil into the nest entrance.

(i) A succession of cells may be prepared from one burrow. The number is indeterminate, depending in part perhaps on success in hunting, the usual range being from 1 to 4. *Argogorytes campestris* is said to make up to 9 cells per burrow, *Clitemnestra chilensis* up to 10, *Dienoplus concinnus* about 15. The first cell is normally the deepest, later cells being prepared back toward the entrance (exception: *Clitemnestra* spp.)

(j) The burrows are more or less oblique and always shallow, the species collectively exhibiting cell depths within the narrow range of from 4 to 15 cm (some minor exceptions).

(k) Mass provisioning is the rule throughout the complex; but the species of *Clitemnestra* are said to provision a cell over 2 or 3 days (but hatching of the egg occurs only after 5 or 6 days in this genus). The prey is placed in the cell head in, venter up. The number per cell varies from 3 to 30, depending largely on their size (exception: up to 60 *Tettigometra* nymphs may be used per cell in *Hoplisoides punctatus*).

(l) The prey consists of small Homoptera Auchenorhyncha of several families. Many species use either adults or nymphs, while certain species appear to specialize on one or the other. Most species will take a variety of Homoptera belonging to one family; a few species of *Dienoplus* attack members of two families; *Ochleroptera bipunctata* utilizes five families (including Psyllidae, the only record for Sternorhyncha). There appears to be little tendency for one genus to restrict itself to only one family of prey (possible exception: *Argogorytes,* two species of which take only cercopid nymphs).

(m) Prey does not normally recover from paralysis; there may be weak leg movements for several hours, but after the 1st day the prey is usually completely immobile, although it remains relaxed for several days.

(n) Prey is carried to the nest in flight, the wasp using her middle legs to hold the venter of the prey close against her own venter (exception: *Clitemnestra,* said by Janvier to employ the mandibles and front legs). The wasp normally lands at the entrance and scrapes through the closure with her front legs.

(o) The egg is laid on the last prey put in the cell. It is laid longitudinally on the ventral side of the thorax, alongside the coxae. (There are many examples for both these statements, but the species of *Clitemnestra* are reported to be exceptional in both respects.)

(p) The cocoons are ovoid and hard, containing sand grains woven into the walls, but apertures are not constructed in the walls.

These wasps are commonly attacked by miltogrammine flies and by wasps of the genus *Nysson;* the behavior of the species of *Nysson* is summarized in Chapter IV.

Chapter IV. The Nyssonini, A Complex of Cleptoparasites

The genus *Nysson* of Handlirsch's "Monographie" has been divided into several genera which make up the tribe Nyssonini of most contemporary workers. Of the seven genera recognized by Pate (1938), only two have been studied in any detail, the widely distributed genus *Nysson* and the Old World genus *Brachystegus,* while a few notes are available on two other Nearctic genera, *Zanysson* and *Metanysson.* Considering the close structural similarity of the genera of Nyssonini, it seems a safe assumption that all are cleptoparasites in somewhat the manner described for various species of *Nysson* in the preceding chapter. The most recent taxonomic treatment of this group is that of Pate (1938), and the following summary of the more important structural features of the group is based largely on that in Pate's paper.

As compared with *Gorytes,* the species of *Nysson* and the several related genera are relatively short, stocky wasps, the basal portion of the metasoma being especially wide when compared with *Gorytes.* Most of them have a rather coarsely sculptured, apparently thick integument, suggesting that of Mutillidae, Sapygidae, Chrysididae, and other Aculeata which enter the nests of wasps and bees and are presumably "armored" so that they can be stung less readily by their hosts. In color, the Nyssonini are generally black, often with limited yellowish banding, in general not very different from that of their hosts. More specific diagnostic features of the adults are as follows:

Mandibles simple, edentate; labrum not, or barely, exserted; eyes diverging above, the lower front rather narrow, the antennae inserted below the middle of the front; mesosoma short, compact; pronotum very short, transverse; mesoscutal laminae usually squarely truncate and abruptly declivous behind; mesopleura strongly developed, bulging, without oblique or scrobal grooves but sometimes with various secondary grooves or ridges, and with only a small signum, the epicnemial ridge complete, incomplete, or absent; mesosternum and metasternum together forming a continuous plate overlying the bases of the middle and hind coxae, the middle coxae and often the hind ones well separated; propodeum short, bearing a strong spinose projection on each side (with rare exceptions); pecten present or absent; middle tibiae with two apical spurs (rare exceptions); fore wing with the stigma small, the second submarginal cell petiolate, the third submarginal cell sometimes absent; females with a distinct pygidial area; males with the apical sternite simple, but the apical tergite often dentate or carinate (Figs. 46–51).

The larva of *Nysson* is unusual in possessing from one to three prominent mid-frontal protuberances; otherwise it resembles the

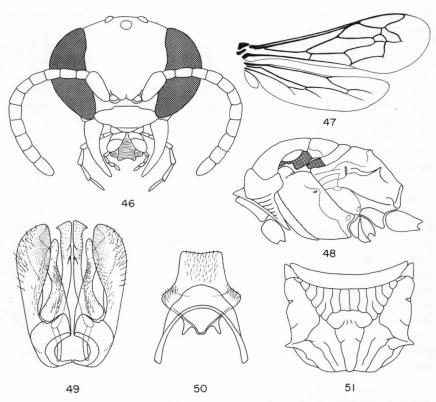

Figs. 46–51. *Nysson lateralis* Packard: 46, head of female, mouthparts extended; 47, wings; 48, lateral aspect of mesosoma of female; 49, male genitalia; 50, sternite VIII of male; 51, dorsal aspect of propodeum of female.

larva of *Gorytes* strongly, although lacking the strong sensory cones present on the labrum of that genus and all other Nyssoninae except *Alysson* (Evans and Lin, 1956b).

A. Ethology of the species of *Nysson*

Shuckard (1837) long ago suspected that these wasps might be parasitic. However, as late as 1929 Arnold stated that these wasps "by some authors . . . have been regarded as parasitic . . . but there is no evidence to support such a view." Apparently Arnold was not persuaded by the several important papers that appeared prior to 1929, including Wheeler's review of the subject in his paper "The Parasitic Aculeata: A Study in Evolution" (1919). Incidentally, Wheeler believed that the adult *Nysson* destroyed the *Gorytes* egg and laid its own in its place (like *Evagetes* in the Pompilidae), but it is now clear that it is the larva of *Nysson* that destroys the host egg (like *Ceropales* in the Pompilidae).

Since I have already discussed the behavior of several species of *Nysson* in connection with that of their hosts, it is necessary to present only a brief summary here. In Table 9 I have assembled what is known of the host relationships of the various species of *Nysson*. It appears from this table that most of them are not completely host specific; in fact, several species are known to attack Gorytini of more than one genus. Probably each species attacks any gorytine of suitable size occurring in the habitat in which the cleptoparasite normally occurs. Wolf (1951) believed that each species-group of *Nysson* restricts its parasitism to one genus of Gorytini (*spinosus* group to *Gorytes*, *niger* group to *Hoplisoides*, *dimidiatus* group to *Dienoplus*). This generalization seems to me unjustified on the basis of available data. It is, however, a matter worth investigation by future students of these wasps.

Whether or not the species of *Nysson* attack wasps other than Gorytini remains a moot point (but see the discussion of other genera below, Chapter IV:B). Pate (1938) cites a record of a *Nysson* entering a nest of *Sphex ichneumoneus*, and Ristich (1953) records two species of *Nysson* entering nests of this same wasp. Hamm and Richards (1930) report seeing *Nysson dimidiatus* around the burrows of *Lindenius albilabris* (Crabronini). One would expect parasites of Gorytini to inspect the holes of other digger wasps in their search for nests of their true hosts, and it seems odd that if *Nysson* does in fact attack various Aculeata, no one has ever found their eggs or larva or reared adults from any Aculeata other than Gorytini. However, we do not know the hosts of many species of Nyssonini, including

TABLE 9. HOST RELATIONSHIPS OF THE SPECIES OF NYSSON*

Species of Nysson	Host species
NORTH AMERICA	
hoplisivora Rohwer	Hoplisoides costalis (Cresson)
bellus Cresson	Hoplisoides tricolor (Cresson)
moestus Cresson	
pumilis Cresson	Hoplisoides spilographus (Handlirsch)
rusticus Cresson	
tuberculatus Handlirsch	Hoplisoides nebulosus (Packard)
daeckei Viereck	
lateralis Packard	Gorytes canaliculatus Packard
fidelis Cresson	
EUROPE	
niger Chevrier	Bembecinus tridens (Fabricius)
	Hoplisoides punctatus (Kirschbaum)
dimidiatus Jurine	Hoplisoides latifrons (Spinola)
	Gorytes laticinctus (Lepeletier)
	Gorytes quadrifasciatus (Perkins)
trimaculatus Rossi	Dienoplus elegans (Lepeletier)
	Dienoplus concinnus (Rossi)
tridens Gerstaecker	Dienoplus tumidus (Panzer)
	Dienoplus lunatus (Dahlbom)
maculatus Fabricius	Dienoplus laevis (Latreille)
	Argogorytes campestris (Mueller)
interruptus Fabricius	Argogorytes mystaceus (Linnaeus)
spinosus Foerster	Lestiphorus bicinctus (Rossi)

* Modified and expanded from a similar table in Olberg, 1959; for references see preceding chapter and also Hamm and Richards, 1930, Wolf, 1951, and Blüthgen, 1952. Heavy lines indicate that the egg or larva of the parasite has been found in the nest of the host; thin lines indicate that the Nysson has been seen entering the nest; dashed lines indicate that the two have merely been found closely associated.

some species (for example, *Nysson plagiatus* Cresson) which are larger than most Gorytini. It is also true that at times certain Nyssonini appear to be far more common than any Gorytini in that habitat. In the light of what is known of the related genera *Brachystegus*

§A. Ethology of the species of *Nysson*

and *Metanysson*, it seems best to remain open-minded with respect to the host relationships of the large genus *Nysson*.

Maneval's (1939) mention of *Nysson dimidiatus* attacking the stizine wasp *Bembecinus tridens* (Fabricius) deserves special comment. Maneval found this *Bembecinus* very common in sandy places along the Loire River, and he screened from the sand a number of cocoons of this wasp as well as several of those of *Nysson dimidiatus*. The *Nysson* cocoons were enveloped in leafhopper remains similar to those surrounding the *Bembecinus* cocoons, and he was able to be certain that they were *Nysson* cocoons by study of the larval exuviae within them. Maneval states that *N. dimidiatus* attacks various Sphecidae that prey on leafhoppers, and its behavior is exactly as it is with respect to *Hoplisoides latifrons*; he cites *Dienoplus laevis* and *Bembecinus tridens* as examples of these other sphecids, but does not indicate how much of the behavior was observed with these species. It seems conceivable that certain Gorytini were nesting within the *Bembecinus* colony. The species of *Bembecinus* practice progressive provisioning, and one would think it improbable that a *Nysson* could successfully attack a *Bembecinus* with the same behavioral repertory used for attacking a gorytine wasp. Nevertheless, there is a distinct possibility that Maneval was correct and that this *Nysson* does in fact attack a species of *Bembecinus*, a genus not far removed from the Gorytini.

The following is a brief summary of some aspects of the ethology of these wasps. For further details, see the preceding chapter and various references cited therein.

(a) Both males and females are sometimes common at honeydew, and certain species have been taken at flowers. Presumably all adult nourishment comes from these sources, as females do not take prey.

(b) The female *Nysson* spends long periods of time in the nesting area of the host species. She flies only a few centimeters high, usually making very short flights of only a few centimeters length, one such flight often following another closely. She also spends much time walking in an irregular course over the soil with her antennae tapping the soil continually.

(c) When the *Nysson* sees the host entering or leaving her nest, she remains motionless on the ground or on a low plant for some time, usually until the host leaves, then approaches the now well-covered nest, which she probably detects partly by odor. There are, however, several records of *Nysson* entering nests while the host was still inside, in which case the host usually merely closes the nest leaving the parasite inside. Sometimes a *Nysson* will enter and leave the same nest several times over a period of time, possibly waiting until several prey have been placed in the cell.

(d) Female *Nysson* also enter various open nests in the sand and probably at times locate the nest of the host species merely by random searching, using primarily olfactory stimuli picked up by the antennae.

(e) When entering the nest, the *Nysson* scrapes open the entrance with her front legs in a manner very similar to that of *Gorytes* and related genera (Fig. 34). When she leaves the nest, she scrapes sand into the entrance to effect a closure similar to that of the host.

(f) The egg is laid on one of the bottom prey in the cell (that of the host being laid later on one of the top prey). It is laid in a concealed location, in the case of adult treehoppers either on the under side of the pronotal shield or beneath the wings (Fig. 35), in the case of the adult leafhoppers beneath the wings. If the prey is a nymph, the egg may be inserted transversely behind the hind coxae, so that it is partially concealed (Fig. 36). Barth found a *Nysson* egg loose in the bottom of a cell, but it is possible that this egg had been knocked off the prey accidentally.

(g) The *Nysson* egg is considerably smaller than that of the host and is duller in appearance because of microscopic surface sculpturing. According to Reinhard, the egg hatches in a shorter period of time than that of the host, and the larva, equipped with long, thin mandibles, seeks out the egg of the host and destroys it, then proceeds to consume the prey in the cell.

(h) The *Nysson* larva spins a cocoon virtually identical to that of its host, though often slightly smaller.

It appears that the host wasp is unable to detect, or at least does not react to, the presence of the *Nysson* egg in the cell, as she continues to provision the cell, oviposit, and close the cell in the normal manner. When the host wasp actually encounters the *Nysson* in or around the nest, she may appear to exhibit "alarm"; according to Hamm and Richards, *Dienoplus* "recognizes its enemy and chases it away at sight." Presumably the coarse integument and compact form of *Nysson* represent adaptations for avoiding the stings of the host. Larval adaptations include the piercing mandibles of the first-instar larva and the frontal prominences of the full-grown larva, the latter of unknown function.

There are several important unanswered questions regarding the behavior of *Nysson*. We do not know where the males and females spend the night and periods of inclement weather. We know nothing about the precopulatory behavior of the males or about mating; presumably there is nothing striking or unusual about reproductive behavior, or it would have been noted. We know nothing of the behavior of the females inside the nests of their hosts, other than the position in which the egg is laid. In fact, there is no aspect of the behavior of *Nysson* that does not require much further study. Needless to say, we also need much more data on their host relationships.

B. Ethology of other genera of Nyssonini

As a result of the work of Deleurance (1943) it seems quite certain that the European *Brachystegus scalaris* (Illiger) attacks a non-nyssonine wasp, the larrine *Tachytes europaeus* Kohl. Deleurance, working in Dordogne in south-central France, found *Brachystegus scalaris* roving about areas where *Tachytes europaeus* was nesting and frequently coming to rest facing open burrows. The *Tachytes* would frequently attack the *Brachystegus,* whereupon the latter would roll in a ball in the manner of chrysidids. Once a *Tachytes* carried off a *Brachystegus* in flight and deposited it at a distance "as it would do with an object obstructing the entrance of the nest" [compare my remarks on *Steniolia obliqua* attacking the chrysidid *Parnopes edwardsii,* Chapter VIII:I,8]. Deleurance found that when a *Brachystegus* approached a nest which had a *Tachytes* inside, she would not enter but would take up a position nearby "in the classic pose of parasites." When the *Tachytes* was absent, the parasite would approach the burrow entrance, palpating the soil and the walls of the entrance with its antennae, and finally enter. Deleurance believed that the *Brachystegus* might be able to detect the presence of the host by odor or by vibrations in the soil.

Deleurance dug out one nest which had been entered earlier by a *Brachystegus.* There was only one cell, and this cell contained two grasshoppers which had been brought in by the *Tachytes,* one of them bearing a small, cylindrical egg attached to the side of the prey near the base of one of the hind legs. This was clearly not the *Tachytes* egg, which is typically laid farther forward and on the venter, and which seemed to be absent. Deleurance believed that it had been destroyed by the adult parasite, although this is contrary to what we know of the behavior of *Nysson.* He was unsuccessful in rearing this egg to maturity, but I see no reason to question that it was the egg of *Brachystegus scalaris.* Deleurance observed that the parasite closed the burrow leading to the cell, but that on leaving the nest entrance she merely raked the soil crudely with her legs.

M. A. Cazier (personal communication) has recently found *Zanysson tonto* Pate in close association with *Tachytes distinctus* Smith at Portal, Arizona. On 4 July 1963 he watched a *Zanysson* enter a *Tachytes* nest several times over a period of about 2 hr, each time remaining in the nest 3-10 sec. This nest was found to contain four completed cells and one incomplete one, but only the eggs and larvae of *Tachytes* were found. On another occasion he observed a female *Zanysson tonto* walking about the nest entrance of a *Tachytes chrysocerus* Rohwer with her antennae constantly tapping the soil. The *Zanysson* inserted her head in the entrance several times but did not enter.

Dr. Cazier has also permitted me to cite some extremely interesting unpublished notes he has made on two species of *Metanysson* at Portal, Arizona. On 19 July 1963 a female *Metanysson arivaipa* Pate was seen to enter a burrow of *Cerceris macrosticta* Viereck and Cockerell and to remain there for 7 min while the *Cerceris* was away hunting for prey. Also, on two occasions during June of the same year, he observed specimens of *Metanysson coahuila* Pate at the nest entrances of *Cerceris conifrons* Mickel. One of them spent 4 min in the *Cerceris* burrow after the latter had been collected for identification. In no case was the egg of *Metanysson* recovered or the presumed parasite reared from the nests of *Cerceris*.

Pate (1938) considers *Brachystegus, Zanysson,* and *Metanysson* to form a single phyletic line starting from a *Nysson*-like ancestor. It seems probable that the more highly evolved Nyssonini have lost their restriction to other Nyssoninae and have come to attack and possibly to specialize in wasps of other subfamilies. Clearly a considerable amount of futher research will be required to clarify the host relationships of members of this complex.

Chapter V. The Cicada Killers, *Sphecius* and *Exeirus*

Members of the genera *Sphecius* and *Exeirus* are the only wasps known to employ cicadas as prey, and as might be expected they are among the largest of digger wasps. Both genera appear, on the basis of structure, to be somewhat isolated from other Nyssoninae as well as from each other. The available information on their nesting behavior, as will be seen, suggests that the two genera have much in common, although they differ in a number of ways from other Nyssoninae.

The two genera are alike in their bulky bodies and relatively small heads, the head being in every case considerably narrower than the thorax; the wings of the two genera are similar in being unpatterned and hyaline or slightly tinged with yellow, also in having the pterostigma extremely small and the marginal cell very long and slender. In both genera the basal vein reaches the subcosta not far from the stigma, and it is for this reason that these wasps are currently usually placed in the Gorytini rather than in the Stizini, although they bear a strong general resemblance to members of the genus *Stizus*. Other features held in common between *Sphecius* and *Exeirus* are to be found in the mesothorax: in both the epicnemial ridge is present, the precoxal ridge is absent but the signum is distinct; also, both genera have the scrobal suture at least moderately strong and of the form typical of *Stizus*, *Bembix*, and most other "higher" Nyssoninae. Beyond this, one is impressed by the many and striking differences between the two genera, which are summarized in part below.

I have presented no summary at the end of this chapter, since only one species of *Sphecius* has been well studied and there is only one species in the genus *Exeirus*.

A. Genus *Sphecius* Dahlbom

This genus falls readily into three subgenera. The most distinctive of these, *Sphecienus*, is Palaearctic in distribution, and so far as I know nothing is known of its behavior except that one species preys on cicadas. The subgenus *Sphecius* occurs in North and South America, the very similar subgenus *Nothosphecius* in Africa, Madagascar, and Australia. It is interesting to note the occurrence of a fairly typical species of this genus in Australia (*pectoralis* Smith), since *Exeirus* is restricted to that continent.

Pate (1936) reviewed the subgenera of *Sphecius*. He believed that the genus belongs in the Gorytini and may be related to more typical Gorytini via the genus *Ammatomus*. The more important structural features of the adults of the genus may be summarized as follows:

> Labrum well exserted, rounded below, broader than long; clypeus contiguous to bottom parts of eyes and extending well up between them, the antennal sockets arising near the middle of the head; front narrow, the eyes diverging above; mesosoma broad and short, compact, all parts smoothly confluent; mesopleura with the epicnemial ridge present but discontinuous below (except continuous across the sternum in *Sphecienus*); mesopleura without other ridges but with a strong scrobal suture that bends downward behind and runs into the suture margining the mesopleurum behind; propodeum without a stigmatal groove; front tarsus of female with a moderately strong pecten; basal one or two segments of the middle tarsus of female asymmetrical, with a strong apical hook; hind tibial spurs of female remarkable broad and flat, the longer one strongly curved (except in *Sphecienus*, where the spurs are unmodified); female with the pygidial area margined by a carina; male with the apical sternite produced into a long spine which superficially resembles a sting. (Figs. 1–7, 52–54).

There is evidence that the modified hind tibial spurs play a role in holding onto the prey (see below, under *speciosus*) and it seems very probable that the modifications of the middle tarsus are also involved in prey carriage. The subgenus *Sphecienus* is less specialized with respect to the spurs; also, the continuation of the epicnemial ridge across the sternum is suggestive of such apparently generalized Gorytini as *Argogorytes*.

B. *Sphecius speciosus* (Drury)

This large and familiar wasp occurs throughout the United States east of the Rockies; the northern limits of its range coincide closely with the northern edge of the Upper Austral Zone as defined

§B. *Sphecius speciosus* (Drury)

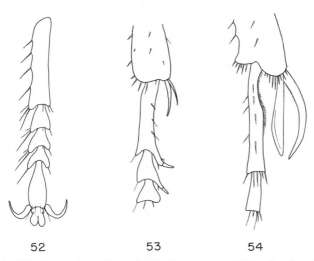

Figs. 52-54. *Sphecius speciosus* (Drury): 52, front tarsus of female, showing pecten; 53, apex of tibia and basal segments of tarsus of middle leg of female; 54, apex of tibia and basal segments of tarsus of hind leg of female, showing modification of the spurs.

by Merriam. Although usually called "the cicada killer," this species has been referred to by various authors under such names as "the larger digger wasp," "the ground hornet," or "the golden digger wasp." There are many popular accounts of the major features of its behavior, beginning with John Burrough's *Pepacton* (1881) and extending down to Edwin Way Teale's *Strange Lives of Familiar Insects* (1962). There are also many brief notes in the scientific literature, but unfortunately only a few of these contribute significantly to knowledge of the species, and some of them contain errors and misconceptions. I have made no effort to review everything that has been published on this species, but I have at least scanned most of the articles appearing in scientific journals. The more important papers are those of Riley (1892a), Howes (1919), Davis (1920), Reinhard (1929), Dow (1942), Dambach and Good (1943), and Lin (1963a). I have studied this species only briefly, once in Pottawatomie County, Kansas, in July 1952, once in Arkansas County, Arkansas, in June 1956, once at Pittsford, New York, in late July 1955, once at Cambridge, Massachusetts, August 1963, and once at Ithaca, New York, September 1964.

Although the observations in Arkansas were made on 8 June, most records for this species are from the months of July and August, and I know of no records from the northeastern or north central states of nesting occurring before about 1 July. Four colonies studied by Lin on the Parade Grounds in Brooklyn, New York, were out of phase, the colony in the most friable soil each year showing the earliest emergence. Initial emergences over various years varied from 26 June to 31 July, while the first nest ever recorded was on 6 July (Lin, 1964). Both sexes are rather commonly seen on flowers, where they apparently take nectar throughout their active season. The Raus (1918) and others have also reported these wasps as feeding at exuding plant sap. Wasps of both sexes are most commonly encountered around the nesting areas, which may remain the same year after year unless there is some change in the drainage or plant cover. Lin (1964) reports that males live a maximum of 15 days, females 33 days.

(1) *Ecology.* Usual nesting situations are at least partially devoid of vegetation, well drained, and with the soil light clay or somewhat sandy. Several authors mention elevated, more or less bare places in lawns and pastures as nesting sites; others mention bare places in gardens, playgrounds, tennis courts, and sandy, gravel, or cinder fill in roadsides or other places. My observations in Kansas and Arkansas were made in fine-grained, pale sand along streams, but in neither case was there a large colony of the wasps. At Pittsford, New York, about 50 nests were found along a sandy road through a broad opening in oak woodland. Nesting aggregations are often found in towns and cities. During some summers cicada killers nest in some numbers in bare places in lawns and beside buildings at Harvard University, and Reinhard speaks of a colony along a flagstone sidewalk at Fordham University "amid the turmoil of the huge metropolis." The colonies studied by Lin (1963a,b) were situated on footpaths along two baseball fields in Brooklyn, New York.

The number of individuals nesting in a particular location during a given season is very variable. I have occasionally found females nesting alone, while Dambach and Good report colonies ranging in size from 2 to 373 burrows. The latter colony was said to be at least nine years old and probably much older: it occupied an area of about 1 acre, the burrows being "grouped in units where they averaged 6.9 per 100 square feet." These authors remark that the distance between burrows "ranges from a little more than one foot to 15 feet or more." In the only sizeable colony I studied, at Pittsford, nest entrances were separated by at least 0.5 m, most of them by more than 1 m.

Norman Lin (1964, and personal communication) states that nest entrances in the colonies be studied in Brooklyn were at times

§B. *Sphecius speciosus* (Drury)

no more than 2 cm apart. He obtained data on total population size over several years by counting emergence holes, and his data show much variation from year to year. The largest colony contained 145 wasps in 1958, 188 in 1959, 241 in 1960, 939 in 1961, 837 in 1962, and 165 in 1963. The increase in population in 1961 and 1962 appeared correlated with precipitation during the breeding season, the greatest rainfall in 41 years having occurred in 1960, the fifth greatest rainfall in 41 years in 1961. Riley (1892a) long ago noted that eggs in dry soil often fail to hatch, and it seems probable that mortality to the immature stages is considerably lower during wet seasons. The colonies studied by Lin have persisted in the same place for ten years, and he has information indicating that they have persisted there for at least 24 years.

(2) *Behavior of males.* The Staten Island naturalist William T. Davis (1920) first presented notes on the behavior of male cicada killers. He observed many males along a grassy lane: "They would station themselves on the tops of plants, or on small bushes," he reports, "and when a female came near they would fly after her. Each male had a particular flower head, or other lookout, to which he would return." Rau (1922) reported males flying about hawthorn bushes: "In their flight to and fro [they] often met in collisions in a manner and frequency that certainly made it appear far from accidental." Reinhard remarks that the male wasps appear before the females and fly back and forth over the area where they emerged. He states that "they chase after every passing hexapod that could be mistaken for one of their own tribe, and come near to assaulting one another in their eagerness." Dambach and Good note that males will pursue inanimate objects tossed near their perches. These authors apparently first described the males as exhibiting territoriality: "There are some indications that a rather well-defined territory is established by the male, within which all intruders are investigated in a vigorous manner." They reported one male pursuing an English sparrow which flew within 25 feet of his perch.

Territoriality in male *Sphecius speciosus* has recently been studied in detail by Lin (1963a). Lin found the territories to consist

of strips of ground ranging from about 16 feet in length and 6 feet in width to about 4 feet in length and 4 feet in width. These areas were vigorously defended by cicada killer males against intrusion by other males, other species of insects, birds, and even thrown or rolled pebbles. The wasps flew, hovered, or perched within their territories. They left frequently, usually to chase intruders, and then returned generally within a matter of seconds . . . The territorial perch was usually an emergence hole, [more commonly next to a hole rather than directly over it].

Other perches used infrequently included a piece of glass, a button, and parts of a black handrail. Individual males maintained the same territories day after day and would return to their territories even when artificially displaced up to 2500 ft away.

According to Lin, the reaction of males toward intruders in the territory may culminate in one of four possible ways: (1) unsuccessful pursuit; (2) threat, that is, the wasp gains on the intruder until within striking distance, then abruptly slows down; (3) butting, in which the territory owner "bangs into the intruder, then turns around and returns to the territory"; and (4) grappling, in which the intruder is actually seized and then released, sometimes after a few seconds of combat. Grappling occurs only with other *Sphecius* males, most commonly those in adjacent territories.

Lin found that the males did not utilize their own emergence holes as perches, but rather set up their territories in areas of high emergence hole density. "Territorial behavior," he says, "enhances mating efficiency by spacing the males and thereby increasing the frequency of male-female interaction. Localizing of territories in places of high emergence hole density fosters male-female interaction because the owners are more likely to contact emerging females. Mating efficiency is also probably enhanced by a reduction in the frequency of male-male interactions, thereby reducing the interference of superfluous males during precopulatory behavior."

Norman Lin states (personal communication) that territoriality is much the same in very small colonies, for example of only 3 males, as in colonies of up to 250 males. However, when colonies are very large and dense, as for example at the Brooklyn Parade Grounds in 1961 and 1962, as cited earlier, there is "a breakdown of territoriality in which large numbers of male wasps were clustered in areas of high emergence hole density . . . Here they perched, hovered, flew about, fought, and chased each other, but they could not establish geographical boundaries."

Lin (1963a) reports that the females are sexually receptive at the time of their emergence and probably mate only once; males are unable to pair successfully with females already just mated or engaged in nesting. The flight of sexually receptive females is said to be straight, without the "jerky, zig-zag" motions of nonreceptive females.

Males leaving their territories to fly after these passing females clasp them from behind in a short, direct, pre-tandem flight which never in-

§B. *Sphecius speciosus* (Drury)

cludes circling, hovering, or butting-like behavior. The pair flies in tandem, facing the same direction, the male above the female; they may land on the ground, or in vegetation, or on some high object such as the branch of a tree. Once a female wasp emerging from the ground within a male's territory was overtaken by the owner and clasped from behind. The pair flew in tandem at about a 45 degree angle to the nearest tree and disappeared among the branches . . . Another female was captured on emergence, marked with paint, and released almost immediately. Before she had flown more than 15 feet she was clasped from behind by a male whose territory she had traversed. The pair flew in tandem approximately 18 feet, landed in a clump of grass, and then copulated.

Supposed mating has been reported by several authors to occur in the air and to be very brief (for example, Davis, 1920), but such reports are certainly erroneous and may well be based upon territorial fights between males. Dambach and Good report one pair "quietly copulating while resting on a tall weed," although they provide no details on the mating posture. N. Lin (personal communication) has observed that copulation is not initiated in the air, but on the ground or some other surface. At first the male assumes a position on the back of the female, facing in the same direction. At this time he may rub the sides of the head of the female with his front legs (as reported by Reinhard) and may lash his antennae against those of the female. The genitalia are soon extruded and are rubbed against the abdomen of the female. As soon as the genitalia make contact with the female orifice, the male dismounts and faces in the opposite direction, so that the two form a straight line or very broad angle, their heads facing away from each other. This posture has not otherwise been reported among the Nyssoninae, although Scullen (1965) has recently published an excellent photograph of a pair of *Cerceris frontata* Say (Philanthinae) copulating in this manner. Lin has also observed that mating pairs of *Sphecius* may take flight, the female taking the lead and pulling the generally smaller male behind her, although the male also moves his wings so as to support himself in flight. Such flight may be elicited by interference by humans or by conspecific males. Eventually the pair lands on the ground or on some object, where mating continues. The total duration of copulation is 45 min or more. Lin has noted that when the female is readily receptive, the male may omit the stroking of the head and antennae and assume the typical copulatory pose very quickly.

I have discussed Lin's work at some length, since his studies of male behavior are the most thorough to have been made on any nyssonine wasp. Further details will, of course, be found in his papers (1963a, and in preparation). It appears at this time that in some features of male behavior the cicada killer is a rather unusual sand wasp, but it is possible that further studies will show that territoriality and a linear copulatory posture are more widespread in the Nyssoninae than is now appreciated.

(3) *Digging the nest.* This wasp is a strong, persistent digger. A female in Pottawatomie County, Kansas (no. HE52) was seen starting her nest at 1750 in sandy soil beneath a hog peanut vine (*Amphicarpa*). At first the sound of her mandibles biting into the soil was clearly audible, but after a few minutes it could no longer be heard. This wasp worked for 1 hr with no pause whatever, even when the vegetation was removed for photography. The forelegs were used for raking back the earth, working simultaneously but much more slowly than in *Bembix*. As the wasp backed out of the burrow periodically, she pushed soil behind her with her hind legs, which opened and closed in a somewhat scissors-like motion. As a result of her manner of backing out the burrow and "bulldozing" the earth behind her, a large trough was maintained through the growing heap of soil at the entrance. This burrow was apparently finished some time that evening, as this female was aready bringing in cicadas at 0850 the next morning.

Reinhard remarks that burrows are often begun in the evening and found to be completed the following morning. His impressions of the digging behavior are very similar to mine; he aptly speaks of the female "diving" into the burrow after pushing out a load of earth, and he also mentions that the female may assume an inverted position in the burrow when rounding off the top. The use of the hind legs for moving soil, most unusual among digger wasps, is also mentioned by Reinhard. The large irregular mound of earth, with a large groove leading from the entrance, has been described by several authors and well illustrated by Savin (1923; see also Figs. 55A,B). The trough and burrow entrance may be as much as 4 cm wide, although the burrow proper is not more than half that width. The mound of earth at the entrance may be as much as 25 cm long by 15 cm wide by 5–10 cm deep. This mound is never leveled off by the wasp, even in small part, although it tends to weather away partially over a period of days. There is general agreement among all who have worked on this wasp that the entrance of active nests is not closed at any time. Reinhard found that provisioning females are unable to find their nests if the entrance is artificially closed, and generally abandon their prey and fly off. However, Lin (personal communication) has found that

§B. *Sphecius speciosus* (Drury)

Fig. 55. Nest entrances of *Sphecius speciosus* showing large, furrowed mound and open burrow: A, nest in sandy loam on lawn in front of Gray Herbarium at Harvard University; B, nest in flagstone walk at the Parade Grounds, Brooklyn, New York. (Photograph by N. Lin.)

some females are, in fact, able to locate nests that have been artificially closed and to dig into them, sometimes without releasing the prey.

Dambach and Good speak of more than one female utilizing the same burrow. One marked female was seen entering several active burrows of other females, and on another occasion two females were seen to pass in a burrow entrance, one entering and one leaving; in no case was a "show of antagonism" apparent. On another occasion two females were found in separate cells of the same nest, each with a cicada upon which an egg had been laid. No other workers have remarked upon the use of a burrow by more than one female, and more data on this point are needed. If Lin (1936b) is correct that density-dependent fighting among females occurs, the multiple use of burrows seems unlikely.

Several workers mention that the female often takes a circling orientation flight upon leaving the nest, but no detailed information is available upon this. Savin speaks of the wasp flying about the entrance "in increasingly larger circles, the outer one having a diameter of about thirty feet."

(4) *Nature and dimensions of the nest.* Most published reports, as well as my own brief observations, indicate that the burrow is at first oblique, at a 35–45° angle with the surface; then, after 10–20 cm, it tends to level off and to remain at approximately the same depth, a rather shallow depth considering the great length of the burrows. Savin remarks that "the termini observed were always slightly nearer the surface of the ground than were the tunnels. This afforded better drainage." I noticed undulations in most burrows, but the cells were often slightly deeper than the main part of the burrow. Each burrow eventually has several major branches, each of which terminates in one or more cells. The depth of the horizontal part of the burrow and of the cells varies from about 12 to 25 cm, the cells of any one nest tending to be at about the same depth (Dambach & Good; also my notes, nos. 1150, 1123, 1124, and 2060). In the one nest I dug out in Arkansas, the distance from the entrance to the cells was well over 1 m (125–140 cm). In five nests in New York the distance varied from 40 to 115 cm. Savin, working in New Jersey, found the burrow length to be from 1 to 4 ft (that is, about 30 to 120 cm), with most burrows being about 2 ft long (60 cm). Various other authors cite burrow lengths varying from 30 to 90 cm. It appears that there is much variation in burrow length and also in the branching of the burrows and the arrangement of the cells, as noted long ago by Riley.

§B. *Sphecius speciosus* (Drury)

The cells are apparently more or less constant in size. Various authors give the size as from 2 to 3 cm in diameter and from 3 to 5 cm in length. Reinhard states that "as the season advances, more cells are built in a sort of semicircular series, each series farther from the opening of the tunnel." Riley says "frequently a number of branches leave the main burrow at about the same point, and terminate, after a length of 6 or 8 inches, in cells . . . More commonly, however, the branches leave the main burrow at irregular intervals." Riley's sketches show one to four cells per burrow, the cells being very widely separated. I have usually found two or three cells grouped rather closely at the end of each branch of the burrow.

Dambach and Good remark that "generally, after a cell has been provisioned . . . it is plugged with a thin wall of earth and a new cell made immediately in front of it. As many as four provisioned cells have been found in one such series. After a cell or series of cells is completely provisioned a new lateral off the main tunnel is excavated and the process repeated." These authors are the only ones of the many who have worked on this species who report cells in close, linear series along the branches of the burrows. I made a special effort to check this point at Ithaca, New York, in September 1964 (no. 2060). In this nest most of the cells did indeed appear to be constructed in series, but I found no more than two in any series; furthermore, the barrier between cells in series was always quite long (3–8 cm). The nest excavated in Arkansas County, Arkansas (no. 1150) also appeared to have its two cells in series, but two nests studied at Pittsford, New York (no. 1123) appeared to have each cell at the end of a short branch of the burrow (see Fig. 56).

Probably many cicada killers dig only one burrow and continue to expand upon it throughout their life. N. Lin (personal communication) observed one female provisioning a nest that had been dug 20 days earlier. However, there is evidence that some females dig more than one nest (Lin, 1964). Burroughs found 9 cells in a nest he dug, I have found as many as 10, and Dambach and Good report an average of 15.8 cells per burrow. Dow dug out 42 cells from the vicinity of two nests about a yard apart, but he believed that more than two nests might be involved. Presumably when a new cell is added to the nest, the soil from it is used to fill the burrow leading to the previous cell. From time to time additional soil is added to the mound at the entrance, presumably when new cells and cell burrows are being constructed. Wasps have apparently not been observed in the act of final closure. Some eight to ten nests

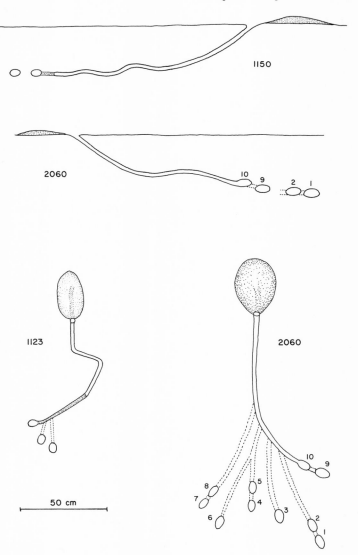

Fig. 56. Three nests of *Sphecius speciosus,* none of them complete: no. 1150 (profile), Arkansas County, Arkansas; no. 1123 (plan), Pittsford, New York; no. 2060 (profile and plan), Ithaca, New York, the profile with some of the cells and cell burrows omitted. Burrows indicated by broken lines had been filled and could not be traced exactly.

§B. *Sphecius speciosus* (Drury)

observed on the biology quadrangle at Harvard in 1963 all remained open until covered by weathering of the surrounding soil well after the disappearance of the females. It appeared that these nests were active until the females gradually died off during cool and wet periods in early September. Lin has noticed that many nests are never closed by the wasps, but some apparently receive a loose plug of soil as a final closure. Perhaps only females that make a second nest close the first one.

The remarks on digging and nest structure in Curran's (1951) popular account of the cicada killer are quite out of line with other published reports and with my own experience. Curran states that the female "digs straight down for a considerable distance—from six to twelve inches or more—then the tunnel curves sharply and is continued for another five to seven inches. Of course, once the shaft has progressed some distance it is no longer possible for the wasp to scratch out the soil; she must carry it to the surface in her mouth and deposit it in a mound away from the entrance." The brood chamber is said to extend "horizontally from the shaft." Curran does not state the source of these apparently erroneous data.

(5) *Hunting and provisioning*. Authentic prey records for this species include several species of the genus *Tibicen* (Cicadidae, Table 10). As Dambach and Good remark, the emergence of *Sphecius* is timed so as to coincide with the height of abundance of these "dog-day" cicadas, but it is too late for early summer cicadas such as *Okanagana* and *Magicicada;* in any event the 13- and 17-year cicadas would scarcely provide a regular supply of food for these wasps. The species of *Tibicen* employed are the ones common in the nesting area. As will be seen from Table 10, much the same species are employed in Massachusetts, New York, and Ohio, while in more southerly localities such species as *T. pruinosa* and *dorsata* are utilized. In Arkansas the six cicadas that I took from one nest were all *T. vitripennis,* and unusually small *Tibicen.* In this case four cicadas were used per cell, as compared with one or two of the larger species, or occasionally three of the more moderate-sized species. Curran's (1951) report of "from eight to more than a dozen cicadas in one brood chamber" is grossly in error.

Riley speaks of the "rapid, strong flight about the trees harboring its prey. The sudden cessation of the regular note of the unsuspecting cicada and in its stead a distressing discordant cry will catch the ear of the observer." Many persons, including myself, have heard this "discordant cry," indicating that a male cicada

TABLE 10. PREY RECORDS FOR SPHECIUS SPECIOSUS

Species	Locality	Note no. or reference
CICADIDAE		
Tibicen canicularis Harris	Mass.	Dow, 1942
	N.Y.	1122, 1123; Davis, 1920
	Ohio	Dambach and Good, 1943
T. chloromera Walker	N.Y.	Davis, 1920
	Ohio	Dambach and Good, 1943
T. dorsata Say	Kansas	HE52
	?	Ashmead, 1894
T. linnei Smith and Grossbeck	N.Y.	Davis, 1920
	Ohio	Dambach and Good, 1943
T. lyricen DeGeer	Mass.	Dow, 1942
	N.Y.	1122, 1123; Davis, 1920
	Ohio	Dambach and Good, 1943
T. marginalis Walker	N.C.	Manee, 1915
	?	Ashmead, 1894
T. pruinosa Say	D.C.	Riley, 1892a
	Ohio	Dambach and Good, 1943
T. robinsoniana Davis	Ohio	Dambach and Good, 1943
T. vitripennis Say	Ark.	1150

has been attacked by a predator. However, nearly everyone who has worked on the cicada killer has noted that more females than males are taken. Davis states that the cicadas are sought "not by the songs of the males as is sometimes stated, but by flying carefully along the limbs of trees and up and down the trunks, and so catching both sexes." Of 66 cicadas taken from nests by Dow, 24 were males and 44 females. Of 703 collected by Dambach and Good, 204 were males and 499 females. Curran's report that male cicadas are used almost exclusively and that these are located by their singing is, as usual, out of line with other published information.

Dambach and Good state that hunting females circle about tree trunks, gradually working their way up through the limbs and branches. These authors provide the following description of the capture of the prey:

> The wasp darted backward and forward in front of the cicada several times meanwhile bending the tip of the abdomen downward and forward. It then hit the cicada viciously and injected the sting between the

§B. *Sphecius speciosus* (Drury) 105

abdominal segments. The cicada buzzed shrilly and immediately ceased struggling. The wasp managed to hold its prey to the limb of the tree and pulled it up on the top side of the limb. After this effort the wasp turned the cicada over on its back, straddled it, grasped the body firmly and flew off with the burden.

Burroughs, in 1881, also noted that a cicada that had been seized was dragged "over on top of the limb."

Although the account of prey capture quoted above mentions that the sting was inserted into the abdomen of the prey, this is clearly not typical. Reinhard took many ("over a score") cicadas from wasps returning to their burrows and found that the sting puncture was always in the membrane at the base of one of the front legs. The puncture is always single, and is long and slitlike, which Reinhard says "testifies to the fact that the sting does not give a simple thrust into the side but bears toward the center, and the deliberate actions of the wasp point to a definite locality which must be reached . . . It appears that the wasp persists with her sting until the victim ceases to struggle." He believed that the venom was injected into or close beside the single large thoracic ganglion of the prey.

Riley's figure of the cicada-killer wasp holding the prey dorsum up is quite erroneous, but unfortunately it appeared in several other publications, including Howard's *Insect Book* (1901, and later editions) before being corrected by Davis (1920) and others. Actually, the cicada is held venter up considerably farther back than figured by Riley (Fig. 57). Dambach and Good describe the wasp as "locking its middle pair of legs about the neck between the eyes and pronotum." According to Reinhard, the claws of the middle legs "are fitted into a pocket at the base of each of the cicada's wings." The painting in Savin's paper is not completely accurate in this regard, and the photograph in Curran's book is clearly posed, but two photographs in Reinhard's book can be considered accurate. (Photographs do sometimes lie. A kodachrome transparency supplied by Ward's Natural Science Establishment is inaccurate and also presumably posed from dead specimens.) Howes presents a photograph of a female dragging a cicada backward into her burrow. This behavior occurs when an unusually large cicada has to be moved over the ground some distance or when the wasp has been disturbed and unable to carry the prey directly into the nest in the usual manner.

Although cicada killers proceeding over the ground or entering the nest hold their prey only with their middle legs, the hind legs also embrace the sides of the cicada during flight. The hind tibial

Fig. 57. Three frames from motion picture film of *Sphecius speciosus* dragging prey to nest (Pittsford, New York, July 1955). Note the trail in the sand made by the wasp and prey, also, in the lowest frame, the movement of the wings for propulsion. The middle legs are being employed to grasp the prey.

§B. *Sphecius speciosus* (Drury)

spurs of *Sphecius* are unusually large and somewhat hooked, and it appears that these hold the sides of the posterior part of the cicada's body, although the details are not clear. When Howes (1919) removed the hind tibial spurs from one individual, she continued to bring in cicadas, but in flight "the cicada was suspended, tail down, in a line perpendicular to the wasp's body; the two insects forming the letter T while in the air."

Balduf (1941) weighed freshly killed wasps and their cicada prey, and found the cicadas to weigh from four to six times as much as the wasps.

I know of no records on the number of cicadas brought in per day by individual wasps. Several authors speak of prey capture and provisioning relatively early in the day, and as already mentioned I found cicadas to be brought in at 0850 in Kansas. However, in the colony at Pittsford, New York, little provisioning was observed until about 1700, and wasps were actively provisioning until dusk, at about 2030.

The cicadas are packed in the cells head in, venter up. When the cell is closed, the earth fill is packed against the abdomen and wing tips of the prey.

Cicadas stung by *Sphecius* remain in deep paralysis but stay fresh and relaxed for at least 1 week. Riley states that "should the egg of the wasp fail from any cause to hatch, the paralyzed victim nevertheless remains in a state of suspended animation, which will last under favorable conditions for a year and how much longer is not known." It is true that uneaten cicadas may sometimes appear relatively fresh if exhumed the following year, but such cicadas have nevertheless been dead for a long time.

Dambach and Good calculated that the colony of 373 burrows which they studied contained an average of 30.4 cicadas per burrow; thus the cicada killers in this colony must have destroyed over 11,000 cicadas.

(6) *Immature stages and development.* The egg is 5–6 mm long by 1–1.5 mm wide, and is laid longitudinally on the ventral side of the thorax opposite one of the middle coxae. The position of the egg is quite fixed, as shown by the figures provided by Riley, Reinhard, and Dambach and Good, and as confirmed by my own observations (Fig. 58). When more than one cicada is placed in the cell, the egg is laid on the last one (Reinhard; my note no. 1150). The egg hatches in 2 to 3 days, and larval development is rapid, requiring (according to Dambach and Good) 4–10 days of feeding. The larva feeds primarily through the coxal cavities, completely

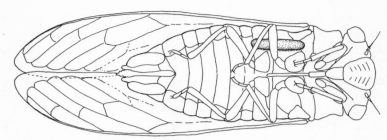

Fig. 58. Adult female cicada (*Tibicen* sp.) bearing the egg of *Sphecius speciosus* (no. 1124, Pittsford, New York).

hollowing out the cicada but leaving the exoskeleton largely intact. Thus Dow was able to sex and identify many cicadas taken from cells in which the larvae had spun their cocoons. The larva has been described and figured by several writers, and recently in some detail by Evans and Lin (1956b:46).

Spinning of the cocoon requires about 2 days and has been described at length by Riley (1892a,b). Particles of earth are incorporated into the walls as in other nyssonine wasps, but in general the particles are smaller than in such genera as *Gorytes* and *Bembix*, giving the cocoon walls a fine texture. *Sphecius* differs from *Gorytes* and its relatives in preparing a series of pores in the wall of the cocoon. Riley (1892a) accurately figured one such pore in cross section. According to Reinhard, there are from 4 to 13 such pores, arranged in an irregular row on only one side of the cocoon, near the middle. The cocoons I have studied contained from 2 to 10 pores, all on the upper side, the larger cocoons tending to have more pores.

Many writers have remarked upon the fact that cells containing a larger amount of provisions produce larger larvae which make a large cocoon and produce female wasps, while the smaller cocoons produce males. Savin provides an excellent photograph of cocoons of the two sizes, and Dow found that the cocoons did indeed fall into two distinct size classes, with no overlap at all. The smaller cocoons, mainly from cells containing a single cicada, produced males; the larger cocoons, mainly from cells containing two cicadas, produced females. Apparently the female wasps lay an unfertilized (male-producing) egg in cells containing a smaller amount of provisions. Further details will be found in Dow's paper.

(7) *Natural enemies and other causes of larval mortality.* Reinhard found the cicada killer to be attacked by the small miltogrammine fly

§B. *Sphecius speciosus* (Drury)

Senotainia trilineata (Wulp). He states that fully half the larvae were destroyed by maggots in the area in which he worked. According to this author, the *Sphecius* egg hatches but the larval food is soon destroyed by the maggots, so the wasp larva dies of starvation. Reinhard found that the flies perch about the nest entrances and trail females carrying prey, sometimes several at a time. As the wasp enters the burrow, the flies larviposit on the posterior end of the cicada.

Dambach and Good also found miltogrammine flies attacking *Sphecius* in Ohio. They found maggots, numbering 16 to 27 per cell, in large numbers of cells in certain localities. They found both maggots and *Sphecius* larvae in certain cells, suggesting that the wasp larva is not directly damaged by the maggots, but merely fails to complete its development because its food has been destroyed. These authors collected three species of flies in or around nests, of which only *Metopia argyrocephala* (Meigen) is known to attack solitary wasps in this manner.

Curran observed flies (not identified) that larviposited on the cicadas as they were being carried in or entered burrows after the wasp had left. "As a rule," he states, "both cicada killers and flies develop to maturity in the same brood chamber, because very seldom are there sufficient fly maggots to devour the entire food supply." It is doubtful if these remarks were based on careful rearing of cell contents (it will be remembered that Curran reported 8–12 cicadas per cell!). In a species such as *Sphecius speciosus,* where the amount of food per cell is graded according to the sex of the offspring, the consumption of part of the food by maggots would appear to be serious. Furthermore, my own observations indicate that the maggots normally destroy the wasp egg before consuming the cicadas.

In the ten-celled nest I excavated at Ithaca, New York (no. 2060), three of the cells contained maggots and two others appeared to have had maggots which had left the cells (although I could not find the puparia). In each case the maggots were feeding at the site of the *Sphecius* egg, the latter having been destroyed. The maggots were reared successfully and found to be *Senotainia trilineata* (Wulp). Both this species and *Metopia argyrocephala* (Meigen) were common in the area.

Nearly everyone who has dug out nests of this wasp has noted many cells in which the wasp larvae have failed to develop. In some cases the presence of mold in the cell suggests that this may be the cause of egg or larval mortality, but of course the mold may have developed after the larva would have completed its development. Some authors have suggested that the cicada killer fails to lay eggs in some of the cells. Dambach and Good found that of 525 cells they examined, only 201, or 38.28 percent, contained viable cocoons; the remainder contained mainly untouched cicadas.

Despite the extensive literature dealing with this wasp, a number of questions obviously remain to be answered.

C. Ethology of other species of *Sphecius*

Very little information is available on the other North American species of this genus. The observations reported by Walsh and Riley (1868) for *Sphecius grandis* (Say) undoubtedly apply to *speciosus*, as *grandis* is a western species and does not occur in Pennsylvania, where their observations were made. Bradley (1920) mentions a colony of *grandis* found at Langtry, Texas; wasps were noted going into their burrows, and a boy questioned said he had seen them with prey. *Mutilla orcus* Cresson (that is, *Dasymutilla klugii* Gray) was seen entering burrows and was believed to be parasitic on the *Sphecius*.

I have encountered *Sphecius grandis* with prey once, at McKinney Lake, Kearny County, Kansas, in August 1952 (no. HE110). At 0900 I heard a loud buzzing near the top of a small cottonwood, about 5 m high. I looked up and saw a female *Sphecius* struggling with a very large cicada, later determined to be *Tibicen dealbata* (Davis). Both fell to a lower branch, and the wasp then tried to carry the cicada back to the top, but instead they fell to the ground. The wasp then tried to carry the cicada up a weed, but without success. Finally the wasp left the cicada part way up in the weed and flew away. When she had not returned 1 hr later, I took the cicada for identification. During this episode I had many opportunities to study the manner in which the wasp held the prey. Invariably the cicada was venter up, held well back beneath the wasp, the middle legs of the wasp passing just behind the eyes of the prey and grasping the dorsal part of the thorax, especially the wing bases. When climbing up trees or weeds, she walked on her front and hind legs and buzzed her wings rapidly for additional propulsion.

The western North American species *Sphecius convallis* Patton is reported by Krombein (1951, 1958d) as preying upon the cicadas *Tibicen pruinosa* (Say) and *Diceroprocta apache* (Davis). In the province of Mendoza, Argentina, Jensen-Haarup (1924) found *Sphecius spectabilis* Taschenberg preying upon the cicada *Tettigades chilensis* Am. and Serv. This wasp was found nesting in considerable numbers in a bare, sunny slope in the foothills of the Andes.

§D. Genus *Exeirus* Shuckard   111

The only member of the Palaearctic subgenus *Sphecienus* to have been studied, and that very briefly, is the south European *S. nigricornis* Dufour. DeGaulle (1908) mentioned this species as nesting in bramble stems, but subsequent authors have considered this to be probably erroneous. Berland (1941) reported that M. A. Devillers, when collecting in Gard, France, had had his attention drawn to a "cri aigu de désespoir" of a large cicada, which had been captured by a *"Sphex."* Berland, after correspondence with Devillers, tentatively identified the wasp as *Sphecius nigricornis* and the prey as *Cicada plebeja*.

Arnold (1929) has reported briefly on the South African *Sphecius (Nothosphecius) milleri* Turner.

This insect preys on large cicadas, such as *Munza furva* Dist., *Platypleura quadraticollis* Butl., *P. lindiana* Dist., and *P. marshalli* Dist. which live largely on small Mopani trees (*Copaifera mopani*). The *Sphecius* circles round these trees and makes a sudden swoop at its prey, bearing it to the ground where it is paralyzed by stinging. It is noticeable that when one of these big wasps comes within a foot or so of the trees, the vociferous din of the cicadas ceases quite suddenly. The nests are at the end of rather long tunnels, often placed in the loose sandy walls of burrows which have been dug by ant-bears.

In Madagascar, *Sphecius (Nothosphecius) grandidieri* Saussure was found by P. Frey to prey upon the cicadas *Poecilopsaltria brancsiki* Dist. and *Platypleura pulverea* Dist. (Handlirsch, 1895).

D. Genus *Exeirus* Shuckard

This genus contains a single species which occurs in eastern Australia. The major structural features held in common with *Sphecius* were summarized at the beginning of this chapter, and I shall here review the many differences. In general, *Exeirus* is a more hairy and much less compact wasp than *Sphecius,* the mesosoma being more elongate and with deep constrictions between many of the sclerites (particularly between the pronotum and mesoscutum). *Exeirus* is also a much more long-legged wasp, so long-legged that it bears much resemblance to a spider wasp (Pompilidae); in fact in some of the earlier literature *E. lateritius* is placed in the pompilid genus *Priocnemis*. *Exeirus* lacks the modifications of the middle and hind legs found in *Sphecius* and presumably assisting in carriage of the cicada prey. It is probable, however, that the very long legs of this genus represent an adaptation for holding the body sufficiently high off the ground to straddle the cicada.

Other pertinent adult features of *Exeirus* are as follows:

Labrum not much exserted beyond the arcuately emarginate clypeus, especially in the female; clypeus much more transverse than in *Sphecius* and not extending much up between the eyes nor in contact with them; front narrow but the eyes not divergent above; eyes situated well toward the front of the head, the temples wide; scrobal suture weaker than in *Sphecius*, but the oblique groove, connecting the epicnemial ridge to the scrobal suture, very strong; epicnemial ridges discontinuous below; propodeum with a strong stigmatal groove; front tarsus of female not only with a strong pecten, but the segments themselves also produced apically as stout teeth; third submarginal cell of fore wing petiolate; female without a well-defined pygidial area; end of sternite VII of male visible beyond sternite VI; apical sternite of male tongue-like, not in the form of a long spine; genitalia strikingly similar to those of *Sphecius* (Figs. 59–62).

Although Beaumont (1954) notes that the venation of the hind wing of *Sphecius* is much like that of *Stizus*, this is by no means true of *Exeirus*; in the latter genus the submedian cell is very much shorter, the transverse median vein meeting the media close to the origin of the cubitus rather than far beyond. All in all *Exeirus* is rather far removed structurally from *Sphecius*, and it is surprising to find so many similarities in the behavior of these two genera. A more detailed study of *Exeirus lateritius* might well reveal more differences than are now apparent.

E. *Exeirus lateritius* (Shuckard)

This large wasp is apparently common in parts of eastern Australia. Froggatt (1903, 1907) refers to it as *Priocnemis* sp. and *Priocnemis bicolor*, but there is no question that he was dealing with *Exeirus lateritius*. McCulloch (1923) and Musgrave (1925) have presented more accurate and up-to-date accounts of the behavior of the species.

Apparently the adults feed at plant sap in the manner of *Sphecius speciosus*. Froggatt (1903) speaks of them driving cicadas from the trunks of trees, but not stinging them, then drinking the sap exuding from the punctures made by the cicadas. The females dig their nests before hunting for cicadas as prey. The nests are dug in friable soil, including gardens and sandy places. The burrow is said to be "about the size of a mousehole." The soil from the excavation is left in a pile at the entrance. The pile contains "four or five cups" of soil, and judging from McCulloch's photographs is irregular in shape and contains a groove in front of the entrance exactly as in *Sphecius*. McCulloch describes the digging of the burrow as follows:

§E. *Exeirus lateritius* (Shuckard)

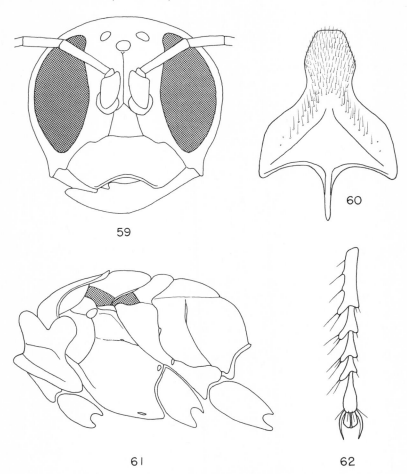

Figs. 59–62. *Exeirus lateritius* (Shuckard): 59, head of female; 60, sternite VIII of male metasoma; 61, mesosoma of female, lateral aspect; 62, front tarsus of female, showing pecten.

Each year, about the middle of November, when the sun has warmed up the earth, they may be seen industriously shovelling sand out of the tunnel mouth and scattering it far behind them. Standing firmly on the two hinder pairs of legs, with the abdomen raised, they turn the front pair inwards till their tips almost touch, and scoop away the sand with quick short strokes, throwing it backwards beneath the body. The amount of sand moved at each throw is, of course, small, so one can imagine what a large amount of energy is expended in digging out several tunnels ten to twenty inches long, from which every grain has to be shifted to the surface with many successive throws, and piled up outside the burrow.

In addition to McCulloch's report on burrow depth, there is also a report by Froggatt that the burrows "vary in depth from 10 inches to 2 feet." McCulloch states that several tunnels connect with the main tunnel, each containing a cicada at its end. Musgrave's paper contains a photograph of a model of the nest made from a plaster of Paris cast of an actual nest. This model shows the nest descending at about a 50–70° angle with the surface and having several turns. There are three cells in this nest, two at the ends of very short side burrows, the third (and apparently most recent) cell at the end of the burrow, considerably deeper than the other two. Apparently the nest is much like that of *Sphecius speciosus* but tends to be more nearly vertical and to descend more deeply into the soil.

The prey consists of large Cicadidae, of "half a dozen species," according to Froggatt, who mentions in another place that one species, *Henicopsaltria perulata* Guer., was being used almost exclusively in one locality. McCulloch lists as prey "common green Mondays" [*Cyclochila australasiae* Donovan] and "black and yellow Fiddlers" [*Macrotristria angularis* Germ.]. Apparently only one cicada is used per cell in most cases. Froggatt states that the cicadas may be "many times" larger than the wasp. McCulloch found a wasp weighing 15 grains carrying a cicada weighing 32 grains. This author found male and female cicadas to be used.

Musgrave states that the hunting female "flies round and round a tree in ascending spirals and on perceiving a cicada darts at it and inflicts a sting in its nerve centres." McCulloch states that the cicadas are apparently not located by sound, since he has seen the wasps bringing in prey when no cicadas were singing. He states that, following stinging, the wasp and prey drop to the ground, "sometimes from a great height, with a thud which provokes a last rattle from the now moribund cicada." The wasp then drags the cicada "though grass and weeds, over sticks, or anything else that may be between it and the burrow. . .Throwing the cicada on its back so that it will glide along easily, and straddling across it, the wasp grips it firmly with the middle pair of legs. Using its long hind legs and shorter front ones, and often assisting itself with its wings, the homeward journey is commenced. The wasp travels along the ground at an amazing rate." Froggatt (1903) had earlier spoken of the wasps holding their cicadas with their hind legs, but this is doubtless an error of observation. Froggatt (1907) also speaks of wasps "riding" the cicadas "down to their tomb," implying that at times the wasp may take flight from a high object with her prey in the manner of *Sphecius speciosus*.

McCulloch mentions that although several wasps may nest in

§E. *Exeirus lateritius* (Shuckard)

close proximity, the females usually proceed without error to their own nests and disappear quickly into them. Apparently the nest entrance is left open as in *Sphecius.*

According to McCulloch, the egg is laid on the lower surface of the thorax near the leg bases. As in *Sphecius* the larva hollows out the inside of the cicada by inserting its slender anterior end through a hole in the body wall.

Chapter VI. The Stizine Wasps: *Stizus, Stizoides,* and *Bembecinus*

The wasps considered here were all formerly placed in the genus *Stizus*. Although they have many similarities, there is no question that three separate genera are involved; the three are quite dissimilar in ethology, as will be seen. The genera *Stizus* and *Stizoides* occur widely in the Old World, where there are many species. In the Americas there are only four species of *Stizus* and two of *Stizoides*, and none of the species are known to range south of Central Mexico. The genus *Bembecinus* occurs throughout the tropics and subtropics of the world.

The Stizini are robust wasps, with a short, compact mesosoma similar to that of *Ammatomus* and *Sphecius*. They differ from all Nyssoninae considered so far in four important features: (1) the epicnemium is completely absent, (2) the metanotum is discontinuous laterally, that is, there is no externally visible connection between the notum and the pleura, (3) the basal vein of the fore wing is less oblique and meets the subcosta far basad of the pterostigma, and (4) metasomal tergite VII of the male has reflexed side pieces, separated from the main part of tergite by a deep incision, containing the spiracles. Other adult features worthy of note are as follows:

Labrum exserted, semicircular; front narrow, eyes subparallel to strongly diverging above; clypeus narrow, extending well up between eyes, antennae arising toward middle of face; ocelli normal; pronotum very short; mesopleurum with a scrobal groove connecting the scrobe with the subalar pit, and with a distinct signum, but otherwise without grooves or ridges; pecten well developed in females; middle tibiae with two apical spurs except with only one in a few Old World species (subgenus *Scotomphales* of the genus *Stizoides*); metasoma robust; female with or without a pygidial

A. Genus *Stizus* Latreille

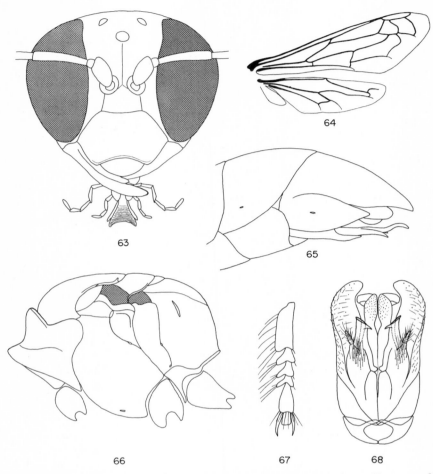

Figs. 63–68. *Stizus brevipennis* Walsh: 63, head of female; 64, wings; 65, apex of metasoma of male, lateral aspect, showing lateral spiracular lobes of tergite VII; 66, mesosoma of female, lateral aspect; 67, front tarsus of female, showing pecten; 68, male genitalia.

area; male with sternite VIII trifurcate apically, genitalia with the basal ring short, volsellae with the digiti much exceeding the cuspides (Figs. 63–68 and 72–75).

A. Genus *Stizus* Latreille

Members of the genus *Stizus* are of medium to large size (up to 3.5 cm); they are often brightly patterned with yellow and red. The mandibles are stout and dentate, and the eyes tend to diverge only rather slightly above; the propodeum is more or less flat behind and is not produced on the sides. The larvae are unusual in possessing rows of large spines on the body and in having a large, blunt tooth on the mandibles close to the apical tooth. The cocoon is also unusual, as described below (Figs. 63–68).

B. *Stizus pulcherrimus* Smith

This East Asian species is the only member of the genus to have been studied in detail (Tsuneki, 1943a,c, 1965b). Tsuneki's first two reports were in Japanese, and hence largely unavailable to Western workers. However, he has now published a translation and expansion of these earlier studies (Tsuneki, 1965b). Since this paper is of considerable significance in the present context, I shall consider it in some detail.

Stizus pulcherrimus is a large wasp, 20–25 mm in length, known to occur from Mongolia to Japan. It is rare in Japan and has not been studied there save for a brief note on the prey by Katayama (1933). Tsuneki observed a female carrying prey at Apaka, East Mongolia, in August 1939, and he excavated the nest of this wasp (first reported in Tsuneki, 1943a). In June 1941 he found several individuals nesting near Seoul, Korea, and was able to study the species in more detail (first reported in Tsuneki, 1943c). In Korea some 20 individuals nested on a hill near the Keijo airport, in reddish brown clay soil containing many stones derived from the underlying gneiss. The place was planted with chestnut trees, but the trees were still very young and gave little shade. In addition to the *Stizus*, species of *Cerceris, Philanthus,* and *Lestica* made their nests in bare places.

(1) *Nesting behavior.* Near Seoul, Korea, Tsuneki found six nests within an area of about 0.5 cm^2. These nests were along the edge of a small mound; the entrances were from 3 to 30 cm or more apart, and in some cases the tunnels of adjacent nests crossed one another, although without actually touching. Later Tsuneki was able to find nine additional nests scattered about the nesting area.

Tsuneki noted that the entrances were always closed with pellets of earth when the wasp was away, so that nests were difficult to detect at first glance. However, he observed various open holes in the nesting area, and it was determined that each nest had one or more "side holes," which were left open by the wasp. According to Tsuneki (1965b):

§B. *Stizus pulcherrimus* Smith

The side holes were two in number as a rule, dug one on each side of the true entrance which was always closed. They were usually 3-4 cm in depth, sometimes only 1 cm and rarely more than 5 cm. They were more steeply [inclined] into the earth than . . . the true burrow. The variation in distribution and in number of these accessory holes was just the same as in *Sphex argentatus* [Tsuneki, 1963a]: sometimes 2 on one side and none on the other, sometimes 2 on each side, or 1 on one side and 2 on the other.

These "side holes" are apparently very similar to those occurring in *Bembix texana, B. troglodytes,* and several other North American wasps. I have called them "false burrows" (see Chapters I:C, 2; XI:E, 6; and XIV:B, 3). Tsuneki found that the single nest he studied in Mongolia lacked a side hole, although these were present in all 15 of the nests studied in Korea.

(2) *Structure and dimensions of the nest.* The burrow of this species is 7 mm in diameter, penetrates the earth at an angle of 25-30°, and attains a total length of 10-50 cm. The main burrow goes in straight for 10-20 cm, reaching a depth of 6-7 cm beneath the surface. Thereafter the burrow gives rise to several branches which vary in length from 5 to 30 cm, thus often exceeding the main burrow in length. These branch tunnels are more or less horizontal and are closed off once the brood cell at the end of each is fully provisioned. The ellipsoidal, horizontal cells measure about 40 mm in length by 15-17 mm in maximum diameter. The upper side of the cell is situated only 4-7 cm below the surface of the ground.

The number of cells (and cell burrows) in the 14 nests excavated in Korea varied from 1 to 9 (two of Tsuneki's sketches are reproduced here in Fig. 69). The wasp, on completing the main burrow,

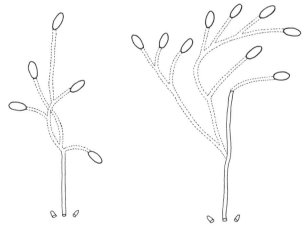

Fig. 69. Two nests of *Stizus pulcherrimus* (plan). (After Tsuneki, 1943b.)

at once makes a cell at its end. She then hunts a grasshopper, places it in the cell, and lays an egg on it. When the necessary number of prey have been stored (up to seven, most commonly four or five), she closes that portion of the burrow permanently. She then digs a new branch burrow in another direction, enlarges the end into a new brood cell, lays a second egg on the first prey in this cell, adds several more prey, and then closes this branch burrow. The process is repeated, there apparently being no particular order in the arrangement of the brood cells.

The nest studied in Inner Mongolia was unicellular; the burrow was 25 cm long, the terminal cell at a depth of about 5 cm. At three separate places in the burrow there were temporary closures of earth, about 7 mm in thickness.

(3) *Hunting and provisioning.* Tsuneki describes the capture and stinging of the prey as follows:

At 10:00, while I was confirming the distribution of their nests I suddenly saw a wasp capture a short-horned grasshopper. It occurred about 2 m in front of my eyes. The wasp caught the insect from the back with her legs and flew to a leaf of the chestnut tree, standing about 3 m apart from there. Instantly she stung the prey twice at the ventral side of the thorax, bending her abdomen from the side of the insect. But the wasp, being frightened by the too close approach of me, flew up with the prey and landed on another leaf 5 m apart from the first leaf. There she hung down from the edge of the leaf with one of her hind legs, turned the prey insect round so as to hold it venter to venter and head to head. She then caught it by the antennae with her mandibles and flew to the trunk of a chestnut tree. She climbed it a little and flew off to go to her nest. Near the ground she flew through the grass leaves and landed on the ground. About 10 cm in front of her several holes were open with the debris by their sides. The wasp proceeded, capturing the prey by the antennae with her mandibles and holding it with her mid legs from both sides. She reached near the holes, but did not enter any of them, but suddenly began to dig the earth in front of them with her front pair of the legs where not the least trace of the burrow was apparent. Soon, however, the entrance to the burrow came to appear and the wasp penetrated into it at once. At this moment the prey carried by her was pushed backwards to the tip of her abdomen and was dragged in the burrow.

Further study indicated that the wasp always grasps the antennae of the prey with her mandibles and the body of the prey with her middle legs; this is illustrated with three photographs (Figs. 4, 5A, and 5B in Tsuneki, 1965b). Sometimes the prey remains in the nest entrance while the wasp enters, turns around inside the nest, then

§B. *Stizus pulcherrimus* Smith

grasps the antennae of the prey with her mandibles and draws it in. During favorable weather provisioning is very rapid; one wasp was seen to bring in four prey at intervals of 5–10 min.

The first prey is always a small nymph and is placed in the innermost part of the cell with its head inward. The egg is laid on this grasshopper and the cell then filled with grasshoppers, all placed head in, either on their backs or on their sides. The prey are deeply paralyzed, but for the first few days movements of the antennae and palpi can be detected and a certain amount of excreta is voided into the cells.

The 40 cells of the 15 nests dug out in Korea contained a total of 118 grasshoppers, about half of each sex, all but 13 of them nymphs. Seven species of Orthoptera belonging to two families were found in the nests, in the following numbers:

ACRIDIDAE
  ACRIDINAE
    Acrida lata Motschulsky   1
    Aiolopus tamulus Fabricius   13
    Parapleurus alliaceus Germar   67
    Stauroderus schmidti Ikonnikov   1
  OEDIPODINAE
    Trilophidia annulata Thunberg   1
  CYRTACANTHACRIDINAE
    Sp. resembling *Oxya vicina* Brunner   32
TETTIGONIIDAE
    Conocephalus maculatus Gouillon   2

The nest excavated in Mongolia contained six nymphs and two adults of *Chorthippus dubius* Zub. (Acrididae, Acridinae). These varied in length from 14 to 30 mm; all were in the cell with their heads inward and most of them were on their backs. In this instance the cell, which was dug out while still being provisioned, contained a small larva attached to a wing pad by its posterior end; the larva was feeding at the base of a foreleg of the same grasshopper. Tsuneki's records indicate that the weather had been good for ten days previously (except for one day), and he therefore feels that the Mongolian population of this species exhibits true progressive provisioning (rather than delayed provisioning as a result of inclement weather, as discussed in Chapter I:C, 3). "However," says Tsuneki (1965b), "the mode of provisioning is very primitive as such, because the waspling was [not half grown] and the prey provided for it were considered already [to have] approximated the full provisioning."

Chapter VI. *Stizus, Stizoides,* and *Bembecinus*

(4) *The egg and development of the larva.* The egg is laid on the fore wing pad of either side of the small nymph that is first placed in the cell (Fig. 70). It is attached with its caudal end to the prey, with its cephalic end free. It is laid across the upper side of the pleurum and reaches with its apex almost to the lower extremity of the pleurum. The egg is milky white in color and measures 4.3 mm in length by 0.9 mm in width. Eggs collected just after being laid by the wasps hatched after 24–30 hr when kept in rearing chambers at room temperature.

The larva, soon after hatching, cuts a small hole in the integument of the prey where it reaches with its mouthparts as it leans over the insect, and at once begins to suck the fluids. As a result the prey becomes emptied of contents in 24–36 hr, the wasp larva by this time measuring 6–8 mm in length. It then perforates the body of another prey lying nearby, inserts its head inside the prey, and devours it. The larval period in this species is about 1 week.

(5) *Cocoon spinning.* Ten larvae were reared in the laboratory in glass vials having a diameter of 15 mm and length of 100 mm. A layer of moist sand, 20–25 mm in thickness, was packed into the vial, the larva or egg with its prey placed next to it, and the tube then closed with a cotton plug. When the larva has eaten the prey it spins silk threads and attaches them to the walls of the chamber, thinly covering the floor with a hammock-like web of silk. Then it gradually adds threads to the upper sides to form an ellipsoidal pouch. The external end of the pouch, that is, the side facing the cotton plug, after gently converging, is gathered into a bundle and suddenly bent downwards and glued to the floor of the bottle. The internal side, on the other hand, is enlarged into a funnel and attached to the wall of the sand plug. At this stage the larva can

Figs. 70 and 71. Immature acridid grasshopper: 70, bearing the egg of *Stizus pulcherrimus* (after Tsuneki, 1943b); 71, bearing the egg of *Stizus fasciatus* (after Ferton, 1902b).

§B. *Stizus pulcherrimus* Smith

be observed through the semitransparent pouch. When it changes direction, it inserts its head beneath the ventral side of its abdomen, pushing its head forward so as to fold the body in two, then by moving its anterior and posterior ends in opposite directions completes its turning. Since the pouch is quite elastic, the larva is able to do this several times until the cocoon is completely spun.

The larva then pushes its head out of the opening on the funnel side, probes the sand with its mandibles, then collects sand grains under its thorax and moves them into the pouch by drawing backwards. At this time a choice of sand grains is made, the larger grains being thrown away from the funnel with the mandibles by quickly moving the head forward. After a certain amount of sand is collected, the larva retreats into the pouch, moistens the sand grains with its secretions, and pushes them to the inside of the pouch and glues them there. When the store of sand in the pouch is used up, the larva again reaches out of the funnel-like end and collects more material, which is in turn added to the walls of the cocoon. In this manner, the inner wall of the pouch is gradually covered with sand grains, eventually forming a complete wet sand cocoon. It is built up roughly ring by ring, the limits of each ring being determined by the ability of the larva to reach out from a given position. When the equator has been passed the larva changes its direction. When it reaches the part near where the pouch converges into a silk bundle, it builds a spherical cap there. It then turns about again, gathers its final mass of sand from the opening, and closes this part of the cocoon. The cocoon becomes dry and hard in one day, but the larva apparently continues lining the inside of the cocoon for 2–3 days.

The cocoon of *Stizus pulcherrimus* differs from that of *Bembix* in having a flatter cap and a nipple-like posterior protuberance. The pores resemble those of *Bembix* in a general way and number from three to six per cocoon. Dr. Tsuneki has kindly sent me a cocoon of this species, and a photograph is included here (Fig. 207A); in this cocoon, the pores appear unusually prominent, as each is surrounded with several large sand grains adhering to the outside.

(6) *Natural enemies.* Tsuneki frequently observed parasitic flies of the genus *Miltogramma* around the nests of this species, and several nests that he excavated contained fly maggots. In one instance, he found several maggots feeding upon the egg of the wasp. He believed that these maggots were brought in on the second prey and quickly sought out the egg and destroyed it. The contents of several cells had been destroyed by mold.

C. *Stizus fasciatus* (Fabricius)

This southern European species has been studied at Bonifacio, Corsica, by Ferton (1899, 1902a,b, 1909). Its behavior is similar in a general way to that of *pulcherrimus*, but the arrangement of the cells in the nest is different and apparently more specialized, and the nest entrance is said to be left open at all times.

In Corsica this species nests during June and July in sandy places, especially along water. The burrows are said to be long, tortuous, and almost horizontal, but Ferton provides no actual figures on their length or depth. The cells are all reported to be in series along the main burrow, separated by barriers of sand about 4–6 mm thick. The first cell is, of course, prepared at the end of the burrow, and cells are added by expanding the burrow slightly and presumably using the soil for the barriers. As Ferton remarks, this type of nest is unusual for a fossorial wasp, but characteristic of wasps that nest in stems and in borings in wood. As cells are added, the burrow becomes shorter. Ferton mentions one nest with three cells, but otherwise provides no data on the number of cells per nest. Because of the brevity of Ferton's account and its lack of quantitative data, it is desirable that the nest structure of this species be restudied.

The prey consists of short-horned grasshoppers (Acrididae), both adults and immatures, five to eight being used per cell. Ferton records three species: *Chorthippus bicolor* Charp. (Acridinae), *Calliptamus italicus* Linnaeus, and *Pezotettix giornae* Rossi (Cyrtacanthacridinae). To this list Bernard (1934) adds *Stauroderus vagans* Eversman (Acridinae). Ferton observed prey capture on one occasion. A female *Stizus fasciatus* was seen hunting in clumps of bushy herbs, where it encountered a grasshopper nymph feeding about 40 cm above the ground. The wasp, which had approached the grasshopper closely, backed up about 20 cm in flight and then plunged forward and seized the grasshopper. She applied the sting for several seconds, curving her abdomen beneath the prey and stinging it on the ventral side. The wasp then flew off with the prey. Ferton was also able to observe stinging by placing grasshoppers outside nest entrances, in which case they would often be accepted by the wasps when they left their nests. In each case the wasp seized the prey with its mandibles and stung it beneath the thorax.

The egg is laid on the grasshopper deepest in the cell, that is, on the first prey placed in the cell. The egg is attached to one of the wing pads of a nymph by its posterior end, its free anterior end extending obliquely forward and toward the venter of the grass-

§D. Ethology of other species of *Stizus*

hopper. According to Ferton, the grasshopper bearing the egg lies on its ventral side, so that the egg is actually directed head down, quite the opposite of the situation in *Bicyrtes* and *Bembix*. However, the remaining prey in the cell are placed sometimes on their backs, sometimes on their sides. The egg is 4–5 mm long and about 0.75 mm wide. Ferton's figure (1902b, reproduced here as Fig. 71) is strikingly like Tsuneki's figure of egg placement in *pulcherrimus*.

Ferton provides no data on the manner of prey carriage, but he does remark that the nest entrance is left open at all times. Provisioning is said to be rapid, "as rapid as permitted by the hazards of the chase," the cell being closed before the egg hatches. In one three-celled nest which he dug out, none of the eggs had yet hatched. Ferton reports that the small larva remains attached for several days to the point at which the egg was laid. He fed the *Stizus* larvae on flies, which they accepted readily and appeared to thrive upon. Ferton found that grasshoppers paralysed by *S. fasciatus* remain alive for several days and often defecate in the cells.

D. Ethology of other species of *Stizus*

The four North American species of this genus are all restricted to the southern Great Plains or to arid parts of the southwestern United States and northern Mexico. None of the species are considered common, although I have taken *S. texanus* Cresson in series on flowers of *Baccharis* in Arizona and *S. brevipennis* Walsh on flowers of *Melilotus* and *Euphorbia* in Kansas. In no case have I been able to locate the nesting sites, and to my knowledge no one else has found the nests of any of the North American *Stizus*. There is one prey record for *brevipennis*. Working in Kansas, Williams (1914) found that this species "hunts in a manner quite similar to [*Tachytes*] *mandibularis*, examining the stems of *Helianthus*, etc., as she flies and finally finds her prey, a large *Xiphidium* [that is, *Conocephalus* (Tettigoniidae); the specimen in question was a short-winged adult female]."

The predilection of wasps of this genus for Orthoptera extends to the South African species as well. Brauns (1911) records four species as using immature Orthoptera. In three cases the nature of the prey is not further clarified, but one of the species, *S. imperialis* Handlirsch, is said to employ mainly Acrididae, with an occasional adult mixed with the nymphs. This species is said to dig its nests deep in vertical banks of rivers. The more or less horizontal burrow often goes 30 cm or more into the bank, widening at the end into an ovoid brood cell. The egg is said to be laid ventrally between the front legs, the cocoon to be oval and hard. This wasp has two or three generations a year, of which the last overwinters; Brauns states that some cocoons may remain for 2 or 3 years without emergence. Parasites include a rhipiphorid beetle and the mutillid wasp *Mutilla merope* Smith.

The remaining species studied by Brauns are all said to make their nests in level places; the species are *S. dewitzi* Handlirsch, *chrysorrhoeus* Handlirsch, and *atrox* Smith (= *pentheres* Handlirsch). The last two are said to prefer saline soil on the coast.

Another African species, *S. marshalli* Turner, is reported by Dow (1935) in a previously unpublished record, to prey upon Mantidae. This record is based on an observation by Captain R. H. R. Stevenson, who observed the wasp and its prey in Southern Rhodesia.

Mantidae are also reported as prey of the first *Stizus* to have been studied anywhere, the European *S. ruficornis* Fabricius (Fabre, 1886). According to Berland (1925), Fabre was probably working with *distinguendus* Handlirsch rather than with *ruficornis*. Fabre reports the prey as *Mantis religiosa* Linnaeus, as well as other mantids, three to five fairly large nymphs being used per cell. This wasp is said to dig its nests in pulverized sandstone and to make a cocoon resembling that of *Bembix* but having a rough protuberance on its side. According to Fabre, the cocoon is constructed in a manner similar to that of *Bembix*, but initially there are two openings, a large one at one end and a small one on the side. Most of the sand is drawn in through the large opening at one end (see Tsuneki's description of cocoon construction in *Stizus pulcherrimus*). Finally this opening is closed, but the wasp requires a small amount of additional sand "for the final retouching," and this is drawn through the opening on the side. This opening is then closed with an application of material from the inside, resulting in a roughened elevation on the outside, which Fabre states is characteristic of *Stizus ruficornis*. This roughened elevation may be homologous to the terminal nipple-like protuberance occurring in *pulcherrimus*.

Deleurance (1941) has also studied *S. distinguendus*, and his observations bear sufficient resemblance to those of Fabre to suggest that the latter was, indeed, working with *distinguendus* rather than the rarer *ruficornis*. Deleurance found *Mantis religiosa* to be used as prey, less commonly the mantid *Empusa egena*, from 4 to 12 being supplied per cell (usually immatures, but sometimes adults). The egg is laid on the first mantid placed in the cell; it is attached to the dorsal side of the pronotum by its posterior end, its anterior end extending free and obliquely backward toward the base of the anterior coxae, where the larva commences feeding. The nests are apparently unicellular and are left open during provisioning. The prey is carried by the middle legs, which grasp the side of the prothorax of the mantid. Larvae development is said to require only 5 to 7 days, construction of the cocoon, 2 days. Deleurance says that the cocoon is comparable to that of *Bembix,* but his Fig. 3 shows a somewhat more pointed cocoon with a rounded structure near one end.

E. Genus *Stizoides* Guérin

These wasps have slender, edentate mandibles and eyes that diverge rather strongly above (Fig. 72); the integumental sculpturing tends to be stronger than in *Stizus,* with which this genus is unquestionably closely related. The larva has not been described. Gillaspy (1963a) has recently reviewed the two Nearctic species and presented notes on the Old World species. His paper includes figures of the terminal abdominal segments of the males of the Nearctic species. The genitalia and apical sternite are so similar to those of *Bembecinus* (Figs. 74 and 75) that they are not figured here.

F. *Stizoides unicinctus* (Say)

The range of this species includes most of temperate North America west of the Mississippi, with a few isolated records from the eastern United States. It is a relatively common wasp, being especially characteristic of open prairies and semidesert country. Despite its abundance, there are no records of females being taken with prey, and there is now considerable circumstantial evidence that this wasp is cleptoparasitic in behavior. Williams (1914) and

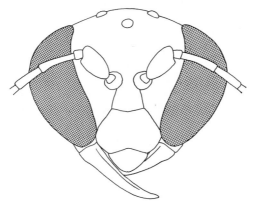

Fig. 72. Head of *Stizoides unicinctus* (Say), female.

the Raus (1918) studied *unicinctus* briefly in Kansas. These observations, plus a still briefer original observation from Kansas and a note by Smith (1915) from New Mexico, all indicate a relationship with sphecine wasps of the genus *Prionyx* (subgenus *Priononyx*). Although none of these notes is conclusive by itself, together they present a reasonably convincing picture.

Gillaspy (1963a) has reviewed much of what is known of the ecology and behavior of this species. He presents a long list of flowers known to be visited, a list that includes some 25 genera of 13 families. As he indicates, these wasps are known to sleep on vegetation somewhat gregariously. Rau (1938) found six adults "asleep, close together, on one head of teasel-grass at the edge of a field of wheat stubble." Gillaspy found two sleeping aggregations of *unicinctus*, both in Idaho. The first consisted of two female and four male *unicinctus* together with two *Prionyx atratus* (Lepeletier) on a live tumble mustard (*Sisymbrium*). The wasps "were in a loose gathering with no individuals in physical contact." The second aggregation consisted of eight *unicinctus* and three *atratus* "atop a plant belonging to the goosefoot family (Chenopodiaceae), at a height of $\frac{1}{2}$ meter, at 1900 hours, in the shade . . . Again the aggregation was loosely formed with no individuals in actual contact." Gillaspy notes that side-by-side sleeping of parasite and host contrasts with the behavior at the nesting sites of the host, where the Raus indicate that *Prionyx* may vigorously attack *Stizoides*.

Evans (1958a) reviewed what is known of the nesting behavior of the North American *Prionyx* (*Priononyx*). These wasps prepare very simple unicellular nests in bare places and provision each nest with a single short-horned grasshopper (Acrididae) which they capture first, then hide in vegetation while they dig the nest. The hopper is placed in the cell head in; the egg is "glued to the membrane just above one of the hind coxae, and curves backward over the base of the femur." The nest is filled rapidly right after oviposition, the *Prionyx* using her head to pack the soil into the burrow. These wasps are not highly gregarious, but they are often plentiful in areas of grasshopper abundance, and females tend to dig a series of nests in the same small area.

The report of Williams (1914), being the first indication of cleptoparasitism by *Stizoides unicinctus*, may be worth quoting in its entirety. Williams' observations were made in western Kansas, where these wasps were

sometimes seen flying low over the ground, alighting now and then as if inspecting the locality for a nest of some sort. In Stanton county one of these wasps was observed to hover about a freshly made tunnel, apparently that of a *Priononyx* [*atratus*], which it entered while the sphecid was away. The latter had brought an *Aulocara* near this burrow, which being occupied by [*Stizoides*], was finally deserted by the disgusted *Prionomyx*.

In Morton county, July 11, 1911, I came upon [*Stizoides*] *unicinctus* engaged in smoothing over a spot with her feet. I unearthed what proved to be a filled-up burrow, which in form and dimensions resembled that of a *Prionomyx*. In the single cell lay a *Melanoplus*. But where the *Prionomyx* egg was to be expected on this locust was only a small bit of soft matter, probably the remnant of the sphecid egg destroyed by the [*Stizoides*], while cephalad of these remains was a short wasp egg, doubtless that of [*Stizoides*].

Williams' second observation is strikingly like an observation reported here for the first time, made in Pottawatomie County, Kansas, 28 June 1953, by C. M. Yoshimoto (my note no. 433). A female *Stizoides unicinctus* was seen scraping soil as if covering a nest entrance. The wasp was captured and the area dug out, revealing a burrow 10 cm long leading to a cell about 7 cm deep containing a paralyzed immature *Spharagemon collare* (Scudder). There was no question that the nest was that of *Prionyx atratus*, as we had been studying this wasp in the same area and had found it making identical nests and using this grasshopper, among others, as prey (reported by Evans, 1958a). This particular grasshopper bore no *Prionyx* egg, but did bear a slightly smaller and less strongly curved egg just anterior to the base of one of the hind femora, the egg extending obliquely toward the thorax rather than across the femur. Unfortunately we failed to make a sketch of this egg, which deteriorated the following day in a rearing tin.

Smith (1915) reported *Prionyx atratus* as a probable host of *Stizoides unicinctus* in New Mexico, and the Raus (1918) reported both this species and *P. thomae* (Fabricius) as probable hosts in Kansas. The latter authors watched one *P. thomae* digging her nest; a *Stizoides* approached the nest and "poked her head in and walked away without further intrusion." Later, when the *Prionyx* had left to move her grasshopper, the same *Stizoides* returned and "wandered about the vicinity, with a 'looking for something' manner." After the *Prionyx* had returned and was digging again, the *Stizoides* returned and actually entered the burrow "where, after a brief tussle, she was chased out by the owner and angrily pursued for a distance of three feet." The *Stizoides* did not return again until after the *Prionyx* had closed her nest and left. The *Stizoides* approached the nest

by walking zigzag, following almost arcs of circles around the hole, with her head close to the surface and antennae vibrating, until she came to the nest . . . She kicked away much of the filling . . . then she worked her way into the nest at the right side of the hopper, remained in for about five minutes and emerged . . . She went in again and this time we removed a little of the soil, yet could not see fully what she was doing; but this much was clear—that she was standing on the hopper with her head near the fore part of the insect, quietly sitting there with the abdomen pulsating for over five minutes. She came out and kicked in all of the loose dirt, then dug up more with her mandibles, kicked it under her body into place, and picked up and placed a few more bits until the hole was once more nicely covered.

The Raus excavated this nest and found that the egg of *thomae* was in its usual position, but that it was "only an empty sac." They were unable to find the egg of the *Stizoides* and believed that they had dislodged it accidentally.

On another occasion the Raus watched a *Stizoides* female

flying lightly from place to place, peeping and examining every spot where the [ground] was a little roughened, such as by ant-hills, or little breaks . . . made by hoof or heel. She made no pretense at alighting on smooth or unbroken spots, but flew lightly from one to another. We followed her and watched her for six minutes; in this time, she alighted ninety-one times on the roughened spots . . . but only twice did she alight on the smooth, unbroken earth, although by far the greater portion of the area was smooth.

Another individual behaved similarly, but paused over "a certain inconspicuous pile of loose earth . . . Instantly her manner changed. She became greatly excited, nervous, quivering with eagerness." This wasp had located a filled *Prionyx* nest, and she dug into it "with the furious eagerness of a dog digging out a rabbit." She disappeared into the nest, only to come out immediately and fly away. The Raus dug out this nest and found the cell was filled with dipterous maggots feeding on the grasshopper.

The Raus believed that the parasite finds the nests of its hosts in two ways: by examining breaks in the surface of the soil and by "shadowing" the *Prionyx* while she is nesting. Their observations suggest that odor as well as vision is involved in nest-finding. The behavior of *Stizoides* strongly suggests that of *Nysson,* as well as of *Evagetes* in the Pompilidae, but the manner of destroying the egg of the host is apparently like that of *Evagetes* rather than the more closely related genus *Nysson.*

In contrast to the remarks of the Raus that *Stizoides* closes and conceals the nest of the host, Smith (1915) states that *Stizoides* "is not particular about refilling the burrow" and may "leave the nest when the burrow is not more than half refilled with soil." Smith found that occasionally the *Prionyx* "will be driven from her nest by [*Stizoides*] while in the act of filling up her burrow."

G. General remarks on the ethology of *Stizoides*

Unfortunately this is very nearly all that is known regarding the behavior of species of *Stizoides*. Two observations remaining to be noted are those of Deleurance (1944) and those of Arens and Arens (1953); both reports pertain to species belonging to the same subgenus (*Tachystizus*) as *unicinctus*.

Deleurance, working on the south coast of France, made some brief but important observations on *Stizoides crassicornis* (Fabricius) which are here quoted nearly in full (in free translation).

On June 17, 1943, I saw a female scraping soil and palpating it with her antennae. Soon she dug with an astonishing precision into a nest of *Sphex albisectus* Lepeletier [that is, *Prionyx kirbyi* Linden]. She removed the two small stones which closed the shaft and then penetrated the nest. But she came out shortly after and flew away permanently without closing the nest. The latter, when dug out, was found to contain an immature *Sphingonotus coerulans* (Linnaeus) (Acrididae) without an egg. The attitude of the wasp led me to believe she was a parasite. The fact that I have never found a female with prey, despite its abundance, appears to confirm this.

Arens and Arens studied *S. tridentatus* (Fabricius) in Russia. These authors found this *Stizoides* associated with four different colonies of *Sphex maxillosus* (Fabricius), a member of the same tribe as *Prionyx* and like that genus a predator on grasshoppers. The *Stizoides* flew about the colonies in a low, searching flight, sometimes remaining about *Sphex* burrows and entering them in the absence of their owners. Upon leaving the burrows they would close them incompletely. Returning *Sphex* would often drive *Stizoides* from their burrows or from the vicinity of their nests. No eggs of the parasite were found, and the parasites were not reared from the cells of their apparent hosts.

These reports might be disregarded as of little significance if they did not resemble so closely the somewhat more complete picture of *S. unicinctus* in North America. Personally I am fully convinced of the cleptoparasitic relationship of the species of *Stizoides* (at least those of the subgenus *Tachystizus*) with grasshopper-predators of the tribe Sphecini. The other two subgenera of *Stizoides* recognized by Gillaspy (1963) appear more specialized structurally than *Tachystizus*, and it would be surprising if they, too, were not cleptoparasites.

One further point needs to be made. Cleptoparasites are typically closely related to their hosts and very much like them in appearance as well as in many features of their gross behavior. For example, this may be said of *Nysson* and the gorytine wasps discussed earlier; of the pompilid genus *Evagetes* and its hosts in the genera *Anoplius, Pompilus,* and *Episyron;* of the bee genus *Coelioxys* and its host *Megachile;* and so on (Wheeler, 1919). However, it most definitely cannot be said that *Stizoides* is closely related to *Prionyx* or *Sphex* or even superficially like those genera. The fact that both *Stizus* and the Sphecini prey upon grasshoppers may be significant, for *Stizoides* is most certainly closely related to *Stizus*. This has led me to suggest (1955) that *Stizoides* arose as a cleptoparasite of *Stizus* and that some species have undergone enough modification of their responses to certain stimili to enable them to attack Sphecini either instead of or in addition to *Stizus*. If this is the case, some species of *Stizoides* should some day be found to attack *Stizus. Stizoides unicinctus* appears much more abundant than any of the North American species of *Stizus,* and its range is broader than that of the four species of *Stizus* together. In acquiring the ability to attack Sphecini, these wasps may have been able to achieve a much higher level of abundance than would otherwise have been possible.

H. Genus *Bembecinus* Costa

These are small, very compact wasps with dentate mandibles (as in *Stizus*) and with the eyes strongly divergent above (as in *Stizoides*). There are two specialized features: the sides of the propodeum are protuberant behind, and the second submarginal cell of the fore wing is triangular or nearly so, often petiolate. The apical sternite is three-pronged, as in *Stizus,* and the genitalia differ from those species of *Stizus* studied only in having much more slender and attenuate digiti and aedoeagus. The larvae resemble closely those of *Gorytes* and related genera, differing mainly in having the two teeth on the inner margin of mandibles quite close together. Willink (1949) has reviewed the eight Neotropical species of this genus, Krombein and Willink (1950) the six Nearctic species. (Figs. 73–75).

Information is available on more than 15 species of this large, nearly cosmopolitan genus. The information varies from mere scraps to fairly complete studies of a few species, most particularly the Nearctic *neglectus* (Cresson) and the Palaearctic *tridens* (Fabricius). The available data together present a remarkably consistent picture, demonstrating that this genus is highly distinctive in its ethology,

§H. Genus *Bembecinus* Costa   133

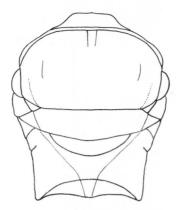

Fig. 73. *Bembecinus neglectus* (Cresson), mesosoma of female, showing the angulate lateral processes on the propodeum.

but suggesting that there are few notable specific differences in behavior within the genus.

I have already presented a review of the ethology of this genus in connection with my studies of *neglectus* (1955). Therefore it seems unnecessary to present a detailed account of members of this genus here. Rather, I shall attempt to present a synthesis of available data on all known species under the following headings: general behavior of adults, nesting behavior, oviposition, provisioning, development, and natural enemies. My review of the literature in

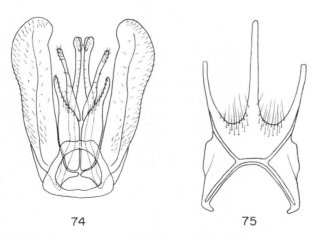

Figs. 74 and 75. *Bembecinus neglectus* (Cresson): 74, male genitalia; 75, apical sternite of male.

1955 was by no means complete, and I have here included published data on several Asiatic and African species omitted from my earlier account. Also included are new data on two North American species: *g. godmani* (Cameron) and *mexicanus* (Handlirsch).

(1) *General behavior of adults.* The species of this genus tend to be very localized in distribution, remaining close to the nesting sites, which are generally in small to fairly large bare, sandy areas, often along water courses (but not usually in dunes). Females apparently do not wander far in their search for prey, and adults of neither sex visit flowers for nectar in any appreciable numbers. Nesting aggregations are often very dense, the nest entrances only one or a few centimeters apart. The size of colonies appears very variable, probably depending in part on the available space, in part on success in nesting in that area, and perhaps in part on specific differences. From my rather limited experience with several North American species, I would judge the very small species *godmani* to be the most highly gregarious, the somewhat larger species *neglectus* and *mexicanus* to form relatively less populous and more diffuse colonies. However, the last two species may well be found to form large and compact colonies under some conditions, and I have described considerable variation in colony size from year to year in *neglectus* (1955). In this species the colonies tend to be spread over a considerable area, but within the area the nests are often grouped in small clusters where the entrances are only a few centimeters apart (sometimes only 2–5 cm). I have described aggressive behavior between females nesting in close proximity.

My studies of *mexicanus* (Handlirsch), reported here for the first time, were made on 22 July 1962, along the highway, 8 km north of Muna, Yucatan, Mexico (no. 1907). Here there were two large, somewhat eroded piles of ground limestone, the result of road-building operations several years earlier. There was an open space about 2 acres in extent around these piles, and the clearing was surrounded by tropical deciduous forest. *Trachypus mexicanus* and *Bembix multipicta* also nested in small numbers in the sandlike soil in the piles. About 20–30 female *Bembecinus mexicanus* nested here, but their nests were well spread over the two piles, no two nests that I found being closer together than 0.5 m, and most of them separated by at least 1 m. These females were actively provisioning their nests in the midmorning hours, when it was exceedingly hot. In my studies of *neglectus,* I noted that these wasps tend to be active in the middle of very hot days, when many other digger wasps are relatively inactive.

§H. Genus *Bembecinus* Costa

It is known that several species of *Bembecinus* do not spend the night in their nests or within the soil in the nesting area, and I fell that it is probable that all the species cluster on vegetation in the manner described for *godmani* and for certain African species. Records for *godmani*, which I summarized in 1955, indicate that both sexes form large balls on vegetation, with much bodily contact between individuals. These balls are apparently formed in the evening and disperse with increasing temperature and sunlight in the morning. They are formed on the top or close to the bottom of various herbs and shrubs in protected places in the general vicinity of the nests; clusters have been reported on *Pluchea*, on *Baccharis*, and on *Argemone*. Apparently these clusters may contain nothing but males early in the season, mostly or entirely females later in the season. Unfortunately little is known of the behavior of wasps at the cluster or of the exact composition of the clusters (for example, whether or not they tend to be layered as to sex as in *Steniolia*).

Brauns (1911) found that certain South African species of this genus cluster in vegetation. *B. cinguliger* Smith (=*clavicornis* Handlirsch) is said to nest in diffuse pseudocolonies on overgrown sand flats. The nights are spent away from the nests, chiefly on stalks and dried seed-heads of *Datura* and on other low plants and bushes in protected places. Here one may find clusters varying in size from only a few individuals up to enormous clumps the size of the head of a baby and containing more than 1,000 individuals. The clusters studied contained mostly females, and Brauns thought these were recently emerged, unfertilized females. Brauns also found other species clustering on *Datura* seed pods, notably *B. rhopalocerus* (Handlirsch), *B. rhopaloceroides* (Brauns), and *B. oxydorcus* (Handlirsch), the last-named in smaller numbers.

In Argentina, Willink (1949) found *B. consobrinus* (Handlirsch) in compact clusters on vegetation at about 1000 in the morning. One cluster, shaken in its entirety from a *Baccharis* branch, was found to contain 43 individuals, including both sexes, as well as a single female *Bembecinus bicinctus* (Taschenberg). Willink states that he found a nesting aggregation of *consobrinus* along a river bank; the colony was very dense and populous, over 100 individuals being taken by passing a net over the nesting area.

Little is known regarding the precopulatory behavior of the males or regarding the mating act. In my studies of *neglectus*, I reported males

walking rather slowly on the surface of the soil, their wings still and either lying flat on the back or held at a slight angle. They tended to walk in small circles in close proximity to one another, and would occasionally take short, somewhat hopping flights . . . Occasionally one was seen to pounce upon a female suddenly and to roll over with her on the earth . . . Males would sometimes pounce upon one another and once a ball of four males was found rolling over and over on the sand. No males were seen after July 1, although females were present at least until July 13, when I left the area.

(2) *Nesting behavior.* The digging of the nest in *g. godmani* (Cameron) has been described by Strandtmann (1945) as follows:

> In starting a nest the loose surface sand is scratched away with the fore legs, the wasp throwing a steady stream under her body for a distance of 2–3 inches, but when moist sand is reached, it is carried out in a lump. The wasp backs to the entrance with a pellet of sand held under her head, poises there a moment, then teeters the back end up and simultaneously tosses the little pellet back underneath her body, this sometimes as far as 3 or 4 inches. This done, she immediately pops back in and shortly reappears and repeats the performance.

In my study of *neglectus,* I also noted that moist sand is removed in pellets; the females "were seen backing out with small pellets of earth in the mandibles; these were then deposited from one to four centimeters from the entrance." I found that about 2 hr are required for completion of a nest; during this time the wasp does not dig steadily, but flies off at intervals and then returns to resume digging. Following completion of the nest, the female *neglectus* lays her egg in the empty cell, then comes out and levels the mound of earth at the entrance.

First the entrance is closed from the outside by scraping sand from the periphery over the entrance and also by loosening sand with the mandibles from around the entrance and packing it in with the tip of the abdomen. The pile of earth . . . is then leveled off by working into it in a straight line kicking earth in the direction of the nest entrance or in various other directions. This was observed on several occasions, and always the movements were in more or less of a straight line, with little tendency toward zigzagging. Having leveled the pile of earth, the female flies away, but may return once or twice for further inspection and further moving of sand over the entrance. At least an hour is required for the completion of these activities of concealing the nest entrance.

Evidently leveling movements of this nature are not characteristic of all species of *Bembecinus.* In *mexicanus* the mound at the entrance is left more or less intact, and this is apparently also the case in

§H. Genus *Bembecinus* Costa

hungaricus (Frivaldsky) (Iwata, 1936)[1]. It is my impression that the mound is leveled in *godmani*, but I have no precise data on this point. In the Palaearctic *errans*, Ferton (1911) reports that the sand is scraped away from the nest entrance, but in *fertoni* Handlirsch the mound is said to be left partially intact (Ferton, 1908).

I have found that in *neglectus*, *g. godmani*, and *mexicanus* a temporary closure is maintained at the entrance at all times when the female is not actually inside the nest. Rodeck (1931) dissents with respect to *godmani*, stating that (at Roggen, Colorado) the wasps did not close the burrows upon leaving; but Strandtmann's observations (1945; made near Fort Davis, Texas) agree with mine (made near Alpuyeca, Morelos, Mexico) that an outer closure is maintained. In *neglectus* I found that "closure is effected by loosening sand from the periphery of the entrance with the mandibles or front legs and scuffing it over the entrance. Most of the soil is dug from certain specific spots surrounding the entrance; hence each nest entrance tends to have a ring (4–5 cm. in diameter) of slight depressions around the entrance. Some wasps take most of the soil from a single point, forming a depression or short 'false burrow' up to 1 cm. deep." These are reminiscent of the side burrows described by Tsuneki in *Stizus pulcherrimus*.

Richards (1937) states that in *B. agilis* (Smith) (=*cingulatus* Smith), studied in British Guiana, "the entrance of the burrow is filled in with sand" while the female is away from the nest. This is also the case in *hungaricus* and *prismaticus*, studied by Iwata (1936, 1964a), and in the European species *errans*, *tridens*, and *fertoni* (Ferton, 1908, 1911). So far as I know, no one has described an inner closure, separating the cell from the burrow, in any species. I watched for inner closures with some care in *neglectus* but found none, even in nests containing eggs. Final closure consists of a complete filling of the burrow, making much use of the tip of the abdomen for packing the soil, followed by fairly elaborate movements of concealment over the former nest entrance (described for *neglectus* and in less detail for *agilis*, *errans*, and *tridens*).

Nest structure shows little variation among the species studied, except that certain species are known to make unicellular nests only, while others make either unicellular or bicellular nests (Table 11). It is very probable that some of the species make more cells per nest than they are presently known to do. Without exception the nest is

---

[1] Iwata studied this species under the names *formosanus* (Sonan) and *japonicus* (Sonan). Tsuneki (1965a) has recently clarified the systematics of the East Asian *Bembecinus*; he considers *japonicus*, *formosanus*, and *formosanus quadrimaculatus* (Sonan) (a form studied by Iwata, 1939) all to be no more than color forms or at most subspecies of *hungaricus*.

TABLE 11. NEST DATA FOR SEVERAL SPECIES OF BEMBECINUS

Species	Area of study	Burrow length (cm)	Cell depth (cm)	No. cells/ nest	Note no. or reference
neglectus	Kansas	5-18	4-13	1 or 2	Evans, 1955
g. godmani	Colorado	15	—	1	Rodeck, 1931
	Texas	12.5	—	1	Strandtmann, 1945
	Mexico	5-9	4-7	1	1593
mexicanus	Mexico	6-9	5-6.5	1 or 2	1907
tridens	France	10-15	—	1	Ferton, 1908
	Italy	6-20	4-13	1	Grandi, 1961
fertoni	France	—	6	1	Ferton, 1908
hungaricus	Japan	12-17	3.5-13	1 or 2	Iwata, 1936
prismaticus	Thailand	6.5-20	—	1	Iwata, 1964

a simple oblique burrow, often somewhat tortuous, especially if there are stones in the soil. Certain species are not included in the table because no quantitative data are available, but in these cases, too, it seems evident that the nests are simple and quite shallow (*agilis, errans, godmani, bolivari*). The cell is broadly elliptical, horizontal or nearly so, and measures about 8 by 10 mm in the small species to 10 by 20 mm in larger species such as *mexicanus* (Fig. 76).

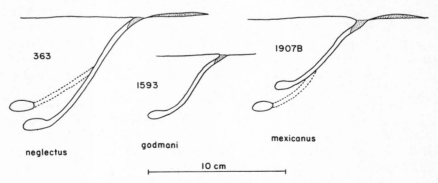

Fig. 76. Typical nests of three species of *Bembecinus: neglectus*, Pottawatomie County, Kansas; *godmani*, Alpuyeca, Morelos, Mexico; *mexicanus*, vicinity of Muna, Yucatan, Mexico. Burrows indicated by broken lines had been filled and could not be traced exactly.

§H. Genus *Bembecinus* Costa

(3) *Oviposition.* As already indicated, oviposition occurs immediately after completion of the nest, before preparation of the first outer closure. The egg has been described for six species of *Bembecinus*, and in every case the situation is similar. The wasp prepares a small pedestal consisting of sand grains that have been glued together, presumably with secretions of the wasp. The pedestal is 1 or 2 mm in height by about twice that in width; it is located centrally on the floor of the cell, but closer to the apical than to the basal (open) end of the cell. The egg is about 2.5–3.0 mm in length and is glued to the top of the pedestal by its slightly broader posterior end. Commonly the egg is not erect, but slightly inclined in a direction away from the open end of the cell. Iwata figures the egg of *hungaricus* as departing from the vertical by no more than 5 to 15°, but in *tridens, errans,* and *neglectus* the egg is more strongly inclined (about 45° in *neglectus,* 15 to 45° from the vertical in *errans*). According to Iwata (1964), the egg of *prismaticus* is fully erect. No prey is typically placed in the cell for at least 24 hr after oviposition, the female spending this period away from the nest. The egg apparently hatches within 2 days (Figs. 77 and 78).

(4) *Provisioning.* In all species in which this aspect of the behavior has been studied, provisioning has been found to be progressive (*neglectus, godmani, nanus, mexicanus, agilis, tridens, errans, prismaticus,* and *hungaricus*). Several of these species are known to provision in the manner I have described for *neglectus*, with minor variations:

The first leafhopper is brought in the day after the egg is laid and is placed in the cell beside the mound on which the egg has been laid; the head of the leafhopper is directed inward and the ventral side is up. Before the egg hatches another leafhopper is brought in and placed on the opposite side of the mound in exactly the same manner; in one case a third leafhopper was found in a cell in which the egg had not yet hatched. When the egg hatches the larva remains in a semierect position glued to the lump of earth. By tipping to either side it is able to reach the center of a leafhopper, which it begins to feed upon by cutting a hole in the

Figs. 77 and 78. Typical position of egg: 77, in cell of *Bembecinus hungaricus* (after Iwata, 1936); 78, in cell of *B. neglectus* (after Evans, 1955).

abdomen. By the second day the larva loses its attachment to the lump of earth and feeds freely upon the prey in the cell.

Once the larva is a day or two old, the female brings in leafhoppers at a rapid pace and soon fills the cell. The larva may be no more than half grown when the cell is fully provisioned and closed off. Iwata's figures of cells of *hungaricus* in various stages of provisioning (1936:243) are in striking agreement with my observations on *neglectus*. All authors who have noted the condition of the prey state that paralysis is very profound or that the prey may actually be killed by the sting. Also, there is general agreement that the prey is carried in flight with the middle legs, much as in *Gorytes* and its relatives. The female typically lands directly in front of the well-concealed nest entrance, scrapes it open with her front legs, then enters carrying the leafhopper, which tends to slide backward and follow the wasp down her burrow, presumably being held at this stage by the hind legs (Fig. 79).

In *neglectus*, I found that females typically flew into the nesting area each morning and entered their nests without prey, presumably to "inspect" the nest and determine the status of the egg or larva. The stimuli received during this inspection trip doubtless determine the intensity and duration of hunting and provisioning activities thereafter, although further stimuli received when each prey is added to the cell may well play a role in accelerating or "shutting off" provisioning.

Fig. 79. *Bembecinus neglectus* female arriving at the covered nest entrance with a leafhopper held by the middle legs (Pottawatomie County, Kansas). (From Evans, 1955.)

§H. Genus *Bembecinus* Costa 141

The number of prey per cell is quite variable, depending in considerable part on the size of the prey. In one cell of *mexicanus* I found more than 50 early instar nymphs of a treehopper, but the usual number in most species is much less than this; I recorded only 10 to 15 in *neglectus*. Although most species take a considerable variety of prey, several authors have noted that individual nests frequently contain only one species. All known species of *Bembecinus* employ Homoptera Auchenorhyncha as prey, virtually all types except Cicadidae and the larger Fulgoroidea being employed. There are two isolated records of psyllids also being used as prey (Ferton, 1908; Iwata, 1936); the family Psyllidae belongs to the Homoptera Sternorhyncha. Most species employ either adults or nymphs. It is not feasible to review fully all prey records for members of the genus. The following is a summary of prey records, with references which may be consulted for specific details.

B. neglectus (Cresson). Evans (1955) recorded as prey in Kansas three species of Cicadellidae, all adults, two of them in considerable numbers, and two records of nymphs of the genus *Scolops* (Fulgoroidea).

B. g. godmani (Cameron). Rodeck (1931) recorded two species of Cicadellidae as prey in Colorado, while Strandtmann (1945) recorded four species of Cicadellidae from western Texas. Along a stream a few miles north of Alpuyeca, Morelos, Mexico, in March 1959, I found two nests provisioned with a single, unidentified species of leafhopper (Cicadellidae) (no. 1593). All records are for adults except for two nymphs taken by Strandtmann.

B. godmani bolivari (Handlirsch). This form is also known to prey upon Cicadellidae of several species (record from British Guiana, Vesey-Fitzgerald, 1956).

B. n. nanus (Handlirsch). Shappirio (1946), at Washington, D.C., found this species preying upon the leafhopper *Graphocephala versuta* Say and an unidentified fulgoroid. Bridwell's (1937) report of *moneduloides* (Smith) at Vienna, Virginia, undoubtedly applied to *nanus*, the only species known to occur there. Bridwell found eight species to be used as prey, four of them Cicadellidae, two Fulgoroidea, one Cercopidae, and one Membracidae; all but 1 of 15 individuals were adults.

B. mexicanus (Handlirsch). In July 1962, I dug out three nests of this species near Muna, Yucatan, Mexico. The three nests (four cells, only one of them fully provisioned), contained roughly 100 rather small, unidentifiable membracid nymphs, apparently all one species (no. 1907).

B. agilis (Smith). Richards (1937) records this species (under the name *cingulatus* Smith) from British Guiana and Panama with nymphs and adults of five species of Cicadellidae. Callan (1954) records the species from Trinidad and Venezuela and states that it preys upon nymphs and adults of Cicadellidae, but he may merely be repeating Richards' statement.

B. tridens (Fabricius). Grandi (1961) has recently summarized his prey records from Italy; his list includes 20 species or varieties, of which half are Cercopidae (mainly *Philaenus*), the remainder Cicadellidae and several families of Fulgoroidea. Ferton (1908), in France and Corsica, found this species employing Cicadellidae and a single adult Psyllidae.

B. errans (Kohl). Ferton (1911) found the prey of this species to consist of 15 species of the families Membracidae, Cercopidae, Cicadellidae, and five families of Fulgoroidea. Most specimens were adults.

B. fertoni (Handlirsch). In France this species is known to prey upon adults of one species of Cicadellidae and adults and nymphs of one species of Fulgoroidea (Ferton, 1908).

B. argentifrons (Smith). Bridwell (1937) reported this South African species as preying upon adult *Bythoscopus olivaceus* Stål (Cicadellidae).

B. hungaricus (Frivaldsky). Iwata (1936) studied this species in Japan, under the name *japonicus* Sonan. Both adults and nymphs were found to be used as prey. The following genera were found to be employed: *Nephotettix, Parabolocratus, Penthimia, Eutettix, Jassus,* and *Platymetopius* (all Cicadellidae). According to Tsuneki (1943b), Iwata also reported Psyllidae and Fulgoroidea as prey in this same paper (which is in Japanese). Iwata (1939) also studied this species on Formosa and found the prey to consist of Cicadellidae and Fulgoroidea of the genera *Tartessus, Tambinia,* and *Dictyophara.*

B. prismaticus (Smith). Iwata (1964a) found the prey of this species, in Thailand, to consist of adult and immature Cicadellidae of the genera *Parabolocratus, Penthimia, Nephotettix, Deltocephalus* and *Minodrylix.*

(5) *Development.* As already noted, the larva remains attached to the pedestal for the first day or two, then loses its attachment and feeds freely on the prey in the cell. The female often completes provisioning and closes the cell when the larva is no more than half its full size (*neglectus, hungaricus*). However, one of the nests of *mexicanus* which I dug out contained a virtually fully grown larva, although this nest was still being provisioned. In this case the wasp was using very small treehopper nymphs, and the labor involved in collecting enough of these minute insects to feed the larva (more than 50) may

§H. Genus *Bembecinus* Costa 143

have caused the female to take longer than usual for provisioning.

Growth of the larva is rapid, and my records for *neglectus* indicate that as little as 4 days may be required for the larva to reach maturity (6 days from oviposition). The full-grown larva of *neglectus* was described by Evans and Lin (1956), and Grandi (1961) has provided a detailed description and excellent set of figures of both the young and the mature larva of *tridens*. The cocoon is hard-walled like that of other members of this complex. In *neglectus*, I found that there were 5 or 6 small pores "located somewhat irregularly around the middle of the cocoon."

(6) *Natural enemies.* It is a curious fact that no dipterous parasites have been recorded attacking the species of *Bembecinus*. Grandi (1961) has recorded the mutillid wasp *Smicromyrme viduata* Pall. attacking *B. tridens* in Italy, and Maneval (1939) has reported *Nysson dimidiatus* acting as a cleptoparasite of *B. tridens* in France (see Chapter IV:A). *B. tridens* is also said to be the host of two chrysidid wasps, *Hedychrum chalybaeum* Dahlbom and *Holopyga chrysonota* Foerster (Olberg, 1959, after Haupt and Trautmann).

Chapter VII. *Bicyrtes,* A Genus of Stinkbug Hunters

This genus contains about 30 species which collectively range throughout tropical and subtropical America, two of the species as far north as southern Canada. This is the first of the genera considered here that is assigned to the tribe Bembicini. The standard taxonomic works on the Bembicini are those of Parker (1917, 1929) and of Willink (1947), the latter covering the Argentine fauna. The Bembicini differ from genera treated up to this point in the following characters: (1) the shorter spur of the middle tibia is absent; (2) the labrum is prolonged and is transversely convex, forming the roof of a large chamber into which the elongate tongue may be withdrawn; (3) the apical parts of the maxillae and labium are greatly elongate, forming a tongue capable of probing deep corollas for nectar; (4) the ocelli are more or less distorted and even reduced to the point of being nearly absent (in *Carlobembix* Willink, 1958, the ocelli are only very slightly distorted); (5) the submedian cell of the hind wing is short, not reaching the origin of the cubital vein. Among the Bembicini, the genus *Bicyrtes* is characterized as follows:

Adults: anterior ocellus reduced to an elongate, transverse, sinuate groove; posterior ocelli in the form of arching lunules, each containing a slender, curved lens; mouthparts moderately elongate, the palpi showing no reduction in segmentation. Mesosoma robust and essentially like that of the stizine wasps, although the pronotum and mesoscutum are more smoothly confluent than in *Stizus brevipennis;* propodeum concave behind, with the posterior angles forming projecting flanges, much as in *Bembecinus.* Female with or without a distinct pygidial area; tergite VII of male simple, without evidence of separated, reflexed sidepieces; male genitalia with the basal ring small, the digiti much longer than the cuspides and strongly separated from them (Figs. 80–85).

§A. *Bicyrtes quadrifasciata* (Say)

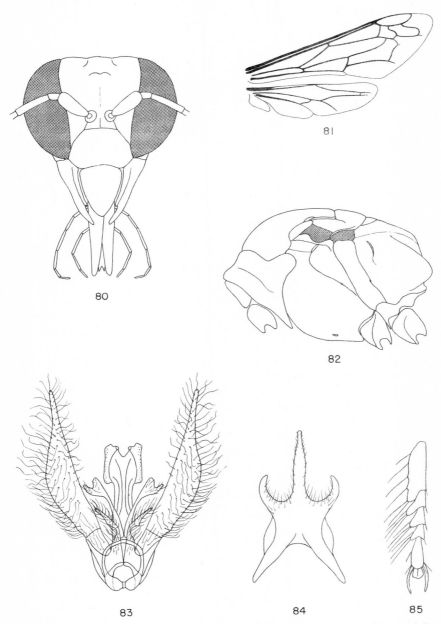

Figs. 80-85. *Bicyrtes quadrifasciata* (Say): 80, head of female, mouthparts fully extended; 81, wings; 82, mesosoma of female, lateral aspect; 83, male genitalia; 84, apical sternite of male; 85, front tarsus of female, showing pecten.

Larvae: mandibles with only a single tooth on the inner margin, this tooth rather broad and blunt; body essentially without setae, and head setae minute; labrum with a bristly apical margin; epipharyngeal sensory areas connected by a bare but pigmented band; inner margin of the maxillae and oral surface of the labium both densely spinulose.

Three species of this genus are discussed in some detail below; these species are *quadrifasciata, ventralis,* and *fodiens.* Following this I have reviewed briefly what is known of other species. The chapter concludes with a brief summary of the ethology of the genus.

A. *Bicyrtes quadrifasciata* (Say)

This wasp occurs throughout much of temperate North America east of the Rockies, but is more common toward the South. It is one of the best studied of bembicine wasps, and my own studies add little that is new, although adding more details on prey, nest structure, and certain other aspects of the behavior. The first studies on *quadrifasciata* were made by Hartman (1905) in Texas. Hartman called it *Bembix belfragei,* "the big bug-hunter," but there are several clues in his report that make it fairly certain that he was working with *Bicyrtes quadrifasciata.* Later studies were made by Parker (1917) in Ohio and Maryland, by the Raus (1918) in Kansas, by Smith (1923) in Mississippi, by Davis (1926) in New Jersey and New York, and by Rau (1934) in Missouri. The most recent and most important studies of the species are those made by Krombein (1953, 1955, 1958b, 1959) at Kill Devil Hills, on the coast of North Carolina. His 1955 paper is the most detailed and contains photographs of females digging and provisioning.

The 36 original observations reported here were made as follows: 8 at Blackjack Creek, Pottawatomie County, Kansas, during June and July 1952, 1953; 3 in dunes 5 mi north of Hutchinson, Reno County, Kansas, in July 1953; 5 in three other localities in central and western Kansas in July and August 1952, 1953; 4 in Highlands and Glades counties, Florida, in April and May 1955, 1961, 1964; 1 at Granby Center, Oswego County, New York, in July 1956; 10 in a sand pit at Van Natta's Dam, Ithaca, New York, July-September 1953-1956; and 5 at three localities near Boston, Massachusetts, July-August 1963. I have also included a few notes contributed by C. S. Lin from observations made at Austin, Texas, in the summer of 1957.

(1) *Ecology and adult behavior.* In Florida this species is active as early as April, and doubtless there are several generations a year

§A. *Bicyrtes quadrifasciata* (Say) 147

at this latitude. Throughout much of its range the species is characteristic of midsummer (mid-June to early September) and probably has one or two generations depending upon the latitude. A Kansas specimen which spun its cocoon on 19 July 1952 (no. CMY39) did not emerge as an adult until the following June. However, in North Carolina, Krombein found that cocoons spun in early July gave rise to adults that same August, the total period from egg to adult being about 7 weeks. In Mississippi, Smith found that cocoons spun in late July gave rise to adults 2 to 6 weeks later.

This species is a frequent visitor to flowers of many kinds, and it is probable that both sexes take nectar throughout their lives. *Melilotus alba* and *Daucus carota* are especially visited. Males are on the wing throughout much of the active season of the females and are usually to be found flying back and forth over the nesting sites of the females. Krombein's remarks confirm my own impression that the flight of the males tends to be mostly back and forth rather than in irregular patterns as in a typical "sun dance" of *Bembix*. However, this may be a result of the fact that both of us were working in areas where the nesting site was itself quite elongate, tending to cause the males to repeat a linear path. In suitable areas as many as five to ten or more males may be found flying 15 to 30 cm high most of the day, now and then resting on the soil or pouncing upon one of the females.

I observed one mating pair of this species, at Granby Center, New York, on 19 July 1956 (no. 1190). The pair rested on a *Daucus carota* flower head at about 1100, the male astride the female. Unfortunately I made no detailed notes on the posture of the pair, and the wasps separated and flew off after a few seconds.

The females nest in fine-grained to fairly coarse sand varying from moderately moist to very dry and powdery on the surface. Favored sites are sandy roads, blowouts, artificial sand pits, small dunes and the periphery of larger dunes, bare places in sandy fields and meadows, and alluvial sand along streams. Common associates are *Stictia carolina*, certain species of *Bembix* (*spinolae, sayi*), and various gorytine and philanthine wasps. *Bicyrtes quadrifasciata* is not highly gregarious, and the nest entrances are usually separated by 1 m or more. Krombein found several dozen females nesting for about 100 m along a sandy road. I have never found more than a few females (up to about ten) in any one sandy area at a time. I have found that individual females tend to make a series of nests within 1 or 2 m of one another.

Adults tend to be active throughout the day when the sun is shining. Much digging occurs in the morning, much provisioning, in the late morning and afternoon. However, there seems to be no strict daily regimen, and new nests are often started in the afternoon and not provisioned until the following day. As Krombein remarks, the female usually requires all of one day and part of another for digging, provisioning, and closing a single nest. It is not known where the adults spend the night.

(2) *Digging and orientation.* Females may start to dig briefly in several places before finally remaining in one spot and completing a nest. Smith reports that the mandibles are used for loosening soil. As Krombein says, the female "inclines her body at an angle of about 30° to the horizontal with her head near the sand, and supporting herself on the mid and hind legs begins to dig out the sand with her forelegs, flinging the loose sand beneath and behind her body some 22 to 30 cm." Smith states that the sand falls "anywhere from six to 18 inches [that is, 15 to 45 cm] from her," and my own notes are in agreement with those of Smith and of Krombein. Hartman states that "each time the head goes down, a single stroke of the leg is given and not several as is the case with *Bembex texanus.*" Evidently there was some confusion in Hartman's notes (he misidentified this species as well as *B. texanus*). In fact, *Bicyrtes* is distinctive in that several spurts of sand are thrown out each time the head goes down. These wasps exhibit unusually pronounced, jerky tilting movements while clearing the entrance. As the head goes down the abdomen is elevated high; several strokes of the front legs are made while the abdomen moves up and down only slightly, the sand being thrown very far, as already noted. Then, the body quickly assumes a horizontal position as the wasp pauses briefly before once again tilting and digging (Fig. 86).

"As the burrow deepens," says Krombein, "the loose sand accumulates in the burrow behind her body and is pushed out to the surface periodically by her abdomen and hind legs and then is dispersed over the area behind the entrance." I have noted that when dispersing the sand at the entrance the wasp turns a little to each side, so that the sand is sprayed over a wide arc. As noted by several authors, and many times by myself, the sand is so well dispersed that no mound accumulates at the nest entrance as occurs in so many digger wasps. My records indicate that from 1 to 2 hr are required for completion of a nest, and published observations are in general agreement with this.

Some females appear to dig more or less continuously until the nest is finished, while others interrupt their digging at times to fly obliquely upward, facing the nest entrance, to a height of from

§A. *Bicyrtes quadrifasciata* (Say)

Fig. 86. *Bicyrtes quadrifasciata* clearing the nest entrance. This wasp is in "head-up" position; she will shortly dip her head down and throw out several jets of sand before reassuming the "head-up" position.

1 to 3 m or more, then descend slowly to the nest and resume digging. In any event, when the nest is complete the female comes out head first, makes a closure, then makes several flights of this nature. Krombein states that following closure, "usually she flies up several times . . . returning to scratch more sand over the entrance from several directions until the surface is completely smooth and with no trace of the entrance. Then she makes one or several short reconnaissance flights in the immediate area, ascending slowly in irregular spirals to a height of about 3 meters, and finally disappears to hunt for her first prey." The Raus say "she rises straight up in the air, so high that it is difficult to follow her with the eye in the dazzling sunlight, then darts away horizontally above the heads of most living things." I have observed this behavior many times and have noted much variation in details, not only in the number and height of the orientation flights but also in the angle of ascent and in whether the ascent is straight or more or less circling. One female ascended to an estimated height of 8 m, then flew off horizontally, as described by the Raus. In one unusual case (no. HE5) the wasp flew from plant to plant nearby, each time landing and facing the nest; then she circled about, landed

directly in front of the nest, and flew off rather close to the ground. Hartman speaks of a female making a locality study by "walking around on the sand in the neighborhood of her nest." However, an ascending flight is much more typical and sets the pattern for the unusual return to the nest, described below.

(3) *Nature and dimensions of the nest.* The nest is a simple burrow which forms an angle of from 30 to 60° with the horizontal (usually close to 45°). The burrow may be straight or nearly so, or there may be one or more gentle lateral bends. Burrow diameter is 6–8 mm; the terminal cell measures 10–15 mm in diameter (commonly 12–13 mm) and 20–35 mm in length (commonly about 25 mm). Most workers are in general agreement on these figures, but Smith speaks of the cell as being "about two inches long" (that is, about 50 mm, considerably longer than that mentioned in any other report). The cell is horizontal or slightly oblique, never as strongly inclined as the burrow. Burrow length varies over the broad range of from 8 to 43 cm, cell depth from 5 to 27.5 cm, but such a range in depth is not found in any one locality (Table 12). In general, nests in very dry soil or in sand dunes (such as those from Reno County, Kansas) tend to be deeper than those in heavier and/or moister soil (such as those reported from New York and Massachusetts). Krombein (1958b) found that in North Carolina, nests dug in September

TABLE 12. NEST DATA FOR BICYRTES QUADRIFASCIATA

Locality	No. of nests	Burrow length (cm)	Cell depth (cm)
Texas (Hartman)	1	25.0	12.5
Kansas (Reno Co.)	3	32.5 (27–35)	24.0 (19–28)
Kansas (4 other localities)	14	21.3 (12–32)	15.0 (10–23)
Missouri (Raus)	2	36.5 (30–43)	24.0
Mississippi (Smith)	several	— (15–20)	15.0
Florida	3	14.8 (12–17)	7.8 (6–9)
No. Carolina (Krombein)	16	—	— (6–13)
Ohio (Parker)	several	— (15–20)	15.0
New York	10	15.5 (8–30)	9.7 (5–18)
Massachusetts	5	15.5 (11–20)	9.0 (6–11)

§A. *Bicyrtes quadrifasciata* (Say) 151

(in which the larvae will overwinter) tend to be somewhat deeper than those dug in midsummer, which give rise to adults the same season. Two September nests measured 11.4 and 12.7 cm in cell depth, while 14 cells from June through August measured from 7.5 to 10.8 cm in depth. Some of the variation shown in Table 12 may be the result of seasonal variation. However, the three very deep nests in Reno County, Kansas, were dug in mid-July, and New York records actually show a slightly greater mean depth for July than for late August and September. The two very deep nests reported by the Raus were apparently in rather dry soil, but the Raus provided no dates. An east-west trend toward deeper nests, similar to that noted for several other species, may be noted (see, for example, *Microbembex monodonta*, Chapter XII:A, 4).

Of eight nests studied in coastal North Carolina in 1954 (Krombein, 1955), one contained two cells, and of five studied in the same area in 1955 and 1956 (Krombein, 1958), one had two cells. In the first case, the wasp was seen doing further digging after a period of provisioning, "just as though she was engaged in constructing a burrow." When dug out, this nest was found to contain a fully provisioned cell closed off by a plug of sand 2.5 cm long, plus a second cell at the end of a lateral burrow 10.5 cm long starting just above this plug. In the second nest the second cell was also at the end of a lateral burrow, in this case starting 7.5 cm from the entrance and reaching only 3.8 cm in length, the lateral being at about a right angle to the main burrow.

In my own studies I found three nests containing two cells and one containing three. In the first (no. CMY39), one of eight nests studied at Blackjack Creek, Kansas, the female was seen digging again 2 days after the original burrow was constructed. This nest was dug out 3 days later and two cells were found, one with a cocoon and the other with a larva. They were at about the same depth and separated by several centimeters of sand, but the exact pattern of the burrows could not be traced. The second nest (no. 717), one of ten nests studied at Ithaca, New York, had one new, empty cell at 10.5 cm depth, and a fully provisioned cell containing an egg at 9.5 cm. The new and the old cells appeared to be connected by burrows making about a right angle; the burrow leading to the fully provisioned cell had been closed off with a plug of sand. The third bicellular nest (no. 1963, Bedford, Massachusetts) was marked on 8 August 1963, when the female was seen digging. She was apparently digging a second cell, for when this nest was excavated 2 days later it was found to contain two cells. The first, 15 cm from the entrance, contained a fairly large larva; the second, 4 cm back toward the entrance, contained several maggots feeding on the prey. In this instance the female had apparently closed off the first cell with about 5 cm of soil, then dug a new, short side burrow off the main burrow (Fig. 87).

The single tricellular nest (no. 1977) was found at Palmdale, Glades County, Florida on 27 April 1964. A female was observed bringing in prey 0930–1100, and when the nest was dug at noon that day it was found to contain a cell with an egg and 10 immature bugs at a distance of 12 cm from the entrance. Slightly to the right and at a distance of 14 cm from the entrance was a second cell, containing an egg and 8 immature bugs. Directly behind the first cell, but at a distance of 17 cm from the entrance was a third cell, containing 11 bugs and a larva about 10 mm long. Apparently this nest had been used over a period of about 4 days. The two deeper cells were closed off with a firm barrier of sand.

These six nests are the only multicellular nests reported for this species, of a total of more than 50 studied by various workers. The available data on these six individuals do not suggest any common

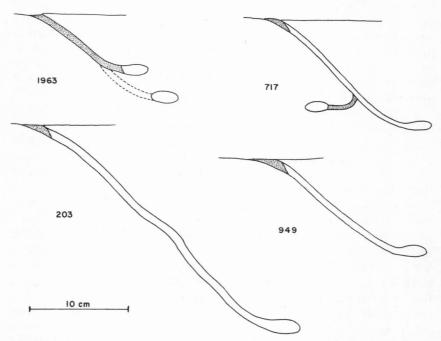

Fig. 87. Four typical nests of *Bicyrtes quadrifasciata:* no. 1963, Bedford, Massachusetts; no. 203, Pottawatomie County, Kansas; nos. 717 and 949, Ithaca, New York. Burrows indicated by broken lines could not be traced exactly.

§A. *Bicyrtes quadrifasciata* (Say)

features that might shed light on the factors causing them to dig another cell from the same burrow instead of digging a new nest. Krombein notes, however, that one of the wasps that made a two-celled nest was also aberrant in that she sometimes approached her nest flying only a few centimeters above the ground.

(4) *Closure.* There is agreement among all who have worked on this species that an outer closure is maintained at all times when the female is not actually inside the nest, but that no inner closure, separating the burrow from the cell, is maintained. The first closure, made after the completion of the new burrow and cell, is often more prolonged than the others, and at this time some use is usually made of the tip of the abdomen for pressing sand into the entrance. I have also observed on two occasions the use of the abdomen for temporary closures during provisioning. Smith describes a typical initial temporary closure as follows:

After digging for an hour, [the wasp] came to the surface and walked around the nest for a few seconds . . . then . . . she began kicking dirt into the entrance to the gallery. While engaged in this she faced away from the nest but was close enough to the entrance so that each spurt of dirt would fall exactly into the right spot. After kicking dirt a moment from one spot she would move to another and repeat the operation, thus completely circling the nest by the time she had finished. Upon filling the entrance hole, she crawled on top of the dirt and lay there, and with her abdomen bent forward under her, used the dorsal side of it in packing the dirt firmly into the gallery to form a plug. This being finished in [a] short while she walked over the nest carefully scratching the dirt here and there that it might not leave any tell-tale trace.

As already noted, an orientation flight follows the initial closure of the nest. Brief orientation flights occasionally occur in the course of provisioning, especially if there has been some disturbance of landmarks or of the soil over the nest.

(5) *Hunting and provisioning.* Krombein found that in three cases the time elapsed between the initial closure and the bringing of the first prey was 18, 15, and 35 min, while one wasp I timed took only 10 min. Subsequent flights to obtain prey timed by Krombein for several wasps varied from 7 to 35 min, with a mean elapsed time of 16.5 min per flight. One wasp which I timed (no. 147) took from 4 to 16 min, but a more typical wasp was no. 521, which took from 20 to 35 min to obtain prey. Wasps usually remain within the nest from 20 to 60 sec when provisioning.

The actual capture and stinging of the prey has not been observed.

The Raus watched one female "climbing in and out among the branches of a cocklebur plant, carrying [her prey] beneath her, resting once or twice on a leaf. All at once . . . she leaped up into the air and flew straight upward, higher, higher, until we lost her in the dazzling sunlight." There is general agreement that the prey is carried to the nest high in the air. Krombein says that the wasp "almost invariably approaches the burrow from two to three meters above the surface, and from a direction opposite to that in which the burrow penetrates the ground. Reaching a point several meters above and behind the entrance she descends extremely slowly in a straight line to the concealed burrow entrance, making a noticeably loud humming noise as she descends. She alights right at the entrance." Hartman, the first to report this striking behavior, says that the wasp does not descend "in a sudden continuous swoop, but in gentle jerks as if she was descending a flight of stairs and had to pause at each step." When about 2 ft high, she "takes a sudden dive" and "lands on her nest." Parker speaks of the wasp poising in the air several feet high and, while producing a humming noise, descending "slowly and steadily" to the entrance of the burrow. The Raus describe a similar descent to the nest, and I have observed this behavior many times and in all the areas of study. Wasps that experience difficulty finding or getting into the nest typically rise high in the air and descend once again with their prey. The descent to the nest is always slow and hovering, and may be either a steady descent or a "step-wise" descent as described by Hartman; the final drop to the nest entrance may be more sudden than the earlier part of the descending flight. As already noted, on one occasion Krombein observed one female that approached the nest with prey only a few centimeters high, although this female usually descended from a considerable height in the usual manner. I also saw one female approach her nest with a low, circling flight (no. 1940, Concord, Massachusetts).

As described and illustrated fully by Krombein, the prey is held venter up beneath the base of the abdomen of the wasp. During flight the bug is held with the middle and hind legs of the wasp, but upon landing the wasp stands on the hind legs, digs open the nest entrance with the front legs, and holds the prey with the middle legs only. According to Hartman, the wasp, when just inside the entrance, passes the prey back to the third pair of legs, "the bug now projecting beyond the tip of the wasp's abdomen."

The prey of this species consists entirely of immature Heteroptera. While members of six families have been taken as prey (Table 13), the vast majority of records are for Coreidae and Pentatomidae.

§A. *Bicyrtes quadrifasciata* (Say)

TABLE 13. PREY RECORDS FOR BICYRTES QUADRIFASCIATA
(All specimens nymphs)

Species of prey	No. of nests	No. of specimens	Locality	Reference or note no.
REDUVIIDAE				
Apiomerus sp.	—	—	Miss.	Smith, 1923
Zelus sp.	—	—	Miss.	Smith, 1923
LYGAEIDAE				
Genus and species ?	—	—	Texas	Hartman, 1905
COREIDAE				
Acanthocephala femorata Fabr.	—	—	Miss.	Smith, 1923
Anasa tristis DeGeer	—	1	Mo.	Rau, 1934
Archimerus alternatus Say	10	60	N.C.	Krombein, 1955, 1958
Archimerus calcarator Fabr.	—	—	Miss.	Smith, 1923
Chariestrus antennator Fabr.	—	—	Miss.	Smith, 1923
	—	—	N.J.	Davis, 1926
	3	12	Kan.	19, 203, CY39
	2	7	Fla.	1037, 1054
Leptoglossus clypealis Heid.	2	14	Kan.	76, 371
Leptoglossus oppositus Say	3	8	N.C.	Krombein, 1955, 1958
Leptoglossus phyllopus L.	—	—	Miss.	Smith, 1923
	1	1	Kan.	751
PENTATOMIDAE				
Acrosternum hilare Say	—	—	Miss.	Smith, 1923
	1	2	N.Y.	959
Acrosternum sp.	1	3	Mass.	1953
Brochymena arborea Say	3	5	Kan.	17, CY39, 532
Brochymena cariosa Stal	3	11	Kan.	17, 75, 116
	—	1	N.C.	Krombein, 1959
	1	8	Texas	CSL-M101
Brochymena carolinensis Westw.	—	1	N.J.	Davis, 1926
	4	13	N.C.	Krombein, 1955, 1958
Brochymena quadripustulata Fabr.	—	—	Miss.	Smith, 1923
	4	25	N.Y.	521, 1306, 1357, 1397
Chlorochroa uhleri Stal	2	7	N.Y.	948, 949
Dendrocoris humeralis Uhler	1	1	N.C.	Krombein, 1955
Edessa florida Barber	—	1	N.C.	Krombein, 1955

TABLE 13. PREY RECORDS FOR BICYRTES QUADRIFASCIATA (CONT.)

Species of prey	No. of nests	No. of specimens	Locality	Reference or note no.
PENTATOMIDAE (CONT.)				
Euschistus tristigmus Say	5	8	N.C.	Krombein, 1955
Euschistus spp.			Miss.	Smith, 1923
	1	5	Mass.	1954
Menecles incertus Say	1	2	Kan.	203
Murgantia histrionica Hahn	—	1	Kan.	541
Nezara viridula L.	—	1	N.C.	Krombein, 1952
	1	4	Texas	CSL-M117
	1	8	Mass.	1963
Nezara sp.	1	2	Ohio	Parker, 1917
	—	1	N.Y.	Davis, 1926
Podisus sp.	1	1	Mass.	1953
Stethaulax marmoratus Say	1	3	N.C.	Krombein, 1958
Thyanta custator Fabr.	3	6	N.C.	Krombein, 1955, 1958
Thyanta sp.	1	1	Mass.	1953
SCUTELLERIDAE				
Homaemus aeneifrons Say	2	3	Fla.	1037, 1054
Tetyra bipunctata H.S.	—	4	N.Y.	Davis, 1926
CYDNIDAE				
Cyrtomenus sp.	—	—	Texas	Hartman, 1905

Most of the nymphs taken are in the last one or two instars, but earlier instars of some of the larger coreids are sometimes used. Some cells contain bugs of only one species, while others contain up to three species, sometimes of two families (probably sometimes more, but this is the maximum recorded). The number of bugs in fully provisioned cells varies from 4 to 14 depending upon their size, the usual number being from 7 to 10. The bugs are placed in the cell head in, venter up.

Provisioning is normally completed on the day on which it is initiated or on the following day, so that there is no contact between wasp and larva. I found no exceptions to this in Kansas, Florida, or New York, but in two of five nests studied in Massachusetts provisioning was completed after the larva had hatched. In one nest (no. 1940, Concord, Massachusetts) the wasp brought in a stinkbug at 0950, then made a final closure of the nest; the cell was found to contain a larva about 12 mm long, probably about 3 days

§A. *Bicyrtes quadrifasciata* (Say) 157

old. The previous day had been clear, but the 3 days before that were rainy; presumably the egg was laid 5 days earlier, but provisioning delayed because of inclement weather and/or scarcity of prey. The second nest (no. 1953, Andover, Massachusetts) was still being provisioned 3 days after I marked it as being dug; this nest contained a small larva still attached to the initial bug, which had deteriorated slightly.

Several workers have noted that the bugs are not killed by the sting of the wasp, but deeply paralyzed. Some twitching of the legs and antennae can sometimes be observed as much as 48 hr after they are stung, or even longer. Krombein found that some bugs remained alive in a rearing tin for 2.5 weeks; they might have remained alive longer, but the sand in the tin was inadvertently allowed to dry out. On the other hand, I have sometimes noted deterioration of bugs stored only 3–4 days earlier, so these bugs must have died shortly after being stung.

(6) *Final closure.* Hartman described final closure as follows: "At 4:10 P.M. the last bug was brought in and the wasp began to permanently close the nest. After remaining inside for seven minutes, she came forth scratching the sand back to fill up the tunnel, biting it loose from the sides, pulling it in from the surface and pressing it down with her abdomen." I have noted that females rake in sand from the front of the entrance and then back in, without making flights as in some related wasps. Much use is made of the tip of the abdomen for packing the sand; the tip is curled under and moved with quick, light blows. Krombein found that females sometimes fly up and hover over the area, returning to back into the burrow and rake in more sand. "After the burrow is filled flush with the surface," Krombein continues, "the wasp scratches sand beneath her toward the entrance from several different spots until the entire area is smoothed over." Krombein found that only the bottom and top few centimeters of the burrow were actually filled, and that in general the sand was not packed tightly but was much looser than the surrounding sand. I found that in about half the nests I dug out after the final closure, only the bottom and top parts of the burrow were filled; in the other half, the burrow was completely filled, although somewhat loosely.

(7) *Immature stages and development.* Several workers have found the egg of this species, and all are agreed that it is laid on the venter of the first bug placed in the cell, this bug being placed at the far end of the cell. The egg measures about 1 mm in width by 4.0–4.5 mm in

length; the slightly broader posterior end is attached firmly to the ventral midline of the bug and the egg extends obliquely upward and forward (Fig. 88). The usual point of attachment is between the fore and middle coxae, but I have found some instances in which the egg was attached between the front coxae and over the second segment of the beak, others in which it was attached between the hind coxae. Krombein mentions one instance in which the egg was laid on the abdominal sternum next to one of the hind coxae. The egg is invariably inclined slightly toward the head, and when it hatches (in 1 to 2 days) the larva remains attached for 1 or 2 days at the point of attachment of the egg, while with its free head end beginning to feed, either around the base of the beak of the same bug or more commonly on an adjacent bug. During the first day or two most feeding is at joints and articulating membranes, but when the larva is larger and has lost its attachment to the initial bug, the prey is extensively macerated. Only 3 to 5 days are required for the larva to consume the prey in the cell and attain full size (Smith records a few larva as taking up to 2 weeks to reach maturity in rearing tins). The cocoon is spun in a manner very similar to that of *Bembix,* and the finished cocoon is indistinguishable from that of a *Bembix;* it measures 6.5–7.0 mm wide by 17–20 mm long and has five to seven pores around its widest part (Krombein, 1955). As already mentioned, Krombein found that cocoons spun in North Carolina in early summer gave rise to adults after 42 days, while

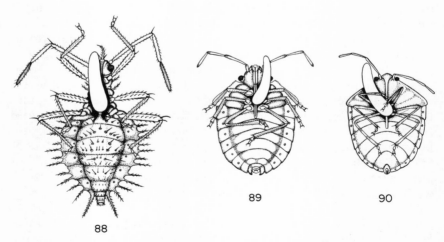

Figs. 88–90. Attachment of the egg: 88, of *Bicrytes quadrifasciata* on an immature coreid bug (no. 538); 89, of *B. ventralis* on an immature stinkbug (no. 430); 90, of *B. fodiens* on an adult stinkbug, *Mormidea lugens* (no. 482). (Drawings by C. S. Lin from material collected in Pottawatomie County, Kansas.)

§A. *Bicyrtes quadrifasciata* (Say) 159

Smith found that cocoons spun in Mississippi in late July gave rise to adults in from 15 to 40 days. Krombein found that cocoons spun in September did not give rise to adults until the following summer.

(8) *Natural enemies.* This wasp is commonly attacked by miltogrammine flies in all parts of its range. Hartman found maggots in two cells, the maggots apparently having destroyed the egg of the *Bicyrtes;* he speaks of them as "muscid" larvae, but doubtless they were Miltogrammmae. Smith found provisioning females being followed by miltogrammine flies on many occasions. The flies, he says, "hover about and to the rear of the wasps, or else sit on stones, blades of grass, or sticks near by, intently watching every move of the wasps." He believed that larviposition occurred on the prey as it was taken into the nest. Maggots found in the nests and reared to maturity proved to be the same species as those taken outside the nests, thus confirming his suspicions. On one occasion he saw a fly dart into a nest after a wasp and then reappear after 3 or 4 sec. The flies were identified as *Senotainia trilineata* (Wulp) and *S. rubriventris* Macq. or a closely related species. Allen (1926) later recorded *S. rubriventris* as being reared from *Bicyrtes quadrifasciata,* and Krombein (1955) confirmed this association in North Carolina. The latter author reared this fly from two of nine cells. In one cell there was only one maggot, and he was able to rear the wasp larva to maturity; the other cell contained ten maggots, and the wasp larva in this cell disappeared, presumably having been consumed by the maggots. Krombein notes that adult *S. rubriventris* do not trail the wasp, but wait near the burrow entrance and dash in the burrow after the wasp carrying prey, reappearing seconds later. Presumably the maggots are deposited as the prey is passing down the burrow.

At Ithaca, New York, on 19 August 1953 (no. 670), a female *quadrifasciata* was observed being trailed by two small flies. Larviposition was not observed, but the cell was later found to contain five maggots. These destroyed the wasp egg and eventually formed their puparia; the following spring (1 May) five adult *Senotainia vigilans* Allen emerged. Another nest at Ithaca (no. 717) contained only two maggots, and in this nest the *Bicyrtes* larva reached maturity (in a rearing tin). The one adult fly reared from this nest was identified as *Senotainia* sp. In Highlands County, Florida, provisioning females were also being followed by miltogrammine flies (no. 1745). These flies would often trail the wasps as they descended obliquely to their nests, whereupon the wasps would often fly about deviously and eventually ascend to a considerable height and reapproach the nest in the usual manner. One of the flies was captured and found to be *Senotainia trilineata* Wulp (complex). Neither of the two nests excavated contained maggots.

Two of five nests excavated in Massachusetts contained maggots. In one case (no. 1940) only two maggots were present, and both maggots and wasp larva reached maturity; the maggots formed their puparia on 26 July, and two adult *Senotainia vigilans* Allen emerged on 17 August. This same species of fly was seen hovering about the wasp while she was making a final closure of the nest. The other nest (no. 1963) contained eight to ten maggots in the cell when it was dug out, and no egg or larva of the wasp could be found, presumably having been destroyed by the maggots. These maggots formed puparia on 14 August, and four flies, *Senotainia trilineata* Wulp (complex), emerged 30 August–1 September.

In one of the nests dug out in Reno County, Kansas (no. 637), the wasp larva developed normally and spun its cocoon on 25 July 1953. On 9 August, only 2 weeks later, an adult chrysidid, *Holopyga ventralis* (Say), emerged from this cocoon. This same species of cuckoo wasp has been reared from cocoons of *Bicyrtes fodiens* (Handlirsch).

One of the five nests studied in Massachusetts was also infested with what I assume to be a chrysidid, but I was unable to rear it successfully (no. 1953, Andover). The larva of the *Bicyrtes* grew normally in a rearing tin, then died of unknown causes when nearly mature; at this time another larva, about 8 mm long, was seen feeding on its dorsum, attached at about the sixth abdominal segment and facing toward the rear of the *Bicyrtes*. The following day the host larva had deteriorated badly, and the parasite had ceased to feed, so I transferred the parasite to a *Microbembex* larva which happened to be available. The parasite larva immediately seized the larger *Microbembex* larva with its sickle-shaped mandibles and began to feed. The following day this host had deteriorated and the parasite was no longer feeding, so I repeated the experiment on another *Microbembex* larva with the same results. Unfortunately this time both host and parasite died, perhaps having pierced each other with their mandibles. It is my belief that the original *Bicyrtes* host died from some other cause, but the cessation of activity triggered the chrysidid into feeding (usually delayed until after the cocoon has been spun and the larva is dormant). The quick deterioration of the hosts was due to the fact that these larvae had not voided their meconium as a diapausing larva would have done.

B. *Bicyrtes ventralis* (Say)

This is a slightly smaller wasp than *quadrifasciata* and has a somewhat more northerly distribution, although the ranges of the two

§B. *Bicyrtes ventralis* (Say)

species overlap broadly. Where the two occur together, they may nest in very similar situations, build very similar nests, and even prey on the same bugs. There are, however, a number of minor although interesting differences between the two species.

Parker (1917) studied *ventralis* at "a pile of clean sand on a vacant lot in the city of Washington [D.C.]." My own studies were made in Pottawatomie County, Kansas, in June 1952 and 1953 (3 notes); Tompkins County, New York, in July 1955 and 1958 and in August 1953–1956 and 1960 (12 notes); and Oswego County, New York, July 1954 and September 1956 (2 notes). I have included two prey records from Groton, New York, contributed by Frank E. Kurczewski.

(1) *Ecology.* Parker remarks that *ventralis* nests in sandy situations similar to those chosen by *Bembix spinolae.* I have also found this to be true, but I have more often found *ventralis* nesting in somewhat coarser sand than that preferred by *Bembix spinolae* and *Bicyrtes quadrifasciata.* In sand pits where the three occur together, one generally finds the second and third species in patches of fine-grained sand and *ventralis* in sandy gravel on the periphery. Most of my studies of *ventralis* have been made in man-made sand fills or excavations, which seem to provide the most suitable nesting areas in the northeastern states. Like other species of *Bicyrtes,* the adults are frequent visitors to flowers.

This species tends to nest solitarily or in small aggregations (up to 20); the nest entrances are typically widely separated and interspersed with those of other digger wasps. I have made no observations on the behavior of the males.

(2) *Digging and orientation.* The nest may be dug in flat soil or (less commonly) in slopes of up to 45°. Large bare places are usually chosen, but I have found some nests in small bare spots surrounded by weeds. Digging is very similar to that described for *quadrifasciata.* There are marked "tilting" or "bobbing" movements in the nest entrance, and the soil is thrown far from the nest entrance (up to 25 cm); by turning slightly from side to side, the wasp sprays the earth in a wide arc. Thus no mound accumulates and no leveling movements are necessary.

None of the wasps were timed, but it seems unlikely that more than 1 or 2 hr are required for completion of the nest. Following completion, the wasp comes out head first and makes an outer closure, scraping in soil from around the entrance and packing it in place with blows of the tip of the abdomen. At times during the closure, and at greater length following the closure, the wasp rises to a height of 10–30 cm and flies circuitously about the nest entrance. One wasp was seen to make a fairly lengthy orientation flight after her second closure, that is, after the cell already contained a bug and an egg. This wasp made five figure-eight flights over the nest, the center of the "8" at the entrance, the whole flight being about 20–30 cm high and covering about 1 m; the wasp then flew off in a straight line rather low. This low, circling orientation flight is very different from that of *quadrifasciata*, as is the return to the nest.

(3) *Nature and dimensions of the nest.* The nest is exceedingly similar to that of *quadrifasciata*, but as might be expected is slightly smaller. Burrow diameter is about 5 mm, and the terminal cell 8–12 mm in diameter and 20–30 mm in length. All nests dug out both in Kansas and in New York were quite shallow, and there was much less variation in depth than in *quadrifasciata* (Table 14). In New York cell depth for *quadrifasciata* varied from 5 to 17.5 cm, with a mean of 9.7 cm, while in the very same areas cell depth of *ventralis* varied over the narrower range of from 4 to 8 mm, with a mean of 6.2 mm. Parker provided no data on the size of the nests in the colony he studied.

Most burrows are oblique and nearly straight, but curves to one side or the other are not uncommon. The cell is horizontal or nearly so. As Parker notes and as I have observed many times, no inner closure, separating the cell from the burrow, is maintained at any time. However, an outer closure is always maintained except during the brief periods when the female is inside the nest.

Parker states that "usually each nest has but a single brood cell, but in the course of my investigations I found two each of which had two brood chambers reached from the same entrance. In each case, however, the first chamber was provisioned and sealed before the second was constructed." One of the three nests I dug out in

TABLE 14. NEST DATA FOR BICYRTES VENTRALIS

Locality	No. of nests	Burrow length (cm)	Cell depth (cm)
Pottawatomie Co., Kan.	3	12.0 (7–15)	10.0 (5–15)
Tompkins Co., N.Y.	11	10.0 (8–13)	6.0 (4–8)
Oswego Co., N.Y.	2	13.5 (12–15)	7.0 (6–8)

§B. *Bicyrtes ventralis* (Say) 163

Kansas had two cells (no. CMY430). This nest was dug out following the final closure; the second cell, containing an egg, was at a depth of 15 cm, while the first cell, containing a partially grown larva, was 3 cm back toward the entrance, on a short side burrow and at a depth of only 12.5 cm. No two-celled nests were found in New York, but 1 of the 11 nests excavated had three cells (no. 1392). In this case the cells were separated by about 5 cm of sand, and the most recently completed cell, containing an egg, was closest to the entrance. Thus it appears that *ventralis,* like *quadrifasciata,* rather infrequently constructs an additional one or two cells from the same burrow (Fig. 91).

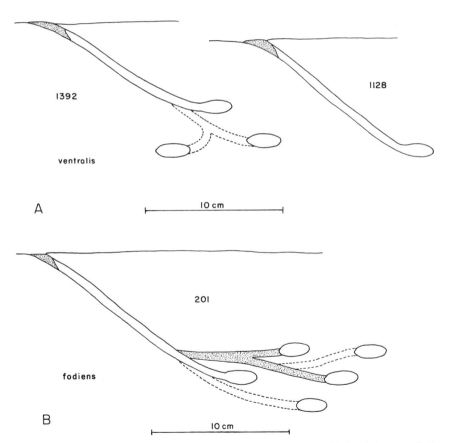

Fig. 91. (A) Typical nests of *Bicyrtes ventralis,* Ithaca, New York; (B) nest of *B. fodiens,* Pottawatomie County, Kansas. Stippling indicates fill that could be traced, broken lines, filled burrows that could not be traced exactly.

(4) *Hunting and provisioning.* When approaching the nest with prey, the female flies quickly, about 10–30 cm high, and either straight to the entrance or in a somewhat sinuate path. She lands directly in front of the entrance, holding the bug venter up with her middle legs, removes the outer closure with her front legs, and enters, remaining inside 30 sec to 2 min. Upon leaving, an outer closure is made, usually with the tip of the abdomen used for pressing the soil into place, and the wasp takes off, flying not more than about 30 cm high and sometimes taking a few loops over the nest before leaving the area. The females observed over a period of time provisioned very slowly, requiring upwards of 45 min to obtain prey. However, provisioning is doubtless more rapid than this at times (Figs. 92 and 93).

Parker reports this species as using the nymphs of stinkbugs (Pentatomidae), but he provides no detailed records. I have found the species using immatures exclusively, usually Pentatomidae but with one record for Coreidae. The three cells dug out in Kansas all contained only *Menecles incertus,* but several of the nests from New York contained as many as three species of bugs. As compared to *quadrifasciata, ventralis* tends to use smaller bugs and to avoid the larger Pentatomidae such as *Brochymena* and various Coreidae which are frequently employed by the former species; however, several species appear on both lists (Table 15).

Parker states that from 3 to 11 bugs are used per cell, fewer bugs being used when the bugs are larger. My own data show a variation in the number of bugs of from 5 to 11, with a mean of 8.2. Parker notes that the bugs are paralyzed but may remain alive for over 1 week.

Of unusual interest in this species is the frequency of delayed provisioning, that is, the rather numerous instances in which the female has been found to be still provisioning the nest after the egg has hatched. Parker states the following:

If unfavorable weather interferes with the work of the wasp before the nest is completely provisioned, she will return to it later and complete the store of food necessary to develop her offspring. Under such circumstances I have observed *Bicyrtes ventralis* carrying bugs into a nest that contained a half-grown larva. On one occasion, after a few days of adverse weather and while the sand was yet wet, I observed a female *ventralis* open and enter a nest but without carrying a bug. After a few minutes spent inside she emerged and sealed up the nest. I at once digged up the nest and found within it a half-grown larva, several untouched bugs, and the remains of several more that had been devoured . . . Here it would seem that the mother wasp, after an absence of two or three days caused by

§B. *Bicyrtes ventralis* (Say)

Figs. 92 and 93. *Bicyrtes ventralis* female: 92 (above), closing her nest entrance (Ithaca, New York); 93 (below), returning a few minutes later with an immature stinkbug held beneath her abdomen by her middle legs. (Fig. 93 from Evans, 1963d.)

TABLE 15. PREY RECORDS FOR BICYRTES VENTRALIS
(All specimens nymphs)

Species	No. of nests	No. of specimens	Locality	Note no. or collector
COREIDAE				
Anasa tristis DeGeer	1	4	N.Y.	953
PENTATOMIDAE				
Banasa dimidiata Say	1	5	N.Y.	1568A
Cosmopepla bimaculata Thom.	1	9	N.Y.	1665
Elasmostethus cruciatus Say	1	2	N.Y.	1128
Euschistus euschistoides Voll.	2	22	N.Y.	1392, 1665
Euschistus tristigmus Say	2	5	N.Y.	947, 1665
Euschistus variolarius P.B.	3	11	N.Y.	953, 1568
Menecles incertus Say	2	21	Kan.	106, 430
Mormidia lugens Fabr.	1	1	N.Y.	1568
Thyanta pallidovirens accerra McAtee	1	1	N.Y.	F.E.Kurczewski
Trichopepla semivittata Say	1	3	N.Y.	953

rainy weather, visited the nest for the purpose of ascertaining whether the larva had been sufficiently provided for."

Of the 14 nests that I dug out during or immediately following the completion of provisioning, three contained larvae: one a newly hatched larva (no. 1665B), the other two larvae that were at least a day or two old (nos. 1214 and 1321). In two instances I have no record of the weather for the preceding 1 or 2 days. In the other (no. 1214), I first observed the wasp digging her nest and bringing in the first bug (on which the egg was doubtless laid) on the afternoon of 11 August 1956. On 12 August she was again seen provisioning in the afternoon. The 13th was a rainy day. On the afternoon of the 14th I did not see the wasp, but it was obvious that she had entered and left the nest, and the nest had received no final closure, although the cell appeared pretty well filled with bugs. The larva was about 6 mm long and was erect, attached to the bug on which it had been laid and feeding on this same bug by bending over to feed at the base of the head. It is interesting to note that this wasp had not completed provisioning at the end of the second day even though the weather was satisfactory up to that point.

(5) *Final closure.* During final closure, the wasp bites sand from the sides of the burrow and scrapes it behind her with her front legs. During the final stages she scrapes in sand from around the entrance, then backs into the burrow and packs it in place with vigorous blows of the tip of the abdomen. She then scrapes sand in various directions over the entrance so that all evidences of the nest are erased. In six nests that I dug out following final closure, the burrow was completely filled with compacted sand, and in one case I was unable to follow it to the cell.

(6) *Immature stages and development.* The egg is from 3 to 3.5 mm long and tapers slightly toward the free (anterior) end. It is laid on the mid-ventral line of the first bug placed in the cell, which is on its back deep in the cell. The actual point of attachment varies from a position between the hind coxae to a position between the front and middle coxae. The free end extends upward and obliquely forward (Fig. 89). After hatching, the larva remains attached in the same position for 1 or 2 days and bends forward and to the sides to feed on the same and on adjacent bugs. Parker states that the egg hatches in from 48 to 72 hr after being deposited, and my own studies substantiate this. The larva requires only from 3 to 6 days to reach full size. One larva spun a cocoon on 30 June in Pottawatomie County, Kansas, and produced an adult on 1 August of that same summer. Doubtless there are two generations a year at least in the southern parts of the range of this species.

(7) *Natural enemies.* Allen (1926) presented a record of *Senotainia trilineata* (Wulp) having been reared from *Bicyrtes ventralis* at Washington, D.C., by J. B. Parker. He also reported a record of *S. vigilans* Allen emerging from a nest of this wasp at the same locality.

C. *Bicyrtes fodiens* (Handlirsch)

This is a species of similar size and general appearance to *ventralis*. It is confined to the southeastern and south central United States. Rau (1922) published a note on the species which is so short that it is here quoted nearly in full. Rau found three females nesting in "sandy clay by the roadside at Lake View, Kansas. The burrows were evidently in course of construction, and went down diagonally for about four inches."

I found this species to be fairly common at Blackjack Creek, Pottawatomie County, Kansas, in June and early July of 1952 and 1953. Adults of both sexes were seen on several occasions taking nectar at *Melilotus alba,* and five nests were located, all in relatively moist, fine-grained sand along the stream. One was in a bank sloping at about 35°, the others in flat soil. Three were in areas partially overgrown with weeds.

(1) *Nesting behavior.* Digging is similar to that of *ventralis,* the wasp bobbing up and down in the nest entrance and, with relatively slow strokes, throwing the soil a considerable distance away (up to 25 cm). The sand, in the area of study, was sufficiently moist that it tended to be raked out and dispersed as small pellets. As in other species of the genus, no mound accumulates at the entrance and no leveling movements occur. Since the nest of *fodiens* is multicellular, periods of digging alternate with periods of provisioning. The five nests that I dug out had a total of 19 cells (several more than the 13 nests of *ventralis* excavated!). Two of these nests had received the final closure (nos. 146 and 200); these contained 4 and 5 cells, respectively. The remaining three were still active when dug out (nos. 201, 444, and 482); these contained 5, 3, and 2 cells, respectively. Number 201 provisioned her five-celled nest over a period of at least 4 days. In any one nest the first cell constructed and provisioned is the deepest one, additional cells being built from short side burrows back toward the entrance, much like the nests of many Gorytini. The cells tend to be separated by 1–3 cm of sand, and in any one nest the cells vary in depth by only a few millimeters (maximum: 4.5 mm). The total variation in all cells of the five nests was barely greater than this, the range being from 8 to 13.5 mm, with a mean of 10.3 mm. Length of the burrow, which is oblique and at a 45–60° angle with the surface, varied from 12.5 to 18.5 mm, with a mean of 16.3 mm (Fig. 91).

As in the preceding two species, an outer closure of the nest is maintained when the wasp is not in the nest, but no inner closure is maintained. At the final closure, the burrow is completely filled with sand that is packed firmly in place with vigorous blows of the dorsal surface of the last abdominal segment. When gathering fill, the wasp walks a few centimeters away from the entrance while shooting sand behind her toward the entrance; then she backs in, raking in the dirt with her forelegs and packing it in place.

(2) *Provisioning the nest.* Females arrive at the nest swiftly, flying rather close to the ground. They land directly in front of the covered entrance, rake it open with a few blows of the front legs, and enter

§C. *Bicyrtes fodiens* (Handlirsch) 169

quickly, holding the bug in the manner described for the preceding two species. Only 30–60 sec are spent inside the nest, somewhat longer if the bug is the first one in that cell and time is taken for oviposition. One female (no. 482) was seen to bring in several bugs in 1 hr, taking a minimum of 4 min between trips.

In the area of study, only two species of small bugs were being used as prey. These were the pentatomid *Mormidea lugens* Fabricius (165 records, all adults) and the scutellerid *Homaemus aeneifrons* Say (66 records, all late-instar nymphs). In some nests the two species were well mixed, but others contained only one or the other. For example, the four cells of nest no. 146 contained nothing but immature *Homaemus,* the five cells of no. 200, nothing but adult *Mormidea.* Presumably there was a rich source of these two species of bugs not far from the nest site. The number of bugs per cell varied from 10–23, with a mean of 15.2. As usual, they are packed in the cell head in, venter up. The bugs are well paralyzed but may show spontaneous movements of the appendages for several days.

(3) *Immature stages and development.* The egg is about 3 mm long and is laid on the ventral midline of the thorax of the first bug placed in the cell; the anterior, more tapered end of the egg extends obliquely upward and forward as usual in this genus (Fig. 90). The egg hatches in approximately 2 days. The larva remains attached at the point at which the egg was attached for a day or two and bends over to feed at the leg bases or base of the beak of the same and adjacent bugs. Eventually it loses its attachment and reduces the bugs to a pasty mass. The larva reaches full size after only 3 or 4 days of feeding and spins a cocoon similar to that of other species of this genus. Three larvae which spun their cocoons 7–10 July 1952, gave rise to adult males 12–14 August the same summer, indicating that there are at least two generations per year.

(4) *Natural enemies.* One cell of nest no. 146 was found to contain two larvae when the nest was dug out. The smaller larva, apparently that of the *Bicyrtes,* was attached to a bug, while a somewhat larger larva was loose in the cell. This larger larva grew rapidly, in a rearing tin, but the smaller larva could not be found again. This larger larva spun its cocoon on 5 July 1952; on 26 July, only 3 weeks later, an adult chrysidid, *Holopyga ventralis* (Say), emerged from this cocoon.

Fly maggots were found in one of the two cells of nest no. 482. These maggots formed their puparia after a few days, the wasp larva failing to develop successfully; three adult *Metopia argyrocephala* (Meigen) emerged from these puparia several weeks later.

Several cells were found to contain nematodes, and in most cases these nematodes multiplied rapidly in the rearing tins and completely destroyed the cell contents.

D. Ethology of other species of *Bicyrtes*

Fragmentary notes are available for three other Nearctic species of this genus and for two Neotropical species. These are useful in further substantiating the generic behavioral features, but they are insufficiently detailed to make clear any specific differences which may exist.

Hartman (1905) presented observations on four individuals that he found nesting near Austin, Texas. One of these was identified as *B. parata* (Provancher), another as "*pictifrons* Smith," a name that properly belongs in another genus. However, it seems probable from Hartman's remarks that he was working with only one species. Although *parata* is a western species, it does occur in eastern Texas, and it is probable that Hartman was in fact working with that species. Unfortunately the specimens on which Hartman based his observations are no longer extant, so it is impossible to check this.

Of the four wasps observed by Hartman, three dug their nests in flat soil, the fourth in the side of a shallow pit. "During the progress of the work of digging," says Hartman, "the wasp makes short excursions, (on foot chiefly), around the neighborhood. [One wasp] continued digging for about two hours and [then] closed the entrance with sand. Before venturing away from her nest for the first time, she made a rather careful study of the locality, flying in and out among the herbs and bushes." On returning to the nest with prey, one wasp "descended slowly toward the nest. When within three inches of the surface, she hovered an instant, then dropped suddenly like a dead-weight and after a moment's pause at the entrance opened it up and walked in. As she entered," continues Hartman, "I could see her pass a very small bug back to her hind legs." Another wasp, when approaching her nest, "flew back and forth above it in nearly parallel lines like a pendulum with ever shortening oscillations. This manner of approach she employed nearly every time. Other individuals . . . showed a habit approaching this, though not quite so marked."

Hartman presented an excellent photograph of one nest. The burrow of this nest curved gradually downward, the cell being about 11 cm beneath the surface. Apparently none of the nests had

§D. Ethology of other species of *Bicyrtes*

more than one cell. Hartman mentions that one wasp provisioned her nest over a period of 3 days, then made a final closure. Eighteen bugs were found in the one cell of this nest. Other nests contained fewer bugs but had not received the final closure. Hartman states that an outer closure of the nest is maintained at all times and that the egg is laid in the same manner as in *quadrifasciata*. The bugs were not identified; presumably they were immature, as Hartman mentions their small size, but there is no indication of the family to which they belonged.

According to Hartman, female *parata* do not spend the night in their nests. He reports that they "dig a shallow nest and crawl into it for the night, closing it from the inside." This remark is significant in that it is one of the only three published notes pertaining to this aspect of the behavior of species *Bicyrtes*. The second and third notes pertain to *B. capnoptera* (Handlirsch) and indicate that these wasps spend the night on vegetation. Evans and Linsley (1960) found both sexes of this species sleeping on a *Melilotus alba* plant in southeastern Arizona along with many other bees and wasps. The *Bicyrtes* were found on this plant 18 of the 23 nights on which observations were made (4–22 August 1959), in numbers varying from one to five (usually two to four). In every instance they were grouped together slightly apart from other wasps, toward the center of the plant. They "oriented crosswise on horizontal stems in what appeared to us to be a precarious balance, held by the middle and hind legs (the front legs tucked under the body)." They tended to line up side by side, but not in actual body contact, somewhat resembling a row of baby birds recently departed from their nest (Fig. 94). The small number of individuals roosting here suggests either that there was only a small nesting aggregation here (actually none were found nesting), or that only certain individuals sleep on vegetation. Linsley (1962) found *capnoptera* sleeping in the same area in 1960, again in small groups of two to five and in the same posture described in the earlier paper (see his Fig. 5 especially). In this case they occupied a dried *Heterotheca* plant and occurred mostly on old flower heads. Again, these wasps were "regular" members of a mixed aggregation, occurring on this plant 15 of 22 nights on which observations were made.

Nothing further has been published on *capnoptera*, a rather common species across the southern tier of states. C. S. Lin has observed the species nesting near Austin, Texas, and has sent me specimens of wasps and their prey. One nest, dug out on 20 June 1957, contained seven small nymphs of the pentatomid *Thyanta pallido-virens accerra* (McAtee), while a second nest dug out in the same area 3

A

B

Fig. 94. (A) two and (B) five *Bicrytes capnoptera* sleeping on *Melilotus alba* in southeastern Arizona. (From Evans and Linsley, 1960.)

§D. Ethology of other species of *Bicyrtes* 173

weeks later contained five adults of this same species of stinkbug.

I am also able to report a previously unpublished prey record for *B. burmeisteri* (Handlirsch), a predominantly Mexican species. J. C. Crawford took specimens of this species at Brownsville, Texas, on 29 September 1906, with adult males of the pentatomid bug *Solubea pugnax* Fabricius. These specimens are in the United States National Museum, and were recorded by Dow (1935) in an unpublished doctoral thesis at Harvard University.

Two South American species of *Bicyrtes* have been studied in some measure; these are *variegata* (Olivier) and *discisa* (Taschenberg), both very widely distributed throughout tropical America. Both Richards (1937) and Callan (1954) indicate that these species may nest in close proximity; Callan sites as a typical locality "bare sandy ground exposed to the sun, such as the dry sand on a beach above [the] highwater mark." Richards found *discisa* nesting solitarily in sandy soil in small clearings at the Essequibo River, British Guiana, and preying upon immature Pentatomidae. Brèthes (1918) studied this species near Buenos Aires, where the females nested along a sandy path. The burrow was found to be oblique and about 10 cm long, leading to a horizontal portion terminating in a cell. The prey consisted of immature Hemiptera of three species: *Spartocerus brevicornis* Stål, *Pachylis argentinus* Berg (both Coreidae), and *Edessa meditabunda* Linneaus (Pentatomidae). In Trinidad, Vesey-Fitzgerald (1940) found *discisa* nesting in sand, the burrows being about 12 cm long and stocked with immature Pentatomidae; the egg is said to be "plastered along the proboscis."

The second species, *variegata*, has been studied in moderate detail by Janvier (1928) in Chile. This author found a colony of about ten females nesting in sand along the banks of a river. The nests contain several cells constructed at the ends of short side burrows, the first cell being closest to the entrance and additional cells added by extending the burrow; the cells vary in depth from only 1–2 cm to 10–15 cm.

According to Janvier, the nest is closed and well concealed while the female is away. Carriage of the prey is similar to that of other species of this genus. In the area of study, the prey consisted entirely of nymphs and adults of the coreid bug *Leptoglossus chilensis* Spinola. Janvier did not find the egg.

Poulton (1917) found *variegata* preying upon immature Pentatomidae west of São Paulo, Brazil. On the Paria Peninsula, Venezuela, Callan (1954) took a female *variegata* with a pentatomid nymph of the genus *Edessa*. Vesey-Fitzgerald (1956) found a colony of this wasp on the south coast of Trinidad. The prey consisted of immature bugs belonging to five genera of Coreidae and Pyrrhocoridae. Vesey-Fitzgerald dug out several nests, but he makes no mention of any of the nests having more than one cell.

E. Summary of the ethology of the species of *Bicyrtes*

The following twelve features appear to characterize the behavior of this genus so far as the presently available data go.

(a) Adults are frequent visitors to flowers for nectar and are not known to feed at the prey.

(b) Adults do not spend the night in the brood nests. One species (*parata*) is reported to dig short burrows in the sand in which to spend the night, while another (*capnoptera*) is known to sleep on vegetation.

(c) Females nest in fine to coarse-grained sandy soil, often along watercourses, sometimes in fairly moist situations.

(d) The species are not highly gregarious nesters, tending to form small colonies with the nest entrances well scattered, often interspersed with the nests of other digger wasps, sometimes other species of *Bicyrtes*.

(e) When digging, the females show strong tilting motions, and with strong thrusts of the front legs throw the earth far behind them over a considerable area, so that no mound of soil accumulates at the entrance

(f) The nests are simple and quite shallow, the cells most commonly being between 5 and 15 cm deep (maximum: a record of 27.5 cm for *quadrifasciata*). Nests of up to five cells are the rule in some species (*fodiens, variegata*), while other species commonly make unicellular nests but occasionally make bicellular or tricellular nests (*quadrifasciata, ventralis*).

(g) An outer closure of the nest is maintained at all times when the female is not in the nest, but no inner closure (separating cell from burrow) is ever made.

(h) The prey is carried in flight with the middle legs, clasped tightly beneath the base of the wasp's abdomen, and is not normally deposited on the ground at any time.

(i) The prey consists of immature or (less commonly) adult Heteroptera of the families Pentatomidae and Coreidae, with some records also for Pyrrhocoridae, Scutelleridae, Cydnidae, Lygaeidae, and Reduviidae.

(j) The prey is well paralyzed but usually remains alive and slightly responsive to stimuli for several days.

§E. Summary of the species of *Bicyrtes* 175

(h) The egg is laid in a semierect position on the mid-ventral line of the first bug placed in the cell.

(1) Mass provisioning is the rule, although several species are known to provision the same cell for more than 1 day, and delayed provisioning (that is, provisioning after the larva has hatched) is known in *quadrifasciata* and is not uncommon in *ventralis.*

It is probable that when the ethology of more species of this genus has been studied in detail, specific differences in ecology and nesting in behavior will become apparent. At the present only a few such differences can be pointed out, and these need fuller documentation.

(1) One species, *quadrifasciata,* makes a much higher orientation flight than is recorded for any of the others, and females return to the nest high in the air and descend slowly and obliquely to the entrance. There are brief suggestions that other species may be distinctive in this regard (for example, *parata*), but several species merely fly swiftly to the nest rather close to the ground (*ventralis, fodiens*).

(2) A large number of prey records for *quadrifasciata* and *ventralis* indicate specialization on immature bugs alone, as do less complete records for *parata* and *discisa.* Many records for *fodiens* show use of both adults and immatures, as do the few records available for *variegata* and *capnoptera; burmeisteri* has been recorded with adults only.

(3) As mentioned under (f), some species appear to show a greater tendency to build multicellular nests than others. Only *quadrifasciata, ventralis, fodiens,* and *variegata* have been fairly well studied in this regard.

Chapter VIII. Three Genera of Bembicini with Recessed Ocelli: *Stictiella, Glenostictia,* and *Steniolia*

The wasps covered in this and the following chapter were included by Handlirsch and other early authors in the genus *Monedula* Latreille, a preoccupied name for which Illiger proposed the new name *Stictia*. An exception to this statement is provided by the genus *Steniolia*, which was recognized as a separate group by Thomas Say in 1837. It was Parker (1917, 1929) who recognized that *Monedula* contained numerous diverse elements and proposed several new generic names for these elements.

It is both convenient and logical to divide these segregates of the old genus *Monedula* into two major complexes: (1) those genera having the anterior ocellus flush with the surface of the head or on an elevation, also having tergite VII of the male strongly excised on each side, as well as several other features in common (*Stictiformes* of Lohrmann, 1948); and (2) those genera having the anterior ocellus depressed and surrounded by a rounded elevation (*Stictielliformes* of Lohrmann). The members of the first group have their center of distribution in South America, and are treated in Chapter IX. The members of the second group have their center of distribution in the southwestern United States and in Mexico, and are considered in the present chapter.

In splitting off the genus *Stictiella* from *Monedula*, Parker pointed out the similarity of these wasps to *Steniolia*. Subsequent work has shown that some species of *Stictiella* in Parker's sense are very close indeed to *Steniolia;* in fact, certain species described since Parker's time tend to occupy an intermediate position between these two genera. At the present time the Bembicini with recessed ocelli are placed in five genera, two of which (*Microstictia* and *Xerostictia*) are of unknown ethology (Gillaspy, 1963b).

The genera considered in this chapter are generally smaller and more slender wasps than *Stictia* or *Bembix*. Many of them also tend to have unusually short wings, a fact doubtless correlated with the whining sounds produced by the females when hunting or when flying about the nest. Certain Old World species of *Bembix* are also relatively short winged, apparently paralleling the development of shortened wings in *Stictiella, Glenostictia,* and *Steniolia.* The genera with recessed ocelli also differ consistently from *Stictia* and its allies in several features of the male terminalia: tergite VII is unmodified except for strongly separated lateral spiracular lobes; sternite VIII is produced into three apical prongs and sometimes also a fourth, ventral prong; and the genitalia have the cuspides prolonged so as to rival the digiti in length.

My treatment of the three genera considered here is uneven. Accounts of the ethology of *Stictiella* and *Steniolia* have recently been published elsewhere, and it seems sufficient to summarize rather briefly what is known of those genera. In the case of *Glenostictia,* I have taken this occasion to present considerable new information. A brief review of the ethology of the three genera is presented at the end of the chapter.

A. Genus *Stictiella* Parker

In this genus the mouthparts are only moderately prolonged and are capable of being fully or almost fully retracted beneath the clypeus; the palpi show no reduction in segmentation. The anterior ocellus is truncate in front and is located in a rather broad depression. The arolia are small, often nearly absent. The middle femora of the male are serrate beneath, and the basal segment of the middle tarsus of the male is strongly curved and beset with several strong spines. These modifications of the middle legs apparently represent devices for holding the female during copulation. Sternite VIII has a ventral prong in only one species, where it is small (Figs. 95–102).

The larvae of *Stictiella* have the body very weakly setose, the spiracles all subcircular; the mandibles have only three teeth; the spiracular atria are lined with spinose ridges; and the epipharynx resembles that of the Gorytini (but differs from that of all Bembicini save *Glenostictia*) in having the median area wholly spinulose.

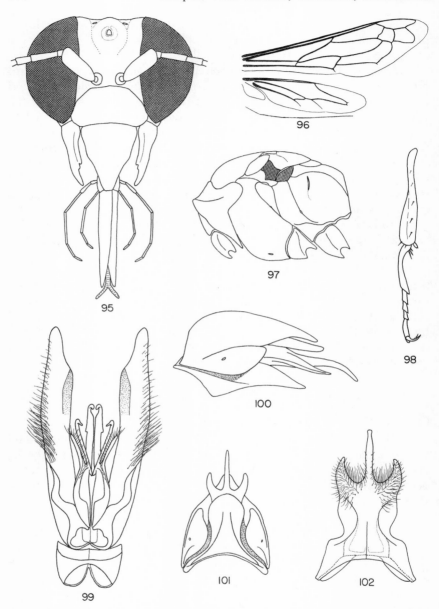

Figs. 95–102. *Stictiella formosa* (Cresson): 95, head of female, mouthparts fully extended; 96, wings; 97, mesosoma of female, lateral aspect; 98, tibia and tarsus of middle leg of male; 99, male genitalia; 100, apex of metasoma of male, lateral aspect, showing lateral spiracular lobes of tergite VII; 101, apex of metasoma of male, ventral aspect; 102, sternite VIII of male.

B. Ethology of the species of *Stictiella*

Gillaspy, Evans, and Lin (1962) have recently discussed the available data on members of this genus, and I shall present only a brief review of their paper. I have included here a few additional notes gathered since this paper was published, also some recent additions by Krombein (1946b) to knowledge of *serrata*. Four species have been studied in some measure: *formosa* (Cresson), *serrata* (Handlirsch), *evansi* Gillaspy, and *pulchella* (Cresson) (= *melanosterna* Parker; see Gillaspy 1963c). Prey records are available for two others: *callista* Parker and *emarginata* (Cresson).

(1) *Ecology and general features of behavior.* The species of this genus typically occur in areas of rather dry, sandy soil. Both *formosa* and *pulchella* are known to nest along watercourses in Texas, tending to nest in somewhat looser soil than *Glenostictia scitula*, discussed in a later section of this chapter. I have found *evansi* nesting in tracts of fine-grained sand behind coastal dunes in Sinaloa, Mexico, *serrata* nesting in broad expanses of light sand in central Florida. Nests are typically dug in flat or slightly sloping sand. These wasps appear to occur in very small aggregations, their nests more or less solitary or at least very widely spaced. Not more than three nests of any one species have as yet been found in any one locality.

I have at times collected both sexes of *evansi* Gillaspy, *plana* (Fox), and other species on flowers in considerable numbers. The flowers visited include Compositae, Labiatae, and Asclepiadaceae. Data on museum specimens indicate that many other types of flowers are visited.

There is no evidence that members of this genus spend their inactive periods inside the nest. Nests of *pulchella* and *serrata* dug out in the evening did not contain adults. Gillaspy found that males and females of *pulchella* cluster on vegetation. Near Bingham, New Mexico, he noticed the wasps flying about *Dalea* plants in the late afternoon

in a slow, searching manner, and finally settling for the night . . . Foci of aggregation were usually places of branching in the upper central part of the plant. Wasps rested close to one another, heads inclined upward to at least some extent but never with the body vertical. Layering and physical contact were scarcely at all present, although one case was noted where individuals were in a partial double layer around a place of branching. As darkness set in as many as possible of the wasps were collected, this being easy to accomplish because of their considerable torpor. Fourteen clusters or aggregations comprising a total of 78 wasps (8 ♂ ♂, 70 ♀ ♀)

were taken over an area some hundreds of feet across in which *Dalea* was the dominant plant growth, but six of these were solitary individuals, either on the same or on different plants from other wasps. The average cluster then consisted of nine wasps, disregarding solitary individuals. (Gillaspy, Evans, and Lin, 1962).

Females flying about the nest often produce an audible sound with their wings, described by Gillaspy, in *formosa*, as "a loud, high-pitched whine."

(2) *Nesting behavior.* Few detailed observations have been made on digging. A female *formosa* observed by Gillaspy near Lajitas, Texas, was adding a new cell to her nest in the late afternoon. This wasp came out the entrance backward at intervals, digging.

She was followed through several cycles of operation as she dug within until the entrance was blocked, then pushed her way out backward and continued backward to various distances up the mound in front of her burrow entrance. She would then begin throwing the soil behind her as she returned and entered the tunnel, holding her antennae close down to the soil as she progressed over it. There was no deviation from this pattern except once, when in place of entering she turned and went up the other side of the mound, then made her way down again throwing soil away from the entrance as before.

After a short flight the wasp made a quick closure, then took another short flight, progressing gradually from the nest, probably for orientation, before flying off to take her first prey for the new cell. There was no true leveling of the mound.

Observations made by me at the Archbold Biological Station, near Lake Placid, Florida, indicate that leveling does occur in *serrata*.

One female was seen digging several short "trial" burrows before finally completing a nest, a feat which took her about an hour. Following completion of the nest, the mound of sand at the entrance was leveled and the surface smoothed off. The wasp first landed at the far end of the mound and worked toward the entrance, zigzagging and kicking sand. This was repeated for 2 or 3 minutes, and the wasp then turned about and began going out in various directions from the nest entrance kicking sand toward the entrance. After about 10 minutes the surface was quite smooth and the entrance well concealed. The wasp then made a brief orientation flight and flew off to catch her first moth (Gillaspy, Evans, and Lin, 1962).

In the case of *pulchella*, Gillaspy made no note of mounds at nest entrances observed near Lajitas, Texas, and a nest excavated by me

§B. Ethology of the species of *Stictiella*

near Presidio, Texas, in May 1963, showed no evidence of a mound at the entrance. The nest of *evansi* which I reported upon in 1962 showed no mound at the entrance. More recently, in August 1962, I found an additional nest of *evansi* in the same area, near Mazatlán, Mexico. The female was observed digging at 0900 in a sloping sandbank in the midst of a dense growth of Acacias. I returned to the spot at 1130 and found the nest entrance to be well concealed and with no evidence of a mound at the entrance. I dug out the nest at that time and found in the cell a moth bearing the egg of the wasp. Thus it appears that leveling of the mound occurs in at least three species of the genus, and so far as is known it occurs following completion of digging (in contrast to *Glenostictia*).

(3) *Nature and dimensions of the nest*. The nest is a short, oblique burrow terminating in an almost horizontal cell. In *formosa*, a relatively large species, the cell is reported to be 35 mm long, in small species such as *pulchella* and *serrata* only 20-25 mm long. However, the nests of *formosa* are not notably deeper than those of the other species, as shown in Table 16. Of the nests recorded in the table, one of *evansi* and one of *pulchella* have been added since the publication of our paper in 1962, also Krombein's data on one nest of *serrata* from the same area in Florida where our studies were made.

Although Lin found a nest of *formosa* near Austin, Texas, and a second at Lake Texoma, Oklahoma, both with only one cell, Gillaspy located two multicellular nests of this same species at Lajitas, in western Texas. One of these nests had 17 cells, a remarkable number for any digger wasp. Lin's nests were dug out early in the season (June and July), and it is possible that more cells would

TABLE 16. NEST DATA FOR SPECIES OF STICTIELLA

Species	Locality	No. of nests	No. cells per nest	Length of burrow (cm)	Depth of cell (cm)
formosa	E. Texas; Okla.	2	1	20	11
	W. Texas	2	5, 17	—	8-14
serrata	Florida	3	1	14-28	8-11
evansi	Sinaloa, Mexico	2	1	15-24	11-16
pulchella	Lajitas, Tex.	3	1	—	12-14
	Presidio, Tex.	1	2	25-32	16-20

have been added to these nests. Gillaspy's nests were dug out in September. Gillaspy found the cells to be rather far apart, but presented no precise data on this point.

Three nests of *pulchella* excavated by Gillaspy at the same time and locality were all unicellular. However, a nest of this species that I dug out near Presidio, Texas, on 3 May 1963, appeared to contain two cells. This female was seen provisioning in the early afternoon, and I dug the nest out at 1900. The burrow was traced to a cell at a depth of 16 cm containing three moths, one of them bearing an egg. Several centimeters beyond, at a depth of 20 cm, I located another cell which contained several moths in a rather poor state of preservation. I could find no trace of wasp egg or larva. It is possible that the wasp had failed to oviposit in this cell or that the egg had been destroyed. Since the cells were reasonably close together and both stocked with the same species of moth, it seems probable that they belonged to the same nest.

When Krombein dug out a *serrata* nest, he found a second cell 15 cm away, with no evident connection with the first cell. It seems possible that this species may also prepare more than one cell per nest, at least at times.

The four species of this genus which have been studied all maintain an outer closure of the nest at all times when the female is not actually in the nest. However, no one has reported an inner closure for any of the species.

(4) *Hunting and provisioning.* Both *formosa* and *serrata* have been reported hunting about vegetation. C. S. Lin (Gillaspy, Evans, and Lin, 1962) watched a female *formosa* attack a butterfly on flowers of *Lippia:* "She followed it within $\frac{1}{3}$ m. and as it perched on the flower, the wasp flew closer and then made a sudden strike and carried the prey off on the wing." I watched female *serrata* "flying about low herbs and tufts of grass. Generally they would approach these plants from the downwind side, swerve back and forth close to them, then dart off to an adjacent plant. Undoubtedly they were hunting for moths. One wasp was seen carrying a pyraloid moth to its nest."

The prey is carried to the nest in flight, the middle legs providing the major support as usual in this subfamily. During transport, the wings of the moth or butterfly are folded close against the body or slightly spread. The wasp typically lands directly in front of the entrance, opens the burrow in the usual manner, and enters, generally remaining within less than 1 min.

Records for six species indicate use of various moths and butter-

§B. Ethology of the species of *Stictiella* 183

flies as prey. The detailed records presented by Gillaspy, Evans, and Lin (1962) need not be repeated here. *S. formosa* has been found to use the butterfly *Libythea larvata* Strecker (Nymphalidae), as well as smaller butterflies of the families Nymphalidae and Lycaenidae and several species of skippers (Hesperiidae). *S. callista* has been taken with a lycaenid butterfly as well as a noctuid moth, *pulchella* also with a lycaenid butterfly as well as with moths of the families Noctuidae and Pyraustidae. The other species have been taken with small moths only: *emarginata* with a noctuid; *serrata* with Phycitidae, Olethreutidae, Epipaschiidae, Pyraustidae, and Crambidae; and *evansi* with Noctuidae, Pyraustidae, and unidentified gelechioid. The two nests I have excavated since the 1962 report was written have added nothing to the spectrum of known prey. Krombein (1964b) listed five species of moths taken from a *serrata* nest in Florida; these belong to three families and mainly to different genera and species than I recorded earlier.

The number of prey per cell varies from 7 to 11 in *formosa*, from 12 to 21 in *serrata*; a fully provisioned nest of *pulchella* contained 19 moths, a similar nest of *evansi*, 15 moths. Gillaspy reported that in *formosa* the butterflies were placed in the cell in a "single, overlapping series extending the length of the cell, all head-in, venter-up, wings folded ventrad. It seemed evident the wasp larva could thus progress readily down the line of contiguous prey bodies, the inedible wings out of the way to the side until finally . . . incorporated like overlapping shingles into a bagworm-like . . . outer sheath of the cocoon." In *serrata, pulchella,* and *evansi* the moths are placed in the cell head in and mostly venter up, but in a more or less layered pile.

Prey taken from cells of these wasps usually appear deeply paralyzed or dead. However, Gillaspy found one fully stocked cell of *formosa* in which none of the prey were completely paralyzed; when he dug it out the butterflies "created pandemonium by flopping aimlessly but vigorously around on the ground." He believed that the wasp associated with this nest was senile and behaving erratically.

Both Lin and Gillaspy found fully provisioned cells of *formosa* containing an egg, thus indicating mass provisioning. Two nests of *serrata* that I dug out both appeared to have been mass provisioned. However, the nest studied by Krombein contained a wasp larva one-third grown, although this nest was still being provisioned. Krombein reports that the weather had been favorable for several days, suggesting that this was true progressive (rather than "delayed") provisioning. In the case of *evansi*, I reported digging a nest still being provisioned and finding a small larva. Another nest of this species, dug out in August 1962 shortly after it was prepared, contained a single egg attached to the initial prey. Since this nest had been closed very carefully and the wasp had not returned by 1130, it seems unlikely that it would have been fully provisioned that same day. In the case of *pulchella*, Gillaspy's observations suggested mass provisioning. A nest of this species dug out in May 1963 near Presidio, Texas, was being provisioned from 1230 to 1600, and when it was excavated at 1900 there were only three moths in the cell, one of them bearing the egg. Thus this cell would clearly have been provisioned over more than 1 day. On the basis of the limited available data, it appears that *Stictiella* resembles *Bicyrtes* in that mass provisioning is the rule, but that at times a cell may be provisioned over more than 1 day, and on occasion the egg hatches before provisioning is completed (delayed provisioning).

(5) *The egg and development of the larva.* The egg has been found in *formosa*, *serrata*, *evansi*, and *pulchella*, and in each case it has been found to be attached to the side of the thorax of the first moth or butterfly placed in the cell. The egg extends more or less vertically upward, in contrast to *Bicyrtes* but resembling those *Bembix* that oviposit on the prey. The egg hatches in about 2 days, and the small larva remains attached at the same point for a day or two. One larva of *serrata* was found to reach maturity after only 4 days of feeding, while Krombein records 6–7 days for this species. The cocoon of *serrata* is very similar to that of *Bicyrtes* and *Bembix*. Gillaspy has indicated the presence of an outer sheath surrounding the cocoon of *formosa*; presumably this is similar to that occurring in the genera *Glenostictia* and *Steniolia*.

Final closure of the nest has not been described for any member of this genus. Nests of *serrata* and of *pulchella* that were dug out after the final closure had the burrow solidly filled with sand and the nest entrance well smoothed off.

(6) *Natural enemies.* I found female *serrata* being followed by miltogrammine flies (*Senotainia* sp. nr. *rubriventris* Macq.) in Florida. One cell was "swarming with maggots" and eventually yielded 26 flies of this same species. A nest of *evansi* was also infested with maggots (unidentified), as was one of the *formosa* cells dug out by Gillaspy. Gillaspy also found mites in one *formosa* cell.

§C. Genus *Glenostictia* Gillaspy

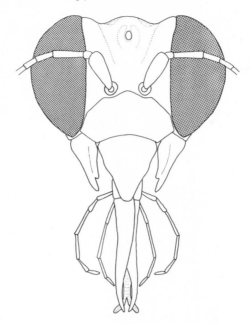

Fig. 103. *Glenostictia scitula* (Fox), head of female, mouthparts fully extended.

C. Genus *Glenostictia* Gillaspy

These are small wasps having many features in common with *Stictiella*. However, the arolia are of normal size, and the anterior ocellus somewhat elongate and placed in an elongate depression. The mouthparts are similar to those of *Stictiella* except that the clypeus is more flattened. In the male the middle femora are not serrate beneath, and sternite VIII has a ventral prong (rarely reduced to a carina). The larvae differ from those of *Stictiella* in having the spiracular atrium lined with nonspinose ridges and the galeae nearly or quite as long as the maxillary palpi (Figs. 103 and 104).

Fig. 104. *Glenostictia scitula* (Fox), sternite VIII of male in lateral aspect, showing the four prongs.

D. Ethology of the species of *Glenostictia*

Erection of this genus was suggested primarily by ethological data, since it appeared that (in contrast to *Stictiella*, from which the genus was split off) these wasps all practice progressive provisioning and utilize Diptera or Hymenoptera as prey. Further studies of the genus have reinforced our belief in its integrity, although it now appears that *G. scitula* (Fox) employs an even broader spectrum of prey than we had appreciated. The present account includes a much more detailed study of *scitula* as well as a few notes on a previously unstudied species, *gilva* Gillaspy, and a few additional notes on the type species, *pulla* (Handlirsch).

Although these three species are the only ones to have been studied in the field, study of museum material indicates that two others employ Diptera as prey. These are *clypeata* Gillaspy and *pictifrons* (Smith). The former has been recorded with flies of the families Apioceratidae and Syrphidae, the latter with Bombyliidae (for details see Gillaspy, Evans, and Lin, 1962). *G. pictifrons* has been reported sleeping on vegetation in a mixed aggregation of wasps and bees (Evans and Linsley, 1960). Actually only one female was found, and on only one night, so its appearance in this mixed aggregation may have been exceptional.

E. *Glenostictia pulla* (Handlirsch)

This species is widely distributed in the Upper Sonoran and Transition zones west of the Rockies. It has been studied by La Rivers (1942) in Nevada, by Gillaspy in Idaho and California, and by myself in Utah (Gillaspy, Evans, and Lin, 1962). The species is characteristic of small to large bare areas where the soil is a light, powdery sand or sandy loam.

Gillaspy found both sexes clustering during July on live Russian thistle (*Salsola*) in Jerome County, Idaho, only a few meters from their nesting area. One cluster contained about 20 wasps, another about 40; the clusters were "densely formed, with many of the wasps in physical contact, often one above the other." On 10 August the two clusters were still about the same size, but on 31 August there were only 25 wasps (16 ♂♂, 9 ♀♀) in the two clusters together. The presence of so many males late in the season suggests that the males may be longer lived than in some other bembicine wasps.

La Rivers watched a female dig briefly in four places before staying in one place and making a burrow. The burrow was started at 0910 and apparently finished at 1027. During digging, the wasp

§E. *Glenostictia pulla* (Handlirsch)

reappeared at the entrance from time to time and "turned about and scattered the pile of debris in all directions . . . meanwhile pivoting swiftly about . . . Each time the wasp backed into the open to clean the tunnel, she rose into the air for a short observation flight, which generally consisted of flying directly up from the opening, hanging motionless in the air for half a second, then dropping back to disappear down the hole." When the burrow was finished, the wasp "moved about over the ground at the entrance of the tunnel, and repeatedly scattered dirt in all directions."

I watched a female starting a nest in a sandbank near Cornish, Utah, at 1030 on 21 August 1961. During digging, this wasp emerged periodically from the burrow and leveled the accumulated sand by rotating to the left or to the right while kicking sand, thus dispersing it widely and allowing no mound to accumulate (Fig. 105). My brief observations and those of La Rivers suggest that leveling is interspersed with digging and resembles closely that of *G. scitula*, described at length under that species below.

The nest I observed in Utah was excavated at 1230 and found to contain a single fly bearing the egg of the wasp attached upright to the side of the thorax. The burrow was straight, 11 cm

Fig. 105. *Glenostictia pulla* leveling the soil at the nest entrance (no. 1831, Cornish, Utah). Note that the wasp has turned to one side so that her body is oblique with respect to the burrow (opening at extreme left).

long, the cell 6 cm beneath the surface; the cell measured 11 × 25 mm. There was no inner closure, and the female was not inside the nest when it was dug out. In Nevada, La Rivers dug out 25 nests and found the burrow length to vary from 25 to 29.5 cm (mean 27.3 cm). Both La Rivers and Gillaspy showed that the nests are provisioned progressively.

The one nest I studied contained a bombyliid fly, *Lepidanthrax* sp. (erroneously reported as *Villa* sp. in 1962). Gillaspy reported Bombyliidae, Therevidae, Syrphidae, Muscidae, Sarcophagidae, and Tachinidae as prey in Idaho, as well as a tachinid in Alameda County, California; La Rivers found certain Syrphidae and Sarcophagidae to be used in numbers in Nevada (for detailed records see their papers).

La Rivers found two species of miltogrammine flies resting about the nest entrances and occasionally being driven off by female *pulla*. He found no evidence that they successfully attack this wasp.

F. *Glenostictia gilva* Gillaspy

This species is closely related to *G. pulla*; it was called *exigua* by Parker (1917), but Fox's *exigua* is quite a different species, now assigned to the genus *Microstictia* (Gillaspy, 1963b). I found two individuals of this species nesting on the periphery of the *G. scitula* colony 1.5 km west of Lajitas, Texas, described under that species below. One female was seen bringing flies to her nest on 24 April 1963. She was captured and the nest dug out on that date. On 30 April a female was seen digging a nest about 2 m away; she brought in a fly later that day, but this nest was accidentally destroyed a few days later. These were the only two females of this species seen nesting in this area.

The following account of digging and leveling is drawn from the second of these wasps (no. 1874). This wasp was digging at 1130 at the edge of the *scitula* colony, in soil that was somewhat more powdery at the surface than that species seemed to prefer. She kept the entrance cleared most of the time by scraping sand while standing just outside the nest and facing the burrow. Occasionally she turned toward one side or the other, much less frequently than in the case of *scitula* or *pulla*. While digging at the entrance, she threw the sand 5–7 cm behind her, mostly to form a small, diffuse pile a short distance in front of the entrance. When the nest was finished, at 1300, the wasp made a partial closure from the outside, then proceeded to the far end of this small mound and worked forward in an irregular zigzag pattern, scattering the sand widely. From time to time, between the leveling movements, she added more soil to the closure. After several minutes the mound was completely dispersed and she flew off, to return after only about 10 min with a bombyliid fly on which the egg was undoubtedly laid.

The story may be continued by turning to the first nest (no. 1855). This wasp was seen bringing in a fly at 1125, and the nest was dug out 20 min later. The cell contained a very small larva, probably hatched earlier that day, still attached erect to the side of a small fly. There were four other flies in the cell. All of the flies were bee flies (Bombyliidae); one was identified as *Aphoebantus interruptus* Coquillett, the other four as *Aphoebantus* sp.

The nest was a simple, straight burrow 11 cm long, terminating in an oblique cell 7 cm beneath the surface. There was no inner closure, but both nests had an outer closure at all times when the female was not actually inside the nest.

G. *Glenostictia scitula* (Fox)

This species is characteristic of Lower Sonoran deserts from western Texas to southern California. It is apparently a vernal, univoltine species, the adults appearing in April and being active for about 2 months. Collections made within its range during the summer and autumn months do not normally include the species; however, Gillaspy (in Gillaspy, Evans, and Lin, 1962) mentions one melanic specimen taken in August in mountains surrounding the Colorado Desert of California.

Gillaspy (in Gillaspy, Evans, and Lin, 1962) briefly reported on sleeping, feeding, and nesting behavior of populations of this wasp studied in 1961 at several localities in western Texas. His observations suggested several unusual features in the behavior of this wasp. Since further study of the species seemed desirable, Gillaspy and I proceeded in late April 1963 to the locality 1.5 km west of Lajitas, Brewster County, Texas, where Gillaspy had worked in 1961. We found the species to be very abundant here, and took 26 field notes, including data on 34 nests. On 2 and 3 May we found the species nesting in smaller numbers about 5 km east of Presidio, Texas, some 70 km west of the first locality; here we made four notes, including data on 12 nests. Finally, on 4 May we made one field note at McNary, Texas, roughly 300 km northwest of Lajitas.

(1) *Ecology and general features of behavior.* The three localities in which the 1963 studies were made were all within a short distance of the Rio Grande, the Lajitas colony actually only about 30 m from the

river bank. Two additional localities visited by Gillaspy in 1961 were also close to the Rio Grande, a third locality along Terlingua Creek only a few kilometers north of the river. The wasps were encountered in greatest numbers in the flat sand or clay-sand of the river valley, above the banks of the river and its tributary draws but in places that have undoubtedly been flooded at brief intervals in the past. All nests were found in soil with a slight crust on the surface. Near Lajitas the major nesting places (Fig. 106) were in slightly encrusted, sometimes pebble strewn, but otherwise very powdery clay-sand in bare spaces among creosote bushes (*Larrea divaricata*). Near Presidio the soil in general was more sandy and friable, but the wasps nested only in certain localized spots which were more stony and firm. At McNary we encountered the species in an expanse of partially stabilized sand dunes. Here the species was less abundant and appeared to nest only in the bottoms of certain blowouts among the dunes, again places which were stony and slightly encrusted on the surface.

Both sexes of *scitula* visit certain flowering shrubs for nectar, often in great numbers. Gillaspy speaks of them as producing a "pronounced hum, somewhat like a swarm of bees"; he notes that the hum of *scitula* is higher in pitch than that of worker honeybees but

Fig. 106. Nesting area of *Glenostictia scitula* in Rio Grande Valley near Lajitas, Texas. The plants in the foreground are creosote bush (*Larrea divaricata*); bushes in the background include *Acacia, Prosopis,* and yucca.

§G. *Glenostictia scitula* (Fox)

lower in pitch than that of *Steniolia duplicata,* "correlating inversely with the apparent ratio of wing area to body bulk."

Catclaw (*Acacia greggii*) was found to be a favored nectar plant at several localites. At Lajitas, both sexes swarmed about catclaw all day long (0900-1800), to a lesser extent around tamarisk (*Tamarix gallica*). Near Presidio, desert willow (*Chilopsis linearis*) and a species of *Croton,* as well as catclaw, were being visited in numbers. At McNary screw bean (*Prosopis pubescens*) appeared to be the major nectar plant, and Gillaspy mentions mesquite (*P. juliflora*) and bluethorn (*Condalia lycioides*) at other localities. The active period of *Glenostictia scitula* coincides with the time of flowering of these trees and shrubs. Gillaspy visited Lajitas on 31 May 1961, when there was little bloom of any kind left, and was unable to find any specimens of *scitula.*

As noted by Gillaspy in 1961 and by the two of us in 1963, the males sleep on vegetation but the females typically do not (specimens can be sexed very readily in this species, as the color pattern is radically different). We did, on one occasion, fine one female and one male sleeping side by side at the top of a creosote bush, but on no other occasions did we find females spending the night elsewhere than in their nests.

In 1961 Gillaspy located a sleeping cluster at the mouth of Maravillas Canyon, Brewster County, Texas. He noted males flying around stems of globemallow (*Sphaeralcea angustifolia*) at 1730, later forming a cluster there about 75 cm above the ground. "The cluster was compact, all individuals in physical contact and surrounding some two or three wasps as a nucleus." The cluster was found to consist of 20 males. About 2 weeks later, Gillaspy found a cluster of four males at Lajitas, at about the same height from the ground but on creosote bush (*Larrea*). These individuals were "in close proximity, but not in physical contact."

At Lajitas in 1963, we found four clusters of males, three on the top branches of creosote bushes, one on a cholla cactus (*Opuntia* sp.). The clusters were all small (four to ten individuals) and all were about or slightly less than 1 m from the ground. In each case the wasps were close together but with little or no bodily contact; commonly the wasps occupied the tips of several closely adjacent branches of the bush. The bodies of the wasps were horizontal or more often with the anterior end of the body tilted upward slightly (Fig. 107). The largest of these clusters was in place each of the 8 nights we were in the area, though varying slightly in numbers from night to night; the others were of more temporary nature, persisting for only 1 to 4 nights. Considering the abundance of

Fig. 107. A sleeping aggregation of 11 male *Glenostictia scitula* on top branch of a creosote bush (in the area shown in Fig. 106).

males at catclaw during the day, there must have been many more sleeping clusters than we were able to find. It is noteworthy that none of the plants used for clustering were visited for nectar during the day (although the *Larrea* was in full bloom). Bushes selected for clustering were generally somewhat isolated ones. The clusters began to form each day shortly before dusk. Each morning the males became restless about 0830 and flew off by 0900.

Since Gillaspy had earlier reported females spending the night in their burrows and since such behavior is otherwise unknown in this section of the Bembicini, we devoted considerable attention to this problem. As already mentioned, one female was found sleeping on vegetation, but this individual did not return a second night. It is possible that early in the season, before the females have established nests, the two sexes may cluster together, but during the period of study all but this one female were found sleeping in their nests.

Just before dusk each evening (1700–1830) females were seen entering their brood nests and closing the entrance behind them. Actually, these closures were rarely complete; often a small hole was apparent through the top of the closure. Since there was of course no smoothing over of the closure from the outside, after some prac-

§G. *Glenostictia scitula* (Fox)

tice we were able to spot these nighttime closures readily. A nest closed in this manner could be counted upon to contain a resting female between dusk and 2 or 3 hr after sunrise.

In order to ascertain the position of the female in the nest during the night, we dug out 19 nests in the early morning hours (0700–0800). In all but one of these nests we found a torpid female about halfway down the burrow, facing toward the entrance. Some of these nests were new the previous day and contained an egg; others were older nests containing small to nearly fully grown larvae. The one nest that contained no female (no. 1892E) had a larva about two-thirds grown in the cell; possibly this female had been overtaken by darkness before reaching her nest. The position of the females in the burrow showed no important variation: without exception the females were about halfway down the burrow, roughly equidistant from the cell and the entrance. The 18 females varied in depth (that is, distance from the surface directly above) from 3 to 5 cm (mean 4.3 cm).

After the females emerge from their nests in the morning (0830–0930), they first clear the entrance of soil, then make a closure from the outside. Provisioning may occur at any time of day, but is at its height in the morning (0830–1300) and late afternoon (1600–1800). Wasps feeding a large larva bring in prey sporadically all day long except during cloudy intervals. Final closures are made during the morning hours (0830–1130); in fact, none at all were observed in the afternoon. A period of digging of new nests follows that of final closures; most new nests were found to be dug between 0930 and 1230. A few females were seen starting new nests in the late afternoon. In at least one case a female is known to have slept in an uncompleted nest and then to have finished it the following morning, but this behavior appeared to be uncommon.

In the center of the major nesting areas near Lajitas and near Presidio, nest entrances were often separated by no more than 2–5 cm. At the periphery of each area, the nests were more scattered. In an area 2 m^2 in the center of the greatest concentration of nests at Lajitas, we counted 25 nests. Several marked females at Lajitas prepared a second nest within a short distance (up to 2 m) from a completed nest. In all three localites, nests were grouped in several small to fairly populous colonies separated by fairly wide intervals with no nests at all, soil conditions apparently determining the distribution of nesting sites. We observed no aggression among females, even those nesting in close proximity, and in no case did we observe females attempting to steal prey from one another.

(2) *Reproductive behavior.* Males were extremely abundant at catclaw and other flowers, and as already described were found clustering at night, chiefly on creosote bush. However, they were notable by their absence within the nesting areas of the females. It seems possible that most mating occurs at the very beginning of the active season, and it may occur at the clusters, as described for the genus *Steniolia*, the females sleeping in their nests only after they have mated and begun their nesting activities.

We did observe brief male-female contacts on two occasions. Once we saw a pair come together at flowers of catclaw, but they separated almost immediately. On another occasion, near Presidio, we saw a male enter a small nesting area and attempt to mate with several females in quick succession. In one case the pair remained together for several seconds, rose into the air about 1 m, and then separated. Evidently none of these were successful matings.

(3) *Digging the nest.* Females often bite into the soil at several different points before remaining in one place and constructing a burrow. Now and then burrows are abandoned after reaching a depth of several centimeters.

In the initial stage of digging much use is made of the mandibles in breaking through the surface crust. At this time the wasps produce a series of fairly loud "chirps." The body is turned from side to side, the head sometimes being completely inverted, as the wasp bores a circular hole through the crust. As the burrow deepens, the female backs up the burrow periodically scraping back loosened soil with her forelegs. Frequently she backs out the entrance, takes a short flight, lands in the entrance, and levels the soil which has accumulated at the entrance. She does this by remaining directly in front of the entrance, facing to the right or left, turning in an arc, and scraping soil with her forelegs in such a way that it lands 5–8 cm behind her (Fig. 108). After a few seconds she takes another short flight and then re-enters the burrow. This is repeated over and over until the nest is completed, no mound of soil being allowed to accumulate at the entrance at any time.

As documentation of this unusual leveling behavior, the following extract from my notes (no. 1891), dictated into a tape recorder on the morning of 26 April, is presented. This wasp began her nest at 1020; the burrow was 1–2 cm deep when this extract of a 10 min period begins:

1043.30. Wasp came out, made a very short flight; back at entrance, leveling while facing to right at about a 90° angle to the burrow; very

§G. *Glenostictia scitula* (Fox)

Fig. 108. Female *Glenostictia scitula* leveling the soil at the nest entrance (Lajitas, Texas). The body is turned at an angle to the direction of the burrow, the entrance of which is seen behind the wasp (*cf.* *G. pulla,* Fig. 105).

short flight; leveling facing right again, 90–100°; very short flight; back, entered nest again and is digging.

1044.10. Came out, flew in two big loops close to the ground; back in 5 sec, leveling facing right, 90–120° from direction of burrow; short flight; leveling facing right again 45° down to 10°; short flight; leveling facing right again 90°, down to 0° (that is, facing the burrow); then into the nest again.

1045.25. Out, flew in two loops; back in 5 sec, leveling facing left, 10° around to right at 90°; short flight and back in again.

1045.15. Short flight; back, leveling at about 180°; short flight; leveling facing right at about 150°; short flight and back into nest at 1046.45.

1047.15. Flight of three large loops; back, leveling facing right 90° to 120°, then up to 160° and back to 90°; short flight and into nest again.

1048. Short flight; leveling facing left at 90°, around to 0° and remained there for several seconds; short flight; back into nest.

1048.50. Short flight; leveling facing right 120–160°, back to 90°; short flight (only one loop); leveling facing right again 90–160°; very short flight (two loops); leveling facing right again 20–120°; short flight and into nest again.

1050. Short flight; leveling facing left, 90° up to 120°, back to 45°, back to 90°; short flight and into nest at only 1050.15.

1050.45. Short, looping flight; leveling facing left at 90°, all the way around to the right at 120°; short flight; back into nest at 1051.

1052. Short flight; leveling facing left at 120°, worked over to right at 120°; flight of one large loop; back, leveling facing right 20-90°; short flight and into nest. [Nest estimated to be 2-3 cm deep.]

This behavior continued in similar fashion until 1157, when closure and concealment began, as described below. It will be noted that on some occasions the wasp made two short flights separated by a few seconds of leveling (as at 1050 above), at other times three or even four short flights separated by short periods of leveling. These flights were always low, 4-8 cm above the soil. In some cases the wasp merely made a single loop, the far point of the loop no more than 1 or 2 m from the nest; at other times she made two loops in the form of a figure eight, the nest entrance at the center of the "8." Once she took several loops, but some of these were occasioned by a lizard near the nest, which she succeeded in driving away by swooping close to it and hovering about it. Since these wasps make no separate orientation flights, I assume these flights interspersed with digging and leveling serve this function.

Each session of leveling took only a few seconds and involved no more than 6-30 scuffs of the front legs. During the first 30 min this wasp leveled facing right 30 times, facing left 16 times; during the last 30 min she leveled facing right 21 times, facing left 11 times. (The only other wasp followed through the entire digging and leveling period, no. 1860, also had a right:left ratio of about 2:1.) The time spent in the nest between leveling sessions increases gradually as a result of the increasing depth of the nest, starting out about 30 sec and increasing to an average of about 2.5 min. While considerable variation occurs in time spent in the nest or leveling, the length of the flight and number of loops, and so forth, as indicated in the above extract, much this same irregularity appeared characteristic of all wasps observed. Brief observations on several females near Presidio showed their behavior to be very similar.

Following completion of the nest, the wasp remains inside for a short time, then makes an outer closure and undertakes movements of concealment. This behavior was also observed many times, but is best illustrated by resuming the account of no. 1891. This wasp did some rotational leveling at 1156.45, then re-entered her nest. Extracts from my notes beginning at this time follow:

1204. She came out head first (after remaining in nest 7 min), scraped some soil into the entrance, and made a short flight; leveling facing right

§G. *Glenostictia scitula* (Fox)

at 90°; short flight; leveling facing right at 90°; short flight; leveling at 180°; flight of two loops; leveling on right again at 10–20°. Now she took a longer flight, with several figure eights over the entrance; landed and added some more to her closure by working away from the entrance scraping soil behind her.

1206. Still adding to her closure by working away from entrance (facing roughly 180° from direction of burrow); going up to 3 cm from entrance, taking short flights between each trip from entrance. Behavior repeated until 1209.45.

1209.45. She has moved to other side of entrance (side toward direction of burrow), where she has not been before, going out 3 cm from entrance kicking sand toward entrance; short flight; behavior repeated in a slightly different direction; short flight; now off to side up to 5 cm from entrance; short flight; now on a different line in general direction of burrow, up to 4 cm . . .

[This behavior continued until 1212, the wasp describing a series of radiating lines, 3–6 cm long, chiefly in the same direction as the burrow; these lines were not straight, but weakly zigzagged. The making of each line was separated by a short flight.]

1212. Wasp made a few big loops and then hovered over the thoroughly concealed entrance; more loops and more hovering; at 1212.30 much more hovering, only about 2 cm high, actually proceeding in small, slow jerks in an irregular pattern over and about the entrance. Now another big loop, and flew away. Not seen again.

Rotational leveling and associated behavior occurs not only in the case of new nests, but also when a more advanced nest is re-excavated after a cave-in. Concealment by radiating lines followed by hovering occurs also following final closure of the nest, but temporary closures other than the initial one do not involve these elements. Excavation of several nests immediately after the initial closure revealed that the egg was already in place. Thus a portion of the several minutes between completion of the burrow and cell and closure and concealment (that is, between 1156.45 and 1204 in the above example) are clearly spent in oviposition.

(4) *Nature and dimensions of the nest*. All nests dug out were unicellular, and study of marked individuals indicated that the wasps always prepare a new nest after the old one is fully provisioned and closed. The burrow is 4–5 mm in diameter and descends at a 30–50° angle (most commonly about 35°) to a nearly horizontal cell measuring 10–12 mm in diameter by 16–20 mm in length (Fig. 109). Burrow length and cell depth showed little variation in the three localities (Table 17). At Lajitas nearly all burrows were straight or almost so. However, one unusual nest started out at about a 20° angle, then dipped down at a much sharper angle and finally made

Fig. 109. Profile of a nest of *Glenostictia scitula* (Lajitas, Texas, April 1963).

a U-turn into the cell. Gillaspy found one S-shaped burrow in 1961; the five cells he excavated at Lajitas were all between 5 and 7 cm deep. Near Presidio, where the soil contained more pebbles, many burrows curved or angled to the left or right or were somewhat tortuous. None of the 47 nests excavated showed any evidence of an inner closure. However, an outer closure is maintained at all times when the female is away from the nest (no exceptions whatever noted in 1963, although Gillaspy reported females leaving the entrances open in 1961). As already noted, a rather incomplete outer closure is made from the inside when the female is inside during the night.

TABLE 17. NEST DATA FOR GLENOSTICTIA SCITULA

Locality	No. of nests	Burrow length (cm)	Cell depth (cm)
Lajitas, Texas	34	12.2 (10-15)	6.3 (5-8)
Presidio, Texas	12	12.5 (10-15)	6.5 (6-7)
McNary, Texas	1	15.0	6.0

§G. *Glenostictia scitula* (Fox)

(5) *The egg.* As already mentioned, the egg is laid immediately after completion of the nest and before the initial closure. Eggs were found in more than ten nests, invariably newly completed nests containing no prey. The egg is slightly curved and measures about 0.8 mm in thickness by 4.0 mm in length. It is laid longitudinally, rarely somewhat obliquely or even transversely, on the floor of the cell a short distance from its apex. Since this is near the point where the cell floor slopes up to the rounded apex, the egg is actually horizontal or nearly so even though the major part of the cell is somewhat oblique.

During the day following oviposition, the female remains outside the nest much or all of the day, re-entering at night. On the second day a few small prey are placed in the cell close beside the egg. The egg hatches in about 48 hr, and the larva begins to feed on the prey beside it.

(6) *Hunting and provisioning.* Females were observed hunting about creosote bushes and other flowering shrubs and trees, especially *Tamarix* at Lajitas and *Chilopsis* at Presidio. Hunting females were also seen flying about herbs, including a small composite at Lajitas and *Lepidium* at McNary. The hunting flight is rapid and irregular, with moments of hovering and quick dashes at prey, quite different from the behavior of adults seeking nectar, flying more slowly and landing on blossoms and protruding their mouthparts. At Presidio I saw one female snatch a small insect from a *Larrea* blossom, hover in the air, and curve her abdomen beneath and forward to sting it while still in the air. I caught the wasp and her prey; the latter was a small bee of the genus *Perdita* which had not yet succumbed to paralysis and flew away when released. Prey taken from wasps at the nest entrances often showed movements of appendages, but prey taken from nests a few hours after they were captured were invariably dead and often quite stiff.

The prey is carried to the nest by the middle legs, which hold it close beneath the wasp's abdomen. Since this species tends to prey on very small insects, it is often difficult to tell when females are carrying prey. If they stand on only the hind legs while opening the nest with the front legs, one can of course be sure their middle legs are employed in holding the prey. When the wasp is carrying larger prey, the hind legs also support the prey in flight, though not on the ground. Females sometimes pause on the ground at some distance from the nest or just in front of the entrance. At these times one can observe rhythmic extensions and contractions of the abdominal segments, associated with respiration.

Females flying in with prey approach the nesting area rather low (3–10 cm high), with some whining of the wings. They tend to hover a bit over the entrance, then drop abruptly to it. The nest entrance is opened with a few scuffs of the front legs, and the female remains inside for only a few seconds (5–25 sec), closing the nest with a few scuffs of the front legs as she leaves and flying off directly. At times females bring in prey very rapidly, at only 3–5 min intervals, such rapid provisioning occurring chiefly in the morning hours or in the late afternoon. Because of the small size of most prey, large numbers are required. However, it was impossible to make an accurate estimate of the number or prey used per cell, since provisioning in this species is fully progressive. That is, the mother provisions periodically as food is needed by the larva, never permitting a large accumulation of uneaten prey in the cell. It appears that prey is brought in almost up to the moment when the larva is ready to pupate. In no case did we find more than five fresh prey in a cell, and usually only one to three. Nests dug in the early morning hours never contained intact prey. Because of the delicate nature of some of the prey, study of debris in cells or surrounding the cocoons was not an accurate method of determining the numbers employed. Some cocoons had wings and exoskeletons of as many as 40 small bees and flies adhering to them.

Since Gillaspy had discovered in 1961 that this species employs both small flies and small bees as prey, we devoted considerable attention to the nature of the prey. We confirmed Gillaspy's observation that bees of the genus *Perdita* provide the most important single type of prey. To a lesser extent other bees, small wasps, male ants, and even parasitic Hymenoptera are employed. Various small to medium-sized Diptera are also used in some numbers. We were surprised to find still a third order of insects utilized: the order Hemiptera, represented by both the suborders Heteroptera and Homoptera, the latter chiefly by Psyllidae, which are exceedingly small insects to be preyed upon by a digger wasp of this size. There was no important difference in prey type in the three localities studied (Table 18).

While some nests contained mostly or entirely *Perdita* bees, and a few contained only flies, the vast majority of nests contained prey well mixed as to taxonomic affinities, several cells having all three orders represented. Thus it appears that the females fly about vegetation, snatching small insects from flowers and foliage almost without discrimination. We were able to sweep the psyllids and some of the *Perdita* bees in numbers from *Larrea*, and since we often saw

§G. *Glenostictia scitula* (Fox)

TABLE 18. PREY RECORDS FOR GLENOSTICTIA SCITULA

Species of prey	No. of specimens collected		
	Lajitas, Texas	Presidio, Texas	McNary, Texas

ORDER HEMIPTERA			
CICADELLIDAE			
Genus and species ?	1		
PSYLLIDAE			
Heteropsylla texana Crawford	14		2
MIRIDAE			
Neurocolpus arizonae Knight	1		
Psallus sp.		1	
ORDER DIPTERA			
CHIRONOMIDAE			
Procladius sp. nr. bellus Loew	4		
STRATIOMYIDAE			
Dicyphoma schaefferi Coq.	2		
Zabrachia sp.	2		
TABANIDAE			
Silvius quadrivittatus Say	3		
BOMBYLIIDAE			
Mythicomyia intermedia Melander	1		
Oligodranes sp.	1		
Phthiria sp.	1		
ASLILIDAE			
Holopogon phaeonotus Loew	1		
SCENOPINIDAE			
Brevitrichia griseola Coq.	6		
SYRPHIDAE			
Allograpta obliqua Say	1	3	
Mesograpta marginata Say		1	
OTITIDAE			
Euxesta magdalenae Cresson		1	
Euxesta nitidiventris Loew		4	
TEPHRITIDAE			
Trupanea bisetosa Coq.		1	

TABLE 18. PREY RECORDS FOR GLENOSTICTIA SCITULA (CONT.)

Species of prey	No. of specimens collected		
	Lajitas, Texas	Presidio, Texas	McNary, Texas
CHLOROPIDAE			
Conioscinella sp.	2	2	
AGROMYZIDAE			
Melanagromyza sp.		1	
ASTEIIDAE			
Astiosoma sp.		1	
MILICHIIDAE			
Milichia aethiops Mall.	1		
ANTHOMYIDAE			
Calythea micropteryx Thomson		2	
Hylemya platura Meigen		1	
Pegomya longimana Pokorny	1		
MUSCIDAE			
Haematobia irritans L.		1	
Musca domestica L.		3	
SARCOPHAGIDAE			
Eumacronychia sp.	1		
TACHINIDAE			
Siphophyto setigera Coq.	1		
ORDER HYMENOPTERA			
BRACONIDAE			
Apanteles sp.		2	
TORYMIDAE			
Torymus sp.		1	
FORMICIDAE			
Iridomyrmex pruinosus Roger		4	
SPHECIDAE			
Lindenius sp.	2		
Tachysphex sp.		1	
COLLETIDAE			
Colletes sp.	1		

TABLE 18. PREY RECORDS FOR GLENOSTICTIA SCITULA (CONT.)

Species of prey	No. of specimens collected		
	Lajitas, Texas	Presidio, Texas	McNary, Texas
ANDRENIDAE			
Perdita exclamans Cockerell	2		
Perdita knulli Timberlake	1		
Perdita marcialis Cockerell	24	3	
Perdita mentzeliarum Cockerell	4		
Perdita larreae Cockerell	3		
Perdita spp.	30	18	
HALICTIDAE			
Dufourea sp.			1
Lasioglossum sp.	2		

female *scitula* hunting on *Larrea*, we assume much of their prey was obtained from that source.

On a number of occasions we observed females entering their nests without prey early in the morning (0815–0900), then later on provisioning their nests. It is possible that the females leave their nests, take nectar, then return to their nests on an "inspection" trip, to determine the needs of the larva. It should be remembered that the females sleep part way down the burrow, facing out, and they may not visit the cell before emerging. Also, in all other known members of this complex of genera the females do not spend the night in the nest and therefore regularly make such an "inspection" trip in the morning.

(7) *Final closure.* As already indicated, final closure is normally delayed until the larva is fully grown and ready to spin the cocoon; in fact, one nest (no. 1856A) dug immediately after final closure already had a fresh cocoon. Final closures are normally made during the morning hours; some were observed as early as 0830, and none later than 1130.

In the initial stages of final closure the nest entrance is found to be open, the female being inside, presumably filling the lower part of the burrow by scraping soil from the walls of the burrow and packing it in place. In a few minutes the female appears at the entrance, comes out 1–2 cm scraping soil behind her into the burrow, takes a short flight, lands in the entrance, rotates 180°, and backs in scraping soil. This behavior is repeated again and again at intervals of 1-4 min, with some variation in the length and timing of the flight (rarely omitted altogether) and in the distance the wasp comes out from the entrance scraping soil (up to 5 cm). When backing in, the wasp often turns from one side to the other reaming the walls of the burrow with her mandibles and front legs. As the burrow becomes fuller, the wasp spends relatively more time outside the nest and may make several short lines from the nest entrance scraping soil before taking a short flight, landing in the entrance, rotating, and backing in. When the burrow is nearly full, it can be seen that the soil is being packed in place with rapid blows of the tip of the abdomen. The filling of the burrow requires approximately 30 min.

When the burrow is full, or very nearly so, the wasp takes another short flight and, upon returning, undertakes movements of concealment on the opposite side of the nest entrance. She lands at the entrance, then proceeds in a straight or slightly zigzag line away from it on the side toward the direction of the burrow. At a distance of 6-10 cm from the entrance she makes a short flight, lands at the entrance, and makes another "radiating line" at a slightly different angle. After several minutes (2-5) of this behavior, she commences hovering only 1-2 cm above the entrance, often moving somewhat jerkily. Hovering may be interspersed with one or two larger, looping flights and with a small amount of additional scraping of soil over the entrance. After only 1 or 2 min of intermittent hovering, the wasp flies away and is shortly to be seen digging a new nest not far off. As a result of the movements of concealment, the original nest is left completely invisible to a human observer, and excavation invariably reveals the entire burrow to be packed tightly with soil.

The movements of concealment (radiating lines, hovering flight) appear virtually identical to those occurring after completion of a new nest.

(8) *Development of the larva.* About 6 days are required for the larva to reach full size. Nest no. 1876A, dug at Lajitas on the morning of 24 April, was provisioned 26-30 April. This nest was excavated at 1030 on 1 May, a cloudy day with intermittent light showers during which little or no provisioning occurred. The larva was found to be about 18 mm long, so presumably would have reached full size the following day. A second nest, no. 1876B, initiated at the same time as the preceding contained a larva 20 mm long when

excavated on 1 May; this larva may have completed its feeding and been ready to spin its cocoon. Neither of these nests contained any fresh prey, and both contained a resting female in the usual position about halfway down the burrow. Data on several other nests indicated 7–9 days to be the usual duration of a nest, the first 2 days for the egg, the next 5–7 for the larva to reach maturity.

The cocoon of this wasp is contained in a silken shroud which has a funnel-shaped opening at the burrow end, as in the species of *Steniolia*. This shroud is typically covered with the remains of the prey, particularly wings, hollowed out head capsules and thoraces, and abdominal terga and sterna.

(9) *Natural enemies*. Nesting females were occasionally seen to drive away lizards and worker *Pogonomyrmex* ants from the vicinity of their nests. A few miltogrammine flies were observed in the nesting area at Lajitas, but on only one occasion (no. 1864B) did I see a fly following a female *scitula* carrying prey. This fly was captured and found to be *Senotainia* sp. None of the many nests dug out contained maggots or other parasites or inquilines.

At Presidio I saw a female *Parnopes concinnus* Viereck (Chrysididae) flying about the nesting area and landing here and there on the sand. She was not seen to enter any nests. *Steniolia duplicata* was nesting in this same area, and it is possible that the *Parnopes* was attacking this species (rather than, or in addition to, *Glenostictia scitula*). The species of *Parnopes* seem especially to attack Bembicini, and no other bembicines were nesting here at this season. No Mutillidae were observed in any of the nesting sites of this wasp.

H. Genus *Steniolia* Say

Wasps of this genus are well known for their remarkably long mouthparts. When retracted, the maxillae and labium extend far beyond the end of the clypeus, reaching at least as far as the middle coxae; when extended, the proboscis is capable of reaching the base of the abdomen. The palpi are much reduced, the maxillary palpi having three segments, the labial palpi one or two. The clypeus is somewhat swollen toward the base. Although these mouthparts appear to contrast strongly with those of *Stictiella* and *Glenostictia*, the genus *Xerostictia* Gillaspy (1963) is somewhat intermediate, having a slightly shorter proboscis than *Steniolia* as well as four-segmented maxillary palpi. In virtually all basic features other than the mouthparts, but including the ocelli and the male terminalia, *Steniolia* closely resembles *Glenostictia*. The larvae are similar to those of

206 Chapter VIII. *Stictiella, Glenostictia,* and *Steniolia*

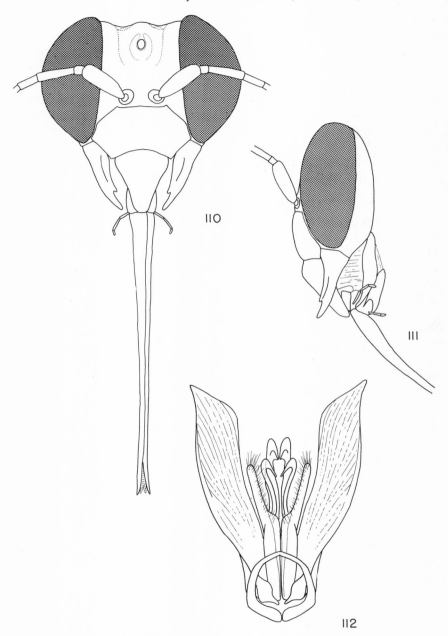

Figs. 110–112. *Steniolia obliqua* (Cresson): 110, head of female, anterior aspect; 111, head of female, lateral aspect, apical two thirds of proboscis omitted; 112, male genitalia.

Glenostictia with respect to the galeae, and also have four-toothed mandibles like some species of that genus. The spiracles are, however, of the type found in the *Stictiella,* and the epipharynx has a bare streak medially, as in *Bicyrtes* and *Bembix* (Figs. 110–112).

Gillaspy (1964) has recently revised the 15 known species of this genus. His paper includes a long discussion of structure and color pattern, illustrated with six plates, as well as a discussion of the ecology and phylogeny of the genus.

I. Ethology of the species of *Steniolia*

Evans and Gillaspy (1964) have recently published an account of what is known of the ethology of members of this genus, and Evans (1963c) has presented a brief popular account of clustering in *obliqua.* The present account adds additional data on *obliqua,* including experimental work on clustering, but is otherwise a review of the 1964 paper. Only *obliqua* has been well studied, but some data are available for seven other species: *duplicata* Provancher, *elegans* Parker, *eremica* Gillaspy, *longirostra* Say, *nigripes* Parker, *tibialis* Handlirsch, and *scolopacea albicantia* Parker.

(1) *Ecology.* Although some of the species of this genus are characteristic of desert and semidesert regions, an approximately equal number are of montane distribution. Females commonly do not nest in pure sand, but in various types of powdery, arenaceous earth, often with a high content of loam, manure, pulverized rock, or alluvial pebbles. Males of this genus appear relatively long lived, their numbers often declining only 1 or 2 weeks before those of the females.

Both male and female *Steniolia* are commonly observed taking nectar from flowers, and it is assumed that they take nourishment throughout their lives from this source. Evans and Gillaspy presented much data on the types of flowers visited, obtaining their data both from museum specimens and from their own observations. Of a total of 280 records (for all species), 170 (61 percent) are for flowers of the family Compositae. For the next family on the list (Leguminosae), there are only 30 records (11 percent); 21 other families make up the remaining 28 percent of the records. It seems apparent that the elongate mouthparts of these wasps represent an adaptation for probing deep corollas, especially those of Compositae (Fig. 113).

208   Chapter VIII. *Stictiella, Glenostictia,* and *Steniolia*

Figs. 113 and 114. *Steniolia obliqua:* 113 (above), female taking nectar from *Erigeron;* 114 (below), nesting area along a dirt road paralleling the Snake River (Jackson Hole, Wyoming, July 1961). (From Evans and Gillaspy, 1964.)

§I. Ethology of the species of *Steniolia*

(2) *Clustering.* One of the most characteristic features of this genus is the fact that adults of both sexes form dense sleeping clusters at night and during periods of inclement weather. These clusters are often at some distance from the nesting sites of the females. The clusters form in the late afternoon and break up 1 to 4 hours after sunrise unless the weather is unfavorable. We have presented data that suggest that temperature is important in inducing and terminating clustering. Evidently montane species such as *obliqua* cluster at considerably lower temperatures (55–60°F) than do xerophiles such as *duplicata* (65–75°F).

S. duplicata has been found clustering on Russian thistle (*Salsola*) and on star thistle (*Centaurea*) in California, also on creosote bush (*Larrea*) and burro bush (*Hymenoclea*) in western Texas. All clusters were on or near the tops of the plants, at heights of from 0.8 to 1.5 m. The larger clusters contained 55–73 wasps, of both sexes, and showed considerable body contact among the wasps. Other clusters were of small size and with little or no bodily contact. Two other species, *elegans* and *scolopacea albicantia,* have been found forming small, bisexual clusters on herbaceous plants.

Most of our knowledge of clustering is based on the widely distributed, montane species *S. obliqua*. This species has been found clustering at the ends of pine branches (1–2 m high) in Jackson Hole, Wyoming, and in the same area also on various herbs, including *Helianthus, Potentilla,* and *Pterospora* (generally less than 1 m from the ground). The clusters varied in size from a few individuals to large masses containing perhaps as many as 500 wasps. One cluster at the Jackson Hole Biological Research Station persisted on the same pine branch for at least five successive summers, but in 1964 it had disappeared. Various counts of the wasps in this cluster showed it to contain from 162 to 298 wasps. In the early part of the season the sex ratio tends to be close to 1:1, but later in the season the females outnumber the males. Although the sexes are fairly well mixed, I have found that the females tend to form tight balls within the clusters, while the males make up the majority of specimens on the outside. (Figs. 115 and 116).

Some of the *obliqua* clusters on herbaceous plants lasted only a few days, but the major clusters persisted for several weeks. By marking all individuals in certain of the clusters it was learned that individuals do not necessarily join the same cluster each night. The major clusters were all situated on trees or herbs in open pine groves but facing an opening directly to the east, permitting the wasps to fly in and maneuver on the downwind side. Most marked females were found to nest 0.3 to 1.0 km downwind from their

210 Chapter VIII. *Stictiella, Glenostictia,* and *Steniolia*

Fig. 115. Sleeping cluster of *Steniolia obliqua* on lodgepole pine (Jackson Hole, Wyoming). (From Evans and Gillaspy, 1964.)

clustering points. This suggested that a social pheromone might be produced and either deposited on the branch itself or released by the first wasps to arrive each evening (generally males), this hormone being dispersed by the wind in the manner of sex pheromones of certain moths. During the summer of 1964 I attempted several experiments at the Jackson Hole Station in an effort to clarify this point. These experiments, reported here for the first time, were not entirely conclusive because of the low population of *obliqua* that year. Only one relatively small cluster was found. This was

§I. Ethology of the species of *Steniolia*

discovered in a small clearing in a pine grove about 1 km east of the station on 20 July, when it contained nine males. The size of the cluster increased gradually until 25–30 July, when it contained an estimated 80 wasps, about half of each sex. Thereafter the number declined steadily, perhaps partly as a result of my manipulations; by mid-August only 2–4 remained.

This cluster was 1.2 m above the ground and occupied the tip of a horizontal pine branch (here called A1) which protruded well into the clearing from the west side. As the wasps assembled each evening, I noted that a few of them flew about the most prominent branch (1.5–2.0 m high) on two other trees, one (X1) 5 m to the north and the other (Y1) 7 m southeast; the latter branch actually extended into the clearing from the south rather than the west (windward) side. However, the cluster formed each evening on A1.

Fig. 116. Sleeping cluster of *Steniolia obliqua* on flowerhead of *Helianthus* (Jackson Hole, Wyoming). Virtually all of the wasps visible in this photograph are males, which typically form the outside of the cluster. (From Evans and Gillaspy, 1964.)

There were several other branches close beside A1, very similar in height and configuration and extending nearly as far into the clearing; these were marked A2, B1-3, and C1-2, depending on the nature of the branching; all arose from the same major branch as A. The following manipulations were performed prior to clustering each afternoon:

25 July. I displaced A1 0.5 m to the left, using a set of ropes and braces, at the same time pulling B1 into the position of A1. Result: wasps clustered readily at A1 and showed no attraction to B1.

26 July. I displaced A1 another 0.5 m to left, pulling C2 into original position of A1. Result: wasps clustered readily at A1, now 1 m to the left, showing no confusion or attraction to C2.

29 July. I cut off A1 and tied it 3 m to the right on a different branch (same height). To A2 (closest to A1) I added nothing; to B1 a brei of female mouthparts; to B3 a brei of female stings and associated glands; to C1 a brei of male mouthparts; to C2 a brei of male genitalia. Each brei was freshened just as the wasps began to assemble. Result: wasps milled about spot where A1 used to be, showing no attraction to A2, B1-3, or C1-2, or to A1 in its new position. Eventually 6 clustered at X1 and 30 at Y1.

1 August. Cluster at X1 has not persisted; all are at Y1. I reattached A1 to its usual place, and some males flew around it, but all settled at Y1. I brought another branch of the same tree (Y2) into position close below Y1 but such that it extended 12 cm farther into the clearing ("supernormal stimulus"). Result: usual clustering on Y1.

3 August. I cut off Y1 and attached it much closer to the trunk of the tree, at a height of 1.2 m, leaving Y2 in place at the usual height of Y1 (2 m). Result: wasps clustered on Y2, but one male went to A1 and one to X1.

No further experiments were performed since the cluster was quite small by this time. No further wasps clustered at X1, but two males were found on A1 on 10 August, although the needles on this branch were then brown, the branch having been cut off 2 weeks earlier. Females were still nesting on 14 August, so there were unquestionably other clusters in the area, but none could be found.

In summary, I was unable to demonstrate that clustering is controlled by a pheromone. It appears (though cannot be considered proved) that the wasps are attracted to the most prominent branches (at heights of 0.5–2.0 m) extending into clearings on the west and south sides; that they fly to several such branches in the evening but eventually cluster together on one; that they readily use other prominent branches when their initial one is unavailable in its usual place; that they orient to these branches each evening by visual cues relating to the configuration of the trees and the branches

§I. Ethology of the species of *Steniolia* 213

(in much the way the females orient to their nesting sites and their nests). The fact that certain branches are used for clustering for more than 1 year may mean merely that these branches are especially well situated and are rediscovered each year.

(3) *Reproductive behavior.* Knowledge of mating is based entirely upon notes I made on *obliqua* at Jackson Hole, Wyoming, in July 1961 and 1964. During the day, the males visit flowers and spend much time flying low over the nesting areas or landing on the ground or low objects near the nests of females. I saw one attempted mating at a nesting site in 1964, but most mating appears to occur at the clusters.

Males begin to arrive at the clustering sites about 2.5 hr before sunset, when shadows are long but the sun still far from the horizon. They approach from the downwind side of the branch used for clustering and fly about facing the branch but not landing. From time to time males land on the ground in sunny spots, extending their front and middle legs rigidly in a manner suggesting the precopulatory pose of many other Bembicini. After about 30 min males may be quite numerous and a few females begin to arrive. The females also approach from the downwind side at a height at 2-4 m. Notes made in 1961 indicate that each female is joined by a male at the cluster or a short distance downwind from it; the pair drops close to the ground and proceeds along in a strongly undulating flight, dropping to within 10 cm of the ground, then rising to about 30 cm, then dropping again. The pair may cover as much as 3 m in this way. Eventually the male flies off to rejoin the growing swarm at the clustering point, while the female usually rests on the ground for 1 or 2 sec before flying off.

Observations made in 1964 on a different cluster revealed a somewhat different pattern: the mating pairs joined at the cluster and remained quite high, sometimes even ascending gradually (up to 6 m), as they flew off a distance of 5-10 m before separating in the air. The usual undulations were noted, and it was remarked that each pair emitted a double buzz, that of the female being higher pitched than that of the male.

Sixty to 90 min after the arrival of the first males, roughly an equal length of time before sunset, some of the females settle on the branch and participate no further in mating. The females gradually build up the closely packed nucleus of the cluster, while the males continue to fly about restlessly, beginning to settle only after most of the females are already quiet. No further matings occur within 30-45 min before sunset, and by sunset the cluster is essentially complete, the remaining males settling on the outside of the mass. The presence of clouds in the west, or of unusually cool temperatures, may cause the clusters to form earlier and may reduce the length of the period during which mating occurs, or even eliminate mating altogether that evening. I noted no tendency for pairing in the morning; the wasps merely became restless and finally flew off one by one after the sun was fairly high.

(4) *Nesting behavior.* As already noted, the nests may be constructed at some distance from the clustering sites, and are typically dug not in pure sand but in some type of alluvial earth or pulverized rock. So far as is known, the females tend to form small, scattered nesting aggregations in areas of suitable soil, where 5 to 20 or more females may nest in close proximity (Figs. 114 and 117). Nest entrances are sometimes only 2–5 cm apart.

In a rather dense nesting aggregation of *obliqua* in 1964, I noted considerable aggression among the females. When two females found themselves close together, they would frequently butt each other in flight, actually striking their heads together. As they did this repeatedly they would rise slowly in the air, up to 1 m or somewhat more, then eventually (after as long as 20 sec) break off and return to their nesting. I could observe no true territoriality of nest space; rather each female appeared to attack others as they entered that section of the nesting area (which often contained several nests).

I (Evans and Gillaspy, 1964) have described digging in *obliqua* as follows: "Every 1–2 min the wasp backs out of the burrow, each time scraping the soil well away from the entranceway. Nearly every time, before re-entering, she makes a short flight, only 10–15 cm high, in the form of several loops just in front of the entranceway . . . When digging, the wasp uses the mandibles to free and to remove any small pebbles encountered. When the entrance is cleared, the mound is spread slightly, but not leveled in any real sense."

As digging proceeds, the flights become somewhat longer. Completion of the nest requires 1–2 hr. Eventually the wasp emerges, closes the entrance, then alternates several orientation flights with further visits to the nest, during some of which she adds more soil to the closure. The mound remains largely intact, although it is a small mound because of the shortness of the burrow. Females appear to revisit their nests fairly frequently, often merely flying about them with a pronounced humming of the wings, sometimes landing briefly, then flying off. They spend no time inside the nest when not actually bringing in flies, except that each morning, and perhaps sometimes later in the day, the female enters the nest

§I. Ethology of the species of *Steniolia*

Figs. 117 and 118. *Steniolia obliqua:* 117 (above), female entering her nest without prey in the morning, front legs thrust forward prior to scraping sand with a backstroke; 118 (below), typical nest, with contents removed (Jackson Hole, Wyoming, July 1961). (From Evans and Gillaspy, 1964.)

for an inspection prior to provisioning. The outer closure is maintained at all times.

I found *longirostra,* in Morelos, Mexico, nesting in a slope along a road fill, in powdery crushed rock. Much use was made of the mandibles for removing stones from the burrow. The females maintained an outer closure, but I have no data on whether or not the mound is leveled. Gillaspy found *nigripes* nesting in the Borrego Desert of California in "a level area of bare soil about three meters above the floor of a narrow, rocky valley." He found many empty cocoons in the soil, indicating that the wasps had occupied the same nesting site for several years.

Steniolia duplicata has been found nesting in several localities in California and in Texas, generally in coarse, sandy clay or gravelly soil. Gillaspy speaks of one female showing motions of "a peculiarly abrupt rocking nature when throwing dirt." Another emerged from the nest from time to time and cleared the earth from the entrance, "sometimes also leveling part of the mound by moving in broad arcs toward the entrance, or sometimes at right angles to the entrance." He found that each time the female emerged she made one or more short flights, as in *obliqua*. This species makes somewhat deeper nests than *obliqua,* and apparently several hours are required for completion of the nest. The mound is small and is partially or almost wholly dispersed by simple leveling movements interspersed with digging and clearing.

(5) *Nature and dimensions of the nest.* So far as is known, the nests of the species of *Steniolia* are always relatively shallow and are simple and unicellular. The cell measures 12–16 mm in diameter by 25–35 mm in length; it tends to be oblique although slightly less strongly oblique than the burrow. An inner closure has not been reported for any of the species, but all species studied maintain an outer closure at all times except during the brief intervals when the female is inside the nest.

Nest data for four species are summarized in Table 19. All species make unusually shallow nests for wasps of this size, those of *obliqua* being especially shallow (Fig. 118).

(6) *Hunting and provisioning.* Prey records for six species were presented by Evans and Gillaspy (1964). The flies taken are mainly flower-inhabiting species (Table 20). Bombyliidae are reported for five species and in most cases appear to make up the bulk of the prey. Syrphidae have been found in the nests of four species. Tabanidae have not been reported.

§I. Ethology of the species of *Steniolia*

TABLE 19. NEST DATA FOR SPECIES OF STENIOLIA

Species	Locality	No. of nests	Length of burrow (cm)	Depth of cell (cm)
obliqua	Jackson Hole, Wyo.	23	7.5 (6-11)	4.0 (3-5)
longirostra	Alpuyeca, Morelos	2	11.0	10.0
nigripes	San Diego Co., Calif.	several	—	8.8 (6-11)
duplicata	Antioch, Calif.	1	17.5	—
	Candelaria, Texas	1	6.5	—
	Lajitas, Texas	2	10.5 (10-11)	7.5 (7-8)
	Presidio, Texas	5	14.0 (11-18)	8.4 (7-10)

Hunting behavior in *elegans* has been described by Gillaspy as follows:

The wasp [hovered] head-on before its apparent intended prey and its high-frequency wing-beat [produced] a shrill sound varying in pitch, as though preparing for the final, high-velocity dart forward. The final lunge was observed on a few occasions without being able to ascertain whether the prey was actually caught, the wasp continuing on her way in rapid flight. It seems possible that the prey is caught in the air as it leaves its resting place, rather than being taken directly from the substrate.

TABLE 20. FAMILIES OF FLIES USED AS PREY BY SPECIES OF STENIOLIA

(For details see Evans and Gillaspy, 1964)

Species	STRATIO-MYIDAE	BOMBY-LIIDAE	ASILIDAE	SYRPHIDAE	MUSCOIDEA
obliqua		X		X	X
longirostra	X	X		X	
nigripes		X			
duplicata		X		X	X
tibialis		X		X	
eremica			X		

In the case of *obliqua,* I reported that "the flies are apparently captured on the ground, low vegetation, and flowers in and around the nesting area. On several occasions wasps were seen hovering near bombyliid flies, with a considerable buzzing, then striking at them suddenly." In *obliqua* the flies appear to be killed by the sting of the wasp, as none of those found in cells showed movements of body parts, and often they were somewhat dessicated. It is my impression that this is also true in *duplicata.* Actually, one rarely finds fresh flies in the cells of species of *Steniolia* unless he digs the nest shortly after they were brought in, for provisioning is fully progressive and not more than a very few flies are allowed to accumulate in the cell.

The egg has been found in *obliqua* and *duplicata;* in both cases it is glued erect to the side of a fly which is placed head in deep in the cell (much as shown in Fig. 164). The next fly is brought in on the day of eclosion. A total of 18 or more flies may be required per larva. Females laden with prey produce a loud whining sound when they approach the nest. The nest does not receive the final closure until the larva is nearly ready to start spinning its cocoon. The cells are not cleaned by the mother, the larva resting upon a mat consisting of the remains of the flies which have been consumed.

(7) *Final closure.* I described the behavior of a female observed at Presidio, Texas, 3 May 1963, as follows:

This wasp would make a short flight, land in the entrance, rotate 180°, then back in; in a few moments she would come out, walking up to 2 cm in front of the entrance, scraping sand into the hole, then take a short flight and repeat the process. After a time she began biting and scraping soil from around the entrance, eventually making a small depression about 3 cm. in diameter and .3 cm deep, using the soil for fill. When the burrow was full she went out in various directions, up to 5 cm. from the entrance, scraping sand over this depression. When the surface was completely flat and smooth, the wasp flew away. The entire final closure required about 30 min. This nest was dug out immediately, and the larva was found to be full-grown and feeding on a fly brought in about an hour earlier; there were no intact flies in the cell (Evans and Gillaspy, 1964).

In the case of *obliqua* I reported that "final closure of the nest apparently differs little from a temporary closure except that the burrow is completely filled with soil and somewhat more soil than usual is therefore scraped from the surface into the burrow."

The cocoons of *Steniolia* are distinctive in that an ovoid capsule of the usual bembicine type is surrounded by a shroud-like structure which has a funnel-like extension on one end (see Fig. 208).

(8) *Natural enemies*. The following notes pertain to *obliqua*, the only species for which any natural enemies have been reported. At Jackson Hole, Wyoming, I found miltogrammine flies (Sarcophagidae, Miltogramminae) to be fairly common early in the season. Wasps being trailed by flies would commonly not enter their nest immediately, but would fly off and return again, sometimes several times, before entering. No maggots or puparia were found in any of the numerous nests excavated, and the number of successful larvipositions may not have been great. Two of the flies following *Steniolia* females laden with prey were captured. One was found to be *Taxigramma heteroneura* (Meigen), the other *Hilarella hilarella* (Zett.).

The chrysidid wasp *Parnopes edwardsii* (Cresson) was very common in this same area and was seen entering *Steniolia* nests on several occasions. The *Parnopes* females dig very effectively, and only a few seconds are required for them to dig through the outer closure of a nest. Female *Steniolia* occasionally return to their nest while the *Parnopes* is inside or just outside the entrance, and when this occurs the wasps attack the parasites vigorously, even attempting to sting them. On one occasion a *Steniolia* was seen to seize a coiled-up *Parnopes* and carry it 30 cm. No *Parnopes* were actually reared from the seven cocoons held over the following winter, but one cocoon accidentally broken in the field was found to contain a *Parnopes* cocoon inside. The *Parnopes* females appeared to be most active in the morning hours (0900–1100), before their hosts reached the peak of their provisioning activities. Apparently they dig short burrows in the soil in which to spend the afternoon and night.

As Evans and Gillaspy reported in 1964, Dr. G. E. Bohart found, in Bonneville County, Idaho, a cocoon of *Steniolia obliqua* containing a small chrysidid larva feeding on the thoracic area of the prepupa of the wasp. Within 3 days the chrysidid larva was about half grown; 13 days later the cocoon had been spun. An adult *Parnopes edwardsii* was reared from this cocoon.

Although no mutillids were observed in Jackson Hole, *Dasymutilla vesta* (Cresson) was common in the Idaho locality. Dr. Bohart found that 5 of the 25 *Steniolia* cocoons he collected contained mutillid cocoons, but unfortunately no emergence was obtained from these cocoons.

J. Summary of the ethology of the genera with recessed ocelli

Although the wasps considered in this chapter show much diversity, both structure and behavior suggest that they form a single phyletic stock. In this brief summary I shall try to point out the common features of the three genera as well as evidence of progressive evolution within the complex.

(a) These wasps nest in diverse situations and only rarely form large, expansive colonies as in some other bembicine wasps; some species of *Glenostictia* and *Steniolia* do form fairly large aggregations, but the nests tend to be grouped in localized areas.

(b) Adults visit flowers for nectar and are not known to feed upon their prey. Flowers with deep corollas are commonly exploited, especially in *Steniolia*.

(c) There is evidence that males of some species are relatively long lived. No one has reported male "sun dances" or territoriality. Some *Steniolia* are known to mate at the clusters.

(d) Both sexes spend the night and periods of unfavorable weather clustering on vegetation (exception: females of *Glenostictia scitula* spend these periods part way down their burrows). Larger, ball-like clusters, are more especially characteristic of *Glenostictia* and *Steniolia*.

(e) Females of all three genera tend to produce a high-pitched whine when hunting or working about the nest; this is doubtless correlated with the relatively short wings of most species.

(f) Nests are of simple structure, multicellular in some species of *Stictiella*, otherwise unicellular. They are especially shallow in *Steniolia*.

(g) No inner closures have been reported, but the females of all species studied make an outer closure when they leave the nest.

(h) In *Stictiella* and *Steniolia* the mound may not be leveled or may be partially or completely dispersed by leveling movements following completion of the new nest. In *Glenostictia* three species are known to exhibit distinctive rotational leveling movements interspersed with digging (but some terminal leveling also occurs at least in *G. gilva*).

(i) When hunting, the females maneuver rapidly on their short wings and strike suddenly at their prey, which is then carried away directly in flight. Paralysis is usually profound, and in *Glenostictia* and *Steniolia* the prey is usually killed and may appear dessicated the following day.

(j) The species of *Stictiella* prey upon Lepidoptera, either small to medium-sized butterflies, skippers, or small moths. The species of *Glenostictia* and *Steniolia* prey upon Diptera, bee flies (Bombyliidae) being especially common prey of most species (exception: *G. scitula* takes a variety of small Hymenoptera, especially bees, as well as Hemiptera in small numbers, in addition to flies).

(k) In *Stictiella* mass or delayed provisioning appears to be the rule. In the other two genera, provisioning is fully progressive, in fact continuous, until the larva is nearly ready to spin; few fresh prey are normally found in the cells at any one time.

(l) The egg is laid erect on the side of the first prey placed in the cell (exception: *Glenostictia scitula*, which oviposits flat in the empty cell).

(m) The cocoons of *Glenostictia* and *Steniolia* have an outer shroud with a funnel-like opening at one end.

(n) All three genera are subject to attacks by miltogrammine flies. Species of the chrysidid genus *Parnopes* have been found associated with species of *Glenostictia* and *Steniolia*.

It is apparent from the above summary that *Glenostictia scitula* is an exceptional wasp in several ways (notably d, j, and l); it is in several respects more specialized than the species of *Steniolia*.

Much more needs to be learned about members of this complex. At the present time nothing at all is known of the ethology of two genera recently described by Gillaspy (1963b), *Microstictia* and *Xerostictia,* and several species of each of the other genera remain completely unstudied.

Chapter IX. *Stictia* and Some South American Genera

Stictia is one of the most characteristic genera of solitary wasps in tropical America. One species, *carolina*, occurs widely in the southern United States, but the other dozen or so species occur in South and Central America and in the Antilles. No less than seven other genera of Bembicini, which may roughly be described as "*Stictia*-like," occur in South America. For the most part these are large wasps, robust and as much as 3 or 4 cm in length. Several of the genera are poorly known, and only three, besides *Stictia*, have been studied in the field. The nesting behavior of *Zyzzyx* and *Rubrica* is fairly well known, but only a few scraps of information are available regarding *Editha*. Only these four genera will be considered in this chapter; the structure of the other four (*Trichostictia, Selman, Hemidula,* and *Carlobembix*) will be treated briefly in Chapter XIII.

Aside from their generally large size and similar facies, the wasps of these eight genera have a number of features in common. In all of them the prolongation of the mouthparts is rather moderate, and the palpal segmentation shows reduction only in *Zyzzyx*, where it is slight. None of them shows as much reduction in the ocelli as in *Bembix, Bicyrtes,* and *Microbembex*, and in no case is the anterior ocellus sunken as in the genera considered in the preceding chapter. In the male, tergite VII has strongly separated sidepieces bearing the spiracles and in all the genera save *Hemidula* the apex of tergite VII is strongly notched on each side, therefore somewhat trifid. In six of the genera the middle femora of the male have a strong preapical tooth and notch (the exceptions being *Hemidula* and *Trichostictia*).

§B. *Stictia carolina* (Fabricius) 223

A. Genus *Stictia* Illiger

The following summary of the important structural features is provided.

Adults: Eyes somewhat convergent above, vertex much depressed below tops of eyes; anterior ocellus without a visible lens, but position clearly marked by a groove which forms somewhat more than half a circle (incomplete in front); within the groove the integument slightly convex, polished, and without setae (forming a blackened covering of the lens called a *cicatrix* in the taxonomic literature); posterior ocelli also forming semicircular cicatrices marked by a groove on the inner margin; mouthparts moderately prolonged, but proboscis capable of being fully retracted beneath the labrum; palpi showing no reduction in segmentation. Mesosoma differing in no important way from that of other bembicine wasps; propodeum flattened or slightly concave behind, but without projecting flanges; wings essentially as figured for *Bicyrtes* except for minor details; female with a strong pecten (very much as in *Bicyrtes*, Fig. 85), but without a pygidial area. Male with a preapical notch and tooth on middle femur; tergite VII strongly modified, as described above; sternite VI sinuate apically and often with a central swelling; sternite VII with a pair of membranous areas which are covered by the large lateral lobes of tergite VII; apical sternite forming a single strong, spinelike process; male genitalia of the same basic pattern as in *Bicyrtes* (Fig. 83), although differing considerably in details (Figs. 119-126; Willink, 1947, has figured the male terminalia of several South American species).

Larvae: body with an unusual number of strong setae and with the more anterior spiracles elliptical, distinctly higher than wide; anus situated above the middle of the anal segment; labium with two patches of spinules separated by a papillose streak.

The ethology of six species of *Stictia* has been studied in some measure, but only one species, *carolina*, is known in detail. *Carolina* may be one of the more specialized species of the genus, so it is important that other species be investigated more thoroughly.

B. *Stictia carolina* (Fabricius)

This large and showy insect is one of the best-known North American digger wasps. It is especially common in the states bordering the Gulf of Mexico and in the southern Great Plains; it is not known to occur north of New Jersey and Illinois, or west of New Mexico. Throughout the South it is commonly known as the "horse guard," because the females are often seen about livestock, where they capture horseflies in considerable numbers. So far as I know,

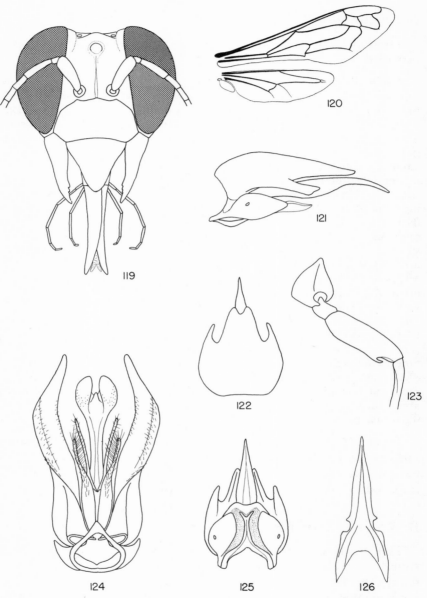

Figs. 119–126. *Stictia carolina* (Fabricius): 119, head of female, mouthparts extended; 120, wings; 121, apex of metasoma of male, lateral aspect; 122, apex of metasoma of male, dorsal aspect; 123, middle coxa, trochanter, femur, and base of tibia of male; 124, male genitalia; 125, apical segments of metasoma of male, ventral aspect; 126, sternite VIII of male.

§B. *Stictia carolina* (Fabricius)

Bryant (1870) was the first to call attention to this behavior and to employ the name "horse guard" in print. Ashmead (1894) reported watching *carolina* preying upon *Tabanus atratus* in Florida; "a singular peculiarity of this insect," he says, "is its ability to fly backwards in front of a moving horse while watching the opportunity to suddenly pounce upon and seize one of these flies." More detailed studies of the species were made by Hartman (1905) in Texas, by Hine (1906, 1907) in Louisiana, and by Krombein (1958b, 1959) in coastal North Carolina. Taken together, these reports provide a fairly complete picture of most aspects of the nesting behavior of this wasp.

The present report is based on studies made nine years apart and in two widely separate localities. From June to September 1952, C. S. Lin, C. M. Yoshimoto, and I made over 50 field notes on a colony of these wasps nesting near Blackjack Creek, Pottawatomie County, Kansas, about 7 mi east of Manhattan. During May 1961, I made 14 field notes on two nesting aggregations in Highlands County, Florida. These observations largely confirm those of earlier workers but provide further detail on several points.

(1) *Ecology and general behavior of adults.* This species is characteristic of sandy soil that is either bare or covered with scattered herbs or grass. It does not normally occur in large dunes or blowouts with such species as *Bembix pruinosa*, but it may occur on the periphery of such dunes. In coastal areas the species is absent from the beach front but may inhabit rolling, partially vegetated sand behind the beach. The colonies studied by Hine were situated "on dry ground where the sand was loose and easy to dig, and on the extent of such ground in one place depended the size of a particular colony." Krombein found females nesting on sandy barrens behind the beach, one at the edge of a wheel rut on a sand road. The Kansas colony we studied was situated in a sandy slope along a corn field where the soil was dry and very friable; there were many herbs growing here, and the wasps occasionally nested among the corn plants (Fig. 127). In Highlands County, Florida, *carolina* occurs widely in sandy uplands, including citrus orchards and open places in pine woods. Two nests at the Archbold Biological Station were situated along wheel ruts in an orchard. At Highlands Hammock State Park a very large colony of these wasps occupied a parking area used by persons visiting a cypress swamp (Fig. 128). The wasps seemed to thrive here even though cars frequently drove or parked over their nests. The soil here was sandy but very compact and difficult to dig in with a trowel. A large colony of *Bembix texana* occupied this

Fig. 127. Nesting area of *Stictia carolina* in partially bare slope along a field (actually planted to corn during the period of study; Pottawatomie County, Kansas).

same parking lot, as well as numerous individuals of *Bicyrtes quadrifasciata* and a species of *Cerceris*. The nests of all four species were well intermingled.

Hartman states that this wasp may "begin her nest in the morning before any digger-wasp was astir . . . *Carolina* is, moreover, least susceptible to the influences of the weather; for, while other diggerwasps will lie listlessly about on a cloudy day . . . she may be as busy as ever." I found these statements to be essentially true. In the Highlands Hammock colony females were often active as early as 0800, though few other digger wasps made their appearance for 1 hr afterward. At the Archbold Biological Station females were observed hunting and provisioning on cloudy days when wasps such as *Bembix sayi* were inactive and within their nests. However, *Stictia carolina* is strongly influenced by sudden drops in temperature. In the Kansas colony, on 7 July 1962, many individuals of both sexes were active 0800–0900. Shortly before 0900 it clouded up and became windy, and within a short time the temperature dropped from 86 to 73°F; at 1100 it was 66°F. At 0910 the last male was seen, and a few minutes later the last female; there was no further activity that day.

§B. *Stictia carolina* (Fabricius)

In Pottawatomie County, Kansas, the males appear in numbers during the last few days of June or first few days of July (records for 3 years); after about 2 weeks they begin to decline rapidly in numbers, disappearing completely in middle or late July. The females begin to appear a few days after the males and remain active for several weeks. The latest date on which we saw a female was on 5 September 1952. Thus I would estimate that males usually live 2–3 weeks, females 6–8 weeks. There is clearly only one generation a year in Kansas.

In central Florida both sexes are common in early May (records for 2 years), and females are nesting as early as 5 May. In the colony at Highlands Hammock, the number of males declined steadily throughout May, and none were noted after 20 May (although I did see males in other localities not far away). Only a few females were active on 5 May, but by 15 May perhaps as many as 200 were active. Records suggest that this species is multivoltine in the southern states. Hine's studies were made in June and July, those of Hartman and Krombein in August.

Fig. 128. Nesting area of *Stictia carolina* in a parking lot (Highlands Hammock State Park, Florida). A large colony of *Bembix texana* was also located here, as well as a few *B. belfragei, Cerceris,* and other digger wasps.

Both sexes visit flowers for nectar and will sometimes fly considerable distances to find suitable flowers. In Kansas, I observed a female taking nectar from *Polygonum*, males visiting *Cephalanthus* and *Melilotus*. In Florida males were seen in numbers on the flowers of *Bumelia tenax*. Nights and periods of cold or rainy weather are spent away from the colony. I have often seen the wasps fly into trees, and I believe they spend these periods resting on the leaves. Females do not remain in the nest at night or more than briefly during unfavorable weather, and males do not dig burrows in the sand as occurs in *Bembix*.

(2) *Reproductive behavior.* Males of this species are easily recognized, as the yellow markings on the abdomen are confined to the basal segments. Their flight is also characteristic, as they tend to fly about rather slowly and circuitously with the tip of their abdomen curved down and the legs partially extended downward, as if in readiness to seize a female. Most flight of the males occurs in and around the nesting site in the morning hours. The flight of the males bears much resemblance to the "sun dance" of most species of *Bembix*, except that it is executed rather high above the ground (between 0.5 and 1 m). As mentioned above, the males appear a few days before the females; for several days the males therefore "patrol the air" over the nesting site by themselves.

Although the males often remain in flight for considerable periods in their characteristic weaving patterns, they also select perches in the nesting area where they pose at certain times, alert and with their wings partially extended. In the Kansas colony herbs 5–20 cm high or occasionally horse droppings or lumps of earth were selected as perches. In the Florida colony the soil was flat and devoid of suitable perches; here the males landed on the ground occasionally, but seemed to spend much of their time in flight.

Males flying in the "sun dance," or flying off quickly from their perches, butt or pounce upon various insects flying about the nesting area, particularly other males. On one occasion a large pompilid wasp flew across the area and was pounced upon by three male *Stictia* before it reached the other side. On several occasions I have seen *Stictia* males begin to pursue dragonflies. These encounters with males of their own species or with other insects are exceedingly brief. Probably individual males patrol a territory and defend it against intrusion by other males, as described for *Sphecius speciosus*, Chapter, V:B, 2 (see also *Stictia vivida*, Chapter IX:D). Unfortunately I failed to mark any males or to study this behavior in adequate detail. Males sometimes swoop down upon females digging

§B. *Stictia carolina* (Fabricius) 229

their nests, but these contacts are also very brief. I believe that mating is successful only with females that have just emerged.

When a male contacts a receptive female, he grasps her from above, and the pair fly together obliquely upward to a height of several meters (several records). After a certain amount of flying about, they settle on the leaf of a tree, where mating occurs. I observed mating only once, in Kansas (no. 370). The pair was on the leaf of a catalpa tree 1.5 m above the ground, in the shade. The male was astride the female, with the tip of his abdomen twisted to one side and the genitalia inserted between the apical tergite and sternite of the female. Both rested quietly with their wings partially extended and still. After 20 sec the two separated and flew off; since they were *in copula* when first seen, I do not know the total duration of mating.

(3) *Nesting behavior.* The Kansas colony contained an estimated 50 females which inhabited, at the beginning of the season, an area about 5 by 20 m. The entrances of active nests were always well separated, generally by 30 cm or more. As the season progressed, the wasps tended to radiate out into adjacent sandy areas, eventually nearly abandoning the original nesting area. The colony at Highlands Hammock, Florida, contained at least 150 nesting females, but even here the nest entrances were usually well spaced. On a few occasions I found nests no more than 8 cm apart, but most nests were separated by 25 cm or more. At the beginning of nesting, in early May, most nests were situated in a parking area, but already by late May the wasps were spreading over the roads leading in and out of the parking lot. This tendency for females to move away from the center of the colony late in the season and to exploit new areas in the vicinity has been noted in certain species of *Bembix* (Evans, 1957b:198); probably in this manner new colonies are formed.

However, *Stictia carolina* does not always occur in large aggregations. In central Florida one often finds widely scattered nests in citrus orchards or along sandy roads. Krombein found only scattered individuals nesting on sand barrens in North Carolina. These may, of course, represent individuals that have recently colonized the area from nearby aggregations; some of these individuals may be successful founders of new colonies.

This species is a vigorous and powerful digger (Fig. 129). Most of the work is performed by the front legs, but some use is made of the mandibles in breaking through hard soil or removing pebbles from the burrow. A mound of loose soil rapidly accumulates behind

Fig. 129. Female *Stictia carolina* digging her nest (Highlands Hammock State Park, Florida).

the wasp as she digs diagonally into the sand. One wasp was observed to dig 17 cm in 2 hr. During the digging of the nest, the wasps often leave for considerable periods of time, returning later to dig further. Whenever the wasp leaves, she scrapes sand into the entrance, often even pounding it in place a bit with the tip of her abdomen. The most prolonged periods of digging occur in the morning and late afternoon; during the middle of the day the female may visit the nest occasionally or even dig a short distance further, but it is rare to observe persistent digging at midday. A common pattern is for a female to begin a nest in the morning, break off several times in the morning, each time closing the nest, then return in the late afternoon and finish or nearly finish the nest; however, in such a nest leveling and oviposition would not likely occur until the morning of the second day. Nests started in the late morning or middle or late afternoon (as they less commonly are) would be completed the following day, with leveling and oviposition occurring late in the day. One female (no. L27, Kansas) did not complete her nest until 54 hr after it was started. Hartman has noted the ex-

§B. *Stictia carolina* (Fabricius)

tremely irregular "working hours" of digging females. He states that 3 or 4 days are required to dig the nest. In both the Kansas and Florida colonies I studied, most nests were completed 22–30 hr after they were begun.

While the female is digging (in a manner very similar to *Bembix*), she periodically sweeps the sand back from the entrance. Eventually a large mound of sand accumulates here. The soil used for making the closure, whenever the female stops digging and leaves the nest, is scraped chiefly from the surface of the soil on the sides away from the mound. Only when the nest and cell have been completed is the mound actually leveled. Leveling is preceded by a closure of the entrance similar to those made earlier; some movements of closure may also be interspersed with leveling movements. When leveling, the wasp backs up quickly to the extremity of the mound and works forward scraping sand vigorously. As she works forward she twists from one side to the other so that much of the sand is thrown to the sides. Having reached the entrance, the wasp often adds a bit to the closure, then backs up and repeats the leveling. Eventually the mound is well spread, and there is a characteristic pattern of irregular lines down the center, as seen in Figure 130.

Fig. 130. A newly completed nest of *Stictia carolina* (Highlands Hammock State Park, Florida). Note the irregular pattern of lines on the partially spread mound, also some additional lines radiating from the nest entrance. The wasp has just re-entered the nest for oviposition, hence the open entrance.

There is also a pattern of radiating lines around the entrance associated with the closure; some of these lines may be up to 25 cm long, but usually they are much shorter, 3–6 cm long. The spread mound is only a fraction of 1 cm deep, and measures from 45 to 60 cm long and from 15 to 25 cm wide. Leveling requires at least 30 min.

Following leveling, the wasp enters the nest again either right away or after a few minutes to 1 or 2 hr. Presumably the egg is laid in the cell at this time; at least I have never found an egg in nests dug out before or during leveling. After oviposition, the wasp leaves the nest and again makes a careful closure, making much use of the tip of the abdomen in pressing the earth into the entrance.

(4) *Nature and dimensions of the nest.* The nest of this species is of simple structure, the oblique burrow tending to level off somewhat before reaching the horizontal cell (Fig. 131). A short spur can be found in most nests, representing the point at which soil is dug for making the inner closure. Nearly all nests have a weak to strong lateral bend part way down the burrow, as noted by Hine. The diameter of the burrow is about 13 mm; at the entrance it is often as much as 20 mm. There is much variation in burrow length and cell depth even within a single colony. Cell depth in the colony at Highlands Hammock, Florida, where the soil was very firm, averaged somewhat less than in colonies in looser sand (see Table 21).

TABLE 21. NEST DATA FOR STICTIA CAROLINA

Locality	Type of soil	No. of nests	Burrow length (cm)	Cell depth (cm)
Pottawatomie Co., Kansas	Loose sand	32	47.5 (37–61)	24.0 (17–30)
Lake Placid, Florida	Loose sand	2	51.5 (46–57)	22.5 (20–25)
Highlands Hammock, Florida	Compact sand	13	35.0 (26–45)	18.5 (13–26)
Kill Devil Hills, N.C. (Krombein)	Moderately loose sand	3	36.0 (32–39)	21.0 (20–22)
Austin, Texas (Hartman)	"Sand"	?	45.0	24.0

§B. *Stictia carolina* (Fabricius) 233

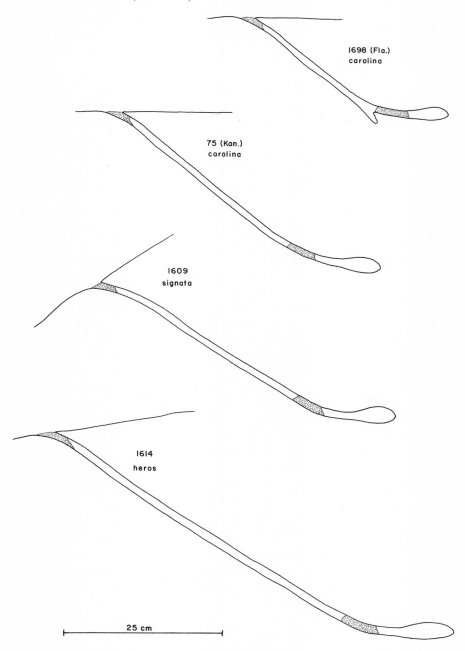

Fig. 131. Typical nests of three species of *Stictia*. Note the presence of an inner closure in all four.

The cell, when new, measures about 22 × 40 mm, but in more advanced nests the cell tends to be lengthened and may measure up to 60 mm. This lengthening probably occurs when the inner closure is removed during provisioning.

The nests of this species are characterized by the presence of a strong inner closure in the more horizontal part of the burrow just outside the cell. In a new nest containing an egg, this closure may be as much as 15 cm long, though more often it is 6–10 cm long. In more advanced nests the inner closure tends to be relatively strong at night, but during periods of provisioning it may be quite weak. On a very few occasions nests that were dug out during a period of active provisioning appeared to lack any noticeable inner closure.

As already pointed out, a strong outer closure is prepared by the female after she lays her egg. This closure is maintained for about 3 days, that is, while there is an egg or very small larva in the cell. In the earliest stages of provisioning the female scrapes open the entrance upon entering with a fly and closes it carefully upon leaving. In later stages of provisioning, a strong outer closure is maintained at night and during periods when the female is not actively provisioning. Between trips for flies, there is much variation with regard to whether or not an outer closure is maintained. Hartman also remarked on this variation. There seems to be a tendency for females to omit the closure more commonly on the last day or two of provisioning, when the larva is nearly mature, but sometimes closure is omitted when the larva is rather small or made when the larva is large. No explanation for this variation is presently available (Table 22).

(5) *The egg.* Hine, in his second report on this wasp (1907), stated that the egg is laid in the empty, newly made cell and that no prey is brought in until it hatches. We recovered three eggs in the Kansas colony and six in Florida, and in every case the egg was found to lie transversely on the floor of the cell near the center. In no case were there any flies in the cell. One other nest (no. 1694, Archbold Biological Station, Florida) contained a newly hatched larva lying on the floor of the cell with its head elevated, but there were no flies in the cell; the wasp was digging at the entrance but had been unable to find the burrow, as the nest was in a wheel rut and had been run over several times by cars. Another nest (no. 1680, Highlands Hammock, Florida) contained a larva only 9 mm long lying on the floor of the cell with its head elevated, feeding on the throat of a small fly, the only one in the cell. Apparently the female places

§B. *Stictia carolina* (Fabricius)

TABLE 22. SELECTED RECORDS OF OUTER CLOSURE
DURING PROVISIONING IN STICTIA CAROLINA

Nest no.	Date	Nature of closure	Size of larva in cell
166 (Kan.)	3 July 1952	Prey at 1642, 1658, 1713... Each time closed on leaving	Two-thirds grown
L 5 (Kan.)	9 July 1952	Prey at 1207, 1244, 1250, 1313, 1317, 1600, 1605, 1610, 1616, 1621... No closure any time	Nearly full grown
L 6 (Kan.)	9 July 1952	Prey at 1223, 1249... Closed both times	
	11 July	Prey at 1455... Closed	
	12 July	Prey at 0845... Closed Prey at 1247, 1252, 1257, 1311, 1322, 1328, 1452... Left open each time	Nearly full grown
L 26 (Kan.)	12 July 1952	Prey at 0836, 0847, 0905... Left open first two times, closed the third time	
	13 July	Prey 1740... Closed	
	14 July	Prey 1415, 1453, 1503, 1517... Left open first three times, closed the last time	
	15 July	Prey 1744, left open Prey 1753, closed	
	16 July	Prey 0953, 0957, 1001, 1014... Closed each time	Nearly full grown
1690 (Fla.)	8 May 1961	Several prey; always closed on leaving	Nearly full grown
1714 (Fla.)	17 May 1961	Several prey; always left nest open	Nearly full grown

the first fly against the anterior end of the egg just as it is about to hatch or beneath the elevated head of the larva soon after it hatches.

The egg of *Stictia carolina* measures 6–7 mm long by 1.7 mm wide; it is curved, slightly thicker at one end than the other. The egg hatches in from 1 to 3 days.

(6) *Hunting and provisioning.* As already stated, the females do not spend the night or periods of inactivity inside the nest. Therefore inspection trips to the nest form an important element in the behav-

ior. Relatively early each morning (0800–1000, unless the weather is unfavorable), each female enters her nest, remains a few moments, and then leaves, usually closing the entrance. The results of her inspection largely determine her behavior for the day, or at least for the next few hours. We have seen some apparent inspection trips later in the day, even in the late afternoon, generally in nests containing an egg or small larva.

As is usual in wasps exhibiting progressive provisioning, the number of flies brought in is determined by the size of the larva; the 1st day one may suffice, the 2nd day three or four, and so forth. Hartman states that the size of the fly is not graduated to the size of the larva, and this is largely true, although nests containing very small larvae in our experience contained only small flies. Hartman states that the larva is fed for 11 days. We found much variation in this regard, with 11 days being about the maximum, 6 about the minimum (both records from Kansas). If we assume the average to be about 8, with the first fly brought in on the 2nd day after oviposition, then about 10 days are required for a single nest.

The practice of females of this species of catching flies about livestock has been known for many years and has, of course, earned it the name "horse guard." I have observed this behavior several times. Near the Kansas colony two horses were pastured, and every day several *Stictias* could be observed flying about them closely. On one occasion a *Stictia* tried to grasp a horsefly feeding on the flank of a horse, but without success. Possibly they are able to capture flies only when they are in flight or have just landed. Not far from the Archbold Biological Station in Florida, I discovered that a calf was being "guarded" by several of these wasps. Although members of the large colony at Highlands Hammock used many horseflies, mostly females, I did not observe where they were taking them. There were no cattle or horses nearby, but there were many deer in the park. As Hine noted, the animals seem to pay no attention to the wasps, even though the wasps buzz loudly and fly about very close to them: in spite of the fact that livestock are sometimes much annoyed or even panic stricken by the presence of certain flies.

Besides horseflies, these wasps also sometimes take stable flies and screwworm flies, as well as various other muscoids which occur about livestock and their droppings. However, not all hunting is done about horses and cattle. Krombein noted wasps coming to the stems of bunch grass, which were also being visited by calliphorid flies of the same species he found in the *Stictia* nests. The fact that male horseflies are occasionally taken, and also flower-visiting flies

§B. *Stictia carolina* (Fabricius)

such as *Volucella*, suggests that these wasps actually hunt rather widely. Doubtless once a female discovers a good source of flies she returns again and again to that source. Krombein found 16 *Cochliomyia macellaria* in one nest, 21 in another. In the Kansas colony I studied, *Tabanus sulcifrons* was being taken in great numbers, nearly all of them females. On several occasions we found ten or more of these large horseflies in a single cell. In Florida I took a remarkable array of horseflies from nests or from provisioning females (Table 23); less than 10 percent of these were males.

On the final day of provisioning as many as 18 flies may be

TABLE 23. PREY RECORDS FOR STICTIA CAROLINA

Species of prey	No. of specimens *				
	Kan.	Fla.	N.C.	La.	Tex.
CULICIDAE					
Psorophora ciliata Fabr.				x	
TABANIDAE					
Chrysops dimmocki Hine		3			
Chlorotabanus crepuscularis Beq.		25			
Hybomitra hinei wrighti Whit.		3			
Tabanus abdominalis Fabr.		2			
T. americanus Forst.			1		
T. atratus Fabr.	2	3	2	x	
T. bishoppi Stone		1			
T. cheliopterus Rondani		1			
T. coarctatus Stone		1			
T. endymion O.S.		1			
T. fumipennis Wied.		4			
T. imitans imitans Walk.		2			
T. lineola Fabr.		17			
T. melanocerus lacustris Stone		6			
T. mularis Stone		4			
T. nigripes Wied.	1	4			
T. nigrescens Beauv.		2			
T. petiolatus Hine		1			
T. sparus Whit.	1				
T. stygius Say		3			
T. sulcifrons Macq.	82				
T. trijunctus Walk.		16			
T. vittiger schwardti Philip	1				
T. spp.	12	20	x	x	
STRATIOMYIDAE					
Odontomyia cincta Oliv.				x	

TABLE 23. PREY RECORDS FOR STICTIA CAROLINA (CONT.)

Species of prey	No. of specimens *				
	Kan.	Fla.	N.C.	La.	Tex.
SYRPHIDAE					
Volucella mexicana Macq.					x
V. nigra Greene		3			
MUSCIDAE					
Graphomya maculata Scop.		1			
Musca domestica L.			1		x
Orthellia caesarion Mg.				x	
Stomoxys calicitrans L.	4				
SARCOPHAGIDAE					
Amobia erythrura Wulp			1		
Sarcophaga spp.	2		6		
CALLIPHORIDAE					
Calliphora vomitoria L.					x
Cochliomyia macellaria Fabr.			37	x	
TACHINIDAE					
Genus and sp.?			x		

* North Carolina records from Krombein (1958, 1959), Louisiana records from Hine (1906, 1907), Texas records from Hartman (1905). The last two authors did not indicate the number of specimens taken.

brought in (no. 1736, Highlands Hammock), probably more under some circumstances. Of course, the number of flies brought in on a particular day, as well as the total of all flies brought in, depends greatly on the size of the flies. In the Kansas colony the total number of flies per cell varied from 15 to 23; many of them were *Tabanus sulcifrons,* a very large fly. In Florida, where the flies were more variable in size and type, the usual total seemed to vary from about 25 to about 35. One nest dug out by Krombein in North Carolina contained 25 flies, mostly muscoids, but this nest was not fully provisioned. Hine found large numbers of flies in the nests he dug out, the maximum being 63, all but 1 of them horseflies (presumably small species).

Hartman notes that the flies are normally killed by the sting of the wasp. In Florida I had several occasions to compare flies stung

§B. *Stictia carolina* (Fabricius) 239

by *Stictia carolina* with those stung by *Bembix texana*. In each case the flies stung by *Stictia* showed much movement of the legs and mouthparts on the evening of the day they were stung, and on the following day all seemed fresh and some still showed some movements. On the other hand, flies stung by *Bembix texana* rarely showed any movements on the evening of the day they were stung, and the next day often appeared dessicated.

The manner of prey carriage in this species does not differ from that in other bembicine wasps, that is, the prey is held venter up by the middle legs and transferred to the hind legs immediately upon entering the nest (Fig. 132). Large horseflies often become stuck in the entrance, in which case the wasp enters, turns around inside the nest, and pulls them in with her mandibles. On one occasion I turned one such fly around while it was in the entrance. The *Stictia* was unable to pull it in because the wings caught on the

Fig. 132. Female *Stictia carolina* taking a *Tabanus* into her nest (Highlands Hammock State Park, Florida). She is scraping open the entrance with her front legs while holding the fly with her middle legs and standing on her hind legs.

sides of the burrow, but after a few moments she came out, straddled the fly in the usual manner, took a brief flight, and then entered the nest in the usual way. However, wasps that are unable to enter the nest because of damage to the entrance normally abandon their flies, and when they succeed in digging into the nests they fly off and catch another. We noted no instances of females picking up flies from the ground, nor did we see any females trying to rob one another. Female *Bembix texana* in Florida rather commonly tried to steal flies from female *Stictia carolina*, but they were not often successful.

If there is some disturbance in the nesting area, or if a human happens to be standing or sitting near the nest, a female *Stictia* laden with prey will normally make a circle near the nest and then fly off a considerable distance, often returning only after several minutes have elapsed.

(7) *Final closure*. Although Hartman says that "each individual performs the final close with scrupulous care, the whole tunnel being filled with sand," our experience with 20 nests dug out in Kansas after the final closure does not support this statement. Indeed, I would characterize the final closure as little more than an unusually thorough inner and outer closure; without an exception a considerable section of the burrow is left unfilled (Table 24). Hartman is correct, however, in stating that the surface of the soil is smoothed off over a considerable area. The outer part of the closure is made by scraping sand from the mouth of the burrow and by going out from the entrance a distance of several centimeters while kicking sand toward the hole. An unusual amount of pounding with the decurved tip of the abdomen is done on the sand in the burrow, and after it has been filled flush with the surface the wasp pounds gently over the filled burrow and the area immediately surrounding it. Any remaining irregularities around the entrance are swept away, and the wasp may return several times to the place, each time smoothing the sand off a little more. In most cases no evidence of the nest remains, but some individuals do leave evidence of a radiating pattern of lines around the entrance (Fig. 133).

The final closure in this species is not made until the larva is very large (30 mm or more) and ready to spin its cocoon within 12–24 hr. Thus larval development can be assumed to take about as long as provisioning (6–11 days, as noted earlier). Within a few hours after making a final closure (or the next morning), the female begins a new nest. I would assume that a female is capable of raising only 6–10 larvae in her lifetime.

§B. *Stictia carolina* (Fabricius)

TABLE 24. DATA ON FINAL CLOSURE IN STICTIA CAROLINA

(Pottawatomie Co., Kansas)

Nest no.	Length (cm)		
	Inner closure	Unfilled burrow	Outer closure
L2	25	10	5
L3	22	25	5
L5	15	18	5
L6	15	12	5
L9	22	21	5
L12	20	13	7
L17	13	15	5
L18	16	20	6
L26	17	23	8
L29	10	17	5
L30	20	8	8
L33	15	22	4
L35	15	15	5
Y1	27	20	6
Y4	20	12	4
Y8	23	20	4
Y9	10	32	10
Y10	20	13	10
Y24	23	10	6
Y27	10	22	9
Mean	18	17	6
Range	10-27	8-32	4-10

(8) *Natural enemies.* Hine reported finding occasional nests containing dead larvae; "in at least one instance," he remarks, "it appeared as though a certain species of little red ant was responsible for this condition, for numbers of them were feeding in the nest, both on the dead larva and on the remains of insects found with it." The ant was identified by W. M. Wheeler as *Solenopsis geminata* Fabr. Krombein found the larva of a miltogrammine fly in one of the nests he dug out.

A dead full-grown larva was found in one cell (nest L2, Kansas). This cell contained numerous small dipterous larvae which fed on the decaying larva and the detritus in the cell for 5 days before pupating. Four weeks later 12 flies of the family Phoridae emerged. These were identified as *Dohrniphora cornuta* (Bigot). There is, of course, no assurance that these flies caused the death of the larva; they may have been developing on the detritus in the cell and

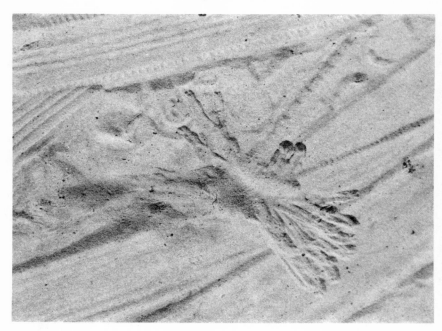

Fig. 133. Nest of *Stictia carolina* that has just received a final closure (Highlands Hammock State Park, Florida). The radiating lines represent pathways followed by the wasp when scraping sand over the nest entrance.

turned to the larva when the latter died of some other cause, or they may have been deposited in the cell after the death of the larva. This nest had received the final closure when it was dug out, so it is evident that the larva died a short time before it would have spun its cocoon.

Miltogrammine flies were common in the areas studied both in Kansas and Florida. In Kansas, *Stictia* females laden with prey were sometimes followed by a small cloud of six or eight flies. The flies would sometimes larviposit on the prey as it was being taken into the nest entrance; at other times they appeared to follow it down the burrow a short distance before quickly reappearing at the entrance. Actually, only 4 of 32 nests dug out were found to contain maggots, though it is possible that some were overlooked. In all 4 nests the larva was either fully grown and about to spin up, or had already spun the cocoon. The maggots merely consumed the considerable mass of fly remains in the cell and could not possibly have harmed the wasp larva at this stage. The maggots in one cell were saved, and several adult flies emerged about 2 weeks later;

they were determined as belonging to the *Senotainia trilineata* Wulp complex.

At Highlands Hammock, Florida, miltogrammine flies of this same complex were even more commonly observed trailing female wasps laden with flies. However, only 1 of 13 nests dug out was found to contain maggots. These maggots were very small, but the wasp larva was very large and nearly ready to spin its cocoon, so again it is unlikely that the maggots would have harmed it in any way. These maggots developed on the fly remains in the cell and formed their puparia 3 days later; 12 days later four flies of the *Senotainia trilineata* Wulp complex emerged.

It should be noted that since these flies larviposit on the prey, their opportunity to enter a given nest is greatest the last day or two of provisioning, when flies are being carried in in the greatest numbers. Presumably *Senotainia* occasionally succeeds in larvipositing on prey being taken into a cell containing a very small larva. In this case, it is possible that the maggots destroy the wasp larva.

C. *Stictia signata* (Linnaeus)

This species occurs throughout tropical South America and ranges north through the West Indies and Central America, barely reaching Florida and southern California. Since there are several records of this wasp capturing horseflies from livestock in the manner of *S. carolina*, the species might appropriately be called the "tropical horse guard." There have been no detailed studies of *signata* although there are several brief reports. Bates (1863) was the first to publish on the species, while the report of Richards (1937) is the most recent and most detailed. I have encountered the species at several localities on the coast of Mexico as well as on the Caribbean coasts of Costa Rica and of the island of Dominica in the Lesser Antilles. However, all of my studies were very brief.

On Dominica, *Stictia signata* nests in both black and white sand beaches, well behind the high-water mark, as well as in sand along the major rivers. In Mexico I have found the species only in dry, rolling sand behind sea beaches. At Puerto Juárez, Quintana Roo, numerous males were observed flying 0.5–1 m high around small hillocks in sand behind the beach. At Mazatlán the two nests located were both in sloping banks behind and facing away from the sea beach. Wolcott (1923) reports the species nesting in sandy soil in Puerto Rico. Bates (1863) studied the species on the upper Amazon, and found that it "sometimes excavates its mine solitarily on sandbanks recently laid bare in the middle of the river."

According to Richards (1937), typical nesting sites in British Guiana are river banks and small, sandy islands in rivers, again places flooded in the rainy season. Willink (1947) indicates that such situations are also usual in Argentina.

In contrast to the numerous records of this species nesting in bare, relatively friable sand, the colony I studied in Costa Rica (no. 1967, February 1964) occupied a grass-covered landing strip for planes. There were between 100 and 200 nests scattered along the narrow landing strip for about 200 m. The nests showed some tendency to be clustered in spots where the grass was sparse, but many of them were amid dense grass, and throughout the area the soil, although sandy, was very firm and filled with grass roots. I attempted to dig several nests but found it most difficult without making a major excavation, which was impossible in this situation. This landing strip was immediately behind the beach (about 50 m from the high-water line), and I concluded that the colony might have occupied this site before it was converted to a landing strip and packed down and planted to grass. The attachment of these wasps to their nesting site may have been such that they remained there even though they were forced to dig through much firmer soil. There was a strip of bare, friable sand between the air strip and the beach front, but it seemed to be unoccupied by these wasps.

This appears to be the largest colony of this species ever reported (although the nests were widely dispersed, the entrances usually separated by 1 m or more). I found only isolated nests in Mexico and no more than ten nests in any one place on Dominica. Bates speaks of the species nesting solitarily. Bodkin (1918) states that in British Guiana the species usually occurs "in colonies of twenty or thirty." Earlier, Bodkin (1917) reported a mixed colony of this species and *Rubrica denticornis* on a foot path. Richards noted aggression between two females nesting in close proximity.

There seem to be no published reports on the nature of the nest of this species. The two I excavated at Mazatlán in July, 1959, were both simple and unicellular, the burrow being initially at nearly a 90° angle with the sloping sandbank, then leveling off well before the horizontal cell. In one nest, the burrow was 30 cm long, the cell 30 cm deep as measured from a point directly above; in the other, the burrow was 40 cm long, the cell 35 cm in vertical depth. The cells measured 20 mm in diameter by 42 mm in length. Both nests had an inner closure 2–3 cm long in the horizontal part of the burrow, just outside the cell (Fig. 131, no. 1609). In contrast, four nests excavated on Dominica were all very shallow, burrow

§C. *Stictia signata* (Linnaeus)

length varying from 24 to 35 cm (mean 29 cm), cell depth varying from 14 to 20 cm (mean 16 cm; nos. 2061 and 2066, February-March 1965).

Bates noted that the nest entrance is closed when the female leaves to hunt for prey. This was true of the nests studied on Dominica, all of which had large larvae when dug out. One of the two wasps I watched at Mazatlán made a lengthy outer closure; this nest was recently completed and was found to contain an egg. The other nest contained a partially grown larva and was left open for several hours while the female brought in flies at a fairly rapid pace. Both nests had some evidence of a mound of sand at the entrance, indicating that leveling is incomplete in this species.

At Limón, Costa Rica, I noted that virtually all nests that were being actively provisioned lacked an outer closure. Many other nests had an outer closure, and I assumed that these contained eggs or very small larvae. I observed much digging of nests after 1400, and most females appeared to disperse the mound only partially, much as in *S. carolina*.

The single egg located (at Mazatlán) was about 6 mm long and was attached to the side of a small syrphid fly (*Allograpta* sp.) This fly was dorsum up; because of its small size the egg seemed to be poorly supported and was oblique rather than fully erect.

The second nest at Mazatlán contained a larva 8 mm long and 16 flies belonging to six species, as follows:

TABANIDAE
    Tabanus truquii Bellardi   1 ♀
    Tabanus sp.   1 ♀
STRATIOMYIDAE
    Hedriodiscus dorsalis (Fabr.)   1 ♂
SYRPHIDAE
    Allograpta sp.   9 ♂♂
MUSCIDAE
    Musca domestica L.   3 ♀♀
TACHINIDAE
    Sitophaga sp.   1 ♀

On Dominica the prey consisted largely of muscoid and syrphid flies. Of 23 flies taken from four nests, 15 were *Cochliomyia macellaria* (Fabr.), the remainder Syrphidae of the genera *Eristalis* and *Volucella*. On Puerto Rico, Muscoidea are also used as prey. Dr. Paul J. Spangler recently sent me a larva and some prey of *S. signata* which he took at Lago Tortuguero, Puerto Rico. The three flies taken were each a different species of Muscoidea: *Morellia scapulata*

(Big.) (Muscidae), *Cochliomyia macellaria* (Fabr.), and *Phaenicia cluvia* (Wlk.) (Calliphoridae).

At Limón, Costa Rica, this species also preyed upon various muscoids (*Sarcophaga* spp. and an unidentified tachinid), although the major prey consisted of the horsefly *Diachlorus curvipes* (Fabr.) (Tabanidae). Females laden with prey approached their nests swiftly, close to the ground, producing a low-pitched hum. Most females plunged into their nests quickly, there being no closure at the entrance.

In the Amazon Basin, Bates reported that *signata* attacks the Motúca (cited as *Hadaus lepidotus*, now known as *Lepiselaga crassipes* Fabr.), a tabanid fly which is a vicious biter; he remarks that the wasps had to fly "at least half a mile" to procure this prey. Bates noted that the wasps sometimes hover around humans. One day, he comments, "I was rather startled when one out of the flock which was hovering about us flew straight at my face: it had espied a Motúca on my neck, and was thus pouncing upon it. It seizes the fly not with its mandibles, but with its fore and middle feet, and carries it off tightly held to its breast." The prey may be seized by the fore and middle legs, but it is carried to the nest by the middle legs alone, as I have noted in Mexico.

Bodkin (1918) notes that in British Guiana these wasps "may frequently be seen about mules and cattle in the pastures waiting to carry off the Tabanidae which are always present about these animals." Wolcott (1923) reports them in Puerto Rico "chasing *Chrysops costatus* Fabr. [Tabanidae] on horses" and "chasing flies attracted to molasses." Richards reports a muscoid fly as prey.

One curious record is that of Howard, Dyar, and Knab (1912) of these wasps capturing mosquitoes and consuming them directly. Although there are isolated reports of bembicine wasps resting on vegetation and feeding at their prey, these wasps were taking prey very much smaller than would normally be employed in provisioning. Their report is as follows:

Mr. H. J. Browne . . . at Calapatch Island, 70 miles south of Batabano, Cuba, and 30 miles east of the Isle of Pines, found the large wasp, [*Stictia*] *signata*, in great numbers. They seemed to make their practical diet on the yellow-fever mosquito, which they seized upon the wing and devoured in large numbers. Mr. Browne says that he has seen one of these wasps seize and devour upon the wing 20 mosquitoes in the space of five minutes.

Richards reports that when the nest is finished, the soil in the entrance is "rammed down with the tip of the abdomen."

According to Bodkin, a bombyliid fly, *Anthrax* sp. (probably actually a species of *Villa*), is often found around the burrows of *Stictia signata* in British Guiana. In the same country, Richards found otitid flies, *Oscinella columbiana* End., perching around nest entrances while the wasps were digging. He believed that they might slip into the nest and lay their eggs among the prey of the wasp. Two of the four nests excavated on Dominica contained maggots. In one instance the maggots appeared to belong to the family Phoridae, while in the other case they were acalyptrates, probably of the family Chloropidae or Otitidae. In each case there were a large number of very small maggots living among the fly remains. The wasp larvae appeared healthy and in no way disturbed by the scavengers.

D. *Stictia vivida* (Handlirsch)

This brilliantly patterned *Stictia* is apparently confined to sea beaches on the east coast of Mexico and extreme southern Texas. In 1957 I published a brief note on observations made in Cameron County, Texas, and also presented records for the states of Tamaulipas and Veracruz, Mexico (Evans, 1957a). In July 1962 I spent several days camped on the beach near Progreso, Yucatan, where I was able to study the species further. However, the population of the wasp was not high, and I was able to find only ten nests.

At Port Isabel, Texas, I collected several males on the flowers of black mangrove, *Avicennia nitida,* growing along the protected beach of a bay. All other specimens were found on unprotected sea beaches, more particularly in rolling sand somewhat above the maximum level of high tides. At Boca Chica, Texas, two nests were found in "fairly hard-packed sand well back from, but facing, the gulf shore" (Evans, 1957a). At Progreso, Yucatan, I observed an estimated 20 females and perhaps that many males in an area about 40 m long and 10 m wide, paralleling the waterfront (Fig. 134). For the most part they were confined to smooth, flat areas, especially blowouts, where the sand was more firm than elsewhere, although still fine-grained and friable. The wasps were also seen on the beach about 2 km away, and it is probable that there were small nesting aggregations in suitable areas for many kilometers along the beach.

At Progreso males were commonly observed flying back and forth and hovering over small plots of sand, usually sections of blowouts or flat areas, or the sides of sandbanks. The same male apparently occupies the same area day after day, and neighboring males appear to maintain quite rigidly fixed territories, estimated to be about

Fig. 134. Beach just west of Progreso, Yucatan, Mexico, July 1962. Nests of *Stictia vivida* were scattered about in the depression between the two rows of small dunes.

3 m in average diameter. The males often remain in the air for many minutes at a time, flying 30 to 60 cm high (Fig. 135), now and then landing on the sand but remaining in an alert posture, the wings half extended, the antennae extended rigidly forward. Males in adjacent territories often butt one another and pursue one another short distances. Also they dart at any moving object within their territory. I threw pebbles and other objects near males and without exception elicited a following response. Presumably the males attempt to mate with females that emerge within their territories or later enter it, but I did not observe any contact between males and females.

The nests at Progreso were well dispersed, no two being closer together than about 1 m. Most new nests appear to be dug in the morning hours (0800–1100). Digging appears to differ in no way from that of *carolina*. When the nest is complete, the entrance is closed and the mound of sand in large part leveled, the leveling movements appearing identical to those of *carolina*. After leveling, the female flies off to capture a fly on which the egg is laid. Upon leav-

§D. *Stictia vivida* (Handlirsch)   249

ing the nest after oviposition, she makes a lengthy outer closure, using the tip of her abdomen for pounding soil in the entrance and scraping sand over the nest from several directions. Apparently the females spend the night and periods of inactivity away from the nest.

The burrow is oblique, at a 45–60° angle with the surface, often slightly less than this near the entrance; deep in the soil, approaching the cell, the burrow levels off, the level portion being from 4 to 10 cm in length (Fig. 136). The cell is horizontal and measures 20–25 mm in diameter by 45–55 mm in length. In the ten nests excavated in Yucatan, burrow length varied from 33–53 cm (mean 47 cm); cell depth varied from 15 to 25 cm (mean 22 cm). The two nests excavated in Texas were considerably deeper, burrow length being 60 and 68 cm, cell depth 33 and 45 cm.

All females observed maintained an outer closure at all times when not actually in the nest. Several nests had an inner closure in the lower, horizontal section of the burrow near the cell. This closure varied in length from 2 to 4 cm. In nests containing fairly large larvae the inner closure was generally absent. One nest containing an egg lacked an inner closure, but this egg was nearly

Fig. 135. Male *Stictia vivida* hovering over restricted portion of nesting area near Progreso, Yucatan, Mexico. Note that the legs hang somewhat loosely, a typical precopulatory posture of male Bembicini.

Fig. 136. Nest of *Stictia vivida*, with contents removed (no. 1911A, Progreso, Yucatan, Mexico).

ready to hatch, and the female was provisioning when the nest was dug out. Probably the inner closure is maintained except during periods of prey carriage.

Eggs were found in three nests, in each case glued to the side of the first fly brought in. This fly has one wing slightly extended, and the egg is attached to the wing base and side of the thorax as in many *Bembix*. In one instance the fly was venter up, as usual in *Bembix*, but in the other two cases the fly was dorsum up, as shown in Fig. 137. Apparently the fly on which the egg is laid is not eaten, and it may be rather dry and hard by the time the egg hatches.

The first fresh flies are brought in shortly before the egg hatches and are placed close to the egg. Prey carriage is similar to that of *carolina*, although the flies employed are generally much smaller. The total number of flies employed per cell was not determined, but it appeared to be very large. Most prey carriage occurred in the afternoon (1400–1700). In Texas I found only *Tabanus texanus* Hine in the two nests studied. In Yucatan one nest contained the remains

§D. *Stictia vivida* (Handlirsch) 251

of about 20 *Tabanus* sp., but the other nests contained more calliphorids than tabanids. The intact flies from the nests at Yucatan were identified as follows, with the numbers of each taken:

TABANIDAE
    Tabanus schwardti schwardti Phil.   2 ♀♀
    Tabanus vittiger guatemalanus Hine   1 ♀
    Leucotabanus itzarum Beq.   1 ♀
CALLIPHORIDAE
    Cochliomyia macellaria Fabr.   28 ♀♀, ♂♂
SARCOPHAGIDAE
    Sarcophaga effrenata Walk.   1 ♀

I did not observe final closure of the nest of this species. One of ten nests excavated in Yucatan contained four miltogrammine maggots which unfortunately were not reared successfully. These maggots were nearly as large as the small wasp larva, but the latter

Fig. 137. Egg of *Stictia vivida* on a fly, *Cochliomyia macellaria*, which is dorsum up (no. 1911B, Progreso, Yucatan, Mexico).

appeared healthy and might possibly have reached maturity if provided with enough flies. I observed no adult miltogrammine flies in the nesting area.

E. *Stictia heros* (Fabricius)

This species occurs along the west coast of Mexico in much the same situations as *vivida* on the east coast; so far as I know these two species are restricted to their respective coasts. I have collected *heros* at several localities, always in sand behind sea beaches. On 20 July 1959 I found a small nesting aggregation 2 km north of Mazatlán, Sinaloa (nos. 1611–1614). I returned to the same place in late August 1963 but saw only one or two females and found no nests; possibly the species had nearly completed its nesting at that season.

These nests were located in a broad, wind-swept area of sand near the top of a series of small dunes about 20 m from the water and in ready view of the ocean. *S. signata* nested nearby, also *Bembix multipicta* and *Stictiella evansi*, but all three of these species tended to nest in more protected places somewhat farther from the water, chiefly in the slopes of the dunes facing inland. No male *Stictia heros* appeared to be present during the period of study. The number of females did not appear large, possibly 20 to 30. These females were digging and provisioning nests well scattered over a slightly sloping area completely devoid of vegetation. Only four nests were marked and dug out during the short period of study, although doubtless many more nests than this were present.

Digging of the nest was observed several times. The wasp backs out periodically and clears the sand from the entrance, producing a large, diffuse mound of sand. When the nest is finished, a prolonged outer closure is made, and the wasp then levels the mound partially in much the manner described for *S. carolina;* the result is a large, rather low, partially dispersed mound with a pattern of zigzag tracks down the center.

The nest is deep, the burrow being at a 45–60° angle with the surface and then leveling off before reaching the horizontal cell, which measures about 22 × 50 mm (Fig. 133, no. 1614). In the four nests dug out, burrow length varied from 49 to 63 cm (mean 56 cm), cell depth from 30 to 41 cm (mean 33 cm). An outer closure is maintained at all times when the female is away from the nest. All four nests contained an inner closure, 2–4 cm long, a short distance from the cell.

The egg is laid on the first fly brought in. The one egg recovered was laid on a stratiomyid, *Odontomyia discolorata* James (♂); this fly was dorsum up in the cell, the egg erect and attached to the side of the fly.

Prey is brought in swiftly, carried by the middle legs. In several instances other females pounced upon females bringing in flies, but I did not observe any successful cases of prey stealing. In this area the wasps were bringing in large numbers of the brightly colored tropical horsefly *Dichelacera caloptera* Hine (Tabanidae). Fifteen intact females of this horsefly were taken from three of the nests, and there were remains of other specimens. One of the nests contained no other flies, while a second contained numerous *Dichelacera* and one *Odontomyia discolorata* James (Stratiomyidae), the same species found in the nest containing an egg. The fourth nest contained numerous *Dichelacera*, one intact *Cochliomyia macellaria* (Fabr.) (Calliphoridae), and remains of another muscoid.

Final closure of the nest was not observed. No parasites were found in the four nests excavated or seen about the nest entrances.

F. Ethology of other species of *Stictia*

Bodkin (1918), in his "Notes on some British Guiana Hymenoptera," states that *Stictia pantherina* (Handlirsch) is "a fairly common species in some parts of the interior, especially where sandy soils exist. It is quite the largest and fiercest of these insects to be found in the colony. Its burrows are stored with Tabanidae."

Hudson, in his book *The Naturalist in La Plata* (1892), devotes a chapter to a wasp identified as *Monedula* (that is, *Stictia*) *punctata* (Fabricius). Llano (1959) points out that *Stictia punctata* is not prevalent in the area described by Hudson, and also that Hudson's description fits *punctata* poorly but is a fairly accurate description of *Rubrica gravida*. The edition of Hudson's book available to me (fourth edition, 1903) has a sketch of a wasp bearing much resemblance to *S. punctata*, but it is probable that this illustration was prepared much later, using museum specimens of *punctata*. I agree with Llano that Hudson was probably dealing with *Rubrica gravida*, and I have considered his account under that species (Chapter IX:L).

G. Genus *Zyzzyx* Pate

This genus was described by Parker (1929) under the name *Therapon*. This name had been used previously for another group of animals, and Pate (1937) therefore renamed it *Zyzzyx*, perhaps an onomatopoetic word suggesting the buzzing sounds made by these insects, as suggested by Willink. This genus contains but one species, *chilensis* (Eschscholz).

Superficially, *Zyzzyx chilensis* bears much resemblance to the species of *Stictia*, but there appear to be enough differences to justify generic status for this form. The clypeus, labrum, and proboscis are all somewhat more elongate, and the proboscis is incapable of being fully retracted beneath the labrum; the palpi are very slender and show a loss of one segment in each palpus. Although the mouthparts thus appear more specialized than in *Stictia*, the ocelli appear more generalized; while the shape is similar to *Stictia*, the dark cicatrices of that genus are replaced by clear and apparently functional lenses, that of the front ocellus being situated on an elevation. The mesosoma is similar to that of *Stictia* except that the propodeum is more rounded behind. The middle femur of the male has a preapical spine as well as a series of small teeth along the under side, and the middle coxa is provided with a long spine. The terminal abdominal segments of the male are similar to those of *Stictia*, but the spiracular lobes of tergite VII are much smaller and do not overlap sternite VII, which is of much simpler structure; also, the parameres of the genitalia are more broadly expanded apically (Fig. 138; see also Willink, 1947, Figs. 77–82).

H. *Zyzzyx chilensis* (Eschscholz)

This is a common wasp throughout much of subtropical and temperate southern South America, ranging as far south as the state of Rio Negro in Argentina and the Island of Chiloe in Chile. Janvier (1925, 1929), who is the only person to have described the behavior of the species, reports it from coastal dunes in Chile as well as from inland localities including the shores of rivers, sandbanks, and earthen roads. Sand containing a considerable admixture of gravel and/or lime is sometimes utilized for nesting. Colonies vary in size from only a few individuals up to those of very large size; they are said to persist in the same place year after year.

According to Janvier, at Santiago the males appear toward the end of November, to be followed in about 1 week by the females. Both sexes are said not only to visit flowers for nectar, but also to capture flies for feeding purposes. They are said to descend on the back of the fly, grasp it with their front legs, turn it venter up, and plunge their proboscis into the fly just in front of the front legs. The wasps remain immobile for several minutes, apparently resting on vegetation in the manner described by Janvier for *Bembix brullei*, sucking the body fluids of the fly, eventually rejecting it "inert and

§H. *Zyzzyx chilensis* (Eschscholz)

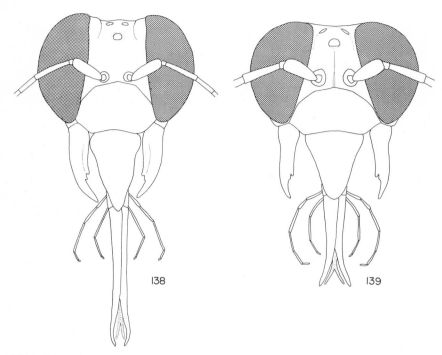

Figs. 138 and 139. Head of female, mouthparts extended: 138, of *Zyzzyx chilensis* (Eschscholz); 139, of *Editha magnifica* (Perty).

half drained." Janvier speaks of the maxillae of *Zyzzyx* as resembling knife blades which are introduced into the viscera of the prey.

A considerable part of Janvier's account of these wasps consists of a description of their sleeping behavior, which is strikingly like that of members of the genus *Steniolia* in North America. Janvier reports that in the late afternoon, well before sunset, the wasps fly from their nesting places to certain trees, there forming dense sleeping clusters which remain intact until 0800 to 0900 the following morning. Many kinds of trees are utilized, including pines as well as broadleafed trees; the branches selected are generally low, pendant ones, at about the height of a man's head. The trees selected are typically in protected places, and the same clustering sites are used night after night. The clusters may contain several hundred individuals, up to 500, with several branches in close proximity sometimes "buried beneath the wasps." Janvier provides a photograph of a cluster on pine.

Janvier explains that wasps coming in to rest fly "in a whirlwind" before settling down:

Some of them settle for a moment, then fly off again, settle once more only to fly off again, and so on ten or twenty times before finally settling down for good. The first ones hook their claws on to the leaves, the next ones grasp the bodies of the first ones, and so on. For more than an hour there is a swarming mass which is augmented constantly by the arrival of newcomers. Towards six o'clock, the latecomers arrive; the agitation diminishes in the clusters; only the extensions and retractions of the mouthparts and the palpitations of the abdomens can be discerned. With the setting of the sun, immobility is complete.

Janvier found that the two sexes cluster together. Some clusters he studied contained as many as 10 males to 1 female, while others contained mostly females; one, for example, contained 212 females and only 9 males. Janvier believed that early in the season the males predominate, "but little by little, their mission fulfilled, they disappear, while the females increase and make up the majority in their turn."

Janvier collected one cluster in the evening and carried it 3 km away. He marked the wasps with red ink and cut off the antennae of some of them. The next morning the wasps flew off in the direction of the locality where Janvier had collected them, flying a straight course although Janvier had brought them by a road that made a right angle. The next evening, the wasps clustered in their original site, including several whose antennae had been cut off. However, some of the marked wasps clustered on another tree 200 m away. He watched these clusters for another week, and found them always on the same branches, although individual wasps did not always cluster on the same branch.

The females often nest at some distance from the clustering sites. Janvier reports much use of the mandibles for dragging out stones and clods. The burrow descends at a gentle angle and terminates in an ovoid cell, the first of several to be constructed per nest. Janvier provides no data on nest depth, and his figures do nothing to clarify this. He reports that an entire week is required to construct the burrow and initial cell, the entrance remaining open most of the time during this period.

The egg is laid deep in the empty cell, glued erect to the bottom. After oviposition, the entrance is closed and smoothed over when the female leaves the nest. Eclosion of the egg occurs on the 5th or 6th day. Close to the time of hatching, the female places a small fly beside the larva. Provisioning is progressive, the larva reaching full size after the 6th or 7th day of feeding. At first small flies are provided, but later on very large flies, including even Mydaidae, are captured. The smaller flies appear dead, but the larger flies are often imperfectly paralyzed; the latter are said usually to have a wound on the thorax, probably the result of having been pierced by the wasp's proboscis. Some 20 to 30 prey are provided per cell. When the cell is fully provisioned, the female is said to dig a new cell and lay her egg, meanwhile leaving the first cell unclosed until the larva has consumed all its provisions; the closure of the first cell usually coincides with the birth of the second larva. Janvier found up to six cells per nest. The cells form a fairly close cluster toward the bottom of the main burrow; the older cells are uppermost, the more recently completed cells slightly deeper. The cocoon is said to have the usual nipple-like elevations around its middle part.

Janvier provides a fairly long list of flies employed as prey, but there seems little point in reproducing his list here. The following families are represented: Tabanidae, Nemestrinidae, Bombyliidae, Therevidae, Asilidae, Mydaidae, Syrphidae, Sarcophagidae, and Tachinidae. He also found one lepidopteran, *Hylephila fasciolata* Blanch., a skipper (Hesperiidae).

Janvier's account is most interesting, but as usual in his case it is necessary to point out certain features that seem somewhat out of line with knowledge of related wasps. When virtually all non-Chilean digger wasps complete their nest in from a few hours to 1 or 2 days, why should so many Chilean species require so much longer? *Zyzzyx chilensis, Bembix brullei, Cerceris chilensis,* and several others are said to require 1 week or more, while *Ammophila rufipes* (which nests in hard soil) is said to require 15 days to 3 weeks! Also, when the eggs of nearly all non-Chilean wasps hatch in from 1 to 3 days, why should the eggs of so many Chilean species take from 5 to 10 days? These questions raise doubts as to the completeness of Janvier's field data and the caution he exercised in writing them up for publication. Such doubts cannot help but extend to other parts of his account. Actually, much of his report seems plausible enough, but those parts that are inconsistent with what is known of other Bembicini should be accepted only tentatively for the present (for example, his statement that the cell is not closed immediately after the completion of provisioning; see also Chapter III:M).

I. Genus *Rubrica* Parker

This genus was erected by Parker (1929) to include three closely related species formerly grouped with the species of *Stictia* but having a somewhat different general appearance. They are rather slender species, with generally short body hairs and more conspicuously punctate integument. All three species are strongly banded with yellow and also have a certain amount of ferruginous coloration. The ocelli are represented by cicatrices, as in *Stictia,* but the anterior ocellus is transverse, the groove along its upper margin somewhat sinuate. The anterior cicatrix gives the appearance of covering the lens more thickly and less smoothly than in *Stictia,* approaching the condition in *Bicyrtes.* The males have a tooth on the middle coxae that is much shorter than in *Zyzzyx,* and the middle femora of the male have a thick subapical tooth that is bifid apically. The structure of the terminal abdominal segments of the male differs little from that of the genus *Stictia,* the spiracular lobes of tergite VII being large and overlapping large membranous areas on sternite VII. Willink (1947, Figs. 43–58) has figured many of the important structural features.

J. *Rubrica surinamensis* (DeGeer)

This is one of the most common and widely distributed of Neotropical bembicine wasps, occurring from central Argentina north to Trinidad and to Central America. It does not seem to be strongly restricted ecologically, occurring in arid as well as humid areas and nesting in various types of soil. Areas disturbed by man are often attractive to these wasps, and several authors report them nesting in paths or roads. Thomas Belt (1874) noted females chasing horseflies in Nicaragua, and Brèthes (1902) made a careful study of the species near Buenos Aires. Bondar's (1930) paper covers several Brazilian wasps, but I judge that his remarks are based mainly on *R. surinamensis.* More recently, Vesey-Fitzgerald (1940) and Callan (1945, 1954) have studied the species briefly in Trinidad and in Venezuela, and Llano (1959) has studied it in Argentina.

This is reported to be a gregarious species, but there are few published notes on the size or density of colonies. Brèthes says that the colonies are "more or less populous"; Callan speaks of a colony comprising "several dozen individuals." Llano reports that colonies may be quite dense; he states that in some nesting areas one may recover 50–60 cocoons from 1 m^2 of earth. His remarks on incipient colonies lead one to believe that the males may be territorial, as in *Stictia.* Vesey-Fitzgerald notes that the males fly around the nesting area in circles and land on objects in the vicinity. "At night," he says, "the males and females segregate and rest on vegetation, the former in groups of many individuals together." Bodkin (1917) also reported sleeping clusters of "at least a hundred" individuals, of

§J. *Rubrica surinamensis* (DeGeer)

undetermined sex, on a "low growing maritime plant" on the shores of British Guiana.

Bare, moderately firm soil is selected for nesting. When digging, Vesey-Fitzgerald reports, "the soil is scraped away with the front legs, any particularly resistant bits being torn loose with the mandibles, while the loosened particles are cast aside with the hind legs to form a fan of excavated material around the entrance of the hole . . . *Rubrica* always chooses the hottest part of the day, the early afternoon, during which to do the hardest work."

The burrow is oblique and terminates in a large, horizontal cell. Brèthes states that the burrow is from 10 to 12 cm long, leading to a cell 4 cm long by 1.5 cm high, the cell being about 6 cm beneath the surface. Bondar presents very similar figures (burrow length 10–15 cm, cell depth 6–8 cm). Copello (1933) gives the burrow length as about 15 cm, cell depth as about 10 cm. In any event the nest is unusually shallow for so large a wasp, and the burrow apparently lacks the horizontal portion characteristic of *Stictia*. Brèthes' sketch of a nest shows a strong outer closure but no inner closure. According to Vesey-Fitzgerald, an outer closure is maintained when the cell contains an egg or small larva, but later on it may be omitted. When making a closure, Vesey-Fitzgerald notes, "the female wasp appears head first in the entrance of the burrow and starts scraping loose particles towards herself. As the entrance fills up she comes farther and farther out, all the time alternating the action of scraping up soil with periods of vigorous tamping with the tip of the abdomen. When the entrance is closed she suddenly flies away without inspecting her work."

Brèthes reports that the egg is laid on a medium-sized fly, one end of the egg being attached to the wing base, the egg standing erect as in many species of *Bembix*. Provisioning is progressive, the prey consisting entirely of Diptera, chiefly species of medium to large size. Brèthes presents a long list of flies used as prey; this list includes members of the following ten families: Tabanidae, Stratiomyidae, Bombyliidae, Nemestrinidae, Asilidae, Syrphidae, Muscidae, Calliphoridae, Sarcophagidae, and Tachinidae. Belt had earlier noted females pursuing tabanids around mules and humans in Nicaragua. Vesey-Fitzgerald reports the major prey in the colony he studied in Trinidad to have been *Volucella obesa* Fabr. (Syrphidae) and *Tabanus occidentalis* L. (Tabanidae). Callan states that in Trinidad the wasp usually takes horseflies and syrphids, but may also use houseflies and stable flies (Muscidae) in considerable numbers, the latter often being engorged with blood. He found one nest provisioned entirely with stable flies, presumably captured from livestock. From the Paria Peninsula, Venezuela, Callan reports *Tabanus sorbillans* Wied. as prey. Willink (1947) reports *Eristalis tenax* L. (Syrphidae) as prey in Argentina. Callan states that the flies are killed by the sting of the wasp.

Brèthes watched one of the wasps attack a large asilid fly of the genus *Mallophora*. The two fell to the ground, and the abdomen of the wasp was seen to double against the venter of the fly. The exact point of penetration of the sting was not observed.

Brèthes devotes a considerable portion of his account to a most unusual behavioral feature of this wasp. He watched a female leave her nest carrying something, fly off several meters, then return to the nest and carry out another object. On the third trip he followed the wasp and saw her drop a part of a dead blowfly on the ground. He then looked about the colony and discovered a great many Diptera on the ground, some intact and some as fragments: one *Sarcophaga* even bore the remains of a desiccated *Rubrica* egg! Brèthes considered various possible explanations for this behavior: perhaps the wasps injured the flies when they captured them, then rejected the more badly damaged ones after they had been placed in the nest; or perhaps neighboring wasps robbed one anothers' nests, discarding some of the booty. He could not fully accept either of these explanations, and he also rejected what was undoubtedly close to the correct answer: "The wasps removed uneaten flies from cells in which the larvae had completed their feeding." Brèthes rejected this explanation because he found flies being removed from cells containing larvae only partially grown. However, species of *Bembix* are now known in which the mother cleans the cell each morning of all uneaten flies and fly remains, then replaces them with fresh ones (see especially the discussion of *Bembix texana*, Chapter XI:A, 8). Brèthes description is so graphic that there can be little question that cell cleaning is well developed in *Rubrica surinamensis*. The possible function of this behavior is discussed by Evans (1957b:230).

Llano (1959) also noted cell cleaning in this species, although of a somewhat different nature. He reported females reutilizing old burrows and carrying out pieces of old cocoons, flying off with them 2 or 3 m, and dropping them on the ground. Apparently cell-cleaning behavior is capable of being transposed to early stages of nesting.

Brèthes reports that the cocoon of this species is 3 cm long, its greatest diameter not less than 1 cm. The outer walls are of earth cemented together with secretions of the larva; the interior is lined with silk. There are two protuberances in the walls, one on each side; the two apertures are covered on the inside with fine silk plugs. Brèthes provides a brief description of the larva and sketches of the larva and cocoon.

Copello (1933) described the structure and behavior of *Hyperalonia morio* (Fabr.), a bombyliid parasite of *Rubrica surinamensis*. The female fly deposits her eggs in the burrow entrances, and the elongate first-instar larva seeks out the wasp larva. When it finds the host, the larva attaches itself to the integument and remains more or less dormant until the following spring; at about the time of pupation of the wasp larva, the parasite suddenly feeds rapidly and kills its host, then pupates inside the cocoon.

Vesey-Fitzgerald and Callan report, from Trinidad and Venezuela respectively, that the fly *Pachygraphomyia spinosa* Malloch (Sarcophagidae) is a constant attendant of this wasp. These flies "land around the vicinity of the nesting holes"; they were observed by Vesey-Fitzgerald to enter nests freshly opened by returning wasps. This author also observed *Liohippelates pusio* (Loew) (Chloropidae) entering and leaving open burrows. He believed that the former might be a parasite of *Rubrica,* the latter a scavenger in the debris in the nest cells. It is interesting that Vesey-Fitzgerald mentions finding dipterous larvae (unidentified) "in the debris composed of excreta and fragments of prey, which collects in the brood chamber." Wasps that clean their cells, as Brèthes indicates this species does, normally leave very little in the way of debris in the cell, and Brèthes indicates that the cells he examined contained very little debris. It is possible that cell cleaning is not characteristic of all populations of this wasp.

K. *Rubrica denticornis* (Handlirsch)

Bodkin (1917) reported briefly on several species of Bembicini in British Guiana which are there called "cowfly tigers," with reference to their striped color pattern and their manner of capturing Tabanidae from cattle. The major part of his report is concerned with *Rubrica denticornis* (Handlirsch), which is said to form mixed colonies with *Stictia signata* in the fairly hard soil of footpaths. The large numbers of humans and livestock passing along the paths do "not appear to incommode the wasps in the slightest." The females "have no difficulty in re-locating their nests despite the fact that all signs of their presence may have been obliterated by the hoofs of passing cattle."

Bodkin describes the nest of *R. denticornis* as being straight and oblique, only about 15 cm long. The nests are apparently unicellular, and the entrance is "not purposely concealed during the absence of the owner as with some species." "Every now and then," Bodkin states, apparently referring to both *R. denticornis* and *Stictia signata*, "two females may be observed engaged in a fierce combat over some prey, both insects falling to the ground locked in a pugnacious embrace." Bodkin states (again apparently referring to both species) that the egg is laid on the first fly placed in the cell.

Bodkin describes the behavior of these wasps capturing Tabanidae from cattle. He believed that they wait until the flies are fully engorged and therefore unable to escape readily. He found many engorged tabanids in the nests. He lists the contents of one cell of *R. denticornis*. Four species of flies were represented, belonging to the families Tabanidae, Stratiomyidae, Syrphidae, and Muscidae; all of the genera are also reported as prey of *R. surinamensis*. Bodkin describes the manner of feeding of the larva and, very briefly, the manner of cocoon spinning. The cocoon is said to have, toward the middle, "four small, rough spots" which "are equidistant from one another and are a regular feature of all such cocoons."

L. *Rubrica gravida* (Handlirsch)

This is the largest species of *Rubrica* and the most southerly in distribution. Both Bondar (1930) and Llano (1959) indicate that its behavior is very similar to that of *surinamensis*. Llano's description of the behavior of this species is graphic and reasonably detailed. As mentioned earlier, Hudson's (1892) discussion of *Monedula punctata* very probably applies to this species. Hudson's chapter, titled "A Noble Wasp," is attractively written but of little scientific value.

Llano reports that this species is common in the province of Buenos Aires, Argentina, occurring in well drained, more or less bare, flat or slightly sloping soil which may consist of fine-grained sand or of more compact, coarse-grained materials. The adults visit flowers for nectar, especially those of Umbelliferae. In a general discussion of Bembicini, based primarily upon *Rubrica,* Llano remarks that the wasps do not spend the nights and periods of bad weather inside the nest, but on branches or in grasslands. Llano describes the digging of the nest as follows:

> The nest is dug with the first pair of legs, which are used in the manner of scoops to extract the earth or sand and throw it beneath the body in a continuous stream. The movements are very rapid and almost mechanical. Little by little the wasp disappears beneath the earth, remaining covered, then backs out scraping sand. The mandibles fulfill the mission of hands,

§L. *Rubrica gravida* (Handlirsch)

being used for drawing out small stones, bits of wood, roots, and so forth. Also, they are used as the compasses in rounding out the straight burrows. I have seen the wasps draw out very large and heavy objects.

The burrow is oblique, more or less straight, about 10–12 mm in diameter and some 17 cm long, terminating in a cell that measures about 25 × 35–40 mm. An outer closure is maintained at all times. Movements of concealment are fairly pronounced, and are described by Llano as follows:

> In closing, the wasp scrapes sand toward the burrow; when this is filled to the surface, she presses her abdomen to the soil and with circular movements and rhythmical blows of its apex, like a mallet, plugs the entrance. Then she . . . rakes and sweeps in all directions, rotating around the nest, which is left disguised and invisible even to the person who is watching. At 30 mm, sometimes up to 70 mm, she makes holes which have the appearance of burrows, which she abandons almost as soon as begun. I suppose that these serve to deceive the enemies which lurk about (parasites) or perhaps serve as reference points or markers to guide the wasps in their goings and comings.

I gather from this that the mound may be well dispersed and that false burrows are constructed. It is interesting that Llano concluded, as Tsuneki and I have concluded independently, that the false burrows may serve to deceive parasites.

Llano believed that the egg was laid on the first fly, but he was uncertain as to the details. Hudson's account leads one to believe that the egg is laid in the empty cell and that prey is brought in only after hatching, but I question the accuracy of this information. Several workers agree that provisioning is progressive and that flies are used as prey. Hudson describes the wasps capturing "stinging flies" from men and horses. He remarks that in addition to flies, this species "kills numbers of fire-flies and other insects." By fire-flies he presumably means nocturnal beetles of the families Lampyridae or Elateridae; or he may have referred to diurnal cantharid beetles of the genus *Chauliognathus,* as suggested by Brèthes (1902). In any event, subsequent workers have failed to confirm the use of insects other than Diptera by *Rubrica gravida*.

Llano found the prey to consist of Stratiomyidae, Asilidae (including *Mallophora* and *Erax*), Syrphidae (including *Eristalis*), Nemestrinidae (*Hirmoneura*), Muscidae (*Musca domestica*), and Tachinidae (*Hypopygia, Rhampinina*). Poulton (1917) mentioned *R. gravida* as preying upon *Volucella obesa* (Fabr.) (Syrphidae) near São Paulo, Brazil. Bondar (1930) also indicates that Diptera are used as prey in Brazil.

Llano believed that adult females live about 120 days and rear about eight larvae during that time. He believed that approximately 50 flies were required for each larva. He describes hunting and capture of the prey as follows:

> Their flight [during hunting] is unmistakable; buzzing, they approach and move away from flowers, heavily, with a swaying or swinging movement, ascending and descending each branch in scrupulous search for their future victims—always Diptera—which are also attracted to flowers. Having selected the prey, they draw back in the air, take aim, then hurl themselves violently forward, quickly and blindly attacking and imprisoning their victims. Not all species succumb immediately, and some of them struggle desperately, offering much resistance and falling and twisting in the twigs and grasses. One may then hear very clearly the angry buzz of the wasp and the anguished cry of the fly. The struggle usually terminates with the victory of the former, who carries the prey to her brood, curving her abdomen about and stinging it in the air.

Llano has provided rough sketches of the egg, larva, pupa, and cocoon. He shows the cocoon as being of the shape typical of the Bembicini, with two large, elevated pores visible in lateral view. He states that these pores "vary in number from 2 to 3 or more." It appears that this species, like the other two species of *Rubrica*, has fewer pores in the walls of the cocoon than one would expect from its size.

M. Genus *Editha* Parker

Parker (1929) described this genus on the basis of the well-developed temples, the barely depressed vertex, and the distinct lenses of the ocelli. The ocelli are, in fact, somewhat convex, although the front ocelli are truncate in front, the posterior ocelli truncate laterally; the front ocellus is flush with the surface of the head rather than being situated on an elevation as in *Zyzzyx*. The middle coxae of the male have a small tooth, and the middle femora have a large apical tooth and notch. The spiracular lobes of tergite VII of the male are relatively small and do not overlie large membranous areas on sternite VII (in this respect resembling *Zyzzyx* but not *Stictia* and *Rubrica*). The mouthparts, male genitalia, and most other features do not differ in any important way from those of *Stictia* (Fig. 139; see also Willink, 1947, Figs. 16–22).

These wasps are black with bright yellow markings on the head and on the abdomen. *E. magnifica* (Perty) is the largest known wasp of the tribe Bembicini, some specimens measuring as much as 4.5 cm.

N. *Editha adonis* (Handlirsch)

Poulton (1917) presented a list of fossorial wasps and their prey, based on a collection made by G. Bondar near São Paulo, Brazil, and given to the Hope Collections at Oxford University. This collection contained two specimens of *Editha adonis*, without definite locality or ecological data. Each specimen had three skippers (Lepidoptera, Hesperiidae) associated with it. The skippers were in poor condition, but they were tentatively identified as belonging to six species of four different genera (*Gorgythion, Staphylus, Megistias,* and *Systasea*). The skippers were believed to have been taken from the cells of the wasp.

Bondar (1930) mentions this species as preying upon small Lepidoptera of several families. The burrows are said to be 20–25 cm in length and to be dug in areas of sparse vegetation. Bondar also mentions *E. magnifica* (Perty) as preying upon leafhoppers of the families Cicadellidae and Fulgoridae. Considering the great disparity in size between wasp and supposed prey, as well as the fact that *adonis* takes Lepidoptera, this record should not be accepted until it has been further documented.

O. Summary of behavior of *Stictia, Zyzzyx, Rubrica,* and *Editha*

Knowledge of species of these four genera is exceedingly uneven, and the following summary is therefore far from satisfactory. A few significant facts do emerge, however, and these are pointed out in the final paragraph.

(a) Adults of the species take nectar from flowers. Two species, *Stictia signata* and *Zyzzyx chilensis,* are reported to capture flies which are used for direct feeding.

(b) Male territoriality has been reported for *Stictia carolina* and *S. vivida,* and may possibly occur in *Rubrica surinamensis;* the other species are unstudied.

(c) Adults of all species of *Stictia, Zyzzyx,* and *Rubrica* that have been studied spend inactive periods away from the nest (*Editha* is unstudied). *Zyzzyx chilensis* and *Rubrica surinamensis* are reported to form sleeping clusters on vegetation.

(d) The nests of the species of *Stictia* are relatively deep and have a horizontal section that typically contains an inner closure. The nests of *Rubrica* are shallow, oblique, and without an inner closure. *Zyzzyx chilensis,* alone of the wasps considered here, is known to make multicellular nests (up to six cells).

(e) An outer closure of the nest is maintained in most species, but may be omitted when the larva is large in *Stictia carolina, S. signata, Rubrica surinamensis,* and *R. denticornis.* Complete leveling of the mound is not reported for any of the wasps considered here; partial leveling occurs in *Stictia.*

(f) Provisioning is progressive throughout (but *Editha* is unstudied in this regard).

(g) Oviposition shows much variation. The egg is laid on the side of a fly in *Stictia signata, S. vivida,* and *S. heros,* the fly often being dorsum up; it is also laid on a fly in *Rubrica surinamensis* and *R. denticornis.* In *Stictia carolina* the egg is laid flat in the empty cell. In *Zyzzyx chilensis* the egg is glued erect in the empty cell. (*Editha* is unstudied.)

(h) The prey consists generally of medium- to large-sized Diptera; most species take horseflies in some numbers, and several species of *Stictia* and *Rubrica* catch them about animals. *Zyzzyx* takes an occasional skipper (Lepidoptera) in addition to flies, and *Editha adonis* appears to specialize on skippers.

(i) Cell cleaning is reported in *Rubrica surinamensis.*

(j) Dipterous parasites and inquilines are reported for several species.

Three features (c, e, and f) are reasonably consistent throughout the members of this complex that have been studied. The variation in (d), (g), (h), and (i) parallels similar variation in other complexes of Bembicini. For example, the genus *Bembix* shows some of the variation expressed under (d), (g), and (i) (see also Chapter XI); while the Diptera-Lepidoptera dichotomy (h) occurs in the *Stictiella-Glenostictia-Steniolia* complex (Chapter VIII).

Chapter X. Some Generalized Species of *Bembix*

Species of *Bembix* occur throughout the world except on many oceanic islands. The total number of species is probably well over 100 in spite of much synonymy. Arnold (1929) recognized about 50 species in the Ethiopian region alone. The genus is represented in South America by only a few species (Willink, 1947, recognized only three in Argentina). About 30 species inhabit North and Central America. Adult *Bembix* may be characterized as follows:

Vertex slightly depressed, mouthparts moderately prolonged but capable of being retracted completely or nearly so beneath the rather long labrum; palpi very short, maxillary palpi with four segments (often only three readily visible), labial palpi with two (often apparently only one); ocelli reduced to mere arcuate grooves with rather heavy integument bearing long setae close beside them (but lenses preserved in a very few species); fore wing with the first intercubital vein possessing a strong bend (except in a few species); propodeum rather flat behind, sides of the declivity roundly prominent but not sharply produced; pecten well developed in both sexes; male often with processes on sternites II and VI; tergite VII with lateral spiracular lobes which are relatively slender and overlie sternite VII slightly if at all, this sternite being rather narrow; sternite VIII in the form of a single median spine; genitalia with the cuspides typically exceeding the digiti (Figs. 140–148, 176).

Several species which appear structurally primitive occur in western North America. In *Bembix u-scripta*, for example, the ocellar lenses are fairly large, the first intercubital vein nearly straight, and the male without processes on sternites II and VI; furthermore, the middle tarsi of the male are modified in exactly the same way they are in *Stictiella*. In the Old World these particular structures show little variation, but the genus has otherwise undergone much

Chapter X. Generalized Species of *Bembix*

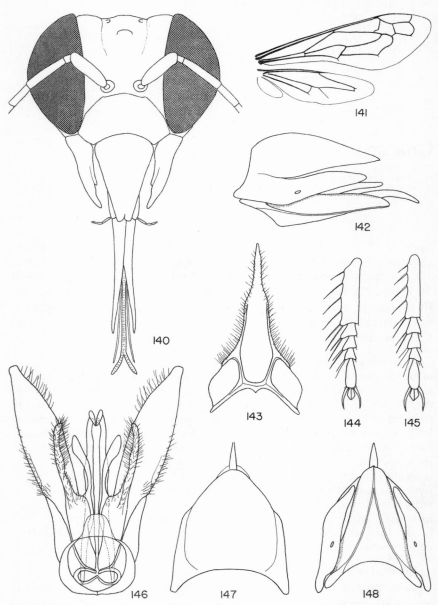

Figs. 140–148. *Bembix spinolae* Lepeletier: 140, head of female, mouthparts extended; 141, wings; 142, apex of metasoma of male, lateral aspect; 143, sternite VIII of male; 144, front tarsus of female, showing pecten; 145, front tarsus of male; 146, male genitalia; 147, apex of metasoma of male, dorsal aspect; 148, apex of metasoma of male, ventral aspect.

radiation, apparently into niches occupied by other genera of Bembicini in the Americas. For example, some species have relatively short wings, simulating *Steniolia* and its relatives (for example, the European species *integra* and *zonata*), while others have become very large, simulating *Stictia* (for example, *monedula* and *bequaerti* in Africa). A few species, in various parts of the world, have developed lateral indentations on the apical tergite of the male, again suggesting the condition in *Stictia* and its allies.

The larvae of *Bembix* also show a fairly broad spectrum of characters. Body setae may be nearly absent (*u-scripta*) or rather strongly developed (*pruinosa*), and there is variation in the mandibular dentition and particularly in the size of the galeae. In general, the larvae are most like those of *Stictia*, differing chiefly in having the body less strongly setose and the anus situated slightly lower on the apical segment.

I treated the ethology of 13 North American species of *Bembix* in 1957, and I shall repeat little of the information covered then (Evans, 1957b). However, enough new data are at hand to justify two chapters, the first dealing with several "generalized" species (defined largely on the basis of the fact that they lay their egg on the first prey), the second dealing with several more advanced species (which oviposit in the empty cell). Some general considerations regarding the behavior of this genus are reserved for Chapter XI. The present chapter includes fairly detailed accounts of three species which were covered only briefly in 1957 (*amoena, sayi,* and *nubilipennis*) as well as brief accounts of five other species (*truncata, cameroni, spinolae, belfragei,* and *u-scripta*). The species are considered in this order; only *truncata* and *cameroni* were wholly unstudied previously.

A. *Bembix amoena* Handlirsch

This species is widely distributed in the western United States, being especially characteristic of semiarid foothills at moderate elevations (2500–7000 ft). Most of my studies were made at the South Entrance of Yellowstone National Park, at about 6900 ft elevation, from 19 July to 15 August 1961. Two colonies were located here, and forty field notes were made. Several other nesting sites were located in Yellowstone Park, and two days in the summer of 1964 (21 July and 6 August) were spent at a large colony in the Lower Geyser Basin, 8 mi south of Madison Junction, Wyoming. A few observations were also made on a colony at Smithfield, Utah, on 19–21 August 1961; this colony had nearly completed its yearly period of activity on those dates, only two worn females being found still active.[1] The brief notes published on this species in 1957 were based on observations made in Los Angeles County, California, by Paul D. Hurd, Jr. I published a short popular account of the Yellowstone colonies in 1962.

(1) *Ecology.* *Bembix amoena* typically nests in flat or sloping areas of coarse, sandy gravel or finely pulverized rock. The nesting area may be completely devoid of vegetation or it may have a sparse covering of short herbs and grasses. The colony at Smithfield, Utah, occupied an artificial excavation made by a sand company into the sides of the Bear River Valley. The soil here was extremely coarse, stony, and dry. The *Bembix* nests were situated in several completely bare areas where there was a larger amount of sand mixed with the gravel than over most parts of the excavation. At the South Entrance of Yellowstone, one colony (hereafter called colony A) occupied a sloping field facing the confluence of the Snake and Lewis Rivers. The vegetation here consisted of a small species of sagebrush (*Artemisia*), a small *Eriogonum,* tufts of grass, and a few small composites. The *Bembix* colony actually occupied only a small portion of a large area of apparently very similar soil and sparse vegetation; parts of the hillside were inhabited by digger wasps of the genera *Philanthus, Cerceris,* and *Tachytes,* but these wasps nested apart from the *Bembix* colony.

The second nesting aggregation at the South Entrance of Yellowstone (colony B) was situated only about 100 m away, and doubtless there was some interchange between members of the two colonies, although none was observed during the period of study. The two colonies were separated by a small hill and a grove of lodgepole pines. The second colony was located on bare soil, with some tufts of grass on the periphery, in an area of hot springs. The soil here was of a very different nature from that in the first colony: it was light in color, rather fine grained and without stones except for occasional chunks of incompletely pulverized geyserite; it was also considerably moister than the soil in the first colony. This area was very typical of what appeared to be the usual habitat of *amoena* in other parts of Yellowstone. I found the wasps in all the major

---

[1]The Smithfield females tended to be more extensively maculated than those from Yellowstone, although both populations showed much variation in color, the more brightly patterned Yellowstone females being similar to some of those from Smithfield. The more fully maculated Smithfield females run to *B. connexa* Fox, and I am convinced that Fox's name should be regarded as a synonym of *amoena.* Fox misassociated the sexes of *connexa,* but his type is a female surely falling within the range of variation of *amoena.*

§A. *Bembix amoena* Handlirsch

geyser basins which were readily accessible by road, and also at Mammoth Hot Springs. In every case they nested in pale, dry to somewhat moist, pulverized geyserite or limestone. Colonies of *amoena* occur along many of the footpaths provided by the National Park Service for access to various geysers and hot springs, and the wasps even nest inside the semicircle of benches around Old Faithful. One wonders if the wasps are able to thrive at this unusual altitude because of the warming effect of thermal action on the soil. That this may be true was suggested by the fact that all colonies in the geyser basins appear to be at their peak in early summer, particularly in July, and by mid-August most activity has ceased. However, colony A at the South Entrance, which was located near but not actually in an area of thermal activity, was only beginning its activity in mid-July, as indicated by the presence of many fresh, unworn males and females. When I last visited this colony in mid-August, a few males were still active and the females were still nesting. This suggests that proximity to hot springs and geysers may influence time of emergence, and it may not be rash to assume that the species is able to exist at all in Yellowstone only because of this warming effect. Colony A was, of course, probably slightly influenced by the nearby hot springs, and in fact a runoff stream from the springs wound its way along the bottom of the hill occupied by the wasps. Most other records for *amoena* indicate that it normally nests at lower altitudes or more southerly latitudes.

Bembix amoena forms distinct and circumscribed colonies. In the center of the colony the nests may be very close together indeed. In fact, I found several nests sharing common entrances, although the burrows diverged immediately beneath the surface. I also found several cases in which two to four nests were clustered in such a way that the soil from each formed a common mound. On 31 July, in the center of colony B, I excavated an area of 1 m^2 to a depth of 30 cm and found 12 cocoons, one cell containing a feeding larva, and many broken cocoons from previous seasons. I had predicted a greater concentration of nest cells than this and of course may have missed some. On one occasion, in the center of colony A, I counted 18 active nests in 2 m^2; since many of these nests had two cells, and other nests were dug in this same area later, the final number of cells per square meter may have approximated 30–40. These figures apply only to the central part of the colony, for nests tend to be more widely separated toward the periphery, and at the extreme periphery they may be separated by several meters. There seemed to be a tendency as the season progressed for colony A to become more diffuse, with females nesting over a slightly expanded area and less close together. This tendency seems characteristic of many bembicine wasps.

I estimated that the population of colony A may have contained between 50 and 100 females and as many males. Judging from the number of visible nest mounds, the population of colony B may have been two or three times that high at its peak, and the colony studied in the Lower Geyser Basin contained an estimated 300 nests. The colony at Smithfield, Utah, must have been a large one, as I was able to turn up 14 cocoons in a short time by random digging (the mounds having mostly weathered away).

(2) *General features of behavior.* Despite its gregariousness, *amoena* appears to be a relatively "unaggressive" species. Wasps sharing common entrances or mounds did not seem to react to one another aggressively except rarely, chiefly when one wasp was making a closure while another arrived with prey. Such episodes were normally resolved by one female flying off long enough for the other to complete her activities at the entrance.

I noted no orientation flights as such in this species, though doubtless some of the flights taken during excavation of the nest serve a function in orientation. On 9 August, after 2.5 days of steady rain, I watched many females emerging from their nests from 1030 to 1130. At first each female groomed and then rested in the sun, perhaps warming her body sufficiently for flight. Then each flew off to take nectar at nearby flowers, returning somewhat later to scrape open the nest entrance prior to provisioning. In no case did I observe females circling the nest entrance in obvious flights of orientation. Apparently they had learned the location of their nests so well that no reorientation was necessary even after a considerable period of inactivity.

These wasps display no strict periodicity in their daily cycle of activities. The males tend to appear about 0900–0930, the females shortly thereafter, although on cool days there may be little emergence before 1030. Both sexes are active until late afternoon (1600–1700), except during cloudy periods. Mating and visits to flowers for nectar may occur at any hour of the day, but both these activities are much more frequent in the morning hours, soon after emergence from the soil. Females with active nests may bring in flies at any time of the day (1000–1600), but the peak of provisioning is at midday (1130–1400). Females beginning new nests usually start in the morning and dig intermittently all day, completing the nest in the late afternoon.

At the South Entrance of Yellowstone the chief sources of nec-

§A. *Bembix amoena* Handlirsch   273

tar were several species of Compositae, particularly a tall *Solidago*. Stands of composites near colonies A and B were being visited in great numbers, while stands at some distance from these colonies were visited much less frequently.

(3) *Activity of males.* I saw no males digging their own holes, but in the late afternoon males were often seen digging into abandoned holes dug by females. On one occasion several males were found spending the night in one such burrow.

During much of the day the males fly in a diffuse "sun dance," generally 3–10 cm above the ground, but now and then rising somewhat higher to pass over tall grasses. Frequently the males rest on the ground for periods up to 1 or 2 min. When resting on the ground, they hold their front legs stiffly toward the front, their middle legs stiffly toward the sides (as in Fig. 160). Frequently the males hover noisily behind females that are digging or closing their nests, and occasionally they land near females and move their antennae up and down rhythmically as if seeking an odor stimulus.

Many attempted matings were observed. In each instance the male descended upon the female and clamped his middle legs over her wings in the manner of some other species of *Bembix* (*belfragei*, for example). The male would then fly a considerable distance (10–20 m) at a height of 1 m or slightly less, holding the female beneath him. The pair then descend to the ground, where mating occurs. The longest coupling on the ground that I observed lasted 20 sec, but it is probable that undisturbed pairs remain together longer than this. The majority of pairs observed broke up immediately upon reaching the ground; I assume these matings were not consummated. The male genitalia are extruded when the pair strikes the ground, but I was unable to tell whether they are extruded during flight.

During 1961 males were active throughout the period of observation at colony A, though some decline in numbers was noted by 10 August. At colony B, and throughout the thermal areas of Yellowstone, males seemed to be scarce after mid-July. However, during 1964, which was unusually cool in the spring and early summer, males were active in many parts of Yellowstone up to 1 August.

(4) *Digging the nest.* As noted above, the digging of the nest requires the greater part of 1 day. For example, no. 1773 began her nest at 1020 on 24 July and finished it at 1500. On several occasions this female interrupted her digging and flew off, each time leaving the

entrance closed. At 1500 she made a brief closure, then at 1520 returned with a fly on which to lay her egg. A few minutes later she was seen outside the nest entrance making a more lengthy outer closure; presumably she re-entered the nest to spend the night, but this was not observed.

In this species the mound of earth is never leveled, and it remains as a conspicuous marker at the nest entrance. A typical mound measures 1.5–2.5 cm deep, 6–10 cm wide, and 9–16 cm long. Since there is much variation in burrow length in this species, it is natural that some mounds are larger than others. Some soil may be added to the mound from time to time, especially when a second cell is constructed from the same burrow, but most mounds undergo no notable increase in size. In both colonies A and B there were numerous holes having unusually small mounds. Such burrows were invariably found to terminate blindly at the depth of a few centimeters. These represented nests that had been abandoned before completion. The unusual number of incomplete burrows in these colonies doubtless reflected the coarseness of the soil and the irregular distribution of rocks in it.

The mandibles of *Bembix amoena* are unusually stout, and much use is made of them for breaking through the soil and for hauling stones from the burrow. On many occasions females were seen backing out of nest entrances dragging stones (Figs. 149 and 150).

(5) *Nature and dimensions of the nest.* The burrow is oblique, forming roughly a 45° angle with the surface, and almost always has one or more lateral curves or angulations before reaching the cell. In some nests the burrow makes a complete circle, so that the cell is more or less directly beneath the entrance. Some of these curves are evidently made around obstacles in the soil, though the burrows seemed no straighter in the less stony soil of colony B (Figs. 151 and 152).

In the three colonies studied in Yellowstone, nest depth appeared to be correlated with friability of the soil. That is, the deepest nests were in colony B, where the soil was relatively easy to dig, the shallowest nests in the colony at the Lower Geyser Basin, where the soil was relatively more firm and contained many large chunks of geyserite. The soil in colony B was distinctly moister than in the other two colonies, and if soil moisture were critical one would have expected the shallowest nests in this colony. Much more impressive than the differences in mean nest depth among the three colonies was the remarkable variation in all three, a variation far greater than one normally expects in one area. In all three colonies the maximum cell depth was more than twice the

§A. *Bembix amoena* Handlirsch

Fig. 149. Female *Bembix amoena* digging a nest in coarse geyserite (Yellowstone National Park). Note that the mandibles are partially extended, the other mouthparts withdrawn; the front legs are in the backstroke and are not visible.

Fig. 150. Female *Bembix amoena* digging at the nest entrance (Yellowstone National Park). The antennae are extended into the mouth of the burrow. (From Evans, 1962.)

276   Chapter X. Generalized Species of *Bembix*

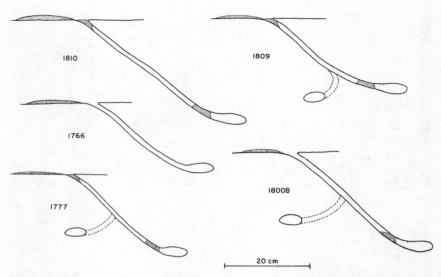

Fig. 151. Five typical nests of *Bembix amoena* (South Entrance, Yellowstone National Park, July 1961). Burrows indicated by dashed lines had been filled solidly and could not be traced exactly.

minimum for that colony; very often adjacent nests were radically different in burrow length and cell depth. It seems possible that nest depth is determined by the amount of labor performed, and wasps that encounter fewer impedimenta may go deeper. In the Smithfield colony, where the soil was of rather uniform consistency, there appeared to be much less variation in cell depth. In this area the depth was considerably less than one would have expected for such friable soil (see Table 25).

Burrow diameter in this species is about 12 mm. The size of newly constructed cells is about 15 × 35 mm, while cells containing large larvae may be as long as 60 mm.

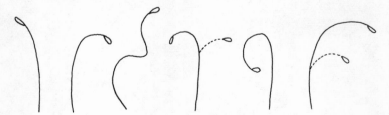

Fig. 152. Several burrows of *Bembix amoena* (plan), showing various types of curvature observed (South Entrance, Yellowstone National Park).

§A. *Bembix amoena* Handlirsch

TABLE 25.   NEST DATA FOR BEMBIX AMOENA

Locality	No. of nests	Burrow length (cm)	Cell depth (cm)
Yellowstone, Wyo. So. Entrance, colony A	31	34.5 (21–57)	16.0 (11–28)
Yellowstone, Wyo. So. Entrance, colony B	10	37.5 (23–57)	20.6 (13–27)
Yellowstone, Wyo. Lower Geyser Basin	9	30.0 (20–40)	12.5 (9–20)
Smithfield, Utah (cocoons only)	15	—	13.0 (11–15)

Nearly all wasps followed over a period of 2 weeks in colony A prepared a second cell after the first was fully provisioned, but none was found to make more than two cells. Less complete data from the other two colonies in Yellowstone tended to confirm the fact that most nests are two celled in their finished form. The second cell is prepared at the end of a fairly long section of new burrow. During this digging the wasp is rarely seen outside the nest, and new soil is rarely added to the mound. Apparently the soil from the new section is merely scraped into the old section and packed in place as final closure of the first cell. For this reason, it is impossible to tell from the outside whether a nest has a second cell or not unless one has been keeping daily records of it (quite the reverse of species such as *belfragei*, where two-celled nests can immediately be identified by the fresh and distinctive new mound). Also, one cannot always be sure whether a nest has a second cell by digging it out; because of the coarse soil, crooked burrows, and variation in depth of cells, it is very difficult to find the closed first cell if one is present. A few extracts from my field notes may serve to illustrate the type of data that convinced me that two-celled nests are the rule in this species.

No. 1811. Wasp seen digging new nest afternoon of 2 August. Nest excavated 11 August and found to contain a fully-grown larva; final closure presumably would have been made today. Total duration of egg and larval stages: 9 days.

No. 1806. This wasp also digging new nest afternoon of 2 August. On 9 August she was digging and adding new earth to mound; excavated at 1600 and found a new cell containing one fly on which an egg had not yet been laid. Failed to find first cell.

No. 1788B. Wasp bringing in many flies in morning of 31 July. Not seen outside nest in afternoon. Excavated nest at 1445 and found that cell had been sealed off with a plug of earth 6 cm long; cell contained a larva 15 mm long plus ten fresh flies and many fly remains. Female was digging a new branch of the burrow from a point 22 cm from entrance, undoubtedly using the soil to close off the first cell. New branch presumably would have contained a new cell.

No. 1800B. Wasp digging new nest 22 July; entrance only 1 cm away from that of 1800A, burrow at about a 45° angle with that nest. Nest remained active through 4 August (13 days), when it was excavated. Found a new cell containing a small larva at depth of 24 cm. At depth of 15 cm found first cell, containing fresh, still rather soft cocoon.

No. 1812. Appeared to be a relatively new nest 26 July. Many flies brought in 31 July. No activity observed 2, 4 August. 9, 11 August much prey carriage. Excavated nest on afternoon of 11 August and found large larva in cell, also a second cell with well-hardened cocoon. Total time for two cells: 16 days.

Five other nests which were marked at the time they were first dug received the final closure 16–18 days later. These nests were not dug out, but evidence such as that cited above makes it clear that these cells were closed after the completion of two cells. Two nests appeared to have been closed after the completion of only one cell. Number 1814C, a wasp which dug her nest on 28 July, was seen bringing in many flies on 2 August, but after this date no further activity was seen. Number 1795B started her nest on 24 July, and was seen provisioning on 28 July. No further activity was seen at this nest on 31 July and 2 August, but on the later date this same female was observed nesting elsewhere. This female was followed through a two-celled nest at her second site. It is probable that some females fail to prepare a second cell for the same reason that some incipient nests are abandoned before completion of the first cell: the wasp encounters many rocks or other impedimenta in the soil.

It should be reiterated that because of the close proximity of nests and the coarse and irregular texture of the soil, it was often difficult to find cells or to be sure which nest a given cell belonged to. In nine nests I was able to find two cells that beyond much question belonged to the same nest. In five of these the two cells were at approximately the same depth, but in the other four the second cell was decidedly deeper (by 5–9 cm) than the first cell. In several other nests I failed to find the second cell after following a burrow deeply into the soil. In general, there seemed to be a tendency for wasps to prepare the second cell at a greater depth than the first. The two cells were invariably widely separated; that

§A. *Bembix amoena* Handlirsch

is, the burrow leading to the second cell was always started at a considerable distance from the first cell and at a considerable angle to the original burrow. It was difficult to discern any particular plan of construction, but in most nests a horizontal projection of the nests might be said to approximate a very irregular Y, or more often the Greek letter ϒ (Fig. 152).

(6) *Temporary closure of the nest.* Most females maintain an outer closure of the nest at all times, even during periods of active provisioning. There is much variation in the nature of these closures, some taking only a few seconds and consisting of only a few scuffs of soil into the entrance, others requiring 1 or 2 min and resulting in thorough concealment of the entrance. A few individuals left the entrance open for long periods, especially during provisioning (no. 1823, for example). Some individuals exhibited much variation in this regard; for example, no. 1813A, provisioning her first cell 24 and 26 July, always closed the entrance upon leaving, but on 31 July and 2 August, when provisioning her second cell, she invariably left the entrance open while away. A few females were even found to omit the outer closure in nests containing eggs or small larvae, while the majority of females always maintained a closure even when the larva was large. I was unable to correlate the variation in closure with any factor in the environment or in the behavioral cycle of the wasp.

Occasional nests undergo cave-ins at the entrance and have to be re-excavated by the wasps. Such nests may have a large hole at the entrance, as much as 3 or 4 cm wide and nearly as deep. The outer closure is made at the bottom of this hole. It is easy to assume that these wasps are omitting the outer closure unless one is observing them closely.

Much variation was also noted in the inner closure. About two thirds of the nests excavated had a strong inner closure; the remainder appeared to have none at all, even though some of these nests contained eggs or small larvae. The inner closures studied varied much in length and in degree of compactness of the soil.

(7) *False burrows, back furrows,* and *back burrows.* Most females prepare the outer closure by scraping a small amount of soil from each side of the entrance. Not infrequently most of the soil is dug from one particular spot, so that a small depression is made. Small "false burrows" of this type were a common sight in all three colonies in Yellowstone. Some individuals made one or two false burrows at the time of the first closure of the nest, right after com-

pletion of digging. For example, no. 1769-22 made a small hole 1 cm deep on one side of the entrance and another 2 cm deep on the other side, both in the course of making the first outer closure of a new nest. Nest no. 1769-1 did not originally have a false burrow, but 4 days after its inception the wasp made a false burrow 1.5 cm deep on the left side. In some cases false burrows persisted for the duration of the nest, but in other cases they were modified by the wasp. For example, no. 1807 (dug 21 July) did not originally have a false burrow, but on the morning of 26 July the female made a lengthy outer closure at 0930, before beginning to bring flies to her nearly full-grown larva, and in the course of this made a small false burrow. On 28 and 31 July there was no evidence of this false burrow, but on 2 August there were two fresh false burrows. On a number of occasions females were seen adding to false burrows or partially filling them accidentally in the course of making a closure. Most false burrows varied from a barely perceptible depth to about 2 cm, but two exceptional ones measured 3 and 5 cm in depth. Approximately half the nests marked in colony A had false burrows at some time or other; about half of this number had two false burrows, the rest only one. There seemed little question that these false burrows served principally as a source of fill for the closures. It is perhaps worthy of note that on one occasion a miltogrammine fly was seen entering a false burrow and on another occasion a bombyliid was seen apparently ovipositing in false burrows (Fig. 153, nos. 1799B and 1807).

In addition to the false burrows, which were lateral or sublateral in position relative to the entrance, a considerable number of nests showed evidence of an elongate groove dug into the mound just outside the entrance (Fig. 153, nos. 1796 and 1807). Observa-

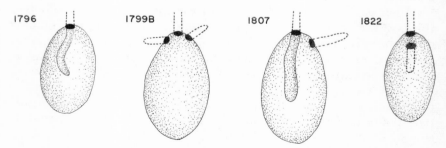

Fig. 153. Nest entrances of four *Bembix amoena* (South Entrance, Yellowstone National Park): false burrows (1799B, 1807); back burrows (1822); and back furrows (1796, 1807). In each sketch the true burrow is shown going directly upward.

§A. *Bembix amoena* Handlirsch   281

tion showed that these grooves were also dug at the time of making outer closures, and the soil used in the closure. Thus they were functionally the same as the false burrows, though different in position and also in that they did not actually penetrate the soil, but merely involved the loose soil of the mound. These grooves varied in length from 1 to 7 cm, the longer ones usually being associated with nests that had received the final closure. Most "back furrows," as these may be termed, were straight or slightly sinuous. In a very few nests (only three specifically noted) there was an actual oblique burrow passing beneath the mound (Fig. 153, no. 1822). One of these, 3 cm long, was in a nest containing an egg. Two others, 1.5 and 2 cm long, were made following final closures (like the "back burrow" so characteristic of final closure in *sayi*). It is probable that the long back burrows of *sayi* are homologous to the back furrows and short back burrows that occur occasionally in *amoena*. Although such excavations into or under the mound often occurred in connection with final closure in *amoena*, they also occurred at times in connection with temporary closures, and final closures did not always involve them.

(8) *The egg.* Eggs were found in several nests. In each case the egg was attached to the wing base of a fly in the manner common to many species of *Bembix*. The flies used as egg pedestals were often of moderate size, and included tabanids, syrphids, calliphorids, and asilids. The egg measures about 1×6 mm. The newly hatched larva remains attached to the fly pedestal by a glutinous thread for 2 or 3 days. The pedestal fly tends to become dessicated and is not ordinarily eaten by the wasp larva. At least two fresh flies may be added to the cell before the egg hatches.

(9) *Hunting and provisioning.* Female *Bembix amoena* were often seen flying rapidly from flower to flower, especially Compositae, apparently hunting for flies. No captures from flowers were actually observed, but many species of flies found in the nests of the wasps were observed on these flowers (particularly Stratiomyidae and Syrphidae). In the Lower Geyser Basin, females were seen to strike at various insects flying over the colony; a skipper and a dragonfly were both knocked off their course slightly, without actually being seized by the wasp. Female *amoena* also flew about us in their search for horseflies; on one occasion a *Hybomitra* was knocked from my arm to the ground, but the wasp failed to immobilize it before it flew away.

The total list of flies found to be used as prey in Yellowstone

282   Chapter X. Generalized Species of *Bembix*

reads like a list of the larger flies of Yellowstone (Table 26). In both areas of study a great diversity of flies was represented, even in individual cells. At the Lower Geyser Basin on 21 July, numerous Asilidae and very few Tabanidae were being employed, and few Tabanidae were employed at the South Entrance. However, on 6 August large numbers of tabanids were present in the Lower Geyser Basin, and these were being used in large numbers by *Bembix amoena;* some cells on this date contained nothing but *Hybomitra,* mainly females. The total list includes members of ten families of flies. The short list of prey from Los Angeles County, California, which I presented in 1957, included four of these same families.

As in most species of this genus, the uneaten parts of the flies are

TABLE 26.   PREY RECORDS FOR BEMBIX AMOENA

(All Yellowstone Park, Wyo., July-August 1961, 1964)

Species of fly	No. of specimens	
	South Entrance	L. Geyser Basin
STRATIOMYIDAE		
Anoplodonta nigrirostris Lw.	4	
Hedriodiscus varipes Lw.	12	4
TABANIDAE		
Hybomitra captonis Mart.	2	
Hybomitra fulvilateralis Macq.	1	
Hybomitra liorhina Philip		16
Hybomitra opaca Coq.	2	
Hybomitra osburni Hine	6	8
Hybomitra phaenops O.S.	1	7
BOMBYLIIDAE		
Poecilanthrax sackeni Coq.	2	
Systoechus fumipennis Painter	1	
Villa alternata Say	20	3
Villa fulviana nigricauda Lw.	8	
Villa harveyi Hine	2	
Villa lateralis Say	3	
Villa sinuosa jaennickeana O.S.	3	
THEREVIDAE		
Thereva sp.		1
ASILIDAE		
Asilus sp.	1	
Cyrtopogon glarealis Mel.		4
Promachus sp.		4

§A. *Bembix amoena* Handlirsch

TABLE 26. PREY RECORDS FOR BEMBIX AMOENA (CONT.)

Species of fly	No. of specimens	
	South Entrance	L. Geyser Basin

SYRPHIDAE

Chrysotoxum ypsilon Will.	3	
Eristalis anthophorinus Fallen	1	
Eristalis barda Say	1	
Eristalis latifrons Lw.	1	
Eupeodes volucris O.S.	3	
Helophilus hybridus Lw.	1	
Metasyrphus lapponicus Zett.	3	
Metasyrphus meadii Jones	3	
Scaeva pyrastri L.	6	1
Stenosyrphus pullulus Snow	1	
Syrphus jonesi Fluke	2	
Syrphus opinator O.S.	1	1
Syrphus ribesii L.	1	
Syrphus vitripennis Mg.	1	

SCIOMYZIDAE

Tetanocera vicina Macq.	1	

MUSCIDAE

Helina punctata R.D.	1	
Limnophora magnipunctata Mall.	1	
Lispe brevipes Ald.	2	
Mydaea persimilis Mall.	1	
Phaonia monticola Mall.	2	
Pyrellia cyanicolor Zett.	1	

CALLIPHORIDAE

Calliphora vicina R.D.	1	
Calliphora vomitoria L.	17	
Cynomyopsis cadaverina R.D.	1	3
Eucalliphora lilaea Walk.	1	2
Lucilia illustris Mg.	1	
Phormia regina Mg.	4	
Protophormia terraenovae R.D.		1

SARCOPHAGIDAE

Sarcofahrtia montanensis Parker	1	
Sarcophaga spp.	4	
Macronychia sp.	1	

TABLE 26. PREY RECORDS FOR BEMBIX AMOENA (CONT.)

Species of fly	No. of specimens	
	South Entrance	L. Geyser Basin
TACHINIDAE		
Acroglossa hesperidarum Will.	2	
Arctophyto sp.	1	
Bonellimyia subpolita Brks.	5	
Eumegaparia flaveola Coq.	1	
Fabriciella rostrata **Tot.**	1	
Gonia porca Will.	2	
Gonia spp.	10	
Melanodexia sp.	1	
Melinocera flavicornis Br.	1	
Mericia alberta Curr.	15	
Mericia ampela Walk.	1	
Mericia **arcuata Tot.**	1	
Mericia bicarina **Tot.**	1	1
Microphthalma disjuncta Wd.		1
Paramuscopteryx sp.	1	
Peleteria iterans Walk.	5	
Peleteria neotexensis Brks.	2	
Ptilodexia sp.	10	1
Rhachogaster algens Wied.	11	
Siphosturmiopsis sp.	1	
Spallanzania sp.	1	
Trochilodes skinneri Coq.		2

allowed to accumulate in the cell. I sometimes found as many as 17 fresh flies in cells containing fairly large larvae, and I believe that the total number supplied to a larva may be in excess of 30.

Although I noted no aggression among females in the colonies at the South Entrance, several females in the dense colony at the Lower Geyser Basin were seen to pounce upon other females laden with flies and attempt to take the flies away from them. One successful attempt at prey stealing was observed, but most ended in failure.

This species is a slow provisioner, females taking several minutes to more than 1 hr to bring in a fly and often remaining in the nest several minutes between trips for prey. One wasp, no. 1807, was watched for several hours during a period of relatively rapid provisioning, with the following results:

§A. *Bembix amoena* Handlirsch

Time entered nest with prey	Time spent in nest (sec)
1013	150
1032	100
1040	90
1100	60
1106	60
1113	45
1135	40
1145	45
1206	55
1216	75
1221	100

This represents, I believe, the total provisions for this individual for that day. It will be noted that the wasp spent less time in the nest between trips during the middle of provisioning than at the beginning or conclusion. The outer closures also tended to be much briefer during this time. Similar figures were obtained for several other nests.

As indicated earlier, the total duration of the egg and larval stages of this species is about 9 days. The cell is normally completely provisioned and closed off only when the larva is approaching full size (7 to 9 days). The slow growth of the larva of these wasps, as compared to most others, is doubtless related to the cool night temperatures at this altitude.

(10) *Final closure.* When the second cell has been fully provisioned (rarely after only one cell has been fully provisioned), the burrow is completely filled up with soil. Many final closures were observed, and seven were studied at some length. In the initial stages the wasp scrapes soil from the walls of the burrow, and can therefore be seen from the outside only to a limited extent if at all. Later the wasp can be seen coming out to the entrance scraping soil, then backing down into the burrow to pack it. When the burrow has been filled to within a few centimeters of the top, the wasp begins coming outside the entrance, walking away from the entrance scraping soil toward the hole. Most of these trips are made on the side away from the mound; they vary in length from 5 to 20 cm. After each trip the wasp takes flight briefly, lands in the entrance facing inward, turns 180°, and backs into the burrow scraping soil. These trips eventually describe a radiating pattern. Fairly large pebbles are sometimes thrown toward the entrance. Not in-

frequently several trips are taken into the center of the mound, resulting in a strong back furrow several centimeters long. Occasionally a short, lateral false burrow may be made instead of or in addition to the back furrow. These may persist or may be filled in by further scraping of soil in the radiating pattern away from the mound. Two of the seven individuals followed in detail made a back burrow after the burrow was completely filled, in one case 1.5 cm deep, in the other case 2 cm deep. These back burrows were oblique, passing into and beneath the mound, and were left unfilled. Slightly over 1 hr is required for those aspects of the final closure that can be observed from the outside (Fig. 154).

The final appearance of a nest that has received the final closure is very variable. The mound may be entirely intact, or some of the soil may have been used up on the side toward the entrance, and there may be evidence of a back furrow or a false burrow; a few nests have an open back burrow directed beneath the mound. The

Fig. 154. *Bembix amoena* female making a final closure of the nest (South Entrance, Yellowstone National Park). The front tarsi are curved in the usual digging position and being used to rake in a stone.

burrow itself is normally packed solidly all the way from the cell to the surface of the ground, but occasionally a slight depression is left. In general, I found it difficult to distinguish with certainty nests that had received the final closure from those that had only a temporary closure. By inserting a straw into the entrance, however, this problem could be quickly solved.

(11) *Natural enemies.* Conopid flies (*Physocephala texana* Will.) were common in colony A from 19 July to about 1 August, also in the Lower Geyser Basin colony on 21 July. After 1 August none were noted in either locality. These flies would often perch on low vegetation in the area of greatest concentration of nests, pivoting about on their perch as if watching for opportunities to pounce upon *Bembix*. On several occasions the flies were seen to dash after wasps (of either sex), and on two occasions they were seen to strike a *Bembix* from above, presumably ovipositing on the wasp's abdomen. The wasps also pursued the conopids on occasion, although these flies were not found to be used in provisioning the nests. In the course of digging out various nests, I dug up five conopid puparia, each of them inside the remains of the abdomen of a male *B. amoena*. All of these were old, empty puparia from previous years. I did not find any conopid larvae or pupae, but there is little question that the flies were maintaining a fairly high population at the expense of the *Bembix*.

Miltogrammine flies of several species were moderately common in all three colonies at Yellowstone. Some of them were seen following females into nests, only to emerge almost immediately. Several of these were captured and found to be *Senotainia trilineata* Wulp (complex) and *S.* sp. nr. *rubriventris* Macq. Other flies perched about on the ground and entered various holes, including one *Bembix* false burrow; these proved to be *Metopia argyrocephala* (Mg.). One specimen of *Hilarella hilarella* (Zett.) was also taken outside a *Bembix* nest. Of the many cells dug out, only one (no. 1777) contained a miltogrammine maggot, and that, only one maggot. The wasp larva in this nest was 18 mm long and probably would have attained maturity in spite of the maggot. The maggot was not reared successfully.

Nest no. 1798B (colony A) was found to contain a large number (20 or more) of minute maggots in the cell. This nest had already received the final closure, and the larva was full grown and began to spin its cocoon in a rearing tin the following day. These maggots grew very slowly and eventually pupated and produced several flies of the genus *Megaselia* (Phoridae).

Although no Bombyliidae were noted at the South Entrance, they were very much in evidence in the Lower Geyser Basin. *Exoprosopa dorcadion* O. S. was common both on 21 July and on 6 August, and females were often seen ovipositing in open *Bembix* burrows and in false burrows. On 21 July the large, dark species *Villa melasoma* (Wulp) was exceedingly common and also engaged in oviposition, but this species had disappeared completely by 6 August. I did not rear these flies from the cocoons collected, but members of both genera have been incriminated as parasites of other species of *Bembix* (Chapter XI:D,2; also Evans, 1957b:213). and there seems little question that they were attacking *B. amoena* with considerable success in the Lower Geyser Basin.

The chrysidid wasp *Parnopes edwardsii* (Cresson) was observed several times in colony A. These wasps were seen entering various holes, including *Bembix* nests, but I have no certain evidence that they were parasitizing the *Bembix*. However, *P. edwardsii* has been reared from cocoons of *B. comata* and *Steniolia obliqua,* so it is not improbable that it also attacks *B. amoena.* Telford (1964) reports that *P. edwardsii* has been taken in close association with *B. amoena* at Walla Walla, Washington.

Fourteen *amoena* cocoons collected at Smithfield, Utah, in August 1961, were overwintered in Massachusetts. Between 10 June and 10 July 1962, eight wasps emerged from these cocoons as well as one female mutillid, *Dasymutilla creusa bellona* (Cresson). I doubt if this mutillid is specifically distinct from *bioculata* Cresson (Evans, 1957b:166), the latter being a well-known parasite of *B. pruinosa.*

On 6 August 1964, at the colony in the Lower Geyser Basin in Yellowstone, *Dasymutilla californica* (Radoszkowski) was fairly common, but I did not actually see any entering *Bembix* nests, and none were reared from the cocoons collected. Several solitary bees, as well as other wasps, nested in this same general area. I saw no mutillids at the South Entrance.

B. *Bembix sayi* Cresson

My 1957 report on this species was moderately detailed but included no information on final closure of the nest. Further studies have shown that final closure is actually relatively complex and unusual in this otherwise rather primitive *Bembix*. The present report is based principally on 32 field notes made between 6 and 27 May 1961 at the Archbold Biological Station near Lake Placid, Highlands County, Florida (where most of my earlier studies of

§B. *Bembix sayi* Cresson

this species were made). One note was also made at Rodeo, New Mexico, in September 1959, and three at Great Sand Dunes National Monument, Alamosa County, Colorado in August 1961.

(1) *Ecology*. As reported in 1957, this species is not highly gregarious, its nests tending to be spread widely over areas of bare sand or sandy gravel, often where other bembicines are nesting. The one nest found in New Mexico was located in a small dune in the midst of a small colony of *Bembix u-scripta*. At Great Sand Dunes National Monument a few nests were scattered over a sparsely vegetated slope away from the dunes proper, an area inhabited by *Microbembex monodonta* and a few specimens of *Bembix spinolae*, although the nests of *sayi* were not close to known nests of those two wasps. In Central Florida, *sayi* was found nesting in small numbers in roadside sand pits, in citrus orchards, and in open pine woods. At Highlands Hammock State Park one *sayi* nested in firmly compacted sand in the midst of a colony of *Bembix texana*—the *sayi* being immediately recognizable because of its very different behavior during final closure. In no case did I find any active *sayi* nests closer together than about 1 m, and in most cases they were separated by several meters. The largest nesting population found was on the grounds of the Archbold Biological Station, where there were an estimated 30 females nesting in four circumscribed areas in a plantation of about 5 acres extent. Each of these areas contained several (3–10) females that tended to remain within that particular area for successive nests. Indeed, nearly all females that were followed through several nests made these nests very close together. For example, no. 1740 dug a nest on 7 May, made a final closure on 13 May, then began a new nest with the entrance only 5 cm from the old one; this second nest was closed on 20 May, and a new nest started only 6 cm away. Since the mound is not leveled and there was little rain during the period of observation, one could often see a number of mounds of varying freshness all quite close together and all the work of one female (Fig. 155). In some cases the mounds of successive nests were partially superimposed upon one another, but in other cases they were separated by as much as 15–30 cm or even up to 1 m. Only in a few cases during the period of observation did marked females move from the immediate vicinity of their previous nest.

No significant observations were made during 1959–1961 on the behavior of the males. At several localities in Highlands County, Florida, individuals of both sexes were found taking nectar at the flowers of *Geobalanus oblongifolius* (Michx.).

290   Chapter X. Generalized Species of *Bembix*

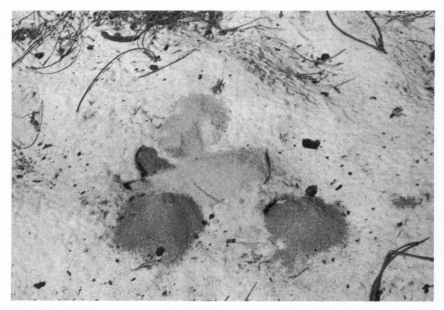

Fig. 155. Mounds of several successive nests of *Bembix sayi* (Archbold Biological Station, Florida). The open burrow and fresh, pale-colored mound in the center belongs to a nest in the course of preparation. The dark mound on the lower right is the one resulting from the back burrow of the previous nest; the other two mounds are those resulting from previous back burrows.

(2) *Nature and dimensions of the nest.* The nests of *Bembix sayi* at and near the Archbold Biological Station averaged somewhat deeper in 1961 than in 1955. This may have been related to the fact that the spring of 1961 was extremely dry, and the wasps may have been digging deeper to seek proper moisture conditions. The dimensions of the few nests excavated in Colorado and in New Mexico did not differ notably from those of the Florida nests. For the sake of comparison I have included all my records for this species in Table 27, including those published in 1957. All nests studied were unicellular (Fig. 156).

In 1957 I reported some irregular leveling movements on the part of one individual, but indicated that the mound of sand at the nest entrance was usually left intact. Further studies have shown that the mound is normally left almost wholly intact, even after final closure. Indeed, the nests in Alamosa County, Colorado, were first discovered and identified as those of *sayi* by the characteristic mound at the entrance. In Florida I was able to maintain a continuing census of the wasps nesting in the plantation of the Arch-

§B. *Bembix sayi* Cresson

TABLE 27. NEST DATA FOR BEMBIX SAYI

Locality	N	Burrow length (cm)	Cell depth (cm)
Lake Placid, Fla., 1955	14	24.0 (19.0-28.0)	13.4 (10.0-16.0)
Lake Placid, Fla., 1961	31	30.5 (22.0-40.0)	16.0 (12.0-22.0)
Gainesville, Fla., 1955	2	27.5 (26.0-29.0)	14.3 (12.0-16.5)
Sun City, Kansas, 1953	3	29.7 (25.5-33.0)	16.2 (15.5-16.5)
Alamosa Co., Colo., 1961	3	27.0 (23.0-32.0)	12.7 (12.0-13.0)
Rodeo, New Mexico, 1959	1	30.0	14.0

bold Biological Station merely by keeping track of the mounds—something that would be quite impossible with a species such as *spinolae* in which mound leveling is a fixed part of the behavior patterns. Fresh mounds of *sayi* measure about 2 cm in maximum depth by 7-9 cm in width and 12-15 cm in length. Over a period

Fig. 156. Typical nest of *Bembix sayi*, with contents removed (Archbold Biological Station, Florida). The rule in the foreground is 15 cm long.

TABLE 28. PREY RECORDS FOR BEMBIX SAYI

Species of fly	No. of specimens	
	Highlands Co., Fla.	Alamosa Co., Colo.
TABANIDAE		
Chlorotabanus crepuscularis Beq.	97	
Chrysops flavida Wied.	4	
Hybomitra hinei wrighti Whit.	2	
Tabanus chellopterus fronto O.S.	1	
Tabanus lineola Fabr.	16	
Tabanus melanocerus lacustris Stone	3	
Tabanus nigripes Wied.	2	
BOMBYLIIDAE		
Exoprosopa fasciata Macq.	2	
Exoprosopa fascipennis noctula Wied.	1	
Poecilanthrax lucifer Fabr.	1	
Systoechus solitus Walk.	8	
Villa cypris Mg.	1	
Villa lateralis Say	1	
Villa molitor Lw.		1
THEREVIDAE		
Furcifera pictipennis Wied.	6	
ASILIDAE		
Ablautus nigronotum Wilcox		1
Erax tabascens Banks	31	
Mallophorina laphroides Wied.	12	
SYRPHIDAE		
Allograpta obliqua Say		1
Eristalis agrorum Fabr.	1	
Eristalis tenax L.		1
Syrphus sp.		1
Volucella sp.	1	
SARCOPHAGIDAE		
Sarcophaga ventricosa Wulp	1	
TACHINIDAE		
Belvosia slossonae Coq.	1	
Fabriciella actinosa Rnh.		1
Fabriciella egula Rnh.		3
Fabriciella latigena Tot.		1
Gonia sequax Will.		1
Juriniopsis sp.		1
Prosenoides flavipes Coq.	1	

of days the mounds tend to settle and to spread out somewhat, especially following periods of rain. As noted above, it is common to find several mounds in varying stages of weathering all quite close together, representing successive nests of one female (Fig. 155).

In contrast to *texana,* and contrary to the impression I obtained in 1955, the *sayi* I studied in 1961 nearly always maintained an inner closure of the burrow, just outside the cell. This closure is at first fairly compact and measures 1.5–4 cm in length, but during periods of active provisioning it may be rather loose and only about 1 cm long, occasionally apparently omitted altogether. An outer closure of the burrow is maintained at night, during cloudy or rainy periods, and when the cell contains an egg or very small larva. Otherwise the outer closure is omitted, and most provisioning females are able to enter the nest very quickly, having no occasion to pause at the nest entrance. These generalizations are based on a fairly detailed study of the nesting aggregation at the Archbold Biological Station in Florida. The one nest excavated in New Mexico contained a very small larva and had both an inner and outer closure, as well as a distinct spur in which the female rested. One of the three nests excavated in Colorado also had a very small larva, and the closures were in place; the other two provided no information on temporary closures.

(3) *Hunting and provisioning.* As noted in 1957, *sayi* preys upon a wide variety of flies, although individual nests often contain only one or a few species. Prey records obtained since 1957 are listed in Table 28. Since cell cleaning does not occur in this species, it is possible to determine the approximate number of flies used per cell by counting the fly remains. The usual number appeared to be about 20. The tendency for females to concentrate on a single type of fly is shown by the following notes (all taken in Florida):

No. 1727. Larva 12 mm long, provisioning not yet complete. Cell contained 5 ♀ ♀ and 9 ♂ ♂ *Chlorotabanus crepuscularis* and one *Systoechus vulgaris.*

No. 1728. Larva only 14 mm long, but cell packed full of flies; nest has received the final closure. Cell contained 5 ♀ ♀ and 11 ♂ ♂ *Chlorotabanus crepuscularis* plus a few unidentifiable fragments.

No. 1729. Larva 25 mm long; nest has received final closure. Cell contained 5 ♀ ♀ and 5 ♂ ♂ *Erax tabascens,* remains of 8 other specimens of this same species, and one *Poecilanthrax.*

On the other hand, some nests in Florida contained a diversity of flies, and one of the nests excavated in Colorado contained an unusual assortment:

No. 1836A. Larva nearly fully grown; nest had received final closure. Cell contained 15 identifiable flies plus a few other fragments. The flies belonged to seven species of four families.

The one nest excavated at Rodeo, New Mexico, contained six bombyliids, *Villa flavocostalis* Painter, with a small wasp larva attached to one of them.

In 1957 I recorded flies of seven families as prey. The present list adds only one family (Therevidae). Although no tabanids were found in the three nests excavated in Colorado, these flies made up about two thirds of the prey in Florida (125 out of 193 records). It is interesting to note that male horseflies outnumbered females about 2:1, suggesting that they were collected by the wasps elsewhere than on animals.

The egg is laid on the first fly placed in the cell, as described in 1957. Under Florida conditions at least, the female brings in a few additional flies on the day after oviposition, and the egg hatches within about 36 hr. Thereafter provisioning proceeds at a rather variable rate, depending perhaps on the ability of the female to locate a rich source of flies. In some cases the cell is packed with flies and closed permanently when the larva is only 2 or 3 days old. For example, no. 1713 was seen starting to dig a nest at 1600 on 12 May; the egg was presumably laid 1 to 2 hr later. When checked on 13 and 14 May the entrance was closed (indicating an egg or small larva in the cell). On 15 May the entrance was open at 1500, and on 16 May the nest had received the final closure at 1600—just slightly less than 4 full days after oviposition. The larva was found to be only 15 mm long when the nest was excavated shortly after final closure; the cell contained 18 fresh flies plus several miltogrammine maggots. Two nests appeared to have received the final closure only 3 days after their inception (in each case several hours more than 3 full days). Data on 26 nests on which records were kept is as follows (for greater detail on selected nests, see Table 29):

Number of days from digging to final closure	Number of nests
3	2
4	4
5	14
6	6
7	1

§B. *Bembix sayi* Cresson

TABLE 29. DURATION OF SEVERAL NESTS OF BEMBIX SAYI

(Highlands Co., Fla., 1961)

Wasp no.	Date nest started	Date of final closure	Total days	Length of larva at final closure
2, nest 1	5 May	10 May	5	22 mm
nest 2	10 May	15 May	5	15 mm
9, nest 1	7 May	13 May	6	—
nest 2	13 May	19 May	6	—
nest 3	19 May	24 May	5	25 mm
22, nest 1	10 May	16 May	6	—
nest 2	16 May	21 May	5	—
nest 3	21 May	24 May	3	18 mm
32, nest 1	13 May	17 May	4	—
nest 2	17 May	20 May	3	14 mm
38	13 May	19 May	6	28 mm

It will be noted that several females showed an increasing reduction in the duration of provisioning in successive nests. Possibly this is a consequence of females learning better places in which to find flies or actually acquiring greater skill in hunting. The actual duration of the larval stage appeared to be about 6 days. Only one of the wasps observed practiced fully progressive provisioning, which would mean closing the nest on the 7th or 8th day after its inception.

(4) *Final closure.* Final closure is elaborate in this species and quite unlike that of any other species of *Bembix* studied. Final closure was observed many times in the Florida location and once in Colorado, and in every case it followed much the same pattern. The following extract from my field notes describes a typical final closure in some detail:

No. 1703, Archbold Biological Station, 14 May. Wasp first seen filling burrow at 1540; apparently started filling 20–30 min earlier, as burrow was already filled to fairly close to surface. She came out repeatedly to just outside the entrance, scraping sand into the burrow, then backed in to pack it in place. As the burrow became nearly full it could be seen that she was packing the sand with light blows of the deflected tip of the abdomen. In the final stages, she reamed out the entrance, making a depression 1–2 cm deep and 3 cm in diameter, using the soil to pack the burrow tightly. Then, at 1600, she began going out in crooked lines away from this depression and kicking sand toward it. She did this for 17 min, making a total of 80 trips from the nest entrance, all on the side away from the mound. The first 70 of these trips varied in length from 5 to 35 cm, most of them being about 20 cm long. In each case, having reached a point well away from the entrance by an irregularly zigzagging path, she would take flight and circle over the mound and back to the depression at the entrance, then work away from it again. After the 70th trip the lines became shorter (only 2–5 cm in length) and were intermingled with some irregular scraping movements around the depression (now well filled).

After the 80th trip she began digging a burrow at the old nest entrance but *toward* the mound, that is, in the opposite direction from the true burrow. She dug this exactly as if she were digging a new nest. She was still digging at 1645 when I left the area. The next morning this back burrow was open; it was measured and found to be 18 cm long, ending blindly at this length. The soil from this burrow formed a small mound opposite that of the mound of the original nest. The entrance of a completely new nest was located nearby; this nest had a typical mound and a closed nest entrance.

This behavior struck me as so unusual (and senseless) that I made many observations on it, but only minor, quantitative differences were found. In every case there was reaming of the nest entrance, resulting in a small pit which was filled by a series of irregular, radiating sand-scraping trips on the side away from the mound (Fig. 157); this was invariably followed by the construction of a back burrow either beneath the mound or at an angle with it of not greater than 90°, this back burrow invariably being left open and its mound intact (Fig. 158). Twenty-three back burrows were measured in the Florida locality; the mean length was 11 cm, the range of variation from 4 to 22 cm. The one back burrow measured in Colorado was 8 cm long.

The most curious feature of this behavior is that the wasp always fills the reamed out old nest entrance more or less completely before starting the back burrow, the digging of which, of course, completely changes the appearance of the nest entrance anyway. The only clue I obtained as to the possible adaptive significance of the back burrows was the observation of a single female mutillid (*Dasymutilla asopus cassandra* Mickel) entering a back burrow briefly and then leaving. Since the true burrow is packed tightly and completely with sand, the back burrow might actually provide an effective device for "duping" mutillids, which normally dig into nests after the final closure and parasitize the larvae inside their cocoons.

§B. *Bembix sayi* Cresson

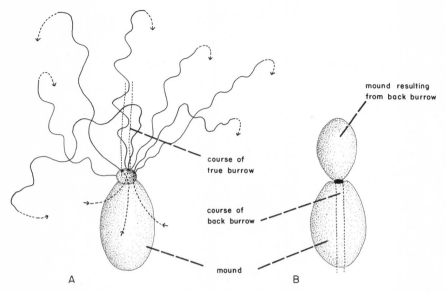

Fig. 157. Final closure and back burrow in *Bembix sayi* (Archbold Biological Station, Florida): (A) a portion of the paths of radiating lines made by the female at final closure; and (B) the final appearance of a typical back burrow and resulting mound.

(5) *Natural enemies.* Although female *Dasymutilla asopus cassandra* were fairly common in the nesting area at the Archbold Station in Florida, the only one actually seen associated with a *Bembix sayi* nest is the one mentioned above as entering a back burrow briefly. I have no definite evidence that this species is a parasite of *sayi*, and since other subspecies of *asopus* have been found associated with bees, it is not probable that *cassandra* attacks *Bembix*.

Miltogrammine flies were seen commonly in the nesting area, and 4 of the 31 nests excavated (13 percent) were found to contain maggots. Two flies captured as they rested near nest entrances and darted after females laden with prey were found to be *Senotainia rubriventris* Macq. The maggots from two of the four infested cells were reared to adulthood and found to be *S. rubriventris*. Another cell, which contained more than 20 maggots, eventually produced two adult *S. opiparis* Rnh. and seven *S. trilineata* Wulp (complex). The other three cells (the contents of one of which was not reared successfully) contained only four to six maggots each. In two cases the wasp larva was practically full grown and doubtless would have

Fig. 158. Female *Bembix sayi* working at the nest entrance, actually making a back burrow, the slope and position of the true burrow being indicated by the base of the stake in the background, which was placed 8 cm behind and at the same angle as the true burrow (Archbold Biological Station, Florida).

spun up successfully in spite of the maggots. In the other two cases the wasp larva was 14–15 mm long, apparently healthy but at least 2 days from maturity. It is possible that these larvae might also have developed successfully if provided by the mother with sufficient food to feed both the wasp larvae and the maggots. I had no difficulty rearing both in tins, but a large number of flies was required.

C. *Bembix nubilipennis* Cresson

This relatively well-known species has been studied by Lafler (1896), Parker (1910), and the Raus (1918). In 1957 I pointed out certain weaknesses and inconsistencies in the published accounts, but I was unable at that time to add anything new. Subsequently I have been able to study in some detail a colony of this species at Versailles, Indiana. These studies comprise 33 field notes made between 16 and 23 July 1957. I am indebted to Dr. Leland Chandler of Purdue University for showing me the location of this colony. During 1958 Dr. Cheng Shan Lin, of Huston-Tillotson College, Austin, Texas, studied about 60 nests of this wasp near Austin, and has kindly placed his notes at my disposal. Altogether these notes add considerably to knowledge of the species.

(1) *Ecology and general features of behavior.* Lafler, Parker, and the Raus all reported this species nesting in bare, compact soil in paths, roadways, and baseball fields. In southeastern Indiana, *nubilipennis* is highly characteristic of baseball diamonds and in fact is something of a pest to ball players at times. The wasps appear to thrive in the hard-packed clay soil of the infield, except along the trampled soil of the base lines. The colony studied by Lin at Austin, Texas, was situated in relatively hard clay soil along the Colorado River.

This species apparently forms large, concentrated colonies wherever it occurs. The major colony studied in Indiana contained at least 200 females. In the center of the colony, around the pitcher's mound, the nest entrances were separated, on the average, by about 5 cm. In other parts of the infield the nests were scattered more sparsely. This ball field was situated in a broad meadow surrounded by woods on all sides. It was situated in the valley of Laughery Creek, about 0.5 mi east of the town of Versailles. The colony studied by Lin, as well as those described by Parker and by the Raus, all contained a large concentration of wasps in a relatively small area.

In the Versailles ball field, *Astata occidentalis* also nested in considerable numbers, and I have reported on this species elsewhere (Evans, 1958b). A few *Cerceris fumipennis* also nested here, and various parasites of these three wasps were very common.

All records indicate that this species is univoltine throughout the central states, the wasps appearing in late June or early July and remaining active into August. However, at Austin, Texas, *nubilipennis* is distinctly bivoltine. Here there is a generation of adults during late May and throughout June, and a second generation in September and October. In 1958 wasps of the two generations exhibited interesting differences in nest structure, type of prey, and aggressive behavior. Dr. Lin found that the second generation was smaller in numbers than the first, and the females did little fighting among themselves and indulged in little or no prey stealing. During the first generation he observed much stealing of prey, as I did on many occasions in Indiana. On one occasion I saw a female lose a fly to a neighbor, then steal one herself and enter her nest (no. 987). On another occasion I watched a female lose two successive flies to robbers (no. 989); each time she entered the nest anyway, remained for a moment, then came out and closed in the usual manner. Stealing of prey generally occurs during a brief struggle on the ground, one female pouncing upon another just as she is about to enter her nest. The wasps make no attempt to bite or sting one another; they merely try to obtain a secure grip on the prey and fly away with it.

In the Indiana colony some active wasps could be found at all times of the day. From 0800 to 0900 wasps could be seen emerging from their nests, the females clearing the entrances and closing from the outside. The late morning hours were filled with activity, the males flying in their sun dance, the females bringing in flies in great numbers. During the afternoon the males tended to be less active, but females could be seen digging or provisioning, at least to some extent, as late as 2000, just prior to sunset.

The males apparently spend the night in holes in the ground, in some cases perhaps in old emergence holes or abandoned nests of females. On one occasion three males were seen emerging from one such hole. The sun dance flight is close to the ground, only 2 to 4 cm high, and is strongest during the midmorning hours, exactly as described by the Raus. Many attempted matings were observed. In each case the male descended upon a female and grasped her with his middle legs, then flew along with her for some distance, eventually descending to the ground. Often two or even as many as six males would form a struggling ball around a female. During mating on the ground, the male holds his front and hind legs rigid while holding down the wings of the female with his middle legs, as occurs in such species as *amoena* and *belfragei*. The longest mating observed lasted about 2 min.

The Raus noted a marked decline in the numbers of males participating in the sun dance after the first 10 days. At Versailles, Indiana, the number of males declined steadily between 17 and 22 July, and after that date no males were seen. Of course, the males had undoubtedly been active for some time before I began to study the colony on 17 July.

(2) *Digging and leveling.* The digging of the nest apparently requires the greater part of 1 day. For example, no. 983 was first seen digging at 0945 on 17 July. At 0958 she closed the entrance, leveled the earth at the entrance, and flew off. At 1056 she re-entered the nest and closed from the inside. At 1155 she was again seen digging

§C. *Bembix nubilipennis* Cresson

and leveling. She left again at 1206 and was not seen again until 1405, when she was digging again. At 1415 she closed the entrance from the outside and spent 10 min leveling the mound. She then flew off, and when she had not returned at 1500 I dug out the nest, finding it complete but with the cell empty.

The practice of digging intermittently, closing, and leveling after each session of digging, was observed many times. Leveling in this species is in the form of fairly regular zigzags across the mound. The wasp first walks quickly backward to the far end of the mound, then works forward to the entrance while making wide sweeps from side to side (though these are by no means as regular as figured by the Raus). The sand is spread widely over the ground, covering an area 30–40 cm in diameter. From time to time the wasp moves laterally away from the closed nest entrance, zigzagging and adding soil to the closure. The longest periods of leveling occur after the completion of the nest (or of a new cell). After the first fly has been brought in and the egg laid, further leveling movements occur, such that the area around the nest entrance is completely smooth. At this time an especially thorough closure is made, the wasp employing the tip of the abdomen for pounding soil in the entrance—even though the wasp shortly thereafter re-enters the nest and recloses from the inside. The Raus describe leveling and closure in this species, but they do not state precisely when this behavior occurs.

It is a curious fact that although leveling movements occur during the digging of the nest and following its completion, they are not necessarily resumed on the following days. Each morning, at least after the egg hatches, the female clears out the entrance after she emerges from the nest and prior to bringing in flies. The small mound that accumulates at this time often remains intact from day to day (but not always; some leveling occurs from time to time in some individuals). Nests having large larvae almost always have a small mound of earth at the entrance. This mound is eventually swept away at the time of final closure or of construction of a second cell.

(3) *Nature and dimensions of the nest.* The burrow forms about a 45° angle with the soil surface and is about 12 mm in diameter. The lower part of the burrow tends to level off and leads to a nearly horizontal cell about 15–18 mm in diameter and about 30–35 mm in length (tending to be lengthened to about 40 mm as the larva matures). The nests are remarkably shallow, and nests from widely separated localities remarkably consistent in depth (Table 30). The

TABLE 30. NEST DATA FOR BEMBIX NUBILIPENNIS

Locality	No. of nests	Length of burrow (cm)	Depth of cell (cm)
Versailles, Indiana	29	17.0 (14-20)	10.0 (8-13)
Austin, Tex. (May-June)	20	24.0 (15-35)	15.0 (9-23)
Austin, Tex. (Sept.-Oct.)	40	16.0 (12-19)	9.5 (5-12)
Kansas (Parker)	several	— (15-20)	—
Missouri (Raus)	several	— (13-15)	—

only exception to this statement is provided by nests of the first generation at Austin, Texas, which were notably deeper than those of the second generation or of colonies where the species has only one generation. I am unable to suggest a reason for this unless it is related to the amount of seasonal rainfall.

Although most of the nests I dug out at Versailles, Indiana, contained but one cell, most of these nests were dug out before final closure. After I had been working in the colony for several days, it became apparent that many females did not make a final closure after the completion of one cell, but rather prepared two or more cells from the same burrow. For example, no. 1456 first dug her nest on 17 July. On 22 July I watched the nest at length, expecting a final closure. Instead, the wasp did additional digging and leveling, then brought in a fly and closed from the inside. This nest was dug out and found to contain a new cell containing an egg on a fly. The old cell contained a nearly full-grown larva still feeding on the last few flies in the cell. The burrow was Y-shaped, the arm of the nest reaching the first cell having been closed off. Two additional two-celled nests and two three-celled nests were found—but none of these had received the final closure. I saw very few final closures during the period of study, and it seemed obvious that most if not all females were making a series of cells from the same burrow. Lin obtained this same impression at Austin, Texas, during May and June 1958. Of 20 nests he dug out (all before final closure), 10 had one cell, 9 had two cells, and 1 had three cells. The cells in multicellular nests are always at about the same depth, at the ends of relatively short branches of the main burrow (Fig. 159).

§C. *Bembix nubilipennis* Cresson

Fig. 159. A three-celled nest of *Bembix nubilipennis* (no. 1473, Versailles, Indiana). The first cell to be prepared (now containing a cocoon) was at the end of a nearly straight burrow; the second (now containing a large larva), at the end of a short side burrow; the third (still empty), at the end of a longer side burrow closer to the nest entrance.

Curiously, all of the 40 nests dug out by Lin during September and October 1958 contained but one cell. Several of these had received the final closure and contained cocoons. The Raus reported only one cell per nest in their studies made in Missouri, while Parker reported that most nests in the area in which he worked in Kansas had four or five cells. Certain parts of Parker's report are probably erroneous, particularly his statement that fully provisioned cells may be walled off and another cell prepared between this cell and the main part of the burrow. However, evidence now seems to point to unicellular versus multicellular nests in *nubilipennis* being a matter of variation in different colonies and different broods.

An outer closure of the nest is maintained at all times, as indicated by Lafler, Parker, and the Raus. During the period of study in Indiana, I noted only two exceptions to this behavior (nos. 988 and 1461, both of which brought in a series of flies rapidly without closures, and then began to close the cell permanently). When closing the nest, the wasp scrapes in soil from the front and sides of the hole, sometimes exhibiting weak zigzag movements. In no case did I find any evidence of a spur or an inner closure, even in nests containing an egg.

(4) *Hunting and provisioning.* This species was reported by Parker and by the Raus to prey upon at least ten species of flies belonging to five families. New records from Indiana and from Texas (Table 31) show the species preying upon nearly 50 species of flies belonging to 12 families, including Nemestrinidae and Trypetidae, families rarely appearing on lists of *Bembix* prey. In the Indiana colony it was common to see the wasps carrying large asilids (*Diogmites*) which they were unable to hold tightly beneath them, but allowed to dangle to some extent. Houseflies were reported as prey by both Parker and the Raus, and were taken in the Indiana colony and by both generations at Austin, Texas. However, Dr. Lin found that while the first generation employed houseflies in small numbers, females of the second generation often took them in great numbers. Some cells contained 24–28 fresh houseflies, remains of a few other houseflies, and nothing else. Doubtless these wasps were taking advantage of the late summer population peak of the housefly in the vicinity of Austin.

The egg of this species measures about 1.2×6 mm. It is glued in the usual manner to the wing base of a small fly, the middle leg of the fly on that side being extended. Flies selected as egg pedestals in Indiana included various small muscoids, especially

TABLE 31. PREY RECORDS FOR BEMBIX NUBILIPENNIS

Species of fly	Versailles, Indiana (H.E.Evans)	Austin, Texas (C.S.Lin)
STRATIOMYIDAE		
Hedriodiscus truquii (Bell.)		3
Nemotelus trinotatus Mel.		4
Stratiomys jamesi Steyskal		1
Stratiomys meigenii Wied.	1	
Stratiomys nigriventris Loew		1
TABANIDAE		
Tabanus sulcifrons Macq.	4	
Tabanus trimaculatus Beauv.	1	
BOMBYLIIDAE		
Anthrax irrorata Say		1
Bombylius io Will.		1
Exoprosopa emarginata Macq.	1	
Exoprosopa fasciata Macq.	1	
Poecilanthrax lucifera (Fabr.)		4
Villa spp.	4	1
NEMESTRINIDAE		
Neorhyncocephalus sackenii (Will.)	2	
ASILIDAE		
Atomosia puella (Wied.)	1	
Diogmites misellus Loew	8	
Diogmites umbrinus Loew	2	
Erax sp. nr. tuberculatus Coq.		5
Proctacanthella cacopiloga (Hine)		1
SYRPHIDAE		
Eristalis arbustorum (L.)	6	
Eristalis tenax (L.)	2	
OTITIDAE		
Chrysomyza demandata (Fabr.)		2
Tetanops luridipennis Loew	2	
TRYPETIDAE		
Paracantha culta (Wied.)		1
MUSCIDAE		
Musca domestica L.	8	120
Orthellia caesarion (Mg.)	7	3
Stomoxys calcitrans (L.)	4	

Chapter X. Generalized Species of *Bembix*

TABLE 31. PREY RECORDS FOR BEMBIX NUBILIPENNIS (CONT.)

Species of fly	Versailles, Indiana (H.E.Evans)	Austin, Texas (C.S.Lin)
CALLIPHORIDAE		
Bufolucilia silvarum (Mg.)	1	
Lucilia illustris (Mg.)	1	
Phaenicia caeruleiviridis (Macq.)	4	
Phaenicia sericata (Mg.)	1	
Pollenia rudis (Fabr.)	39	
SARCOPHAGIDAE		
Sarcophaga derelicta Wlk.	3	
Sarcophaga lherminieri R.D.	2	1
Sarcophaga rapax Wlk.	3	
Sarcophaga sinuata Mg.	1	
Sarcophaga ventricosa Wulp	3	
TACHINIDAE		
Aplomya theclarum (Scud.)	1	
Archytas apiciferus Wlk.		4
Archytas metallicus (R.D.)	1	
Gymnoclytia occidua (Wlk.)	1	
Hyalomyodes triangulifera (Loew)	1	
Phorocera tachinomoides Tns.		1
Trichopoda pennipes (Fabr.)	2	

Pollenia rudis (four records). The housefly was very commonly employed as an egg pedestal in Texas.

Females carrying flies tend to curve the tip of their abdomen down as if to support the fly from behind, a trait also characteristic of *Bembix spinolae* and *B. comata*. During periods of active provisioning, a female may bring in a fly every 5-15 min. As I have already noted, a closure is almost always maintained between trips for flies. This closure increases the amount of time required for females to get into their nests, and therefore provides more opportunities for prey stealing by other females as well as attacks by miltogrammine flies.

The duration of a single cell in this species may be as short as 5 days. For example, no. 1455 dug a second cell in her nest on the morning of 17 July. On 21 July and on the morning of the 22nd she brought in a great many flies, then on the afternoon of the 22nd she prepared a new cell and brought in a single fly on which to lay

§C. *Bembix nubilipennis* Cresson 307

her egg. The nest was excavated late that day, and the larva in the second cell found to be very large, with only one intact fly still to be eaten. Thus development of the larva is very rapid.

(5) *Final closure.* This was observed in detail only once (no. 985). This wasp filled the burrow completely, packing the soil firmly with blows of the tip of her abdomen. Much of the soil for fill was reamed from the walls of the burrow. Later the wasp came out the entrance a distance of 1–3 cm scraping soil, then took flight briefly and landed in the entrance, rotated quickly 180°, and backed in scraping soil. This was repeated many times, much of the small mound of accumulated soil being used up in this manner. When the burrow was full, the wasp walked away from the entrance a distance of 2–4 cm in various directions scraping soil. The final result was an almost completely smooth surface. Some zigzagging occurs as the wasp scrapes soil toward the entrance, but I did not observe the elaborate zigzagging described by the Raus, and in particular I observed no zigzagging toward the entrance (scraping soil away from the entrance) at this time. As I suggested in 1957, the Raus probably observed leveling following the completion of a new nest and believed that their nest was receiving the final closure.

(6) *Natural enemies.* The chrysidid wasp *Parnopes chrysoprasinus* Smith was very common in the Indiana colony from 16 to 21 July, but none were noted on the 22nd or 23rd. These wasps would appear in numbers each day about 0930 and remain active throughout the morning, to a lesser extent throughout the afternoon. They spent most of their time on the soil in the colony, taking short, hopping flights from one *Bembix* nest to another, perching or walking about the nests, and now and then digging into nest entrances. They seemed especially attracted by evidences of fresh digging or leveling. On several occasions female *Bembix* were seen to pounce upon these chrysidids, appear to sting them, and actually begin to carry them to their nest. My impression was that these were not attacks on the chrysidids as parasites; the wasps apparently merely mistook them for flies. In the cases observed, the wasps were not able to paralyze the chrysidids.

Apparently the *Parnopes* females oviposit on growing *Bembix* larvae, but the *Parnopes* larva does not consume the *Bembix* larva until the latter has spun its cocoon. When I dug out nest no. 985, I found an apparently normal *Bembix* larva 22 mm long. This larva spun its cocoon in a rearing tin on the following day. The following spring I reared a *Parnopes chrysoprasinus* from this cocoon. Apparently the *Parnopes* larva was extremely small until after the cocoon had been spun. The percentage of parasitism by *Parnopes* in this colony may have been rather high. Only six cocoons were saved, and of these one was parasitized by *Parnopes*.

The parasitic beetle *Macrosiagon flavipenne* Lec. (Rhipiphoridae) was seen several times in the Indiana colony. These beetles fly sluggishly and land heavily on the ground. They were sometimes seen near *Bembix* nests, but none was actually seen entering them. Barber (1915) reared this species of beetle from a cocoon of *Bembix spinolae*, so it is probably safe to assume that it is also a parasite of *nubilipennis*.

Miltogrammine flies were moderately common in the Indiana colony. They were often seen sitting on the ground near *Bembix* nests and flying up and pursuing wasps laden with flies. One fly taken as she was following a *Bembix* was found to be *Senotainia trilineata* Wulp (complex). Only one of the nests dug out contained maggots (no. 986); this nest contained only two maggots, and these were not reared successfully. *Senotainia trilineata* was attacking *Astata occidentalis* very successfully in this same area, but it did not appear to be as successful in its attacks on the *Bembix*.

Two of the 60 nests dug out by C. S. Lin at Austin, Texas, contained maggots, but these were not successfully reared. Lafler, Parker, and the Raus all reported dipterous parasites of *nubilipennis*, probably *Senotainia* in most cases.

In the Indiana colony I often saw kingbirds (*Tyrannus tyrannus*) perching on the periphery of the colony and making occasional dashes at the milling hordes of wasps and their parasites. I had no way of knowing exactly what they were taking, but it is probable that at least a few male *nubilipennis* were snatched from their sun dance.

D. *Bembix truncata* Handlirsch

This species was described by Handlirsch from Orizaba and Cuernavaca, Mexico. It has remained poorly known, and was omitted altogether by Parker (1929). I encountered the species in small numbers in June 1959, at a locality 5 km northwest of Cuernavaca, Morelos (no. 1606). The species appears to resemble *nubilipennis* in structure and color, although lacking the clouding on the wings of that species, and my limited studies of the behavior also suggest *nubilipennis*.

Although I have collected and observed species of *Bembix* in many places in Mexico, I have encountered *truncata* only this once.

§D. *Bembix truncata* Handlirsch

The locality was an open grassland dotted with low *Acacias,* at an elevation of about 6500 ft, only a short distance below the lower limit of pine forest. Both males and females occurred in small numbers at the flowers of milkweed (*Asclepias* sp.), beginning about 20 June. There seemed to be no large nesting aggregation in the area, but on 26 June I found one female digging in the bare soil of a trail. The soil here was extremely hard, by no means sandy, best described as a heavy clay-loam. It took this female some time to break through the hard surface crust, and the soil beneath the surface was only slightly less compact. This burrow was eventually abandoned, but I did find one emergence hole nearby which led to an old cocoon at a depth of about 10 cm. On 30 June I located two active nests in this same area. It appeared that there was a very small, diffuse colony here, the nest entrances spread over a considerable distance along the trail. Unfortunately I was unable to continue my studies beyond 30 June. It is possible that the species was just beginning to nest at this time. All specimens appeared fresh and unworn. June is the first month of the rainy season in this area, and it is probable that the soil in which these wasps nest is completely unworkable in the dry season.

One of the females observed on 30 June was bringing in flies during the late morning and early afternoon. She flew in with prey only 20–30 cm high, remained within the nest only a few seconds, and closed the entrance each time she left. This nest was found to contain a half-grown larva and ten fresh flies, all of them small assassin flies of an apparently undescribed species of *Atomosia* (Asilidae). The larva was feeding upon a larger asilid (*Asilus* or a related genus) and resting upon a mat consisting of wings and body remains of several Asilidae. There was no inner closure of the nest. The burrow was oblique and nearly straight, 20 cm long, the cell at a depth of 11 cm.

The second female entered her nest without prey at 1330 and threw up a closure from the inside. This nest was found to be of about the same dimensions as the preceding one. The female was inside the otherwise empty cell when the nest was dug out. Although this was apparently a newly finished nest, there was no evidence whatever of a mound at the entrance. This was also true of the first nest.

Characteristics suggesting *nubilipennis* are (1) the use of very hard soil for nesting, (2) the apparently complete dispersal of the mound by leveling movements, and (3) the use of Asilidae in considerable numbers. One would predict multicellular nests in *truncata,* but my studies shed no light on this matter; both nests might have eventually had more than one cell if left undisturbed.

E. *Bembix cameroni* Rohwer

This species ranges from Arizona and New Mexico south through the Mexican Central Plateau. Its nesting behavior has not been studied previously, and my own very brief studies do nothing to distinguish it from other members of the *spinolae* complex, to which it appears to belong. In Cochise County, Arizona, where these studies were made, *cameroni* appears to occur in canyons at about 4500 to 5500 ft elevation. There is no evidence of intergradation with *spinolae*, a smaller wasp represented at generally lower elevations in this area by an extensively maculated subspecies or color form, *similans* (Fox).

I collected *cameroni* at several localities in the Chiricahua Mountains, but found it nesting only in Cave Creek Canyon, near the Southwestern Research Station, 7–29 August 1959. Both sexes occurred in numbers on flowers of *Melilotus alba* at the station. Several nests were located in bare, sandy strips along the stream, never more than one or two nests in any one place. This was an abnormally rainy month, and the stream was in frequent flood. Several nests were destroyed by violent floods which greatly altered the nesting sites, but a few nests which were subject to only slight and occasional flooding survived. The first nest (no. 1615) located on 7 August, was situated in a sand bar that had definitely been flooded the night before. The only other nest (no. 1624) I dug out was found on 29 August, after the floods were largely over.

In all I marked about ten nests, but some of these were destroyed by floods and others were not followed because of the pressure of other work. All nests were in light, alluvial soil along Cave Creek, sandy on the surface but often silty or loamy a few centimeters down. I made no observations on digging or leveling except to note that the mound was always leveled more or less completely, apparently in much the manner of *spinolae*. Without exception, outer closures were maintained. One nest containing a half-grown larva had a strong inner closure, while the second nest, dug out during a period of active provisioning and having a large larva in the cell, lacked an inner closure. The burrow of the first nest was at first oblique, then made a sweeping curve to the right and leveled off; the burrow was 20 cm long, but the cell only 9 cm deep. The second burrow was straight and more oblique, only 16 cm long, the cell 10 cm deep. Both cells measured about 18 × 35 mm and were horizontal.

Both nests contained several fresh flies and many fly remains when dug out. The flies were diverse in kind and size; the largest,

Esenbeckia delta, a tabanid common on flowers in the canyon, was of nearly as great a bulk as the wasp. The following is a list of the flies taken from the two nests and the numbers of each taken:

TABANIDAE	
Esenbeckia delta Hine	1
BOMBYLIIDAE	
Geron sp.	1
Phthiria sulphurea Loew	3
SYRPHIDAE	
Eristalis latifrons Loew	1
MUSCIDAE	
Musca domestica L.	1
Orthellia caesarion Mg.	6
TACHINIDAE	
Peleteria sp.	1
Plagiprospherysa parvipalpis Wulp	1

Both cells contained numerous very small maggots which appeared to feed upon the fly detritus in the cells. In both cases the wasp larva developed normally and spun its cocoon. The maggots formed their puparia at about the same time. Nine flies (*Lasiopleura grisea* Mall., family Chloropidae) emerged from one of the cells 2 weeks later.

F. *Bembix spinolae* Lepeletier

I presented a fairly detailed account of this widely distributed and relatively well-known species in 1957. Since that time I have made occasional observations at various localities where it has not been studied previously. My hope has been to obtain a better impression of variation in some facets of behavior, also to obtain more data on parasitism. I do not feel that I have added a great deal to knowledge of the species, but my data are summarized briefly here for whatever they may be worth.

The notes summarized below were made at the following six localities: (1) Eastern Massachusetts (Andover and Lexington), July 1961-1963; (2) Central New York (three localities), July-August 1958; (3) Presque Isle, Pennsylvania, July 1961 (contributed by Frank E. Kurczewski); (4) Jackson Hole Biological Research Station, Moran, Wyoming, August 1961 and July-August 1964; (5) Cornish, Cache County, Utah, August 1961; (6) Rodeo, Hidalgo County, New Mexico, August-September 1959.

The populations in the first four of these localities all consisted of "typical" *spinolae,* that is, with relatively reduced maculations. Specimens from Cornish, Utah, run to *primaaestate* Johnson and Rohwer in Parker's keys, those from Rodeo, New Mexico, to *similans* Fox. I have already suggested (1957b) that *primaaestate* does not deserve formal separation from *spinolae.* Fox's *similans* averages somewhat smaller than typical *spinolae* and has quite extensive yellow maculations; whether or not it deserves subspecific status is uncertain. There seems no reason to treat it separately here.

(1) *Ecology and general features of behavior.* I have already discussed the ecological versatility of this species. In areas of extensive inland sand, such as those at Andover, Massachusetts, and Cornish, Utah, *spinolae* tends to nest in the periphery of dunes and blowouts, *pruinosa* occupying the more central areas. At Jackson Hole, Wyoming, *spinolae* nests in small patches of fine-grained, pale sand along the Snake River, while other bembicines (*Bembix amoena* and *Steniolia obliqua*) occur in coarser and less sandy soil. Thus *spinolae* might be characterized as less of a psammophile than *pruinosa* (and *occidentalis*), more of a psammophile than *amoena* (and *cinerea*). However, in the absence of competition, *spinolae* may nest either in pure, open sand or in rather coarse soil.

In Massachusetts adults are often common on *Aralia hispida* (Araliaceae) in June and early July. At Jackson Hole adults appeared to be taking most of their nectar from *Solidago,* while at Rodeo, New Mexico, *Baccharis* and *Asclepias* were being visited in great numbers. Kurczewski and Kurczewski (1963) have recently presented several flower records from Presque Isle, Pennsylvania.

At Andover, Massachusetts, in early July, the sun dance of the males was so populous that the humming sound produced was audible from some distance. At Rodeo, New Mexico, there was a large sun dance over a series of sand piles along an irrigation ditch (Fig. 160). Although the dance persisted for more than 1 week, I only rarely saw females in this area. A dense growth of weeds was beginning to cover the piles, and my impression is that they were no longer suitable for nesting and that the females were flying elsewhere. The piles were checked over a period of 3 weeks, but I found no active nests at all. The only nests I was able to find in this area were in a sandy field about 1 km away, where this species formed a small mixed colony with *Bembix troglodytes* and *B. u-scripta.*

Observations made in Massachusetts, Wyoming, and New Mexico confirmed the fact that the mound at the nest entrance is leveled with irregular zigzag movements after the entrance is first closed,

§F. *Bembix spinolae* Lepeletier

Fig. 160. A male *Bembix spinolae similans* in typical position on the sand during a pause in the "sun dance" (Rodeo, New Mexico). The body is close to the ground, the legs and antennae extended somewhat rigidly.

also that an outer closure is maintained at all times. Nests containing an egg usually have a distinct spur, and nests not being actively provisioned generally have an inner closure. Twenty-nine nests excavated in the six localities were all unicellular. All were within the known range of variation in depth (Table 32). The five nests from Massachusetts were all from areas of fairly extensive sand, and this may account for their greater depth as compared to those from most other localities (assuming I am correct in my belief that there is a correlation between nest depth and amount of blowing sand). The two very shallow nests from Pennsylvania were from a generally sandy area, but both were dug in coarse sand which contained bits of stones and rotting driftwood.

(2) *Hunting and provisioning.* In 1957 I presented a list of 52 species of flies of 11 families known to be utilized as prey by this wasp. I am able to add one more family to this list: Sciomyzidae, based on a single *Sepedon fuscipennis* Loew taken from a nest at Andover, Massachusetts. This same nest contained the second record for Otitidae, a single *Delphinia picta* (Fabr.).

Otherwise, recent records add little to the known spectrum of

TABLE 32. RECENT DATA ON NEST DEPTH IN BEMBIX SPINOLAE

Locality	No. of nests	Burrow length (cm)	Cell depth (cm)
Massachusetts	5	20.0 (18-25)	10.5 (8-14)
New York	7	15.5 (11-21)	7.5 (5-10)
Pennsylvania	2	16.5	4.5
Wyoming	12	15.5 (10-20)	8.2 (5-11)
Utah	1	15.0	10.0
New Mexico	2	22.0	15.0

prey of this species. New York records accumulated since 1957 include quite a number of Syrphidae not previously reported: *Eristalis transversus* Wied.; *Mesograpta geminata* Say; *Metasyrphus americanus* Wied.; *Pipiza femoralis* Loew; *Syrphus vittafrons* Shan.

At Jackson Hole, Wyoming, this species was taking advantage of an extremely high population of deer flies, over 80 percent of the flies taken from nests belonging to two species of *Chrysops* (the majority females): *C. furcatus* Walker and *C. noctifer pertinax* Will. The nest at Cornish, Utah, contained one tachinid and six *Lepidanthrax proboscideus* (Loew) (Bombyliidae). Two nests excavated at Rodeo, New Mexico, contained nothing but Bombyliidae, of the following five species: *Bombylius flavipilosus* Cole; *Lordotus pulchrissimus luteolus* Hall; *Villa (Chrysanthrax)* sp.; *Villa (Hemipenthes) chimaera* (O.S.); *Villa (Villa) salebrosa* Painter.

At Jackson Hole I once saw a female land on the sand with a *Chrysops*, insert her proboscis into the neck region of the fly, and remain in this position for 30 sec, apparently imbibing fluids from the gut of the fly.

(3) *Natural enemies.* Although chrysidid wasps have not previously been found associated with *spinolae*, I found *Parnopes chrysoprasinus* Smith to be very common in a small colony at Ontario, New York, during July 1958. The *Parnopes* seemed definitely attracted to the *Bembix* nests and spent much time flying from one nest to another and walking about and sometimes digging into the entrances. Since this chrysidid is a known parasite of *B. nubilipennis,* it is probably

safe to assume that it was attacking *spinolae* in this area, which is well north of the known range of *nubilipennis*.

In Jackson Hole, *Parnopes edwardsii* (Cresson) was very common in the nesting areas of *Bembix spinolae* as well as in adjacent nesting areas of *Steniolia obliqua*. The *Parnopes* was often seen near known *Bembix* nests and often visited the same flowers as *Bembix*. It seems very probable that *P. edwardsii* attacks both *B. spinolae* and *Stenolia obliqua* in this area.

At Rodeo the mutillid wasp *Dasymutilla ursula* (Cresson) was common on the sand piles inhabited by many male *spinolae similans*. At Jackson Hole the bombyliid fly *Exoprosopa dorcadion* O.S. was plentiful in the nesting area, as was the conopid fly *Physocephala texana* Will. The latter was seen pouncing upon adult male *spinolae* and is unquestionably a parasite of that species, as it is of at least three other species of *Bembix*. The *Exoprosopa* and *Dasymutilla* also belong to groups otherwise known to attack bembicines, although in this case their association with *spinolae* is only a surmise.

The Peckhams and others have reported a high incidence of parasitism by miltogrammine flies. These flies were occasionally seen at Jackson Hole, but none of the 12 nests excavated contained maggots. Of 2 nests I dug out in Tompkins County, New York, in 1958, 1 was infested with maggots, and of 5 nests dug out in eastern Massachusetts, 2 had maggots. In one case I was able to rear both wasp larva and maggots to maturity in a tin, and in the other two cases the wasp larva appeared healthy and probably would have reached maturity if supplied with enough flies by the mother. However, the importance of these parasites cannot be discounted entirely. Even if the completion of provisioning is delayed an average of only 1 or 2 hr, the result will be a slight reduction in the total number of offspring produced by a given colony in a season.

None of the maggots from the above nests produced adults in my rearing tins. The miltogrammine fly *Senotainia vigilans* Allen was observed around *spinolae* nests at Andover, Massachusetts, and may have been the major species involved in that locality.

G. *Bembix belfragei* Cresson

My 1957 studies of this species were all made at a single locality in Kansas. Hence it is of interest to examine data from other localities. The following notes, all brief, were made at three localities in Florida in May 1961 and April 1964. A colony of 80–100 nests was located in mounds of sand, mostly covered with grass, along Fisheating Creek at Venus, Highlands County (no. 1972). In sandy roads along the same creek at Palmdale, Glades County, a few scattered nests were located (no. 1742). At Highlands Hammock, near Sebring, two or three females nested on the periphery of the *Stictia carolina* colony described in Chapter IX:B, chiefly in places where the soil was moderately friable and covered sparsely with grass (no. 1732).

In the colony at Venus, nest entrances were fairly well spaced, mostly 30–100 cm apart, a few only 10 cm apart. In all three localities the mound was without exception left intact, conspicuously marking the nest entrances. With rare exceptions, the nest entrances were left open. These exceptions were two nests at Venus which were closed from the inside; one of these was dug out and found to contain the female wasp, but I failed to locate the cell. In both of these nests the female may still have been digging and may have merely failed to clear the entrance.

In all three localities, nest structure was typical of the species, the cell being elevated well above the burrow, as shown in Fig. 161, a two-celled nest dug out at Palmdale. At Venus, as in the Kansas colony studies several years ago, I was able to distinguish between two-celled and one-celled nests by the nature of the mound. I dug out only one nest at each of the three localities; burrow length varied from 25 to 30 cm, cell depth from 11 to 17 cm.

Pooled prey from these three nests consisted of the following flies:

TABANIDAE	
Tabanus spp.	9
SYRPHIDAE	
Eristalis agrorum Fabr.	8
MUSCIDAE	
Orthellia caesarion Mg.	1
CALLIPHORIDAE	
Cochliomyia macellaria Fabr.	15
Phaenicia caeruleiviridis Macq.	4
SARCOPHAGIDAE	
Sarcophaga ventricosa Wulp	1
TACHINIDAE	
Sturmia sp.	3
Genus & species?	2

One cell of the nest at Palmdale contained six small maggots. The wasp larva from this cell was nearly mature and started to

§H. *Bembix u-scripta* Fox 317

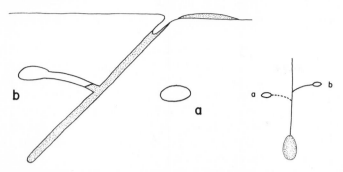

Fig. 161. Typical nest of *Bembix belfragei* (no. 1742, Palmdale, Florida), shown in profile (left) and in plan (right). Cell a contained a cocoon and had been closed off; cell b contained a nearly full-grown larva. This nest was dug out just before the female completed her final closure.

spin up soon after the nest was dug out; the maggots developed on the fly debris for 2 days more and then formed their puparia. Four adult flies of the *Senotainia trilineata* Wulp complex emerged from these puparia 2 weeks later. In the Venus colony *Dasymutilla pyrrhus* (Fox) was very common and was seen entering open nests on several occasions.

Thus these brief studies revealed no important differences in behavior from the Kansas colony studied earlier. The distinctive nest structure and lack of closure and leveling were essentially the same, and several genera of flies appear on both lists of prey. The greater use of Tabanidae in Florida may have merely reflected their relatively great abundance near the nesting areas.

H. *Bembix u-scripta* Fox

As I pointed out in the introduction to this chapter, *B. u-scripta* has a number of unique and apparently primitive structural features, including fairly well-developed and presumably functional ocelli. The species also has one striking behavioral peculiarity: the females provision their nest just before and at dusk, the last females closing their nests at about the time it becomes difficult for a human to observe them without artificial illumination. Since large ocelli are characteristic of nocturnal habits in the Hymenoptera, it seems probable that the preservation of ocellar lenses is correlated with the crepuscular flight period of this species. It is probable that other deserticolous species with well-preserved ocelli (for instance, *stenebdoma* Parker) will also be found to be in some measure crepuscular.

I published a few notes on *u-scripta* in 1957, then in 1961 presented a much more detailed study based on two widely separated colonies: one on the Gulf coast of southern Texas, the other in southwestern New Mexico. Several minor differences in the behavior of the two colonies were noted, and I suggested that further study of behavior and color pattern in this species might be rewarding. A photograph of the egg of *u-scripta,* taken in southern Texas, is included here as Fig. 162.

I have since found *u-scripta* in two other localities, but in each case only one or two females were found nesting. In both cases (as also in New Mexico) the species appeared to be a relatively solitary nester, its nests tending to be interspersed with those of other bembicine wasps. Also, the type of soil selected for nesting appears quite diverse.

On 31 July 1962 I found two females nesting in coarse, moderately compact, sandy gravel at the side of the stream bed in

Fig. 162. Cell of *Bembix u-scripta* showing the egg on a beefly at the apex of the cell (no. 1526, Port Isabel, Texas). This photograph appeared in *Wasp Farm* (Evans, 1963d), where it was incorrectly identified as representing the cell and egg of *B. spinolae.*

§H. *Bembix u-scripta* Fox

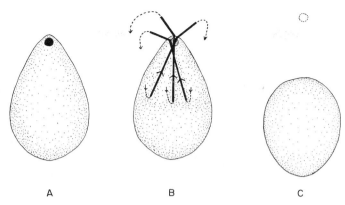

Fig. 163. Mound leveling in *Bembix u-scripta*. In A the burrow has just been completed, and the resulting mound is intact, the entrance open; in B three typical routes of a female leveling the mound are shown, the dashed lines indicating flight; in C is shown the final product, with the nest entrance well concealed, the mound well removed from the entrance. (From Evans, 1961.)

Cañon del Zopilote, Guerrero, Mexico, at about 2,000 ft elevation. One female was seen digging in the early morning (0730); I dug out her nest but found that it had not yet been completed. The second female was seen about 0800 as she entered her nest and closed from the inside. This nest was dug out and found to contain a cell beyond a strong inner closure; in the cell was an egg which seemed to have deteriorated, also two fresh *Ptilodexia* sp. (Tachinidae) and three dry, stiff *Saropogon* sp. (Asilidae). There was a small fresh mound at the entrance of this nest. *Bembix multipicta* and a *Bicyrtes* also nested in this canyon.

Additional observations were made from 26–30 April 1963, 1 km west of Lajitas, Brewster County, Texas, where several females were seen in the nesting site of *Glenostictia scitula* described in Chapter VIII:G. Only one completed *B. u-scripta* nest was located, although several other females were seen hunting around creosote bushes (*Larrea*) in the late afternoon (1715). This one nest was in relatively soft, friable sand, devoid of the surface crust found in the *scitula* colony. It was started on the morning of 26 April. The female dug off and on in the morning, but no activity was seen between 1000 and 1500; after that time she was seen digging again. At 1830 she emerged from the nest. The mound was large, about 3 cm deep by 8 cm wide and 20 cm long. Leaving the entrance open, she leveled this mound by starting near its middle and working forward, exactly as I described in 1960 (see Fig. 163). At 1850 she flew off, leaving the entrance open and the threshold largely cleared of sand, the mound flattened and moved some distance from the entrance.

For the following several days, this nest was seen to be open in the early morning and again in the late afternoon, but it was closed during most of the day. On the 30th, the female was seen provisioning from 1730–1930. I dug this nest out at 1930 and found 11 fresh flies and about 5 partially eaten flies in the cell with a half-grown larva. The burrow curved gently to the right, and there was an inner closure 1 cm long before the cell. Although the sand on the surface was rather powdery, the burrow passed through a layer of rather firm clay-sand and into a layer of coarse, sandy gravel.

The flies in this cell were quite diverse, and included a fairly large asilid as well as an apioceratid. The complete list follows:

ASILIDAE
Efferia sp.   1
APIOCERATIDAE
Apiocera haruspex O.S.   1
BOMBYLIIDAE
Aphoebantus spp.   6
Desmatoneura argentifrons Will.   2
Villa flavipilosa Cole   1

All available nest data have been brought together in Table 33. The Cameron County, Texas, nests were in very compact, somewhat moist soil, and this may account for their lesser depth. Although the very deep Brewster County nest was in soil not noticeably more friable than in the case of those in New Mexico and Guerrero, Mexico, the soil here was very much drier (Fig. 164).

TABLE 33. NEST DATA FOR BEMBIX U-SCRIPTA

Locality	No. of nests	Burrow length (cm)	Cell depth (cm)
Cameron Co., Texas	36	30 (19–40)	15 (10–22)
Brewster Co., Texas	1	52	28
Hidalgo Co., N. Mex.	8	35 (30–42)	18 (16–23)
Guerrero, Mexico	1	34	17

§H. *Bembix u-scripta* Fox

The fragmentary data from Guerrero and from western Texas are in general agreement with my earlier observations with respect to crepuscular provisioning, diversity of prey, nest structure, closure, and leveling. The fact that both nests were unicellular is not significant, since neither nest had received a final closure. The new data are insufficient in quantity to add anything of importance to an understanding of geographic variation in this species. However, confirmation of the unique features of a species from other localities is always worthwhile. I suggested in 1961 that this species is able to nest in diverse situations and to coexist with other bembicine wasps because it exploits its environment differently: that is, the females capture flies, differing widely in size and taxonomic affinities, after they have come to rest for the night (perhaps to some extent also before they become active in the morning).

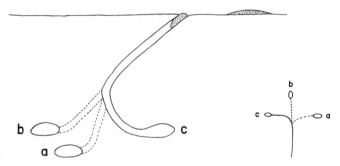

Fig. 164. A typical three-celled nest of *Bembix u-scripta* (no. 1635, Rodeo, New Mexico), shown in profile (left) and in plan (right). Cells a and b had been closed off and contained cocoons; cell c was empty when the nest was excavated. (From Evans, 1961.)

Chapter XI. Some Specialized Species of *Bembix*

This chapter begins with accounts of three closely related species of *Bembix* which lay their eggs glued erect on the floor of the empty cell before bringing in prey. Two of these species, *texana* and *troglodytes*, were treated in 1957, but I am able to add a good deal to knowledge of the former and a few scraps of information to knowledge of the latter. The third species, *multipicta*, has not been studied previously. These species possess several behavioral specializations other than their manner of oviposition: all of them tend to omit the outer closure when the larva is large, two of them make false burrows more or less consistently, and one of them (*texana*) exhibits cell-cleaning behavior. The three species have larvae that are very similar but differ sharply from other *Bembix* in one character: the galeae are very stout and terminate in several sensilla instead of the usual two. The adults all have black on the clypeus and are very similar structurally, but so far as I know there are no adult characters that separate this group sharply from other *Bembix*.

In a later section of this chapter I have presented a few further notes on *pruinosa*, a species that also oviposits in the empty cell but makes a very different type of nest and lays the egg against the wall at the extremity of a very long, curved brood chamber. There are other specializations, in both behavior and adult structure, which I covered at some length in 1957. The present report merely adds evidence from other localities which confirms some of the behavioral peculiarities and adds slightly to knowledge of variation in the species.

A final section of the chapter reviews some aspects of behavior in *Bembix* in a more general way, with special attention to matters studied further since my 1957 paper.

A. *Bembix texana* Cresson

In 1957 I presented a brief discussion of this species based entirely on 3 hr of study in a large colony at Highlands Hammock State Park, Florida. This preliminary study indicated that cell-cleaning behavior was especially well developed in this species. I returned to this same colony in May 1961 and studied the species intermittently for 3 weeks, with special attention to the details of cell cleaning. Krombein (1958b) published a brief report on several nests of this species discovered at Kill Devil Hills, North Carolina. The present report includes these newer observations and is much more complete than that presented in 1957.

(1) *Ecology.* The colony at Highlands Hammock, Florida, has inhabited the same parking lot for several years, along with a large colony of *Stictia carolina* and smaller numbers of several other digger wasps (Figs. 128 and 165). Although I explored other apparently

Fig. 165. Nesting area of *Bembix texana* (Highlands Hammock State Park, Florida). Several nest entrances are visible in this photograph. (See Fig. 128 for a broader view of this same area.)

suitable nesting areas in the vicinity, I was unable to find the species nesting elsewhere. During May 1961 I estimated that there were between 100 and 150 females nesting in this parking area. As I pointed out in 1957, the soil here is sandy but quite hard packed. The nests are continually being run over by cars, but this seems to have no serious effect on the wasps.

Krombein's colony consisted of only about six females nesting along a sand road through the woods; the soil here was a "rather loose dry sand at the surface with very sparse vegetation." Krombein's studies were made in August, and the wasps were in worn condition. In this area, Krombein has taken the species as early as 23 June and as late as 11 August; the same area was examined during mid-September, but no specimens were seen. Probably *texana* has only one generation a year in North Carolina.

A brief visit to the Highlands Hammock colony from 20–28 March 1957 revealed that males were already out in some numbers. In 1961 males were found to be present in small numbers in early May, but later in the month none were noted. The species probably has two or more generations a year in this area, but I have no exact information on this point.

Bembix texana is decidedly more influenced by cool or cloudy weather than is *Stictia carolina*, which nested with it at Highlands Hammock, the nests of the two species being intermingled. Between 0800 and 0900, when many *Stictia* were active, scarcely any *Bembix* could be seen, and not until 1000 could the colony be said to be at the peak of its activity. Also, *Bembix* tended to be much less active in the late afternoon (after 1700) or during cloudy intervals than *Stictia*.

(2) *Activity of males.* The males are active in some numbers before the first females emerge. On 20 March 1957 I found two males flying over the ground about 3–5 cm. high in circles and irregular patterns, a typical incipient sun dance. On 28 March at least 30 males were active, but still no females. The sun dance was now very conspicuous and much like that in species such as *spinolae* and *cinerea*. In the late afternoon (1600–1700) the males dug short burrows in the soft sand along the edge of the road and parking lot—in notably more friable sand than that in which the females nested. These burrows were oblique and only 3 to 7 cm deep (average of nine burrows: 5.4 cm). The males left the small mound of sand intact and made a substantial closure from the inside; at the bottom of the burrow they rested facing the entrance. One of the nine nests dug out had two males in it.

I have not observed mating in this species.

(3) *Nesting behavior.* In the Highlands Hammock colony, the females nested primarily in the roadway and those parts of the parking lot which were most hard packed and most devoid of grass. The nests tended to be well scattered over the available space, with the entrances generally 15–50 cm apart.

Most new nests appear to be started in the morning and to require 1 to 3 hr for completion. At intervals during the digging, the female backs out to the center of the mound and works forward into the hole, often with some zigzagging, so that the mound tends to be removed slightly from the entrance (by 1–3 cm). These actions tend to produce a rather elongate mound, measuring 10–20 cm long by 5–8 cm wide and 1–2 cm deep. The mound is not leveled at any time and tends to persist for the duration of the nest, although it is often much reduced by the action of wind or rain or (in this particular area) by cars or people passing over the nest.

Females very commonly discontinue digging for a period of time, in which case they make an outer closure and either rest within the nest or fly off for a period. Nests begun in the late afternoon may be completed the following morning, the female remaining in the burrow during the night. Four of six nests dug out by Krombein did not contain a cell, the nests not having yet been completed.

(4) *False burrows.* I noted no false burrows in this species in the brief studies reported in 1957, and it was with considerable surprise that I discovered in 1961 that most individuals make not only one but two false burrows, one on each side of the true burrow. These are dug immediately after completion of the true burrow. The nest entrance is first closed by scraping in a little sand from the periphery. The wasp then makes a short false burrow at a right angle to the true burrow, clearing the sand from it and leaving it open at all times thereafter. She then makes another false burrow on the opposite side of the entrance of the true burrow. Some individuals alternate in digging the two false burrows. The soil from the false burrows is scraped away onto the mound. Having completed the false burrows, the female enters the nest, closes from the inside, and lays her egg in the cell. These false burrows persist for a variable length of time, depending upon the amount of wind, rain, or human disturbance (Figs. 166 and 167). On several occasions I observed wasps making a final closure of the nest when evidence of the false burrows was still present.

When I first discovered these false burrows, on 4 and 5 May, I received the impression that they were present in all nests. I later discovered that they were sometimes omitted either on one or on both sides. There seemed to be a general trend for the false bur-

326   Chapter XI. Specialized Species of *Bembix*

Fig. 166. *Bembix texana* female entering her nest with a horsefly (Highlands Hammock State Park, Florida). Note the open false burrow to the right of the wasp.

rows to be shorter and to be more often omitted as the season progressed. My data on the occurrence and depth of the false burrows are summarized in Table 34.

The probable function of these false burrows is discussed in the final section of this chapter. They are not dug as a source of earth for filling, nor are they used by the female for storage or for resting. Occasionally I found pieces of dead flies in them, but I believe that these merely blew in from the debris on the mound following cell cleaning. A possible clue was provided by the fact that I saw several female *Dasymutilla pyrrhus* (Fox) enter false burrows and dig at the bottom of them, but I saw none enter true burrows, which are kept closed much of the time (Fig. 170).

(5) *Nature and dimensions of nest.* The nests are unicellular and of simple structure. Nearly all nests have a strong lateral curve just before the cell. The nests studied in 1961 averaged somewhat

§A. *Bembix texana* Cresson

shallower than those studied in 1957. The two completed nests studied by Krombein were very shallow in spite of the fact that they were in looser sand than the nests in the Florida colony (Table 35).

I did not find a spur in any nest, nor did Krombein. This is related to the fact that there is almost no inner closure of the nest in this species. I found a small inner closure in one nest containing an egg (no. 1677) and in one nest containing a very small larva (no. 1716), but ten other nests lacked inner closures altogether even though one of them contained an egg.

The outer closure is another matter. Nests containing eggs or small larvae (up to 18 mm long) invariably have a strong outer closure—except during the brief intervals when the female is taking a fly into the nest. In 1957 I reported that an outer closure is maintained at all times, but this is not strictly true. On the final day of provisioning, and possibly sometimes on the preceding day,

Fig. 167. *Bembix texana* female opening her nest entrance while carrying a beefly (Highlands Hammock State Park, Florida). Again, the open entrance of a false burrow can be seen in the background.

TABLE 34. LENGTH (IN CM) OF FALSE BURROWS
IN NESTS OF BEMBIX TEXANA

(Highlands Hammock, Fla., 1961)

Nest	Left F.B.	Right F.B.	Nest	Left F.B.	Right F.B.
	5 May			17 May (cont.)	
A	6.0	1.0	N	0.1	0.1
B	1.0	0.5	O	1.0	1.0
C	0.5	2.0	P	none	none
D	0.5	4.0	Q	none	none
E	1.0	4.0	R	0.5	none
F	2.0	0.3	S	none	none
G	0.2	4.0	T	none	none
H	4.0	3.5		27 May	
I	6.0	0.1	U	1.0	1.0
J	2.0	0.5	V	none	1.5
K	4.0	none	W	0.5	0.5
	17 May		X	none	none
L	0.5	0.5	Y	0.5	1.0
M	none	none	Z	0.3	0.5

the outer closure is often omitted. At these times the nest has no closure at all, there being an open tunnel all the way from the surface to the larva (as is common in the related species *troglodytes*).

(6) *The egg.* Eggs were found in two nests. In each case the egg was glued upright to the floor of the cell at about the center of the cell, as reported in 1957.

(7) *Hunting and provisioning.* In 1957 I reported that *texana* employs relatively large flies, chiefly Tabanidae, and takes males and females

§A. *Bembix texana* Cresson

TABLE 35.   NEST DATA FOR BEMBIX TEXANA

Locality	No. of nests	Burrow length (cm)	Cell depth (cm)
Highlands Hammock Fla., 1955	7	23.5 (21-26)	15.0 (12-18)
Highlands Hammock Fla., 1961	14	19.5 (16-25)	11.5 (10-15)
Kill Devil Hills N.C., 1956 (Krombein)	2	15.0	9.0

in about equal numbers. In 1961 Tabanidae again provided the major prey, but smaller species were in the majority and only a very few males were being taken. Of 125 examples of prey taken, 118 (94 percent) were Tabanidae, and of the Tabanidae only 23 (20 percent) were males. Some of the horseflies brought to the nests were freshly engorged, suggesting that they may have been taken directly from the animals on which they were feeding. I did not observe *texana* hunting, but I suspect they may have taken at least some of the horseflies from deer in the park, as deer were plentiful and apparently much bothered by horseflies (Table 36).

In North Carolina, Krombein found *texana* preying upon Stratiomyidae, Syrphidae, Micropezidae, Calliphoridae, and Tachinidae, but not Tabanidae.

Provisioning females pause long enough at the entrance to scuff it open (Fig. 167), except that on the last day of provisioning they may enter the nest swiftly, the outer closure having been omitted. The fly is grasped with the middle legs unusually strongly; even when captured in an insect net, the wasp usually retains her grip on the fly for some time. This is probably related to the fact that an unusual amount of stealing of prey occurs in this species. Indeed, females were even seen attempting to take flies from the much larger species *Stictia carolina,* but without success.

(8) *Cell cleaning.* As pointed out in 1957 and as also noted by Krombein, it is an unusual feature of this species that the cells normally contain only fresh flies and very few fly remains. The females obviously clean the debris from the cells at intervals, thus presumably providing few opportunities for miltogrammine or phorid

TABLE 36. PREY RECORDS FOR BEMBIX TEXANA

(All Highlands Hammock, Fla., May 1961)

Species of fly	No. of specimens
STRATIOMYIDAE	
Hedriodiscus trivittatus Say	1
TABANIDAE	
Chlorotabanus crepuscularis Beq.	6
Chrysops dimmocki Hine	1
Chrysops pudica O.S.	1
Hybomitra hinei wrighti Whitn.	16
Tabanus bishoppi Stone	8
Tabanus coarctatus Stone	1
Tabanus endymion O.S.	1
Tabanus fuscicostatus Hine	2
Tabanus gracilis Wied.	1
Tabanus lineola Fabr.	68
Tabanus melanocerus lucustris Stone	2
Tabanus mularis Stone	5
Tabanus quinquevittatus Hine	1
Tabanus sparus milleri Whitn.	1
Tabanus trijunctus Walk.	4
SYRPHIDAE	
Meromacrus acutus Fabr.	1
CALLIPHORIDAE	
Cochliomyia macellaria Fabr.	2
SARCOPHAGIDAE	
Sarcophaga ventricosa Wulp	2
TACHINIDAE	
Exorista larvarum L.	1

maggots to develop in the cells. During 1961 I studied this behavior in some detail and observed and photographed many females cleaning their nests.

All cell cleaning occurs when the female first opens the nest in the morning. Only 5–20 min are normally required, and since most females in the colony perform this behavior at the same time of day, it is easily overlooked unless one is on the alert during the morning hours (especially 0900–1000). Not every female cleans her nest, however. Nests containing eggs, of course, require no clean-

§A. *Bembix texana* Cresson   331

ing, and nests containing small larvae (up to about 12 mm long, about 2 days old) are not cleaned. Thereafter for the remaining 2 or 3 days of provisioning, the cell is cleaned each morning before any fresh flies are brought in (Fig. 168).

When the female emerges from her nest in the morning, she first clears out the entrance, sweeping the soil back onto the mound. She then goes down to the cell and begins removing the remains of flies from the cell, using both her mandibles and tarsal comb. The remains are moved up the burrow in stages and eventually swept or carried onto the top of the mound. Legs, wings, and other small pieces are normally swept along with the front legs in the same manner as sand. When handling a hollowed-out thorax or a more or less intact fly, the wasp seizes it with her mandibles and walks backward rapidly. On a very few occasions I saw females take flight with pieces of flies in their mandibles, dropping these on the ground. However, I believe this behavior to be unusual, the female perhaps taking flight because of human disturbance. Following cell cleaning, the top of the mound is strewn with legs, wings, heads, and thoraces of flies. The mound itself tends to be flattened and spread slightly as the fly remains are dragged or scraped onto it. In the course of the day the debris may blow about to a certain extent, but one can nearly always recognize a nest that has recently been cleaned by the fragments on the mound. Normally, the cell is cleaned very thoroughly; in some cases even intact flies which were not eaten the day before are cleaned out. Having finished cleaning, the female closes the entrance and flies off to hunt for fresh flies.

The following notes on various nests will serve to demonstrate that cell cleaning begins only after the larva is about 2 days old, also that the outer closure is omitted only when the larva is nearly full grown.

No. 1679. Entrance closed; mound high and fresh, with no evidence of cell cleaning. Cell contained an egg.

No. 1677. Entrance closed; mound high and with no evidence of cell cleaning. Cell contained a very small larva, four fresh flies, and remains of about four flies which had not been removed from cell.

No. 1717. Entrance closed; mound high and with no evidence of cell cleaning. Cell contained a larva 12 mm long, five fresh flies, and remains of several others.

No. 1718. Entrance closed; mound more spread out, and had some pieces of flies on it. Cell contained a larva about 15 mm long, plus nine fresh flies but no fly remains.

No. 1720. Entrance left open during provisioning; many fly remains

Fig. 168. *Bembix texana* female sweeping away the debris cleaned from the cell. At first (above), only a few fly bodies have accumulated, but after a few minutes (below) the remains of many flies appear on the mound.

§A. *Bembix texana* Cresson

on mound. Cell contained a larva 18 mm long, plus five fresh flies but no fly remains (female was still bringing in flies).

No. 1749. Nest dug on morning of 23 May; had been finished by 1100. On 26 May there was evidence of cell cleaning, but the outer closure was maintained during provisioning. Larva found to be 16 mm long; cell contained five fresh flies but no fly remains.

Because of the large amount of human traffic over the colony, it was difficult to follow individual nests to completion. Several nests marked as new on 17 May had received the final closure 6 days later. Two nests marked as new on 23 May (including no. 1749, above) contained medium-sized larvae 4 days later. From these data we can outline the general history of a nest as follows:

First and second days: egg in empty cell (fly probably brought in on second day). Outer closure maintained, sometimes a weak inner closure.

Third and fourth days: larva growing; a few flies brought in each day, but cell not cleaned (possibly some cleaning on fourth day). Outer closure maintained but not inner closure.

Fifth day: cell cleaned out thoroughly in morning; many fresh flies brought in, but outer closure generally maintained.

Sixth day: cell cleaned as above; many flies brought in, outer closure being generally omitted between trips with prey. Final closure begun in afternoon or on morning of seventh day (in the latter case preceded by further cell cleaning).

(9) *Final closure.* Most final closures are made in the morning or in the late afternoon. Before the closure is begun, a certain amount of sand is cleared from the burrow and entrance and swept back upon the mound. Some fly remains are often also swept out at this time. Then the wasp beings filling, first by scraping sand from the walls of the burrow, later by scraping in sand from the periphery. A small pattern of lines radiating from the entrance usually results, and this pattern as well as the mound is usually left intact; that is, no effort is made to level off the surface smoothly over the top of the filled burrow (Fig. 169). This, plus the fact that the lower part of the burrow is usually filled rather incompletely, would seem to render the nests rather easy for mutillids to enter, although I did not actually see any digging into nests that had received the final closure.

(10) *Natural enemies.* The mutillid *Dasymutilla pyrrhus* (Fox) was common in the nesting area, and females were often seen examining

Fig. 169. Final closure in *Bembix texana*. The wasp is scraping soil from the periphery into the nearly filled nest entrance.

holes in the ground, especially the "false burrows" (Fig. 170). However, I have no definite evidence that they were successfully parasitizing the larvae of *Bembix texana*. On several occasions I saw female *Bembix texana* attacking these mutillids vigorously, although not actually attempting to sting them. Miltogrammine flies were also common in the nesting area, and on several occasions specimens of *Senotainia trilineata* Wulp (complex) were observed trailing *texana* females laden with prey or perching on the mounds outside the entrances. However, none of the nests dug out contained maggots. This same miltogrammine was more commonly seen trailing *Stictia carolina*, and maggots were found in the nests of that species. On one occasion a miltogrammine was seen trailing a female mutillid!

Two nests I dug out each had an earwig in the cell, in one case a small nymph, in the other case an adult. Both cells con-

B. *Bembix troglodytes* Handlirsch

tained fresh flies and healthy wasp larvae. Perhaps the earwigs merely made use of the nest as a place to spend the day; there is no evidence that they did any harm in the cell.

B. *Bembix troglodytes* Handlirsch

This species, treated in detail in 1957, resembles *texana* in its manner of oviposition and in its practice of constructing a false burrow immediately after completion of a new nest. It differs in that the nests are deeper and are dug in fine-grained sand, chiefly along watercourses, also in that pronounced zigzag leveling movements are performed after the completion of a new nest and again just before the final closure.

Brief additional studies of this species, made near Rodeo, New Mexico, 9 September 1959, tend to confirm several of these characteristics. However, neither of the two nests studied had evidence of a false burrow, although one of these was a recently completed

Fig. 170. Female mutillid, *Dasymutilla pyrrhus,* entering a false burrow of *Bembix texana* (Highlands Hammock State Park, Florida). A depression marking the closed entrance of the true burrow may be seen in the extreme upper right-hand corner.

nest containing an egg. New Mexico is close to the center of the range of this species, and it is worth noting that another species, *pruinosa,* is known to make a false burrow only occasionally near the center of the range, although always making such a burrow toward the periphery of the range. Kansas, where my earlier studies of *troglodytes* were made, is distinctly peripheral for this species.

The specimens of *troglodytes* studied in New Mexico agree well in size and coloration with the usual concept of this species (Table 37). As discussed below under *multipicta,* there is some question as to whether these two forms are actually distinct species. I have collected "typical" *multipicta* at a locality in Chihuahua, Mexico, only about 300 km south of the New Mexican locality where I studied *troglodytes* in 1959, and *multipicta* has been recorded from Arizona.

The two nests located near Rodeo, New Mexico, were in open patches of sandy soil in a broad, semidesert valley; *B. u-scripta* and *B. spinolae similans* also nested in small numbers in these same sandy places, the nests of the three species being to some extent intermingled. One female was first noticed as she was leveling the mound at the nest entrance. Following zigzag leveling movements as described for this species in 1957, she scraped sand in various directions over the entrance, but failed to make a false burrow. When I dug this nest 2 hr later I found an egg glued erect in the empty cell; there was a weak inner closure, but at this time the female was not in the nest and there was no outer closure. The second nest was being provisioned when it was discovered; the entrance was closed each time the female left. There was no evidence of a mound or a false burrow. When this nest was dug out shortly after being discovered, the cell was found to contain a large larva as well as several fresh flies and some fly remains. This nest had a weak inner closure as well as an outer closure. The two burrows were 19 and 33 cm long, the cells 15 and 18 cm deep, respectively. These figures are within the known range of variation in nest depth of this species, although close to the minimum.

Of the 11 fresh flies in the cell of the second nest, 1 was a calliphorid (*Cochliomyia macellaria* Fabr.), the other 10 were bombyliids (8 *Villa salebrosa* Painter, 1 *Villa chimaera* (O.S.), and 1 *Sparnopolius* sp.).

C. S. Lin has recently sent me 20 prey records for this wasp from Austin, Texas. Most of the species and all but one of the genera are included in my list of prey from Kansas (1957b, Table XVI). The one new genus recorded is *Exoprosopa,* represented by the two species *fascipennis* Say and *iota* O.S. (Bombyliidae). The former is a parasite of several species of *Bembix*. *B. troglodytes* is also known to use several species of *Senotainia* (Sarcophagidae) as prey; again, these are known natural enemies of *Bembix*. In contrast to most species of *Bembix*, *troglodytes* appears to do much hunting around the nest entrances and to use flies that are parasites of various bembicine wasps. One notes also that the flies used by *troglodytes* are mainly small- to medium-sized species of compact body form; absent from the list are any especially large flies or flies of slender body form (for example, Therevidae, Asilidae).

C. *Bembix multipicta* Smith

This species was described from Oaxaca, Mexico, and is one of the most characteristic bembicine wasps of inland localities in that country. I have collected it in many places, at altitudes varying from 1,000 to 5,000 ft, chiefly in sandy areas along streams or dry washes. Some of the diagnostic features of typical *multipicta* are summarized in Table 37.

Unfortunately I have had no opportunity to study the typical form of this species in detail. Rather, my opportunities to study *multipicta* have come in coastal localities, where this species appears to exhibit a mixture of *multipicta* and *troglodytes* characters. The population I studied from Yucatan showed no noteworthy differences in color from *troglodytes* populations from Cameron County, Texas, but averaged somewhat larger. Some males taken in Yucatan proved indistinguishable from males taken in Sinaloa, on the Pacific Coast. However, Sinaloa females differ consistently in color from those from Yucatan and tend to be smaller. Some Sinaloa females will run to *multipicta* in the keys of Handlirsch and of Parker, although most of them have yellow maculations and are small in size for that species.

It should be pointed out that when U.S. specimens of *troglodytes* are compared with inland Mexican *multipicta*, the two give every appearance of being different species. I have seen no intergrades from the southwestern United States or northern Mexico. However, there seem to be no reliable structural differences between the two species, and the genitalia and secondary sexual characters of the male appear identical. Clearly a detailed study of size and color pattern in this complex is in order. For the present I shall treat the Yucatan and Sinaloa populations (which seemed identical in ecology and behavior) under the name *multipicta*. There is only one point regarding the behavior of these wasps that is different in any important way from that of *troglodytes*: they nested in con-

TABLE 37. COMPARISON OF FEMALES OF MULTIPICTA-TROGLODYTES COMPLEX

Locality	Body form	Band on 2nd metasomal tergite	Color of labrum of female	Color of clypeus of female	Color of bands of metasoma	Color of apical tergite
South central and south-western U.S. (troglodytes)	Relatively slender (length 11-16 mm)	Complete, enclosing black spots	Black (may be partly or wholly whitish near Mexican border)	Wholly black (only paired black spots in some So. Texas spec.)	Yellow (whitish in occasional ♀♀, never in ♂♂)	Black or spotted
Yucatan, Mexico	Somewhat larger and more robust (15-18 mm)	As above	Whitish	Whitish, with paired black spots	As above	Spotted
Sinaloa, Mexico	Smaller, moderately robust (14-17 mm)	Interrupted or weakly connected medially, black spots not always completely encircled	Pale, with or without a median black band (rarely all black)	With paired black spots, connected basally, or all black	Usually yellow, but whitish in some ♀♀ and some ♂♂	Black
Inland Mexican localities (multipicta)	Robust (15-19 mm)	Interrupted medially (or weakly connected), usually not fully enclosing black spots	Whitish or with a partial black streak	Whitish with paired black spots connected basally	Whitish	Black

§C. *Bembix multipicta* Smith 339

siderably more compact soil (suggesting *texana*!). I found no false burrows in any Mexican nests, but as already pointed out I also failed to find false burrows in *troglodytes* nests in New Mexico.

My studies in Yucatan were made between 17 and 24 July 1962, on the beach just west of the city of Progreso. I spent parts of 3 days working on this species and dug out seven nests. From 5 to 10 August of the same year I studied a small nesting aggregation just behind the beach at Mazatlán, Sinaloa. I have also included a very few notes on typical *multipicta*, made at localities in Morelos and Guerrero as noted below.

(1) *Ecology and general aspects of behavior.* At Progreso, these wasps nested in large numbers in rather hard-packed sand along several short roads leading from the highway to the beach (Fig. 171). In this same area, *Stictia vivida* nested in the much more friable sand of small dunes and blowouts just behind the beach; there appeared to be no overlap in nesting sites between the two species. At Mazatlán, *multipicta* was fairly common, but I was able to find only eight

Fig. 171. Hard-packed, sandy road leading to beach near Progreso, Yucatan, Mexico, site of a colony of *Bembix multipicta*. Fig. 134 was taken in this same spot, looking off to the right toward the more friable sand just behind the beach, where *Stictia vivida* nested.

340   Chapter XI. Specialized Species of *Bembix*

nests, all along a hard-packed footpath leading up a sandy rise behind the beach (Fig. 172). Again, the more friable sand closer to the beach was inhabited by species of *Stictia*. Where I have found *multipicta* at inland localities, however, I have found it in areas of more friable sand, chiefly in draws or along streams or irrigation ditches.

Males were fairly abundant both at Progreso and at Mazatlán, flying back and forth over the nesting areas and attempting to mate with females. Several males were dug from their burrows in the sand in the late afternoon (1700). One such burrow, shown in Fig. 174 (no. 1923), was 7 cm long, the bottom 4.5 cm beneath the surface. The males rest in these burrows facing toward the entrance; there is a strong closure as well as a very small mound at the entrance. Excavation of several brood nests in both localities showed that the females spend their inactive periods inside the nest, just outside the inner closure.

In the areas of study, *multipicta* seemed to maintain a fairly strict daily regimen. From 0800 to 1100 females were seen opening their nests and then bringing in flies, while males flew actively

Fig. 172. Female *Bembix multipicta* digging nest in a path leading to the beach at Mazatlán, Sinaloa, Mexico, August 1962.

§C. *Bembix multipicta* Smith

about the colony. From about 1000 to 1300 final closures of some nests were observed, as well as the initial stages of digging of new nests. From 1300 to 1500, when temperatures in the areas of study were from 90 to 95°F, very little activity was to be seen. In the late afternoon females having large larvae in their nests may bring in additional prey, while other females resume the digging of their nests, finishing them off and leveling the mound either late in the afternoon or on the following morning. Males are not commonly seen after 1300, and females having eggs or small larvae in the nest may remain inside the nest after about 1300. During the early afternoon one may stand in the midst of a large colony of these wasps and not appreciate its presence.

I estimated 50–100 nesting females at Progreso, most of them along part of one sandy road, where the nests were quite close together, sometimes separated by no more than 5–10 cm. However, there were scattered nests on other roads not far away.

(2) *Nesting behavior.* The better part of 1 day is required for digging the nest, though for considerable periods of time the female appears to rest within the partially completed nest, the entrance closed from within. When the nest and cell are completed, the female emerges from the burrow and, without first closing the entrance, begins to level the pile of sand at the entrance. She takes flight briefly and lands at the far end of the mound, then works toward the entrance in a zigzag pattern, scattering the sand widely. Having reached the entrance, she takes flight again and repeats the pattern over a slightly different path. After about 10 min, when the mound is fully or at least in large part dispersed, she makes an outer closure. In the instances observed, the female made a closure from the outside, then flew off, presumably to obtain nectar (since the egg is laid in the empty cell), then returned, entered, and closed from the inside. As noted above, I observed no false burrows at any of the Mexican localities.

Nests containing an egg or small larva invariably have an outer closure and a strong inner closure, the latter 2–4 cm long. A distinct spur was noted in some nests containing eggs. However, nests containing larger larvae often have neither an outer nor an inner closure during periods of active provisioning (although both are typically present during the night). It appears that once the larva is 2–3 days old, the female opens the entrance wide, producing a small mound which is not leveled, and leaves the entrance open during the period of provisioning each day.

342   Chapter XI. Specialized Species of *Bembix*

The nest is a simple, oblique burrow which tends to become horizontal at the extreme bottom and terminates in a horizontal cell (Figs. 173 and 174). The cell measures 15 × 25 mm at first, but may be lengthened to about 35 mm as provisioning progresses. Data on nest depth at the two major localities are as follows:

Locality	No. of nests	Burrow length (cm)	Cell depth (cm)
Progreso, Yucatan	7	35.0 (28–42)	20.0 (15–25)
Mazatlán, Sinaloa	2	33.5 (32–35)	22.0 (20–24)

I discovered several nests of "typical *multipicta*" 5 km north of Alpuyeca, Morelos, in April 1959, but was unable to study their structure because they were shortly thereafter trod upon by cattle. I managed to locate three cells by excavating a large area of soft

Fig. 173. Nest of *Bembix multipicta*, showing inner and outer closures (no. 1910, Progreso, Yucatan, Mexico, July 1962).

§C. *Bembix multipicta* Smith

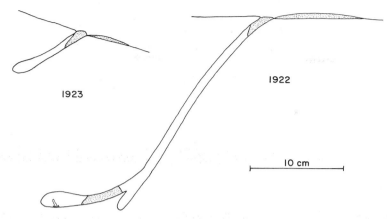

Fig. 174. Resting burrow of a male *Bembix multipicta* (no. 1923), and brood nest of a female containing an egg (no. 1922; Mazatlán, Sinaloa, Mexico, August 1962).

sand where the nests had been located. All three cells were 45–50 cm deep, just within a layer of heavier clay soil. This is very considerably deeper than any of the cells located at Progreso or Mazatlán.

(3) *The egg.* I located eggs in two nests at Progreso and one at Mazatlán. In each case the egg was in the empty cell, glued by its broader end to a few sand grains, and situated slightly closer to the apex than to the base of the cell (Fig. 174). In two cases the egg was on or near the bottom of the cell and extended upright or nearly so, while in the third case (no. 1910, Progreso) it was actually glued to the side of the cell and extended nearly horizontally into the cell. All three cases are illustrated in order to show the variation (Fig. 175).

Fig. 175. Cells of three *Bembix multipicta* nests (cross section), showing variation in the position of the egg. The position shown in no. 1922 is typical of *B. troglodytes* and *texana* and is presumably normal for *multipicta*.

(4) *Provisioning the nest.* Flies are brought in rapidly during the morning hours, the nest entrances typically being left open if the larva is more than 2 days old. The flies are carried with the middle legs in the usual manner of bembicine wasps. The flies utilized are apparently those common in the vicinity. Near Alpuyeca, the cells contained the wings of several muscoid flies as well as two intact and several partially eaten stratiomyid flies, *Hoplitimyia mutabilis* (Fabr.). At Mexcala, Guerrero, on 29 June 1951, I took a female with a fly of the genus *Asilus* (Asilidae). At Progreso, *multipicta* preyed almost exclusively on the blowfly *Cochliomyia macellaria* (Fabr.); more than 30 flies of this species were taken from various nests, along with a few fragments of Stratiomyidae and Syrphidae. At Mazatlán, *Cochliomyia macellaria* (Fabr.), *Musca domestica* L., and various other muscoids were being used, also the stratiomyid *Hedriodiscus dorsalis* (Fabr.). Fragments of flies that have been eaten are not removed from the cells but allowed to form a mat in the bottom of the cell.

(5) *Final closure.* This was observed both at Progreso and at Mazatlán. After first leveling off any accumulation of sand at the entrance, the wasp scrapes sand from the walls of the burrow and from the periphery of the entrance and backs in to pack it in the burrow. About 30 min are required to fill the burrow, and as much as another 30 min may be used for concealment. Having filled all but a small depression at the top of the burrow, the wasp goes out in various directions on the side toward the direction of the burrow (that is, on the side opposite the mound, which is no longer evident). Here she makes a series of as many as 30 radiating lines, some of them up to 15 cm long, each separated by a very short flight. During each movement away from the entrance she throws sand back toward the entrance. Finally, she walks in an irregular pattern over the entrance, kicking sand in various directions, leaving no trace of the burrow. After completion of the final closure, the female may start a new nest only a short distance away.

(6) *Parasites.* At Mazatlán a chrysidid wasp, *Parnopes concinnus* Viereck, was common in the nesting area. These wasps were often seen perching on sticks overlooking the *Bembix* nests. Although none were seen entering the nests, it seems very probable that this *Parnopes* is a parasite of *Bembix multipicta* and probably other bembicine wasps (I have also reported it as a probable parasite of *Glenostictia scitula* in Texas).

D. *Bembix pruinosa* Fox

I regard this as one of the most highly evolved species of *Bembix*, indeed of all digger wasps. Several features of its behavior are unique and were totally unsuspected prior to my 1957 studies. The behavior of the species is of sufficient complexity to justify considerable further study, particularly from the point of view of clarifying some of the variation that has been observed. A carefully planned, long-range research program aimed at gathering quantitative data in nesting aggregations at various points throughout its wide range might prove extremely rewarding.

At the present time I am able to present only a few miscellaneous notes gathered during the past few years. In 1961 I discovered large colonies at two localities near the western edge of the range of the species. One of these, seen on 25 August at Great Sand Dunes National Monument, Alamosa County, Colorado, contained many hundreds of males engaged in their "hopping dance," but so far as I could determine the females were only just beginning to nest. The *pruinosa* population at this locality has bright yellow markings instead of the usual greenish-white, and may prove to show behavioral divergences. The other colony, in a broad expanse of sand near Cornish, Utah, was in full activity from 19–21 August, and most of the notes presented below were made at this colony.

I also made a very few notes at a small colony in a large, man-made sand pit at Andover, Massachusetts, during July 1963. This is the easternmost record of this species known to me. Several additional notes were also made at Granby Center, Oswego County, New York, in the same colony reported on in 1957. The population here seems subject to great fluctuations. In the summer of 1958 there were hundreds of nesting females, many more than I had ever observed in this area before; however, in 1960 the population appeared very much lower. The appearance of the nesting area has not changed noticeably in several years, and the incidence of parasitism appears very low. Very probably weather conditions in this peripheral locality (particularly the number of sunny days during the period of nesting) have a profound effect upon population size.

At Granby Center I made one essentially new observation. On 11 August 1958 I sat in the midst of a very large colony for several hours, during the course of which time I confirmed many behavioral details reported earlier. I also saw several females laden with prey stop on the sand for 10–20 sec and use their mouthparts to probe the oral opening of the fly. It appeared that they were imbibing fluids from the foregut of the flies. They then proceeded to their nests with the flies in the usual manner. The flies at which they were feeding included various muscoids as well as Syrphidae. Similar behavior has been reported for certain other bembicine wasps (for instance, *Bembix niponica*) but not for *B. pruinosa*.

(1) *Nesting behavior.* Nests excavated at Andover, Massachusetts, and at Cornish, Utah, were all of the usual *pruinosa* type; that is, the cell was in the form of an extremely long chamber which in most cases curved strongly to one side or the other, often forming a semicircle. The one nest excavated in Massachusetts was close to the minimum depth recorded for this species; the burrow was 20 cm long, leading to a curved brood chamber 18 cm long and 17 cm beneath the surface. Six nests excavated in Utah showed burrow lengths of from 30 to 40 cm (mean 35 cm); cell length varied from 10 to 16 cm (mean 12.5 cm), cell depth from 18 to 21 cm (mean 20 cm). Thus the Utah nests were also relatively shallow for this species. This may have been related to the fact that the nests were located in three rather small blowouts in an extensive area of partially stabilized sand; the amount of shifting sand did not appear to be great.

At Cornish, Utah, I counted 16 newly dug nests on 21 August. Half of these had false burrows, the other half did not. I previously reported that about 20 percent of the nests studied in Kansas had false burrows, 85 percent in Arkansas, 100 percent in New York. I postulated that this behavior may have "become vestigial in the center of the range due to decreased selection pressure." The 50 percent figure from Utah is not entirely consistent with a central-peripheral pattern. Many more data are needed on this point. In *texana*, I found that false burrows were more often omitted as the season progressed, but the data on *pruinosa* do not take this into consideration.

The nests excavated in Utah had the flies lined up in single file in the cells as is usual in this species. However, none of them appeared to have a cache of fly remains at the tip of the cell, representing the result of the unique cell-cleaning behavior of this species. These nests were being provisioned very rapidly and some of them were being closed when the larva was only partially grown (truncated progressive provisioning). One nest contained a larva only 10 mm long (1–2 days old) and had 16 fresh flies in the cell; another contained a larva 18 mm long with 21 fresh flies; a third nest received a final closure when the larva was 25 mm long, with 10 fresh flies in the cell. I noted in 1957 that cell cleaning may be omitted when

§D. *Bembix pruinosa* Fox

provisioning is very rapid, and the nest closed well before larval maturity.

A record for the number of fresh flies found in a single cell was set by nest no. 1675, Granby Center, New York, 31 August 1960. This nest contained 29 fresh flies of 10 species and 6 families.

At Cornish, Utah, *pruinosa* was employing a remarkable array of flies as prey (Table 38). Although only 90 examples of prey were taken, these represented 37 species belonging to 11 families. Two families (Otitidae and Anthomyidae) were added to the 10 families

TABLE 38. PREY RECORDS FOR BEMBIX PRUINOSA

(All Cornish, Utah, 19-21 August 1961)

Species of fly	No. of records
STRATIOMYIDAE	
Anoplodonta nigrirostris Loew	11
Odontomyia tumida Banks	3
TABANIDAE	
Tabanus productus Hine	7
BOMBYLIIDAE	
Anastoechus barbatus O.S.	1
Aphoebantus sp.	1
Lepidanthrax proboscidea Loew	10
Poecilanthrax sackenii Coq.	1
Poecilanthrax willistoni Coq.	1
Villa faustina O.S.	3
Villa lateralis Say	2
Villa molitor Loew	1
THEREVIDAE	
Psilocephala aldrichii Coq.	2
SYRPHIDAE	
Asemosyrphus polygrammus Loew	2
Eristalis brousii Will.	2
Eristalis latrifrons Loew	1
Eupeodes volucris O.S.	1
Helophilus latifrons Loew	4
Sphaerophoria robusta Curran	1
OTITIDAE	
Ceroxys latiusculus Loew	1

TABLE 38.   PREY RECORDS FOR BEMBIX PRUINOSA (CONT.)

(All Cornish, Utah, 19-21 August 1961)

Species of fly	No. of records
ANTHOMYIDAE	
Hydrophoria divisa Mg.	1
MUSCIDAE	
Lispe cotidiana Snyder	4
Lispe nasoni Stein	1
Muscina assimilis Fall.	2
Muscina dorsilinea Wulp	1
CALLIPHORIDAE	
Eucalliphora lilaea Walk.	1
Bufolucilia silvarum Mg.	3
Lucilia illustris Mg.	3
Phaenicia sericata Mg.	4
Phormia regina Mg.	1
SARCOPHAGIDAE	
Amobia floridensis Tns.	1
Sarcophaga lherminieri R.D.	3
Sarcophaga querula Walk.	4
TACHINIDAE	
Achaetoneura archippivora Riley	1
Paradidyma affinis Rnh.	1
Peleteria eronis Curran	2
Spallanzania sp.	1
Sitophaga sp.	1

previously known to be employed as prey (Evans, 1957b, Table XX). Apparently most hunting is done on flowers, as suggested by my earlier studies.

At Great Sand Dunes National Monument, Colorado, I saw a female pursuing a bombyliid fly across a sand dune. The wasp followed about 5 m behind the fly, gradually closed the gap, then pounced upon the fly quickly. The two descended to the sand, where stinging apparently occurred. The wasp seized the fly with her middle legs and flew off. Shortly thereafter she landed on the sand again and was captured. The fly proved to be *Poecilanthrax sackenii* Coq.

The large horsefly *Tabanus sulcifrons* Macq. was being used as prey at Granby Center, New York, during the summer of 1958. I have elsewhere published photographs of a female entering her nest with this horsefly (Evans, 1959b) and of another female (from the same colony) entering her nest with a drone fly, *Eristalis tenax* L. (Evans, 1963b).

(2) *Natural enemies.* At Cornish, Utah, the conopid fly *Physocephala texana* (Will.) was common around the *Bembix* colony, the females perching on vegetation and flying off in pursuit of the wasps from time to time. Also, the bombyliid flies *Exoprosopa arenicola* J. & J. and *Villa atrata* (Coquillett)[1] were seen apparently ovipositing in holes left open by provisioning female *pruinosa*. The small miltogrammine fly *Senotainia inyoensis* Rnh. was also observed following close behind females laden with prey. In no case was I able to incriminate the flies definitely as parasites of *pruinosa*, but all belong to genera known to attack bembicine wasps, and *Physocephala texana* is a known parasite of this species.

The chrysidid wasp *Parnopes edwardsii* (Cresson) was present in numbers in the Utah colony, and some females were seen digging in the sand near known *pruinosa* nests. This wasp undoubtedly attacks *pruinosa*, as it is a known parasite of certain other species of *Bembix*.

E. General remarks on the ethology of *Bembix*

The literature on this genus is extensive, and I shall make no attempt to review all of it, or even much of it, here. The papers of Nielsen (1945), Tsuneki (1956, 1957, 1958), and Evans (1957b, 1959b, 1961) are most relevant and contain references to most of the earlier papers. Emphasis here will be on behavior patterns peculiar to *Bembix* or subject to unusual variation within that genus. Studies on orientation, such as those of van Iersel (1952, 1965) and others, and of behavior inside the nest (Nielsen, Tsuneki), while of much intrinsic interest, have been conducted on too few species to be of present value in comparative studies.

(1) *Ecology.* Of the species I have studied since 1957, *truncata* may be said to nest in hard-packed earth, *cameroni* in small, sandy tracts, *multipicta* in firm sand near beaches or watercourses, *u-scripta* in friable or moderately firm sand in a variety of situations.

[1]This is apparently the species identified by Bohart and MacSwain (1939) as *Anthrax atrata* Loew and recorded as a parasite of *Bembix occidentalis beutenmulleri* Fox at Antioch, California.

It appears from the literature that the North American species are more restricted as to soil type than are the Old World species. Certain species (most notably *hinei* and *cinerea*) appear rigidly confined to one particular soil type, others somewhat less so; a few species (for example, *spinolae, amoena*) are relatively versatile. In general, one finds more ecological restriction in the southern states, where there are more species of *Bembix*. In the north, where two species of *Bembix* coexist (for example, *spinolae* and *pruinosa*), each seems more restricted ecologically than does either when alone. I believe that the species of *Bembix* (and to some extent local populations) have evolved so as not to compete seriously with others for nesting sites.

Records compiled by Nielsen (1945) suggest that some of the Mediterranean and African species of *Bembix* do, in fact, tend to be somewhat restricted ecologically. Doubtless this is no more than an example of a general tendency for more tropical regions to support a greater number of species, each species occupying a relatively smaller niche (Klopfer, 1962).

The most recent edition of Thorpe's *Learning and Instinct in Animals* (1963) unfortunately still contains the statement that "there seems to be no evidence whatever that the limited distribution of *Bembix* has any ecological basis," in spite of much evidence to the contrary. Thorpe also continues to cite Parker's feeble evidence that odor plays an important role in nest finding, although no recent workers have confirmed Parker's remarks (see especially Tsuneki's experiments with *B. niponica,* 1956:146-150).

(2) *Sleeping*. If I were asked to select the single ethological character most diagnostic of the genus *Bembix,* it would be this: both males and females spend the night in burrows in the sand, the males in short "sleeping burrows" which they usually dig themselves, the females in their brood nests, usually just outside the inner closure. Both sexes close the burrow from the inside. Males (or females without a brood nest) may re-enter old or abandoned nests or emergence holes, sometimes more than one occupying the same hole. Only one other genus of Bembicini approaches this: *Microbembex;* but, as will be seen in the next chapter, the females of this latter genus do not sleep in their brood nests but dig short sleeping burrows like those of the males. No important exceptions are known to these generalizations, even among the Old World *Bembix* which have been well studied. None of the various students of sleeping Hymenoptera have found *Bembix* spending the night on vegetation (the report by Lindauer, 1962, being in error both as to the genus of the wasps and the name of the author).

§E. General remarks on *Bembix* 351

(3) *Nest structure.* Nielsen's description of digging and of the nest at various stages of construction is detailed and probably applicable in a general way to the majority of species (but see under 11, below). The function of the spur, and its presence chiefly in nests containing eggs or small larvae, is explained by Nielsen. Much variation has been noted with respect to the spur. Tsuneki found a spur (= accessory branch) in only 65 out of 143 nests of *B. niponica,* and noted much variation in the length and position of the spur; in this species some individuals apparently maintain the spur even when the larva is in the last instar. Many details of nest structure, including the nature of the spur, the slope and curvature of the burrow, and the depth of the cell are clearly strongly influenced by the nature of the soil, including such matters as friability, layering, size and number of stones and roots, moisture content, and so forth. Again, I feel that there are real differences among the various species, but these are difficult to document because of the great amount of ecophenotypic and individual variation.

That distinctive types of nests occur in such species as *belfragei, occidentalis,* and *pruinosa* cannot be questioned. The first two of these species have been studied in two widely separate localities, the third in eight localities from Massachusetts to Utah. The nests of many species are always unicellular; for example, about 100 nests of *pruinosa,* from these same eight localities, were all unicellular. On the other hand, *belfragei* is known to make bicellular nests in Florida and Kansas, *nubilipennis* to make nests of three or more cells in Indiana and Kansas. At the present time the American species may be grouped as follows with respect to the number of cells per nest:

One cell	Often or always two cells	Three to six cells	Insufficient evidence
sayi	*amoena*	*u-scripta*	*cameroni*
spinolae	*cinerea*	*nubilipennis*	*truncata*
comata	*belfragei*	*hinei*	
texana		*brullei*	
troglodytes			
multipicta			
pruinosa			
occidentalis			

It should be noted that Parker (1925) found up to three cells in nests of *comata,* the Raus (1918) only one cell in nests of *nubilipennis.*

It is easy to dismiss these discrepancies on the basis of the fact that only a fairly long period of study plus numerous careful excavations are sufficient to clarify this point. On the other hand, after extensive studies Tsuneki found that in *B. niponica* the nest is typically unicellular, but "in a colony observed in the neighborhood of the city of Chiba, the majority of the nests contained more than two [up to six] brood-cells." Near Sapporo, Hokkaido, about 4 percent of the nests were compound, the remainder unicellular. Observations on *nubilipennis* at Austin, Texas, cited in the preceding chapter, suggest that the spring generation of this species makes multicellular nests, the autumn generation, unicellular nests. In certain other species (*cinerea* is the best example), unicellular nests seem to be most common, but some females in the same colony make bicellular nests (not very different from the situation in certain species of *Bicyrtes*.) Clearly, this seemingly important behavioral feature is also subject to variation and deserving of much further study.

(4) *Nest closure.* The majority of species of *Bembix* maintain an outer closure of the nest all or most of the time. Nielsen lists several European species that maintain a closure and mentions one (*olivacea*) that does not. Tsuneki indicates that *niponica* makes an outer closure except that during periods of rapid provisioning the closure is sometimes weak or even omitted entirely (as in *spinolae, pruinosa,* and several other American species). Present evidence suggests that the American species may be placed in three groups: those that typically make a closure, those that do not, and a number of species that are intermediate in this respect. The intermediate species are not by any means all alike: in *amoena,* for example, closure seems to be subject to much individual variation, some females omitting the closure even when an egg or small larva is present in the cell, others closing even when the larva is large; but in *troglodytes* and *multipicta* all females close when the cell contains an egg or small larva and omit the closure when the larva is large. The following arrangement of the known American species is, then, somewhat oversimplified:

Outer closure except rarely last day or two		Intermediate		No outer closure at any time
spinolae	*occidentalis*	*amoena*	*u-scripta*	*belfragei*
comata	*texana*	*sayi*	*troglodytes*	
cameroni	*pruinosa*	*cinerea*	*multipicta*	
nubilipennis	*brullei*	*hinei*		
truncata				

§E. General remarks on *Bembix*

The inner closure seems to be a relatively constant feature of nests of most species of *Bembix*, with the reservation that it tends to be weak or absent during periods of active provisioning. Most *Bembix* burrows tend to become nearly horizontal as they approach the horizontal cell. This nearly horizontal section, the "cell burrow," is of course the site of the inner closure. I have noted apparent individual variation in the inner closure in *amoena* and some other species.

In two species of *Bembix, nubilipennis* and *texana*, fairly detailed studies have failed to reveal an inner closure (except rarely). The nests of these species are dug in rather hard soil and lack a spur or any appreciable tendency for the burrow to become more horizontal near the cell. However *cinerea*, another species nesting in hard soil, does usually make an inner closure.

(5) *Leveling*. Some of the most stereotyped species-specific behavior occurs in the manner in which the mound of earth at the nest entrance is leveled. I have noted geographic variation in this behavior in *u-scripta* (Evans, 1961), but most species seem to show little variation. My coverage of this topic in 1957 was fairly complete, and I have since further confirmed the type of leveling in several species (*spinolae, pruinosa*) and the absence of leveling in others (*sayi, texana*). In *amoena* leveling is also omitted. In *nubilipennis* elaborate zigzag movements occur following closure of the entrance (somewhat as in *spinolae*); in *multipicta* rather similar leveling occurs, but while the entrance is still open (as in *troglodytes*). In *u-scripta* leveling movements are also performed while the entrance is open, but they are of a somewhat different nature. Apparently *truncata* and *cameroni* also level the mound, but the details are unknown.

(6) *False burrows*. In 1957 I was familiar with false burrows only in *troglodytes* and *pruinosa*. I have since studied them in three additional species: *texana, amoena,* and *sayi*. In *texana* the females typically prepare two initial false burrows, quite short although relatively persistent; however, I have noted a tendency for them to omit one or both false burrows later in the season. These false burrows are dug after the outer closure is made, so it cannot be assumed that they function as a source of soil for the closure.

In *amoena* false burrows are of irregular occurrence and may be single or paired. In this species they are quite definitely associated with the taking of soil for the closure. Some nests which initially lack a false burrow may acquire one in the course of later closures; false burrows may at times be lost, reexcavated, or modified over a period of days. In *amoena* I have also noted a tendency for females to take soil for closure from the side of the mound toward the entrance, forming a groove or "back furrow" into the mound. The back furrows differ from the false burrows in being dug into the mound rather than into the substrate and in being in front of the entrance rather than beside it, but they appear to be similar in origin and function. Sometimes these back furrows actually penetrate beneath the mound, forming a "back burrow" 1.5-3.0 cm long. The back burrows seem especially associated with the obtaining of soil for the final closure (Fig. 153).

Tsuneki and Yasumatsu have independently observed that in *niponica* some females obtain most of their fill for final closure from a hole in front of the entrance, "resulting at last in a new tunnel that lies in the opposite direction to her closing nest. The wasp carries out the sand from the bottom of her new tunnel and carries it in her old tunnel." This back burrow is abandoned after enough soil is obtained to close the nest, but Tsuneki found one that was extended to form a new nest (Tsuneki, 1956:115). The resemblance of *niponica* to *amoena* in this regard is striking.

In *sayi* there are no initial false burrows, and none are dug in the course of various temporary closures. However, following the completion of final closure, all females observed dug a strong back burrow (2-22 cm long; mean 11 cm) passing beneath the mound. In this case there is no question of obtaining earth for fill, for the old nest has already been closed.

It seems to me possible that the false burrows and back burrows, serving as a source of soil for closure in primitive species such as *amoena* (back burrows also in *niponica*), have in certain more advanced species become ritualized to serve quite a different function. I have presented a number of records of mutillids and other parasites entering these false burrows, but proof that these burrows actually reduce the incidence of parasitism is lacking. Tsuneki has given much thought to this matter with respect to *Stizus pulcherrimus* (Tsuneki, 1943c) and with respect to certain species of *Sphex* (Tsuneki, 1963). He found some evidence that *Sphex argentatus fumosus* Mocsary, which makes two or three strong false burrows, is less heavily parasitized by flies than the sympatric *S. flammitrichus* Strand, which does not make false burrows. However, he felt that not too much significance should be attached to this, since the species without false burrows "are able to survive with no serious inconvenience nor hindrance in the competing world. This may be an instance of the fact that there is sometimes a phenomenon in

§E. General remarks on *Bembix*

nature which has no strong direct connection with the survival values or natural selection."[2]

Both Tsuneki and I are satisfied that false burrows are of little or no significance in orientation. I find it hard to believe that such conspicuous behavioral elements, which require considerable time on the part of the female wasp and must contribute to wear of her mandibles and pecten, are without survival value. If we assume that in species such as *amoena* the false burrows and back burrows, having arisen in the course of closures in relatively intractable soil, cause parasites to drop a small percentage of their eggs or larvae into these burrows, or cause mutillid or chrysidid wasps to spend a considerable part of their time digging in vain at the bottoms of these burrows, we may further assume an increase in the frequency of genes enhancing this behavior. Thus species descending from such a prototype might possess stereotyped false-burrow digging movements disassociated from their original function. These ideas, for whatever they may be worth, might be summarized briefly as follows:

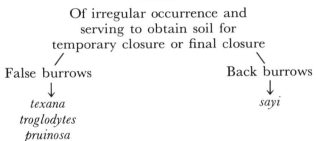

| False burrows of more or less regular occurrence, serving not for fill but presumably for deception of parasites | Back burrows of regular occurrence, serving not for fill but presumably for deception of parasites |

[2]In his most recent paper on *Stizus pulcherrimus,* Tsuneki (1965b) questions my use of the term "false burrow." He states that in his opinion these burrows probably do function to divert parasites, but since there are differences of opinion on this matter it seems best not "to adopt the term which restricts the significance to a single opinion only"; furthermore the term "has a very strong anthropomorphic odour." Tsuneki prefers the strictly morphological term "side holes." I use the word false to mean simply "not true," and in this sense it is widely used in biology: for example, false scorpions, false ribs. However, it is perfectly true that "false" has alternate meanings such as "feigned" and "dishonest." Probably another term should be used, although I question whether the term "side holes" is sufficiently descriptive; these holes are beside the burrow and not from the side of the burrow. Tsuneki has sometimes called these "accessory burrows," a term which I prefer to "side holes."

(7) *Oviposition.* Most species of *Bembix* lay their egg on a fly which serves as a pedestal but is not usually eaten by the larva (the fly is killed and generally somewhat dessicated by the time the egg hatches). The wing and middle leg of the pedestal fly are dislocated on the side on which the egg is laid. Tsuneki (1956) has discussed this behavior at some length and provided a whole plate of figures of flies bearing eggs, showing some slight variation in the disposition of the legs and wings. Similar observations have been made on several European species and on nine American species. Although the species of *Stictia* often place the pedestal fly dorsum up, such behavior only very rarely occurs in *Bembix* (I have noted it once in *u-scripta*).

A few species of *Bembix* lay their egg in the empty cell before any flies are brought in, not always in the same manner. Oviposition in these species may be outlined as follows:

Egg placed erect near center of cell, glued to substrate	*troglodytes* *multipicta* *texana* *brullei* (Chile) *olivacea* (Europe)
Egg placed in semierect position against apical end of cell	*pruinosa*
Egg placed flat on bottom of cell	*occidentalis*

(8) *Paralysis of prey.* Tsuneki (1956, 1958) has discussed hunting and paralysis of the fly in *niponica* at some length, and doubtless his remarks will apply in a general way to most species of the genus. He found that although the pedestal fly is killed, subsequent flies are merely paralyzed, living up to 9 days, but most of them dying 4–6 days after their capture, possibly of starvation. I have gathered no quantitative data on this point, but it is my impression that this would be found true of most American species. I have noted that in *belfragei* and *hinei* paralysis seems unusually light. However, in *texana* the flies are surely killed, as they generally appear dry and hard on the day following capture. There may be a correlation between cell cleaning and the killing of the prey. It would seem worthwhile for future students of *Bembix* to pay more attention to the effects of the sting, as there appear to be specific differences.

(9) *Type of prey.* There are now several thousand prey records for various Nearctic and Palaearctic species, and all indicate a restriction to Diptera. The South American *B. citripes* (Taschenberg) also

§E. General remarks on *Bembix*

takes Diptera, but Llano (1959) found a bee in one nest. ("En una oportunidad encontré entre las presas a una abejita de abdomen rayado, siendo siempre 'moscas' en los demás casos.")

Several African species also take Diptera, but there is one rather indefinite record of an African species taking Orthoptera and one somewhat less nebulous record of an East African *Bembix* using adult Lepidoptera (Wheeler and Dow, 1933). These authors also recorded the Australian species *palmata* Smith using Orthoptera as prey, but their information was secondhand. Wheeler himself found a species of *Bembix* in Western Australia using damsel flies (Odonata) as prey. Wheeler's specimens are in the Museum of Comparative Zoology at Harvard; the *Bembix* (not identified by Wheeler and Dow) runs to *atrifrons* Smith in Parker's (1929) keys.

It is possible that the records for Orthoptera involve confusion with the genus *Stizus*. However, it seems very probable that *Bembix* has undergone a certain amount of radiation with respect to prey in the warmer parts of the Old World. As noted at the beginning of Chapter X, the genus has undergone morphological radiation in this area unlike that occurring in the Western Hemisphere.

Some American species (such as *pruinosa*) appear to do much of their hunting at flowers, while others (such as *texana*) hunt about animals and often take freshly engorged horseflies, much as in *Stictia*. Most species do little hunting around their nest entrances and hence take few parasitic flies (*Exoprosopa, Senotainia*), but here *troglodytes* appears to form an exception. Certain species (*nubilipennis, truncata, u-scripta*) appear to use an unusual number of assassin flies (Asilidae). It is said that those European species with short wings (such as *integra*) prey mostly upon flies with widespread wings, chiefly Bombyliidae (filling the niche occupied by the species of *Glenostictia* and *Steniolia* in North America).

It is evident that female *Bembix* learn certain sources of flies and return to them until the supply is exhausted. If the source is a stand of flowers, the flies taken may be very diverse, but if it is a carcass or an aggregation of newly emerged flies resting on vegetation, the nest cells may be filled with only one or a few species of flies.

Whether or not the accumulation of long lists of prey records for the various species is worthwhile is a moot question. At least one can anticipate records of interest to dipterists, for these wasps have earned a reputation for taking flies not often collected by routine methods. For example, male horseflies may be taken in numbers in the forest canopy (Oldroyd, 1954:35–37). *Bembix pruinosa* is responsible for the first record for *Musca autumnalis* from upstate New

York, where it is now a major pest (Sabrosky, 1956). *Bembix cinerea* collected a new stratiomyid fly, *Eulalia evansi* James, in southern Florida. These are only two of a great many interesting records obtained from the study of *Bembix*.

(10) *Cell cleaning.* Earlier in this chapter I described cell cleaning in *texana* at some length, since this behavior has been little studied previously. Less complete observations suggest that similar behavior occurs in *occidentalis.* Tsuneki found that *niponica* carries out the remains of larval food on rare occasions, and there is evidence that this also occurs occasionally in the European *rostrata.* In *texana* (at least in the one colony I studied), cell cleaning seemed to be characteristic of the entire population. Brèthes observed very similar behavior in *Rubrica surinamensis* (Chapter IX:J).

B. pruinosa exhibits a very different type of cell cleaning (discussed especially in Evans, 1959b). Here the fly remains are not actually taken from the nest but are swept into a small cache at the apex of the cell and closed off. When provisioning is rapid and the cell filled up when the larva is small, as often occurs in this species, cell cleaning does not occur.

(11) *Some unusual features of nesting and provisioning.* Tsuneki (1956) found that in *niponica* the female does not make the spur and cell until after the first fly has been brought in. This is not true in *rostrata,* as Nielsen has shown, and I am not aware that it is true of any of the North American species, though admittedly my studies of some of the species do not cover this point adequately.

Tsuneki also found that females in a colony where compound nests were the rule would sometimes rear more than one larva simultaneously. This is of much interest in the light of Janvier's (1928) earlier report of this in the Chilean species *brullei.* There is no evidence that any of the North American species rear more than one larva at a time.

Tsuneki also found that each time the female brings a fly into the nest she leaves it in the tunnel and enters the cell empty-handed. She then either touches the larva with her antennae or, during periods of rapid provisioning, merely thrusts her head into the cell. When the larval food is nearly gone, she brings one or two flies to the larva. When the larva becomes large, it may move to the pile of food near the entrance, but in this case the wasp carries it back to the middle or back part of the cell. When nests are manipulated experimentally, the wasp often moves the larva into a more favorable location.

§E. General remarks on *Bembix*

To what extent these statements apply to the North American species is unknown. I have in no case seen an accumulation of flies in the cell or burrow not actually continuous with those being fed upon by the larva.

Clearly the behavior of *Bembix* is very much richer in detail than was appreciated only a decade ago. Although each species is in some measure distinctive, and some (most notably *pruinosa*) have a number of apparently unique characteristics, there is a great deal of variation which must be considered. In most cases, we do not know whether the observed variation is genetic or nongenetic, to what extent it is related to soil type, geography, population size, and so on, or to what extent learning is involved. Although *Bembix* is in every sense the best-known genus of Nyssoninae, we may safely say that knowledge of the ethology of the genus is only in its beginnings. At least we have reached the stage where it is possible to ask relevant questions.

Chapter XII. *Microbembex:* A Genus of Scavengers

Microbembex is a genus of some fifteen species which collectively range from Argentina to southern Canada. As the name implies, the wasps resemble diminutive *Bembix,* and it has sometimes been considered that the genus is a derivative of *Bembix.* In fact, I recently surmised that the genus might have evolved from a fairly advanced species of *Bembix* (Evans, 1963d). However, a close study of the structure and behavior of these wasps reveals some important differences from the genus *Bembix,* and I would not now make such a statement.

The following summary of the important structural features is provided.

Adults: body form similar to that of *Bembix,* but generally smaller, usually well under 15 mm, the species extensively ornamented with yellow or greenish-white; eyes prominent, subparallel, the vertex somewhat concave between the eyes; mandibles simple, tapering, without teeth; labrum elongate, tapering; mouthparts, when extended, very elongate, capable of reaching the middle coxae, the sheathlike galeae embracing the slender and delicate glossae and paraglossae; maxillary palpi of four segments, labial palpi minute, of one segment; ocelli completely obliterated, anterior ocellus represented by a transverse groove, posterior ocelli by oblique, arching grooves; thoracic structure differing in no important way from that of *Bembix;* both sexes with a pecten, that of the female very strong, the tarsal segments produced apically; wings much as in *Bembix,* but the marginal cell slightly separated from the wing margin apically and more pointed apically, the second submarginal cell more narrowed above; female without a pygidial plate; male with tergite VII without any indication of lateral spiracular plates; sternite II of male generally with a hooklike prominence, sternite VII simple, rounded apically, sternite VIII produced into a single long, curved spine; male

§A. *Microbembex monodonta* (Say)

genitalia with long, slender parameres and an unusually large, complex aedoeagus, the digiti much exceeding the cuspides (Figs. 177–185).

Larvae: body somewhat setose; head about as wide as high, bearing a number of fairly strong setae; antennal papillae unusually long; mandibles with a single strong, broad and blunt tooth on the inner margin; epipharynx spinulose except with a bare streak between the sensory areas (for figures and further details, see Evans and Lin, 1956b).

A. *Microbembex monodonta* (Say)

This is a common wasp in sandy areas from northern South America to southern Canada. It is rather variable, especially in color, and several of the variants have been given subspecific names, although none of these are currently in use. Typical habitats include dunes, blowouts, beaches, and artificial sand pits; in each case the sand is typically fine grained and friable, devoid of large pebbles and roots, and not excessively dry. Within its very broad range, virtually every area meeting these requirements is inhabited by a good population of this wasp.

As might be expected, there are a number of published reports on the behavior of this species. However, several of the reports are

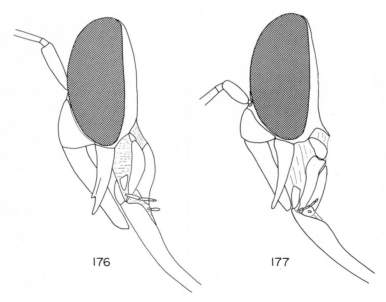

Figs. 176 and 177. Head: 176, of *Bembix spinolae*, in lateral aspect, apical portions of antennae and proboscis omitted; 177, of *Microbembex monodonta*, showing differences from *Bembix* with respect to clypeus, mandibles, and palpal segmentation.

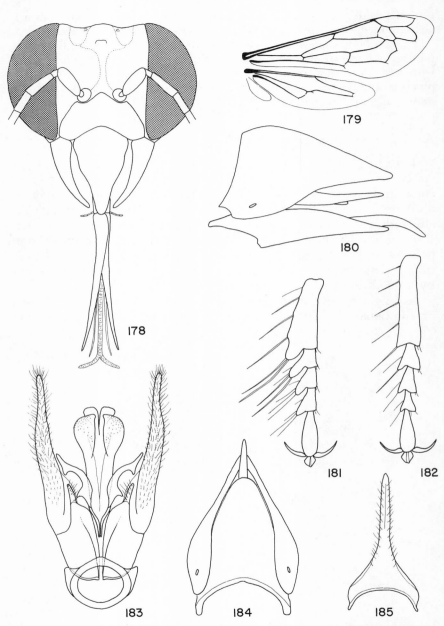

Figs. 178–185. *Microbembex monodonta* (Say): 178, head of female, mandibles spread and mouthparts extended; 179, wings; 180, apical segments of metasoma of male, lateral aspect; 181, front tarsus of female, showing pecten; 182, front tarsus of male, showing weaker development of pecten; 183, male genitalia; 184, apical segments of metasoma of male, ventral aspect; 185, sternite VIII of male.

§A. *Microbembex monodonta* (Say) 363

very brief, and even the longer reports contain less detail than might be desired. *Microbembex monodonta* is one of the most remarkable of digger wasps, and even my own fairly extensive studies leave a number of questions unanswered. This wasp is readily amenable to experimentation and available in many places in great numbers; it is to be hoped that it will some day provide a subject for much more detailed studies than have yet been made.

Hartman (1905) studied this species briefly at Austin, Texas, and was the first to call attention to the fact that this wasp is a general scavenger, taking many types of dead and disabled arthropods as "prey." Parker (1917) studied the species near Sandusky, Ohio, and in the same year Stoehr (who was apparently unfamiliar with the work of Hartman or Parker) published his observations made in Quebec, Canada. Since then there have been several brief reports, including those of Rau (1918 and others) from Kansas and Missouri, Krombein (1953) from North Carolina, and Vesey-Fitzgerald (1940, 1956) from Trinidad. I studied the species in Kansas in 1952 and 1953 (11 notes made at Blackjack Creek, Pottawatomie County; 3 notes made in localities in Reno and Stafford counties), in New York 1954–1960 (9 notes made at two localities in Oswego County), and at Mazatlán, Sinaloa, Mexico (1 note made 20 August). In the summer of 1963 I made a more detailed study of the species in a large man-made sand pit at Andover, Massachusetts; studies in this area involved 22 field notes and the excavation of 40 nests. Frank Kurczewski has also sent me notes which he made at Presque Isle, Pennsylvania, and near Auburn, New York.

(1) *Ecology and general behavior of adults.* I have found this species nesting in open sand behind sea beaches and on the shores of lakes and streams, as well as in dunes, sandpits, and other areas remote from water. Hartman reported the species from "sandy woods," and Parker found large colonies in blowouts in sandy soil along the shores of Lake Erie. Parker says:

> They prefer the open spaces entirely free from vegetation, but their burrows may be found almost anywhere among the clumps of grass and even under the trees wherever the sand is free from leaves or not hidden by foliage. Although a high wind . . . or a violent rain-storm . . . alters in no small measure the surface of the naked sand, nevertheless such changes in no way discourage these energetic little insects or even seriously interfere with their prosperity.

Vesey-Fitzgerald found *monodonta* nesting in Trinidad in sandy soil at the seaside or at the mouths of rivers, in dry soil above the level of high water. On the Paria Peninsula, Venezuela, the species is also reported as nesting in sand above the high-water mark (Callan, 1954).

Chapman *et al.* (1926) found *Microbembex monodonta* to be very common in sand dunes in Anoka County, Minnesota, and studied the effect of physical conditions on this and several other sand dune insects. They found that *Microbembex* is active over a more limited range of temperature than any of the other insects studied. When confined to glass tubes in the laboratory, adult *Microbembex* first became active at about 25°C, assumed normal activities between 30 and 44°C, and were all rendered inactive by 50°C; their effective "zone of activity" was only 14°, as compared to 17° for *Bembix pruinosa* and over 30° for *Ammophila* sp., various Pompilidae, and *Dasymutilla* sp. In nature, the first *Microbembex* usually appeared at air temperatures of 19–24°C, that is, somewhat later in the morning than many other wasps (this was confirmed by me at several localities, the first appearance usually being at 21–24°C). Since the temperature at the sand surface at midday often exceeds 50°C, Chapman *et al* conclude that the wasps penetrate the hot surface layer of sand by "a juggling of time and space during which the wasps alternately dig furiously at the surface for a short period of time and fly about." While no one questions that any insect will be killed by forced exposure to the exceedingly high temperatures of the surface of sand on a hot day, I feel that the conclusions of these authors are weakened by the fact that they failed to describe in detail the normal daily regimen and digging behavior of these wasps. As a matter of fact, the penetration of the surface layer of sand takes at most a few seconds anyway, and all digger wasps spend the greater part of the time either above the surface or below it, where temperatures are very much lower. To speak of this as a "juggling of time and space" verges on the metaphysical.

These wasps seldom fly far from their nesting sites and hence are decidedly local in distribution, although often present in countless thousands in restricted areas. Common associates include particularly certain species of *Bembix,* sometimes also species of *Bicyrtes, Philanthus,* and *Tachysphex.* Hartman reported *Bembix texana* as nesting beside *Microbembex monodonta* in Texas, but his photographs show *B. spinolae,* and I feel that he was surely working with the latter species. In much of the eastern United States *B. spinolae* and *B. pruinosa* are common in the same areas as *Microbembex,* their nests often intermingled; in the southwest *B. troglodytes* is a common associate, in the far west *B. occidentalis,* and in Mexico *B. multipicta.*

In New York and Massachusetts the males appear toward the end of June and are present in large numbers for only 2 or 3 weeks,

§A. *Microbembex monodonta* (Say)

when they disappear almost completely. Females appear a few days after the first males and are active for from 8 to 10 weeks, that is, well into September. Stoehr states that in Quebec the males emerge in mid-July, to be followed in a few days by the females. In the warmer parts of the range there may be two or more generations per year. In Kansas my records begin on 12 June and extend through 9 September. One individual that spun its cocoon in a rearing tin on 18 June 1952, emerged as an adult in August of the same year. In Florida and Texas I have collected the species as early as April.

In the early part of the season, the males may be found visiting flowers in great numbers; females apparently take nectar from flowers throughout the season, but spend much less time at it than do the males. A great variety of flowers are used, generally species growing in the sand of the nesting area or in the immediate vicinity. At Blackjack Creek, Kansas, smartweed (*Polygonum*) was especially favored, while in New York and Massachusetts the flowers of various brambles (*Rubus*) are often visited in great numbers early in the season. I have also taken the wasps in numbers on *Aralia*, *Melilotus*, and several other plants. Kurczewski and Kurczewski (1963) report the species on flowers of *Rubus*, *Asclepias*, and *Solidago* at Presque Isle, Pennsylvania.

It is possible that the females obtain part of their nourishment from the bodies of the insects they capture. Hartman on one occasion "noticed a wasp fly to a weed and hang there by one of her hind feet while with the remaining five she held an apparently dead Syrphid. I could approach very close to her," Hartman continues, "and could see how she held the fly and alternately apply her mandibles and proboscis to the fly's thorax. It is probable that *Microbembex* was this time enjoying a little fly-juice for herself."

Virtually everyone who has worked on *Microbembex monodonta* has remarked upon the large number of open holes in the sand where these wasps occur (Fig. 186). Hartman speaks of the sand as being "riddled with holes," and Stoehr remarks that "one sees hundreds of little openings, in the form of minute horseshoes, which are crowded one against the other." These holes do not represent the true nests of this species; in fact, they are often most numerous somewhat to one side of the main nesting area. I have sometimes counted as many as 50 such holes per square meter. A few of these holes may possibly represent incipient nests which were later abandoned, but there is no question that most of them represent sleeping burrows. These are typically left open during the day, when they are empty, but at night and during inclement weather many of them

Fig. 186. Open sleeping burrows of *Microbembex monodonta* (Pottawatomie County, Kansas, July 1952).

are closed from the inside, the wasp remaining near the bottom of the burrow, the head directed toward the entrance. In the morning, shortly before the wasps become active, they are usually found just beneath the closure; in fact, they may remove the closure and remain looking out the entrance for a short period before taking flight.

Both males and females dig burrows of this type in which to spend their inactive periods. Since the males disappear after 2 or 3 weeks, the number of sleeping burrows drops markedly later in the season, and one rarely finds areas "riddled with holes" except early in the season (the sleeping burrows of the females tending more to be scattered among the true burrows). I have observed wasps reentering old sleeping burrows in the evening, and I have also observed them digging new ones, in the latter case presumably when an old one cannot readily be found open. On windy days the majority of wasps may make new sleeping burrows, as the old ones will largely have been buried.

The sleeping burrows are oblique, like the true burrows usually directed into the slope of the sand; there is no terminal cell and

§A. *Microbembex monodonta* (Say)

little if any mound of sand at the entrance, although the closure thrown up from the inside is sometimes noticeable at least when it is new. Several burrows dug out at Mazatlán, Sinaloa, were all about 3.5 cm long, the bottom of the burrow being only 1 to 2 cm beneath the surface. Five sleeping burrows dug out at Andover, Massachusetts, varied in length from 3 to 6 cm, the bottom being from 2 to 3.5 cm beneath the surface. I have never found more than one wasp in a sleeping burrow (Fig. 189, nos. 1924 and 1931).

I have seen females enter sleeping burrows as early as 1315. Presumably females whose nests contain an egg or very small larva and who therefore are engaged in little or no provisioning spend more time resting. Also, it seems probable that both sexes spend the hotter parts of very warm days in these burrows. Most commonly, the wasps enter their sleeping burrows from 2 to 4 hr before sunset and leave them from 2 to 4 hr after sunrise. There is no evidence that the female ever spends the night in the true burrow or that either sex sleeps on vegetation.

By way of example, my no. 65 (Blackjack Creek, Kansas) started a new nest at 1530, finished the initial closure at 1645, then left and began digging again nearby. She abandoned this hole in a few minutes, leaving it open, then dug two more shallow holes before finally making a burrow 8 cm long and closing it from the inside (1730). She was dug from this burrow, and her true nest about 1 m away was dug out and found to contain an egg but no wasp. Many similar examples could be cited.

This species typically produces no sounds, even when digging or provisioning. The flight is silent, rapid, and generally only a few (1–5) cm above the soil surface. Hartman aptly compares these wasps to birds skimming the water. Stoehr mentions the difficulty in catching them in an insect net, even though thousands may be present. Active nests of this species may be only a few centimeters apart (rarely less than 5 cm); I have counted as many as 15 active nests per square meter. On the other hand, at the periphery of the nesting area or in areas of low population, the nests may be relatively isolated.

Both Stoehr and Parker have noted aggression among females nesting in close proximity. Parker says that they "push and shove and crowd each other . . . One will gain the entrance only to be seized by the wing or hind leg and dragged out by the other . . . Frequently as one of them seeks to enter the burrow the other will pounce upon her back, seize her, and rising on the wing carry her to a short distance and drop her without ceremony upon the sand." Parker believed that these struggles occur when two or more nests are so close together that a wasp "disturbs the entrance to the nest of a neighbor."

(2) *Reproductive behavior.* I have made no observations whatever on this aspect of the behavior and must content myself with quoting others. Stoehr states that the males form a veritable cloud over the sand. "They fly on all sides, going and coming only to go off and return once again. When they settle, their antennae strike the sand, and their tilted head auscultates the soil so as to catch the slightest noise which can be detected within it. As soon as the females appear, they lose no time in paying their court, and unions are soon contracted." Parker is more prosaic but provides more detail:

> Mating occurs immediately after the female has emerged . . . The males are constantly searching the sands for the emerging females, and a female is not long above ground before she is discovered by a passing male and fertilized. The female is usually found and seized by a male before she has made any attempt to fly, and a fierce but brief struggle precedes copulation, the pair rolling about on the sand or in some instances rising into the air. Copulation requires but a brief time, about half a minute, and the male seems capable of fertilizing a number of females. All data secured tend to show that copulation occurs but once. While the females are digging their burrows and searching the sands for food for their young they are continually pestered by the roaming males . . . In the majority of such cases the male retains his hold but momentarily . . . in others a struggle ensues . . . but in all of these cases, and hundreds were observed, not a single mating was effected.

(3) *Digging the nest.* New nests are usually started in the afternoon, most commonly between 1300 and 1600, but sometimes as early as 1030. Several individuals timed at Andover, Massachusetts, required from 1.5 to 2.5 hr to complete a nest. Most wasps dig more or less continuously until the nest is finished. Some individuals dig one or more short "trial burrows" before remaining in one place and completing a nest. One individual at Blackjack Creek, Kansas (no. 65) made more than 12 such holes before completing a nest; after completing this nest, at 1757, she immediately dug another one, this one a sleeping burrow, only a few centimeters away. Parker remarks that when the sand is reasonably moist and firm, as after a rain, the "burrow can be completed in the course of two or three hours or even less," but that if the sand is very dry the wasps may "work for half a day without being able to get out of sight in the sand." I have not been able to confirm this last remark of Parker's.

Hartman described the digging behavior well when he said that "the body of the wasp, balanced, as it were, on the middle pair

§A. *Microbembex monodonta* (Say)

of legs, represents a teeter-totter in miniature. Each time the head goes down, the tail bobs up and a stream of sand pours out from under the wasp, propelled by several smart strokes of the front legs in quick succession. Then there will follow a brief pause while the wasp rests with head in air as if looking around an instant to survey the landscape." Stoehr compares the jerky digging movements to those of a mechanical windup toy.

Movements such as Hartman has described are best observed as the wasp stands in the entrance, as she does at intervals, clearing the sand that has been thrown up during the actual digging. One to several spurts of sand are thrown out each time the head goes down and the abdomen up, although a small additional up-and-down movement of the abdomen can be noted between each stroke of the front legs. The sand is thrown up and back a considerable distance, generally 1–4 cm in the air and 8–20 cm behind the wasp. The sand is so well spread that no mound accumulates at any time, and there is no external indication that the burrow may actually be quite long. This manner of clearing the entrance is very similar to that of the species of *Bicyrtes* (Fig. 187).

Fig. 187. Female *Microbembex monodonta* digging (Andover, Massachusetts, July 1963). This female is in the "head-up" position; shortly her head will go down, her abdomen up, as she scrapes back several jets of sand.

When the nest is complete, the egg is (usually) laid immediately, before the entrance is closed and concealed. The initial closure is fairly lengthy, taking one half to several minutes, during which time sand is scraped from several directions over the entrance. Following this, the wasp goes out scraping sand in a straight line several times, forming a pattern of radiating lines, all or mainly on the side toward the direction of the burrow (that is, usually upslope from the entrance). The number of such "radiating lines" is highly variable, usually from three to ten. During closure and between the radiating lines the wasp flies up briefly and may make a few small loops over the nest; following completion of the last radiating line she flies about and often reapproaches the nest several times, sometimes scraping a bit more sand over the entrance.

I have observed no orientation flights in this species other than those described above. When the initial nest concealment is finished, the wasp revisits the nest only (1) to make an inspection each morning, shortly after emerging from the sleeping burrow, and (2) to bring in food for the larva. Otherwise, the time is spent away from the nest, including some whole days when the weather is poor. Nevertheless, I have never seen females exhibiting the slightest hesitation or difficulty in locating their nests, even though there may be hundreds of nests scattered over a large area of bare sand with few notable landmarks. This matter is discussed further in a subsequent section entitled "orientation."

(4) *Nature and dimensions of the nest.* The nest is a simple, oblique burrow with a small terminal cell. Published observations are in agreement that the nest is unicellular. All of the more than 60 nests I have studied have had but one cell; some of these nests were dug out after the final closure, and some were nests of marked individuals which had been watched over a period of several days. Such marked individuals were generally found to dig a new nest fairly close to the old one, after the final closure of the latter. For example, no. 1959 constructed a second nest only 0.8 m from the first; no. 1932B made a final closure from 1330 to 1420 and then dug a new nest 2 m away from 1430 to 1630 (both Andover, Massachusetts).

The burrow is small, only about 5 mm in diameter. The angle formed with the horizontal varies from 40 to 60°; nests dug into slopes tend to maintain about the same angle with the horizontal, but of course a slightly greater angle with the surface. Where there is a slope, however slight, virtually all nests are dug into the slope. Nearly all nests are straight, or almost so, but I have

§A. *Microbembex monodonta* (Say)   371

found some tending to curve downward at a steeper angle part way down, as well as a few that curve gently to the right or to the left before reaching the cell. The cell is oblique or nearly horizontal, always at a distinctly lesser angle than the burrow. The cell is broadly elliptical and measures from 2–3 cm in length by 1–1.5 cm in diameter at the middle. Typical nests are shown in Fig. 188.

As may be seen from Table 39, there is a considerable amount of variation in nest depth. However, in any one locality the variation is usually not great. Thus, at Andover burrow length was found to vary from 12 to 25 cm, but as a matter of fact all but 7 of 40 nests were between 15 and 20 cm; in this locality cell depth varied from 7 to 15 cm, but all but 6 nests were within the very narrow range of from 9 to 12 cm. This is in strong contrast to the over-all recorded cell depth of this species, which is from 7 to 25 cm.

I believe the variation in nest depth to be attributable to two factors. (1) Smaller wasps tend to make shallower nests, larger wasps, deeper nests. To check a general impression that this was true, I marked the nest of the smallest wasp I could find at An-

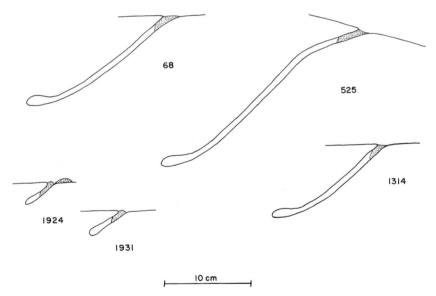

Fig. 188. Burrows of *Microbembex monodonta*: no. 68, Pottawatomie County, Kansas; no. 525, Stafford County, Kansas; no. 1314, Oswego County, New York; no. 1924, Mazatlán, Sinaloa, Mexico; and no. 1931, Andover, Massachusetts. Nos. 1924 and 1931 are sleeping burrows, the others, brood nests.

TABLE 39. NEST DATA FOR MICROBEMBEX MONODONTA

Locality	No. of nests	Burrow length (cm)	Cell depth (cm)
Stafford Co., Kan.	1	32.0	25.0
Pottawatomie Co., Kan.	11	21.5 (10-25)	12.5 (8-19)
Missouri (Rau)	2	— (32-35)	16.0
Ohio (Parker)	several	— (20-30)	— (8-15)
Oswego Co., N.Y.	9	17.0 (14-21)	9.5 (8-12)
Andover, Mass.	40	17.5 (12-25)	11.0 (7-15)
Quebec (Stoehr)	several	15.0	8.0

dover on 12 July 1963 (no. 1933C). This wasp was only 7 mm long and made the shallowest nest of this species I have ever found (burrow 12 cm, cell depth 7 cm). By contrast, no. 1935A, a rather large female nesting nearby and measuring 12 mm in length, made a burrow 16 cm long, the cell 11 cm deep. (2) Toward the north and east, nests tend to be shallower than in Missouri and Kansas. This parallels a trend in several species of *Bembix* and *Bicyrtes* that nest in open sand and is perhaps correlated in some measure with lower temperatures and greater moisture content of the soil toward the north and east. I also believe there may be a correlation with the amount of blowing sand, as I found to be true in *Bembix pruinosa* (Evans, 1957). Unfortunately I dug only one nest in the large dunes in Stafford County, Kansas, but this was much the deepest nest dug anywhere. Presumably selection favors deeper nests in areas of large sand dunes, where there is greater danger of the cells being uncovered by wind action.

An outer closure of the nest is maintained at all times when the female is not actually in the nest (but Hartman noted one wasp that left the entrance open while bringing in prey in quick succession). Since the female is in the nest only at certain brief intervals (digging, making an inspection, bringing in prey, and during the final closure), this means that the outer closure is in place nearly all the time and is always made from the outside. The closure consists of several scuffs of sand made while the wasp is facing away

§A. *Microbembex monodonta* (Say) 373

from the burrow; the initial closure involves movements of concealment, as described above, and the last closure each day may be fairly long and also involve scraping sand over the entrance from the sides. After a series of closures there may be a small depression at the nest entrance, although this depression is not likely to be noticed unless the nest has been marked while the entrance is open. This wasp does not make an inner closure of the nest at any time, nor is there ever a spur or any other branching of the burrow.

(5) *Orientation.* Since these wasps often nest in great numbers in broad expanses of open sand, and since the nest entrances are kept closed while the wasps are away and are not marked by a mound of sand at the entrance, a human observer is at once impressed with the remarkable powers of orientation of these wasps. The wasps always fly low, even when approaching the nest entrance, yet they invariably stop at precisely the right spot, even though they spend the greater part of their time away from their nest. As already noted, there are no conspicuous orientation flights during or following nest construction. Yet they are able to find their nests without hesitation even when the sand surface has been greatly disturbed, for example, by humans or livestock walking through the area. Parker cites a case in which he had been sitting on the sand for almost 1 hr and had thoroughly disturbed the sand over a considerable area; nevertheless, when he moved his foot a *Microbembex* female proceeded to uncover her nest, which was directly beneath where his foot had been. I have had similar experiences many times.

Parker deliberately altered landmarks about nests by throwing buckets of water over the surface, piling sand over the entrance, or placing paper, leaves, and rubbish over the nest. He was unable to prevent the wasps from finding their nests. Feeling that larger landmarks farther from the nest may be important, I experimented by moving some of these (Andover, Massachusetts). In one instance a log lying on the sand was the only major landmark within several meters. I moved this by 1 m, but this caused no disorientation whatever of the *Microbembex* provisioning their nests nearby. On another occasion I moved both a clump of grass and a small oak tree 20 cm in the same direction (again, the only large objects within several meters); several females provisioning nests in the near vicinity continued to find their nests without hesitation.

Parker concluded that these wasps find their nests "through the sense of smell or some power similar to smell." Although his evidence was entirely of a negative nature, of the type indicated above, his remarks have been cited by a number of authors, for example by Thorpe (1963), who believes that if odor plays a role in nest finding in bembicines it may also "be responsible for colony localization year after year."

I am unable to believe that odor plays a role in orientation in this species for the following several reasons. (1) When there are hundreds of nests in close proximity in an expanse of sand, it seems to me highly unlikely that there can be a difference in odor between each nest sufficient to guide wasps to their own nests. (2) All work that has been done on homing in solitary wasps indicates that visual and not olfactory cues are used. This includes not only many observational data, but the important experiments of Tinbergen and others on *Philanthus*, of Baerends on *Ammophila*, and of van Iersel on *Bembix rostrata*, a member of the same tribe as *Microbembex*. (3) I have placed transparent plastic sheets, 1 m^2, over nests of known location which were being actively provisioned. Without exception, the wasps homed in to the exact spot where their nest was located. They would not land, but would fly off again and again and return (at the usual very low altitude) to the same spot, always without important hesitation or circling about in the vicinity of the nest (note no. 1957, Andover).

It is my feeling that these wasps learn the nesting area in great detail, including such major features as the contours of the dunes, distant trees, and other large objects in relation to closer objects, also intimate detail of the soil throughout a wide area. The destruction of a portion of the landmarks has no important effect upon the total picture, as the wasp locates its nest against a broad back-drop of objects, probably including objects on the horizon, as known to occur in *Bembix niponica* (Tsuneki, 1956) and *B. rostrata* (van Iersel, 1965). The orientation of wasps inhabiting broad expanses of open sand is a subject worthy of much further investigation.

(6) *Hunting and provisioning.* This aspect of the behavior of *Microbembex* has received most attention, and deservedly so, for no other wasp is so completely non-host specific or utilizes dead arthropods as "prey."[1] It is commonly stated that *monodonta* uses either dead or living arthropods, following Hartman's account. However, Stoehr found only dried insects and pieces of insects to be used. Parker states unequivocally that this species "feeds its young on dead insects, which it gathers up from the surface of the sands, instead

[1]However, I know of no basis for the assumption of Malyshev (1959) that wasps of this genus may provision their nests with a mixture of nectar and saliva, thus suggesting a stage transitional to the bees.

§A. *Microbembex monodonta* (Say) 375

of capturing and paralyzing living insects by stinging, as do other wasps of similar habits." Later he says, "I did not observe a single case in which a wasp attacked and carried off a living uninjured insect."

After many hours of watching these wasps, I find myself in agreement with Parker rather than with Hartman. Doubtless arthropods that are not actually dead but only greatly disabled are sometimes used, but I am not convinced that healthy, fully active ones are ever attacked. I have several times placed active insects on the sand and seen *Microbembex* females approach them closely, but in no case did they seize or sting them. The wasps hover about them or land on the sand and extend their antennae rigidly toward them, but as soon as the insect moves they fly off. Once I placed several small caterpillars (Geometridae) on the sand. They walked about and were approached but not attacked by the *Microbembex*. Soon the intense heat of the sand began to slow them down, and one by one they curled up and became motionless, only to be seized promptly by a *Microbembex* and carried off; several were found in nests later on, and they had not recovered. At Andover, Massachusetts, the wasps often hovered about *Formica fusca* workers briefly, but never attacked them; this ant was abundantly represented in the cells of the wasp, but only by mangled or dried and broken individuals, probably taken from the outside of the various nests in the vicinity. On one occasion I saw a spider crossing the sand. A female *Microbembex* landed beside it and touched it with her antennae, but the spider defended itself and the *Microbembex* flew off. This was repeated several times. However, spiders are often found in the cells of the wasps, doubtless specimens found dead on the sand.

Hartman's examples of live insects being attacked by *Microbembex* are worth re-examining. Hartman says: "That *Microbembex* does attack the living ants seems probable from a struggle I once saw between a wasp and two red ants, one of which had probably fastened its hold upon the wasp at the start until joined by the companion. The wasp was evidently dead when I took the two into a bottle." This sounds more like a case of ants attacking a disabled wasp; in any case, Hartman's "probable" conclusion is based on something which "probably" happened, and I am not fully convinced. Later on, Hartman mentions that the caterpillars found in the nests are always limp and fresh. I have often found dried or partially decayed caterpillars in nests, but have found fresh caterpillars only under the artificial conditions described above. It is probable that caterpillars do from time to time drop from trees onto the sand and die there from the heat of the sand, thus simulating the conditions in my experiment; perhaps paralyzed caterpillars are also stolen from time to time from *Ammophila harti* and similar wasps (see discussion of prey stealing, below). Even Hartman believed that the taking of active arthropods was exceptional. Pending clear-cut evidence to the contrary, I shall regard *Microbembex monodonta* as strictly a scavenger.

There is also some uncertainty as to whether these wasps sting their "prey." The sting and poison sacs are no smaller than would be expected for a wasp of this size, so it seems probable that they are not completely functionless. I have often handled females in the process of marking them, and they extrude their sting in the manner of many wasps, although even the larger specimens seem unable to penetrate the skin. In describing the response to disabled insects thrown on the sand, Parker writes as follows:

> In seizing such an insect the wasp appears to sting it, but of this I cannot speak with certainty. When one of these insects was seized the wasp invariably bent her abdomen forward, bringing the tip into contact with her victim, thus going through the performances that would be incidental to stinging; but this same performance may sometimes be seen when the wasp seizes a dead and dried insect lying on the sand. I am of the opinion, however, that the instinct to sting still remains, and that whether the prey be a disabled or a dead insect the wasp uses her sting upon it.

Although I have elsewhere suggested (1963d) that stinging in *Microbembex* may be an example of vestigial behavior, following Parker, I now feel uncertain of this. In July 1963 I made a special effort to observe the seizure of prey. It appeared that in many cases the mandibles and front legs were used in picking it up, and that it was then passed back to the middle legs for carriage. Only occasionally did I observe the tip of the abdomen to be curved forward. I also observed an occasional female stopping during flight to the nest, either hovering in flight or dropping the prey to the sand, and then curving the abdomen forward. It seems possible that the tip of the abdomen plays a role in adjusting the position of the prey, which perhaps is especially important in a wasp that takes arthropods of varying sizes and shapes. My present feeling is that these wasps remain capable of stinging but do so only rarely. This matter deserves further study.

Females marked at Andover, Massachusetts, appeared to do most of their hunting within a radius of 3–5 m of their nest. However, under some circumstances they sometimes venture many meters from the nest, chiefly later in the day when the dead arthropods

§A. *Microbembex monodonta* (Say)

have been largely cleared from the nesting area. The females hunt in irregular searching patterns only one or a few centimeters above the sand surface, pausing to hover about objects lying in the sand, sometimes picking them up and dropping them again, sometimes carrying them a short distance only to abandon them or to pick them up again. Hartman speaks of these wasps "calmly flying through the woods much like a dragon fly, steadily maintaining a level of a foot from the ground." I have rarely seen these wasps flying at this high an altitude. There is little question that all or nearly all prey is picked up from the surface of the soil rather than from vegetation. Even bits of plants lying on the sand are investigated, occasionally picked up and dropped, and rarely taken to the nest (see (7) below). By and large, the arthropods used most commonly are those to be found in abundance in and around the nesting area. At Andover, *Formica fusca,* an ant nesting abundantly around the nesting area, was most commonly employed, and in beach areas mayflies and midges which have flown the night before and died on the sand are used in great numbers.

The prey is carried to the nest by the middle legs; at the entrance, the prey is normally held by the middle legs while the entrance is opened by the front legs. Large prey may be carried by a series of short flights interrupted by pauses on the sand, and in any case pauses on the sand with prey are not uncommon. Krombein watched a wasp with a dead phalangid, "the long legs of the harvestman projecting way beyond the wasp's abdomen, the ensemble presenting a very ludicrous appearance as the wasp hovered over the sand." I watched a female carrying a damselfly about three times her own length; she held the thorax of the damselfly and let the abdomen drag along behind, the two resembling a *Pelecinus* flying along. The damselfly was dropped outside the nest while the wasp scraped open the entrance, then took flight briefly, picked up the damselfly by the middle legs, and proceeded into the nest in the usual manner. Parker also notes that large prey may occasionally be dropped outside the nest. Often, also, large prey becomes stuck in the entrance, so that the wasp has to turn around inside the nest and pull it in with her mandibles. I watched one female that had a rose chafer (*Macrodactylus*)—a disabled specimen I had provided—stuck in the entrance. After 10 min of trying to pull it in, she came outside and pulled it out of the entrance, whereupon it was immediately picked up by another *Microbembex* and carried away.

Females that arrive at the nest entrance laden with prey and are unable to enter immediately because of the large size of the prey, cave-ins of the upper part of the burrow, or destruction of the nest entrance by footprints, often elevate their wings obliquely for brief intervals, as is common in *Gorytes* and *Bicyrtes*.

Parker mentions that females often take prey from one another, not only specimens that have been dropped but ones that are actually being carried to the nest. "The struggles at the mouth of the burrow," he says, "are frequent and furious, the contestants grappling and rolling over and over on the sand. Frequently it happens that the prey is dropped in the struggle and while the pair of contestants are rolling on the sand a third wasp comes along and settles the quarrel by quietly carrying off the coveted treasure." I have sometimes seen as many as four other females hovering closely about a struggling pair, and at other times I have seen prey change hands twice before being taken into a nest. During these struggles, the females merely try to grasp the prey; they do not, so far as I have observed, actually try to bite or sting one another.

These wasps will also take prey from other species of wasps nesting in the same areas. I saw a female attempt to steal a cricket from a *Lyroda subita*, in this case without success. I have seen flies abandoned outside their nests by *Bembix pruinosa*, as well as queen *Formica* ants dropped by *Aphilanthops frigidus*, picked up and used by *Microbembex*. Frank Kurczewski (personal communication) has observed several times, near Auburn, New York, *Microbembex* females taking small paralyzed grasshoppers deposited at their nest entrances by *Tachysphex terminatus*. He observed this only in cases where both species were abundant and the nests of the two were intermingled.

Microbembex will also occasionally attempt to take dead arthropods, or pieces thereof, from worker ants. At Andover, Massachusetts, I often saw wasps hovering over *Formica fusca* workers carrying or dragging insects over the sand to their nests, and in at least one case the *Microbembex* was actually able to snatch the insect from the ants and fly away with it. I have also seen instances of ants carrying off insects dropped temporarily on the sand by *Microbembex*, and Hartman describes such cases.

The first trip to the nest each morning is made without prey; the wasp merely enters by scraping her way into the burrow, remains within the nest for 1 or 2 min, and then comes out and closes. Some clearing of the entrance often occurs either before or after the wasp goes down to the cell. These inspection trips often occur between 0830 and 1000 in the morning, but on several occasions I noted that marked wasps having nests containing an egg or very small larva delayed their inspections until somewhat later. The results of this inspection presumably determine the number and to some extent the type of arthropods to be brought in that day. I have noted that nests with eggs just hatched or about to hatch always contain very small and usually relatively fresh arthropods; for example, no. 1933B contained a small male ant pressed against the egg, no. 1948B had a small midge, a small plant bug, and a small leafhopper close beside a day-old larva. Larger and hard, dry pieces of arthropods occur generally in nests containing large larvae. Presumably as the larva grows and requires more food, the thresholds of response involved in hunting are lowered so that almost anything of arthropod (rarely even plant) origin is accepted.

(7) *Nature of food gathered for the larvae.* Since this species is a general scavenger, it seems pointless to present a detailed breakdown of everything I have found to be used or that is recorded in the literature. However, I do wish to present a sufficiently detailed summary to provide documentation of the fact that the species uses almost anything readily available in and around the nesting area. In all, two classes of arthropods and ten orders of insects have been found to be used under natural conditions, and experimentally three other classes of arthropods have been found acceptable (as summarized in a later section).

ARACHNIDA
PHALANGIDA. One record of a dead specimen (North Carolina, Krombein).
ARANEIDA. The Raus record *Xysticus nervosus,* a dry and hard specimen. I have found small, unidentified spiders in seven nests, including nests in Kansas, New York, and Massachusetts, the spiders varying from fresh in appearance to shriveled or rotted. One nest contained six spiders, the others only one or two.

INSECTA
EPHEMERIDA. Mayflies are commonly used along the shores of lakes. Parker found them to be used abundantly at Cedar Point, Ohio, and Frank Kurczewski has sent me a record from Presque Isle, Pennsylvania, of a subimago *Hexagenia limbata* (Serv.), a specimen with portions of the head decayed. I found two mayflies in as many nests on the south shore of Lake Ontario, New York.
ORTHOPTERA. Hartman reports grasshopper legs used commonly in Texas, as well as an "orthopterous pupa-case with dry dead pupa inside." Stoehr reports small, dried grasshoppers from Quebec. Vesey-Fitzgerald (1940) found winged termites to be commonly

used in Trinidad, and also took one adult cockroach (1956). I took eight orthopterans from seven nests in Kansas and Massachusetts, varying from small, whole nymphs to pieces of adults (including two hind legs of crickets); families included Acrididae, Gryllidae, and Tridactylidae.

PSOCOPTERA. I took one adult bark louse from a nest in Oswego County, New York.

HEMIPTERA. Hartman reports treehoppers as well as several species of true bugs; Stoehr reports several tarnished plant bugs (*Lygus*); the Raus report treehoppers. Vesey-Fitzgerald reports stinkbugs (Pentatomidae) from Trinidad. I found both Homoptera and Heteroptera to be used in all localities studied, both nymphs and adults, both fragments and whole specimens; in all some 48 Hemiptera were taken from 20 nests. Homopterous families represented were Cercopidae, Membracidae, Cicadellidae, Fulgoroidea, Psyllidae, and one record for Aphididae. In the Heteroptera, the families represented were Lygaeidae, Tingitidae, and Miridae. One nest in New York contained ten leafhoppers of seven species.

NEUROPTERA. I took one ant-lion larva (Myrmeleonidae), in poor condition, from a nest in New York.

TRICHOPTERA. I saw and photographed a female carrying into the nest a large adult caddis fly (Leptoceridae) on the shores of Lake Ontario, New York (Fig. 189).

LEPIDOPTERA. Hartman and Stoehr both report slender caterpillars. I found eight lepidopterous larvae in six nests, all small, smooth species except for a head of a large caterpillar. One nest in New York contained a pupa of a small moth, and two nests each contained a single small adult microlepidopteran.

COLEOPTERA. Adult beetles, or often pieces of beetles, are used commonly. Stoehr, Rau, and Vesey-Fitzgerald report Carabidae, Coccinellidae, and Curculionidae. My records include these three families and also Anthicidae, Cerambycidae, Chrysomelidae, Dermestidae, Elateridae, Helodidae, Hydrophilidae, Pedilidae, Scarabaeidae, and Staphylinidae. I have also found larvae of Chrysomelidae, Cicindelidae, Coccinellidae, Elateridae, Nitidulidae, Ostomatidae, and several unidentifiable larvae. In all I have records of 49 Coleoptera from 28 nests.

DIPTERA. Stoehr reports shriveled flies and pieces of flies; Hartman reports Syrphidae, Hine (1906), Tabanidae of the genus *Chrysops*, and Parker, midges (Chironomidae) in some numbers. I have recovered specimens of the following families: Anthomyidae, Calliphoridae, Chironomidae, Culicidae, Dolichopodidae, Empididae, Muscidae, Sarcophagidae, Stratiomyidae, and Therevidae. All were

§A. *Microbembex monodonta* (Say)

Fig. 189. Female *Microbembex monodonta* entering her nest with a dead leptocerid caddis fly (Trichoptera) (Oswego County, New York, July 1955). The prey is being held by the middle legs while the front legs scrape open the nest entrance.

adults, except for one empty puparium of a muscoid fly. Total records are 38 flies from 16 nests.

HYMENOPTERA. Nearly everyone who has worked on this wasp has found ants and pieces of ants to be used. Dead worker ants are apparently picked up from the entrances of ant nests in the vicinity, and alate forms are often picked up in numbers following nuptial flights. I have often found several species of ants in one nest. Dead wasps and pieces thereof are also not uncommonly taken. Hartman reports dried specimens of *Polistes* and a mutillid; Vesey-Fitzgerald reports a tiphiid wasp. I have taken specimens of five genera of digger wasps from various nests, including one very large one, a whole *Prionyx atratus* which had apparently been sucked dry by an asilid fly. I have also taken one whole female mutillid, several Ichneumonidae and Chalcidoidea, two sawfly larvae, and five bees or pieces thereof. Total records for Hymenoptera come to over 150 examples taken from 46 nests (over three fourths of the records for ants).

Thus, in my records, Hymenoptera are most prevalent in numbers of individuals found in nests, followed by Coleoptera, Hemiptera and Diptera in some numbers. However, in some situations other orders, such as Ephemerida, may predominate.

As mentioned above, bits of plant material may occasionally be picked up and dropped again. Of the 61 nests I dug out, 2 contained pieces of plants, both at Andover, Massachusetts. One nest, no. 1956A, contained a rather small larva and only three small insects when it was dug out; in addition there was a piece of a dried, brown leaf which was fresh and clean and resting among the insects. The other, no. 1949, contained a nearly full-grown larva, and the cell was packed full with insects. This wasp was seen bringing in prey (a disabled damselfly I had placed on the sand) at 1030. I dug this nest out at 1130. The damselfly was in place near the top of the pile, but above it was half of a maple key (*Acer* sp.), slightly imperfect and somewhat dessicated but clean and lying loosely on top of the pile. Evidently this had been brought in after the damselfly. Very probably these bits of plants are not eaten by the larva (any more than are the carcasses of weevils, beetle elytra, fly puparia, and so forth). At least none of these items were consumed by larvae that I reared; such larvae invariably consumed the fresher specimens and left most of the harder objects intact.

Before leaving this subject, it seems desirable to list the contents of a single nest in order to further demonstrate the variety of arthropods used. Hartman, Stoehr, and the Raus have also done this, and my example is not greatly dissimilar. The nest selected is my no. 1102 (Oswego County, New York); this nest was dug out on 27 June 1955, following the final closure. Surprisingly, the nest contained a small larva, probably no more than 1 day old. The cell was packed full, and very little had yet been eaten by the larva. The cell contained the following:

1 spider (*Icius harti* Peckham, ♂)
1 adult bark louse (Psocoptera)
2 leafhoppers, one adult, one nymph (Cicadellidae)
1 psyllid (*Psylla striata* Patch)
1 plant bug nymph (Miridae)
1 ground beetle (*Tetragonoderus fasciatus* Hald.)
2 helodid beetles (*Cyphon variabilis* Thunb.)

§A. *Microbembex monodonta* (Say)

4 beetle larvae, one ostomatid, one nitidulid, two unidentifiable
1 small moth pupa
1 dried fly, without legs (*Phormia*)
2 anthomyid flies in poor condition
2 midges (Chironomidae)
1 worker ant (*Lasius*)
1 dealate queen ant (Formicinae)

(8) *Experiments on prey selection.* Parker found that *Microbembex* females would accept various insects which he dissabled and threw on the sand. At Andover, Massachusetts, I placed a great many dead or damaged insects on the sand, and all were accepted readily unless they were much larger than the wasps. Accepted insects included small butterflies, beetles of moderate size (*Macrodactylus*), freshly killed *Microbembex* females, and representatives of several groups not found in nests under natural conditions, such as adult lacewings (Neuroptera, Chrysopidae), and damselflies (Odonata, Coenagrionidae). It seems probable that members of all orders of insects could be shown to be acceptable to these wasps.

The question naturally arises as to what other groups of arthropods these wasps will take. I broke a centipede (*Lithobius forficatus* L.) into several pieces, and the pieces were readily accepted by the wasps; one piece was later recovered from a nest (no. 1934). Millepedes (*Julus*), when broken into pieces, were picked up and carried off; several were dropped again at distances varying from 5 cm to 2 m, although all eventually disappeared. I failed to find millepedes in any of the nests in the vicinity, but of course I did not dig out every one. Freshly killed sowbugs (isopod crustaceans, *Oniscus asellus* L.) were also picked up, and one was found in a nest (no. 1945A). However, small pieces of earthworms (*Lumbricus*) were not picked up, nor were pieces of cooked beef and chicken (although all were eventually carried off by ants). In these experiments, the arthropods were interspersed among the bits of meat and earthworms in a small feeding area. The *Microbembex* would examine the various items by hovering closely or landing beside them, often touching them with their antennae. They readily picked out the arthropods and left the others. However, if bits of plants are occasionally picked up, it seems rather likely that the earthworm fragments and other non-arthropod food may be used under some circumstances.

(9) *The egg and immature stages.* As noted by Parker, the egg is fastened in an erect position in the floor of the empty cell shortly after

it is completed (Fig. 190). I dug out several nests shortly after the female made the initial outer closure and (with one exception) found the egg in place. The egg is 4.0–4.5 mm long, slightly curved, and whitish; it is glued to a small clump of sand grains which, however, do not form a noticeable elevation on the floor of the cell. About half the eggs I recovered were near the center of the cell; the remainder were between the center and the apex of the cell, one of them almost against the wall at the extreme tip of the cell. One exception was provided by nest no. 1933D, at Andover, where the egg was erect but not actually in the cell at all; rather it was at the extreme base of the burrow, 1 or 2 mm before it broadened into the cell.

Parker states that no food is placed in the cell until the egg hatches, but this is not strictly true. No food is normally supplied the day following oviposition, but on the following day several small insects may be supplied and placed very close to the egg. For example, no. 1933B contained a very small male ant pressed close against the egg, which hatched later the same day.

The Raus (1918) claim to have found a nest in which the female had provisioned the cell with several dead insects and not yet laid

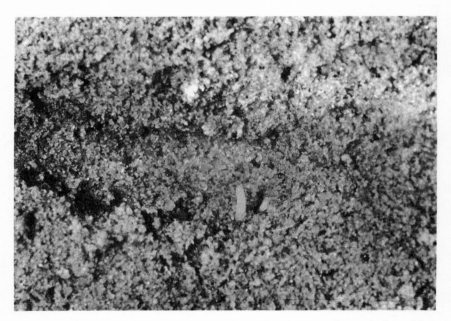

Fig. 190. Nest cell of *Microbembex monodonta*, showing egg in typical position (Andover, Massachusetts, July 1963).

§A. *Microbembex monodonta* (Say)

her egg. Undoubtedly the Raus merely overlooked a small larva, as the many observations of Parker and myself indicate that provisioning is progressive.

Parker carefully reared several larvae from the egg stage and found that from 8 to 9 days were required from oviposition to larval maturity, that is, 6 or 7 days of larval feeding. My data from Kansas agree perfectly with Parker's, from Ohio. However, in Massachusetts several females were found to provision their nests for 8 days, indicating a somewhat slower larval development. Two nests that were marked as being dug on 31 July were dug out 1 week later and found to contain larvae still rather small (8–12 mm in length). A nest marked 31 July while it was being provisioned was still being provisioned 1 week later; the larva in this nest was found to be about 16 mm long; it fed for 3 days more in a rearing tin before spinning up.

As already indicated, only a few small, generally soft-bodied insects are brought in the first day or two, while larger larvae are fed more rapidly and with a great variety of arthropods, often rotted or dessicated. The female spends no time in the nest except when provisioning and during the inspection trip made to the nest each morning. The speed with which the cell is filled up and closed off is very variable, probably depending in part upon the female's success in hunting. No. 1102, the contents of which were listed earlier, contained only a 1 day-old larva when the final closure was made, and several other nests in this locality (Oswego County, New York) were also provisioned very rapidly. None of the many nests I dug out at Andover was filled this rapidly, although much variation was noted. For example, no. 1936A, dug out after final closure, had a larva about 15 mm long (probably 3–4 days old); on the other hand, no. 1950, dug out while still being provisioned, had a larva of essentially full size.

Parker states that larvae are unable to spin in the brood chamber but must advance into the tunnel where the "narrower diameter permits the larva to reach the sand on all sides, or, more rarely, make its way into the sand from the side of the brood chamber." I found larvae spinning up inside the cells in the field, and recovered several cocoons from cells; reared larvae also spun up readily inside artificial cells in sand, without burrowing into the side as described by Parker. The cocoon of this species resembles a small *Bembix* cocoon, but has a single pore in the walls.

(10) *Final closure of the nest.* As mentioned above, final closure may be made when the larva is very small up to the time when it is

nearly fully grown, depending perhaps on the success of the wasp in obtaining dead arthropods to fill the cell. I have observed final closure many times, always in the afternoon (1300–1700). The burrow is filled completely, largely with dry, surface sand. Approximately 30 min are required for completion of the final closure.

When filling the burrow, the wasp is repeatedly seen coming out the nest entrance and up to about 2 cm from it, pivoting about slightly, scraping sand behind her. Each time, when she reaches her maximum distance from the hole, she takes flight briefly, lands in the entrance facing the burrow, then rotates rapidly 180°; she may then back in or come out again scraping, then fly again, rotate, and back in. This is repeated many times, with the result that the burrow is filled at the expense of the area immediately surrounding the entrance. This shallow depression is then filled by a second series of movements. The wasp goes out in more or less of a straight line away from the entrance scraping sand toward it. Most or all of these "straight line" movements are made on the side away from the threshold of the burrow, that is, generally on the upslope side. The number and length of these lines varies greatly. In Kansas I obtained the impression that they were quite numerous (ten or more) and often quite long (up to 25 cm). However, in Massachusetts I observed no final closures involving more than six radiating lines, and none of these exceeded 20 cm in length; three wasps were seen to omit the radiating lines altogether. In this instance the shallow depression was more or less erased by some irregular movements close to the nest entrance, but in at least one instance the depression was still fairly distinct when the wasp flew off. The greater number of radiating lines in the Kansas localities may be correlated with the generally deeper nests there, for the amount of fill needed was of course greater, and the depression around the nest entrance therefore also greater.

(11) *Natural enemies.* Both Hartman and Parker mention that small flies follow wasps carrying prey or perch about the entrance "awaiting an opportunity to dash into the opening behind the wasp as she entered her nest" (Parker). Hartman noted that the *Microbembex* females occasionally dropped their prey and "pounced upon one or the other fly and threw it to the ground," but without injuring it or deterring its activities. Neither author identified the flies or reported finding maggots in the nest cells.

I found *Microbembex monodonta* to be plagued by miltogrammine flies only rather uncommonly. In Stafford County, Kansas, I noted two small flies (not identified) hovering 10–20 cm behind a female

§A. *Microbembex monodonta* (Say)

digging her burrow. At Andover miltogrammines were also noted, but they were not common. One fly was seen hovering behind a female making a final closure, and several were seen following females laden with prey (in each case rather large insects). However, only 1 of the 40 nests dug out contained a maggot, and that one, only one (no. 1955B). This maggot was alone in the cell with a few pieces of insects, apparently having destroyed the *Microbembex* larva. It was not reared to the adult stage, but may have been *Senotainia rufiventris* (Coquillett), the species seen following females on several occasions both here and in a small colony at Bedford, Massachusetts. At Andover, *Bembix spinolae* was attacked more successfully by miltogrammine flies; in fact, two of three nests dug out contained maggots. The reason for the low incidence of parasitism on *Microbembex* may be related to the fact that the prey of this wasp is often so small that it does not project well behind the wasp in flight, thus providing less of a target for larviposition by the flies.

Parker reports the bombyliid fly *Exoprosopa fascipennis* Say ovipositing in the sand at nest entrances of *Microbembex* in Ohio, and he found a pupa of this fly in a cocoon of the wasp. This same fly was observed in Oswego County, New York, and at Andover. In both localities it was seen apparently ovipositing in various open holes in the sand, including open, abandoned sleeping burrows of *Microbembex*, but in no case in true *Microbembex* nests, which are of course closed most of the time. It is possible that these open holes, so characteristic of this wasp, serve an important function in diverting these flies from the true nests. Bombyliid flies are well known to oviposit indiscriminately in open holes. *Exoprosopa fascipennis* is a widely distributed species that probably also attacks *Bembix spinolae*, *B. pruinosa*, and other bembicine wasps.

Parker also took a mutillid wasp, *Dasymutilla* sp., from a *Microbembex* cocoon. At Andover, *D. nigripes* (Fabr.) was seen commonly in the *Microbembex* nesting area, often digging in various places. However, I did not see them digging into known *Microbembex* nests. Mickel (1924) found *D. bioculata* Cresson to be an important parasite in Anoka County, Minnesota. He collected 285 *Microbembex monodonta* cocoons and from them reared 11 *Dasymutilla bioculata* (about 4 percent parasitism). In the same locality, *D. bioculata* also attacked *Bembix pruinosa;* adults reared from the latter host were larger in size, with very little overlapping in the two size categories. The *Dasymutilla* larvae are said to consume the bembicine larvae completely "before they enter the prepupal stage."

Bohart and MacSwain (1940) report the chrysidid wasp *Parnopes fulvicornis fulvicornis* Cameron (= *westcottii* Melander and Brues) as a parasite of *Microbembex aurata* Parker, a species closely related to *monodonta*. At Antioch, California, they collected several hundred cocoons of *aurata* and found them to be about 20 percent parasitized by *P. f. fulvicornis*, the cocoon of the parasite being inside that of the host. They describe the behavior of the female *Parnopes* as follows:

> While searching, the wasps made short flights close to the sand, alighting frequently to run over the surface, tap rapidly with the tips of their antennae, and examine every depression. At frequent intervals the wasps would tunnel their way beneath the loose top sand. The digging efforts ... are interesting because of their ineffectiveness. In contrast to other aculeates these wasps have neither well developed mandibles nor tarsal combs and have restricted use of their front legs. When starting to dig, the body is raised to an almost vertical position from which they often lose their balance and fall over backwards.

Bohart and MacSwain (1940) and Telford (1964) note that *Parnopes f. fulvicornis* has been taken in the same places as *Microbembex monodonta* in the absence of *aurata*. I have taken *P. f. fulvicornis* in numbers in Utah, behaving much as described above in the midst of a large *monodonta* nesting area. Doubtless this parasite attacks several species of *Microbembex*. It is unlikely to attack species of *Bembix*, since it is a small *Parnopes*, and Bohart and MacSwain failed to rear it from *B. comata* or *B. occidentalis*.

Krombein (1958c) described another subspecies, *Parnopes fulvicornis atlanticus*, from Kill Devil Hills, Dare County, North Carolina, with other specimens ranging from Maryland to Florida. Krombein found *P. f. atlanticus* in close association with *Microbembex monodonta* at Kill Devil Hills. He says:

> I watched several *Parnopes* females investigating the *Microbembex* nesting site. Several examined the partially open entrances of some burrows, probably dug by male *Microbembex*, but they did not enter the burrows. Another female *Parnopes* was noted running on the ground and flying low over it in short flights; she dug in the sand several times but did not uncover any burrows. Once I saw a female *Microbembex* chase off a *Parnopes* that was investigating some partially open male (?) burrows.

B. *Microbembex ciliata* (Fabricius)

This is a widely distributed South American species which occurs in much the same situations as *monodonta*. Janvier (1928) has studied the species in several localities in Chile, and Richards (1937) studied it very briefly in British Guiana; both of these authors called the species *sulfurea* (Spinola), a synonym.

§B. *Microbembex ciliata* (Fabricius)

In Chile, Janvier found large colonies in coastal sand dunes and along the banks of rivers; some of the latter were exposed to flooding, and Janvier believed that cocoons might be washed downstream and survive to augment other colonies. He found that the males emerge earlier in the morning than the females and fly about in large numbers just above the surface of the sand, scraping sand here and there and entering burrows likely to contain females. As soon as females begin to appear, about 0900 to 1000, the males attempt to mate with them. Coupling occurs in the air, and the pair fly along and drop to the ground further on. Males and females confined in tubes mated several times in the course of a day. Richards reports that the males spend the night in short burrows in the sand, and it is probable that the females do likewise.

The nests are usually dug into the sides of hillocks in the sand. The burrows are short, but Janvier presents no actual data on their length. He found that each nest contained from four to eight cells that "formed a cluster around the main burrow." He figures a four-celled nest in which the older cells are closer to the entrance than the newest cell. The egg is placed near the apex of the cell, glued erect to the floor of the cell before any prey is brought in; it is said to be placed between several grains of sand. The initial closure of the nest is prolonged, and the wasp returns two or three times "to see if the hole is completely closed. It smooths over the surface of the covering of its nest and departs when no vestige of the burrow remains." The entrance is kept closed while the female is away from the nest. Janvier's figure shows no inner closure of the nest.

Provisioning of the nest is progressive, the first prey being placed close to the newly hatched larva. The first prey is of small size and is said to be completely paralyzed, while the larger prey brought in later may be incompletely paralyzed, especially beetles. The prey consists of arachnids, including a solpugid, *Galeodes variegata* Gerv., and insects of five orders. These include Odonata (damselflies of the genus *Ischnura*), Hemiptera (Pentatomidae, Membracidae), Diptera (Muscidae), Hymenoptera (Formicidae, Sphecidae), and especially Coleoptera (adults and larvae of several families, including even a *Gyrinus*). Apparently stinkbugs (*Podisus chilensis* Spin.) were especially common as prey, occurring in nearly all cells. The prey is said to be captured in flight and to be paralyzed immediately in the air. The prey consists mainly of arthropods common in the vicinity of the nesting sites.

Apparently many of the prey were identified from fragments in the cells, so Janvier had no way of knowing in many cases whether they were paralyzed or merely picked up dead. Richards believes it is probable that the prey are picked up dead, as in *monodonta*.

Janvier presents a brief description of the larva and cocoon of *ciliata*. The larva has long antennal papillae, as in *monodonta*, and the cocoon is said to lack the mammilliform pores characteristic of most bembicine wasps.

C. Ethology of other species of *Microbembex*

The western United States species *M. aurata* Parker was reported by Bohart and MacSwain (1940) to be abundant in the sand dunes of Antioch, California. These authors collected several hundred cocoons and found 20 percent of them to be parasitized by *Parnopes f. fulvicornis* Cameron, as discussed earlier, under *monodonta* (Chapter XII: A, 11). Both host and parasite were found to be common on the flowers of *Croton californicus*.

Willink (1947) found *M. argentina* Brèthes to be abundant in the province of Mendoza, Argentina, digging its nests in loose sand in the slopes of sand dunes. In the province of Buenos Aires, Llano (1959) found *M. uruguayensis* (Holmberg) forming large colonies in sandy ridges. Llano found these wasps to be "very nervous and excitable, going and coming tirelessly and quarreling continuously. The burrows are located very close together, and it is common to see several females working and throwing clouds of sand behind them. At times they walk for a distance, but generally they fly, and anything whatever, a whirlwind, an insect flying past, suffices to bring them up in a swirling cloud."

Llano found the nests of *M. uruguayensis* to descend obliquely some 10 cm, then make a sharp bend and descend another 15 cm to the cell. The egg is laid in the empty cell, and the larva supplied with from 12 to 14 prey. In all cases observed, the prey was found to consist of a single species of small carabid beetle. These beetles were captured on the ground, seized with the middle legs, paralyzed, and carried to the nest in flight.

It appears that this species is not a general scavenger but uses living beetles as prey. Llano's account is brief, and it is important that this point be confirmed by other workers.

D. Summary of the ethology of the species of *Microbembex*

The following summary is based primarily upon *monodonta* and *ciliata*, the only two species to have been studied in detail. Doubt-

§D. Summary of species of *Microbembex*

less some modification will be found necessary as more data on other species become available.

(a) The wasps occur in tracts of open, fine-grained sand and may form very large colonies.

(b) Adults take nectar from flowers in or near the nesting sites, and females of *monodonta* are reported by Hartman to feed also on their prey.

(c) Both sexes spend the night and other periods of inactivity in short burrows in the sand. These "sleeping burrows" often pepper the sand in or near the nesting area; they are not closed during the day.

(d) When digging, the females show strong tilting movements in the nest entrance, throwing out one to several loads of sand while the head is down, the sand landing well away from the nest entrance so that no leveling movements of the mound are required.

(e) The nests are simple, unicellular (*monodonta, uruguayensis*) or multicellular (*ciliata*).

(f) An outer closure of the nest is maintained at all times when the female is not in the nest, but no inner closure is ever made.

(g) Females enter the nest each morning for an inspection of the cell before initiating provisioning for the day. Provisioning is progressive.

(h) The egg is laid erect in the cell shortly after the nest is finished; no prey is brought in until the egg is nearly ready to hatch.

(i) Prey is carried by the middle legs, mainly in flight, but large prey may be carried by a series of short flights alternating with pauses on the sand.

(j) The prey consists of arachnids of 3 orders and of insects of 11 orders. Experimentally, *monodonta* will also use millepedes, centipedes, and terrestrial Crustacea. (However, *uruguayensis* is reported to utilize carabid beetles exclusively.)

(k) In *monodonta* dead or disabled arthropods are collected from the sand, healthy and active arthropods being used rarely, if at all; it is probable but not certain that the sting is used for immobilization of the prey in some cases. In *ciliata*, Janvier reports that the prey is paralyzed, as in other solitary wasps, but this is questioned by Richards. *M. uruguayensis* is also reported to paralyze the prey.

(l) The cocoon is similar to that of *Bembix*, but is smaller and has a single pore (*monodonta*) or none at all (*ciliata*).

Chapter XIII. Fossil History, Distribution, and Comparative Morphology of the Nyssoninae

In this chapter I shall attempt to make a more direct comparison of the structure of the various genera than has been done up to this point, also to consider the distribution of these genera in time and space insofar as this is possible. The contemporary nyssonine fauna of some parts of the world is still poorly understood, only the Nearctic, Palaearctic, and Ethiopian faunas having received anything like a comprehensive treatment in recent years. As in most groups of insects, the fossil record is exceedingly inadequate, the known fossils represented mainly by wings and more or less crushed and distorted bodies. In the case of the Nyssoninae, where many characters are to be found in the structure of the exoskeleton and few characters in the wing venation, the situation would seem to be particularly hopeless. Nevertheless, the fossil record of this group, sketchy though it is, is not without elements of considerable interest. The following account considers only the described fossil Nyssoninae. Possibly there are fossils of this group in certain museums which are as yet undescribed (although I have seen none), and doubtless there are fossils which have yet to be discovered.

A. Fossil record of the Nyssoninae

No fossil wasps, ants, or bees are known from eras prior to the Cenozoic. However, since ants and several groups of wasps are known from the Eocene, approximately sixty million years ago, it is assumed that the Aculeata had their origin somewhat earlier, at least in the Palaeocene and probably in the Cretaceous. The parasitoid Hymenoptera, a large and diverse group presumed to contain the ancestral stock of the Aculeata, date back to the Juras-

§A. Fossil record of the Nyssoninae

sic; the earliest hymenopteron of all, the sawfly genus *Archexyela*, dates from the Triassic.

(1) *Eocene*. Two presumed Nyssoninae have been described from the Eocene. Because of the importance of these records, I have here included the original descriptions in full, with commentary. The presence of this group in the Eocene indicates that these are among the most ancient of wasps, most groups of which are not known prior to the Oligocene (although many major types, including Pompilidae and several subfamilies of Sphecidae, are represented in Baltic amber, of Oligocene age).

Didineis solidescens Scudder, 1890. This is the first presumed nyssonine wasp to have been described. *Didineis* is a genus allied to *Alysson* and was considered briefly in the final pages of Chapter II. Scudder's description is as follows:

> The body of the single specimen known is preserved on a side view but partially dorsal, and though the antennae and legs are destroyed, the wings are tolerably well preserved. There is, however, no sign of any spine on the sides of the metanotum, the thorax here appearing to be well rounded; nor would the abdomen appear to be so closely narrowed at the base as in *Didineis*. The neuration of the wings agrees very closely with that of *Didineis lunicornis* Fabr. sp., except in the very much larger size and subtriangular shape of the marginal cell, the width of which is nearly one-third that of the wing. The middle discoidal cell also is remarkable for its extreme length, being at least three times as long as its basal breadth. The body is not very darkly colored on the stone, being of a rather pale testaceous tint, but the apical half or less of the abdominal segments are paler than the rest.
>
> Length of body, 7 mm.; of wing, 5.25 mm.
>
> Green River, Wyoming. One specimen, nos. 132 and 263 (Dr. A. S. Packard).

This specimen is preserved in the Museum of Comparative Zoology at Harvard University, and it is in better condition than Scudder's description and figure (Pl. 10, Fig. 30 in his *Tertiary Insects of North America*) would suggest. The assignment of this fossil to *Didineis* seems wholly capricious, for it is an altogether more compact, robust wasp, and there are some major venational differences: the marginal cell is very different, as Scudder admits, and the basal vein is long and very strongly oblique, reaching the subcosta very close to the stigma. It is true that the second submarginal cell is petiolate, but this is true not only of *Didineis* but of several genera of wasps of at least four families, including, for example, the vespid genus *Alastor*. The nature of the basal vein certainly suggests a vespid wasp, and the shape of the mesosoma also suggests the Vespidae. In particular, there is a pair of converging lines on the anterior mesosomal dorsum suggesting the posterior margin and the humeral keel of the pronotum. The other visible lines on the mesosoma all appear to represent real structures (for example, the meso-metapleural suture and the epicnemial ridge). The lines on the pronotum are therefore probably not artifacts, and if they are not, this can be nothing but a vespid wasp.

I conclude that this is certainly not a *Didineis* and almost certainly not a nyssonine wasp. I assign it tentatively to the vespid subfamily Eumeninae and suggest a relationship with *Alastor*.

Hoplisus archoryctes Cockerell, 1922. This wasp was described in *Nature* in a short paper titled "An Ancient Wasp." Cockerell wrote as follows:

I have just received from Mr. John P. Byram a small collection of fossil insects which he obtained at the head of Bear Gulch, 12 miles from Una, Colorado. The formation is Green River Eocene, and Mr. Byram states that the material comes from a lower stratum than the insects previously obtained by us. One of the specimens is a beautifully preserved wasp, with wings outspread, belonging to the modern genus Hoplisus. It is 12 mm. long, with a wing-spread of about 19 mm.; the head and thorax are black; abdomen fusiform with narrow base, and the hind margins of the segments broadly pale-banded, as in living species; the legs are colourless, probably yellow originally; the anterior wing shows a pallid stigma, and a strong dusky cloud including the basal part of the marginal cell and the whole of the second submarginal; the venation is essentially that of the modern *Hoplisus quadrifasciatus*, except that the marginal cell is more slender, and in the hind wing the cubitus practically meets the nervulus.

The only fossorial wasp from the Eocene previously described is Scudder's *Didineis solidescens*, which is evidently quite different from the present species, but is too poorly preserved for the accurate determination of the genus. No older wasps are known.

This Eocene Hoplisus, which may be called *Hoplisus archoryctes*, doubtless preyed upon Homoptera, which are so numerous in the same rocks. It is, I think, the most impressive instance of the persistence of type which I have ever seen, when we consider that it belongs to a highly specialized group of insects, and proves that within this group there has, at least in one line, been no change of form or colour in the many millions of years which we now believe to have elapsed since the Eocene. Even the cloud on the wings is as in living species. Could the species be restored to life, *H. archoryctes* would fall into our system, merely forming another species to be added to the many similar ones existing.

Cockerell published a photograph of *Hoplisus archoryctes* in 1924, but his photograph does not adequately show all the remarkable detail which is preserved in this wasp; I have, therefore, included a

§A. Fossil record of the Nyssoninae

better photograph here (Fig. 191). It should be noted that *Hoplisus* is another name for *Gorytes*, the latter name having priority. The ethology of several species of *Gorytes* was discussed in Chapter III.

I have studied the type of Cockerell's *archoryctes*, which is preserved in the United States National Museum. I find myself in essential agreement with Cockerell's conclusions; this wasp is indeed so much like contemporary Gorytini that it could even be (so far as can be judged from what is preserved) a living species. However, since the pleural regions of the mesosoma are not preserved and the venation of the hind wing not distinct, it is impossible to place it exactly in our modern classification. The very small first segment of the metasoma suggests *Harpactostigma* rather than *Gorytes* as that genus is now conceived. The important point is that the Gorytini have apparently undergone no major changes since the Eocene; even the wing color of this fossil is strikingly like that of many living species, as Cockerell pointed out. Cockerell was very probably correct in believing that this was a fossorial species that preyed upon Homoptera.

(2) *Oligocene*. Wasps of many different groups occur in Baltic amber, but no one has attempted to treat them systematically, and most

Fig. 191. Type specimen of *Hoplisus archoryctes* Cockerell (Eocene, Colorado). (Photograph courtesy of the Smithsonian Institution.)

of them remain undescribed. Baltic amber Sphecidae in the collections of the Museum of Comparative Zoology include specimens belonging to, or close to, modern genera such as *Ampulex* and *Dolichurus* (Ampulicinae), *Pemphredon* and *Passaloecus* (Pemphredoninae); there are even specimens assignable to the apparently highly evolved tribe Crabronini. However, the material I have examined contains no Nyssoninae. Brischke (1886) mentions the genus *Gorytes* as occurring in Baltic amber he examined, and that genus (in its broad sense) can be predicted as occurring in the Oligocene. Unfortunately this page must remain a blank for the present.

The only nyssonine wasp described from the Oligocene, so far as I know, is the following species, found in shales at Rott, Germany.

Nysson rottensis Meunier, 1915. I have not seen the type of this species, and the description leaves much to be desired. However, Meunier (1915, Pl. XXII, Fig. 1) provided a very good photograph from which a number of conclusions can be drawn. First of all, one notes a very long basal vein essentially continuous with the median vein and reaching nearly to the stigma; this feature, plus the nature of the branching of the discoidal and transverse median veins and the shape of the marginal, submarginal, and discoidal cells, all place this wasp in the Vespidae. The rather long, slightly thickened antennae, which (together with the triangular second submarginal cell) apparently suggested the genus *Nysson* to Meunier, could equally well belong to a vespid wasp. The spinose processes of the propodeum, so characteristic of *Nysson*, are absent, and the body is much less compact than in that genus.

I conclude that this is certainly not a *Nysson* and almost certainly not a nyssonine wasp. I assign it tentatively to the vespid subfamily Eumeninae. This wasp may be related to Scudder's *Didineis solidescens* from the Eocene, but it differs in having the second submarginal cell triangular, the third, wide above rather than strongly narrowed as in *solidescens*.

(3) *Miocene*. The rich Miocene shales of Florissant, Colorado, contain many aculeate Hymenoptera, and nearly all major groups are represented. Cockerell (1906) reviewed the Hymenoptera known from Florissant, and he and Rohwer later added numerous other species. Several of the Sphecidae were regarded by these authors as belonging to modern genera (*Ammophila, Sceliphron, Passaloecus, Mellinus, Philanthus, Tracheliodes*). Cockerell described two Nyssoninae, both from fairly good specimens which are preserved at the Museum of Comparative Zoology.

Hoplisus sepultus Cockerell, 1906. This wasp is represented by the apical two thirds of the wings, most of the legs, the posterior third

§A. Fossil record of the Nyssoninae

of the mesosoma, and all of the metasoma. The most striking feature, and one immediately suggesting *Gorytes* (= *Hoplisus*), is the banding of the fore wing. As Cockerell points out, this band extends from the base of the marginal cell through the second submarginal and "suffusedly below, still showing brilliant iridescent colors." The fore wing, so far as preserved, is strikingly similar to that of the Eocene *Hoplisus archoryctes*, including the relatively long and slender marginal cell. Cockerell states that the hind wing is reversed and that the cubital vein meets the transverse median vein. This may well be true, although I find the hind wing rather difficult to interpret. The abdomen and legs, so far as preserved, seem similar to those of the genus *Gorytes*, to which genus (in its broad sense) I would unhesitatingly assign this species.

Hoplisidia kohliana Cockerell, 1906. This is a larger wasp than any of the fossils considered up to this point; the body is about 20 mm long, the fore wing about 11 mm long. The entire body is preserved, its ventral side up, only slightly distorted; the legs are in poor condition and mostly missing, but one pair of wings is almost wholly preserved. The head, mesosoma, and even the venter of the metasoma (at least the second segment) bear simple hairs, most of them 0.5–0.8 mm in length. There are many such hairs on the mandibles, and an especially large group on each side of the head, where they may form a psammophore. The mandibles are quite large, but no other mouthparts are visible. The coxae of each pair appear to be contiguous. The abdomen is fusiform and rather slender.

The fore wings might be considered fairly typical of Gorytini except in two details: the basal vein appears to pass upward to meet the subcosta far basad of the stigma, and the basal and transverse median veins are interstitial, the discoidal vein forming a perfectly straight line with the median vein. In the hind wing, the transverse median vein is interstitial with the base of the cubital vein.

All in all this wasp presents a most unusual mixture of characters. It is clearly not a typical gorytine, and one might assume from the large size, hairy body, and junction of the basal vein and subcosta remote from the stigma that this was a primitive bembicine wasp. The fact that the proboscis, ocelli, tibial spurs, and mesosomal pleura are not preserved makes it difficult to come to a decision on this. Several features lead me to believe that Cockerell may have been incorrect in assigning *Hoplisidia* to the Sphecidae (the pronotal lobes and other diagnostic features of this family are not preserved). The nature of the branching of the basal, discoidal, and transverse median veins of the fore wing is suggestive of the family Tiphiidae, and the shape of the head, especially the very large, hairy mandibles, also suggest this family. I have compared *Hoplisidia kohliana* with males of the tiphiid genus *Anthobosca,* and am impressed with the similarity in body form, hairiness, and venation of the fore wing. In *Anthobosca,* however, the cubitus of the hind wing arises well before the transverse median vein. There is much variation in wing venation in *Anthobosca,* even in one sex of one species, and the two sexes are often strikingly different in some details (see Turner, 1912). In some males the second recurrent vein curves back to meet the second submarginal cell, as is true in *Hoplisidia* (and all Nyssoninae). Cockerell described a *Geotiphia foxiana* from these same strata, and Turner has placed this genus in the synonymy of *Anthobosca. Geotiphia foxiana* is a smaller wasp than *Hoplisidia kohliana* and differs in several details; tentatively, I would consider them generically distinct.

I conclude that this is not a gorytine wasp, and if it is a nyssonine wasp it is well off the main line of evolution, although having some superficial resemblances to the Bembicini. I consider it much more probable that it is a tiphiid wasp not unrelated to *Anthobosca.*

(4) *Discussion.* Unfortunately this concludes the inventory of known or supposed fossil Nyssoninae, none being known from epochs later than the Miocene. Only two fossils can be assigned with assurance to this subfamily, the Eocene *archoryctes* Cockerell and the Miocene *sepultus* Cockerell, the two being apparently very similar and both assignable to the genus *Gorytes* in a broad sense. *Gorytes* is also reported to occur in Oligocene amber.

One notes with interest the absence from the fossil record of any Alyssonini, Nyssonini, Stizini, and Bembicini. In the modern fauna, Gorytini are generally uncommon and inconspicuous insects as compared to Bembicini, and since the latter are relatively large and often nest gregariously along watercourses, one would expect them to fossilize readily. One must be cautious in using negative evidence of this nature, but it is tempting to conclude that until quite recent geologic time (that is, after the Miocene), the subfamily Nyssoninae was represented principally by fairly typical *Gorytes*-like wasps, which may have made up a considerably larger percentage of the wasp fauna than they do today.

B. Distribution of the genera of Nyssoninae

In this discussion, as elsewhere in this book, I omit the genus *Mellinus* Fabricius, considered to form the subfamily Mellininae,

§B. Distribution of Nyssoninae

and the genus *Bothynostethus* Kohl, considered to belong to the Larrinae. Also omitted is the poorly known genus *Xenosphex* Williams (1954), described as possibly a gorytine wasp, but more probably a primitive member of the Larrinae.

Most of the distributional data are summarized in Table 40. Doubtless this table contains errors in the Gorytini, since many of the species of the tropics and the Southern Hemisphere have not been studied from the point of view of determining their place in the modern classification. For the sake of reducing the length of this table, I have omitted a number of small genera of Gorytini which are restricted to one zoogeographic region. These include *Handlirschia* (Ethiopian), *Psammaecius* (Palaearctic), *Megistommum* (Neotropical), and the following Nearctic genera: *Hapalomellinus, Trichogorytes, Arigorytes,* and *Psammaletes.* The Nyssonini I have

TABLE 40. DISTRIBUTION OF THE GENERA OF NYSSONINAE

Genus	Neotropical	Nearctic	Palaearctic	Ethiopian	Oriental	Australian
ALYSSONINI						
Alysson			X	X		
Didineis	X	X	X			
Entomosericus			X			
GORYTINI						
Clitemnestra	X					X
Ochleroptera	X	X				
Olgia			X			
Argogorytes	X	X	X		X	X
Exeirus						X
Lestiphorus		X	X			
Harpactostigma	X	X				
Gorytes	X	X	X	X	X	
Hoplisoides	X	X	X	X	X	
Dienoplus		X	X	X		
Ammatomus		X	X	X	X	
Sphecius	X	X	X	X	X	X
Kohlia			X	X		
NYSSONINI						
Nysson (s. lat)	X	X	X	X	X	X

Genus	Neotropical	Nearctic	Palaearctic	Ethiopian	Oriental	Australian
STIZINI						
Stizus		X	X	X	X	
Stizoides		X	X	X	X	
Bembecinus	X	X	X	X	X	X
BEMBICINI						
Bicyrtes	X	X				
Stictiella		X				
Microstictia		X				
Glenostictia		X				
Xerostictia		X				
Steniolia	X	X				
Stictia	X	X				
Trichostictia	X					
Zyzzyx	X					
Rubrica	X					
Selman	X					
Hemidula	X					
Editha	X					
Carlobembix	X					
Bembix	X	X	X	X	X	X
Microbembex	X	X				

treated as a single genus, *Nysson* in the broad sense, since no purpose would be served in treating in detail this relatively homogeneous and obviously sidewise-specialized complex (see Pate, 1938, for a summary of currently recognized genera and subgenera and their distribution).

Distributional data such as those presented in Table 40 are relatively meaningless by themselves, but they form one of several sets of data that must be considered in an attempt to reconstruct the evolution of the group. A genus of worldwide or nearly worldwide distribution may be a very old genus, in which case one expects considerable structural diversity and a tendency to form subgenera or distinctive species-groups in various parts of the world (possibly *Argogorytes* and *Ammatomus* are examples of this). On the other hand, a cosmopolitan genus may be a recently evolved, successful genus with strong dispersal powers, in which case one expects structural advances but relatively little diversity within the genus (*Bembix* and *Bembecinus* may qualify here). A genus of very limited distribution may represent a relict of a previously widespread group (*Harpactostigma* and *Olgia*, perhaps), or it may represent a relatively modern group which has not had an opportunity to disperse into other regions (*Stictiella*, *Zyzzyx*). In each case only a careful weighing of all available evidence can help one in deciding which alternative is correct.

Disjunct distributions are almost always indicative of old and formerly widespread groups. *Clitemnestra* has a marsupial-type distribution (Australia and Chile), to which the related and perhaps slightly more advanced genus *Ochleroptera* adds an element suggestive of the American opossums (South America and southern North America). Beaumont (1953) considers *Olgia* related to these two genera more than to other Gorytini. *Olgia* is a small genus confined to the southern Palaearctic region, and may represent a relict element of this same general complex of presumably formerly widespread distribution.

Among other Gorytini, *Argogorytes* and *Ammatomus* both have a somewhat spotty distribution, and each falls into at least two quite distinctive subgenera. On the other hand, *Gorytes* and *Hoplisoides* have a predominantly Holarctic distribution and show little structural diversity; these are presumably relatively modern elements in the Gorytini.

The Alyssonini are of Holarctic distribution, the one Neotropical species of *Didineis* having been described from Cuba where it is presumably a recent arrival from Florida. While *Alysson* and *Didineis* have a number of species (exhibiting no great structural diversity) in both Europe and North America, *Entomosericus* is a small group restricted to the southern Palaearctic region.

The Stizini have unquestionably undergone most of their evolution in the Old World (probably chiefly in Africa); both *Stizus* and *Stizoides* have only a very few species in (mainly western) North America, and none in South America. *Bembecinus*, however, is cosmopolitan.

Even more clearly, the Bembicini have undergone most of their evolution in the New World. Only the genus *Bembix* itself occurs in the Old World (but the apparently most primitive species occur in the deserts of North America). One whole series of genera centers in the Neotropics (*Stictia* also having a single species in southern United States). Another series of genera, all having recessed ocelli and several other features in common, centers in the southwestern part of the Nearctic region (one of these genera, *Steniolia*, has a species in Central America, with one record from Ecuador). Two other genera, *Bicyrtes* and *Microbembex*, are widespread in North and South America.

In summary, distributional data appear in agreement with the fossil record in suggesting that the Gorytini are an old group: they have had time for the development of widespread, spotty, and discontinuous distribution patterns, while the genera of Alyssonini, Stizini, and Bembicini all exhibit continuous and generally more limited distribution patterns. The Nyssonini are widespread, like their hosts the Gorytini, but the group is a much smaller one, and most of the genera and subgenera appear to show more or less continuous distributions. There is still much to be learned about the generic classification of the Nyssonini and Gorytini, and some changes are inevitable. The classification and known distribution of the genera of the other three tribes appears reasonably well settled.

C. Comparative morphology of the Nyssoninae

A complete comparative morphology of the Nyssoninae would take a volume by itself, as such a study should include a point-by-point description of every significant detail of structure, including internal features such as the digestive tract, musculature of flight and of retraction of the proboscis, ovaries, chromosomes, and so on. Ideally, one should include the detailed structure of the corpora pedunculata of the brain, major nerve tracts, and other features of the nervous system. Such a study might be enormously revealing, but it would require the efforts of a team of workers

TABLE 41A. THIRTY CHARACTERS OF ADULT ALYSSONINI, GORYTINI, AND NYSSONINI *

Genus / Character	1. Ocelli	2. Antennae	3. Labrum	4. Mandibles	5. Proboscis	6. Max. palpi	7. Labial palpi	8. Mesosoma	9. Mesosc. laminae	10. Metanotum	11. Propodeum	12. Epicnem. ridge	13. Oblique groove	14. Scrobal groove	15. Metasternum	16. Tibial spurs	17. Hind femora	18. Pecten	19. Basal vein	20. Discoidal vein	21. Pterostigma	22. 2nd SM cell	23. Recurrent veins	24. TMV hind wing	25. Pygidium ♀	26. Sternite II ♂	27. Tergite VII ♂	28. Sternite VII ♂	29. Sternite VIII ♂	30. Volsellae ♂	TOTAL
ALYSSONINI																															
Alysson	1	1	1	1	1	1	1	1	1	1	1	1	1	3	1	2	3	1	1	1	1	3	1	2	1	1	1	2	2	1	40
Didineis	1	2	1	1	1	1	1	1	1	1	1	1	1	3	1	2	3	1	1	2	1	3	1	2	1	1	1	1	2	1	40
Entomosericus	1	2	1	1	1	1	1	2	1	1	1	1	1	2	1	3	3	1	1	3	1	1	3	2	1	1	1	1	1	1	42
GORYTINI																															
Clitemnestra	1	1	1	1	1	1	1	2	1	1	1	1	1	1	1	1	1	1	1	1	1	1	1	1	1	1	1	3	3	1	33
Ochleroptera	2	1	1	1	1	1	1	2	1	1	1	1	1	2	1	1	1	1	1	1	1	1	1	1	1	1	1	3	3	1	35
Olgia	2	1	3	1	1	1	1	2	1	1	1	2	2	2	1	2	1	1	1	2	1	1	1	2	1	1	1	3	1	1	41
Argogorytes	1	2	1	1	1	1	1	2	1	1	1	1	1	1	1	1	1	2	1	2	3	3	2	1	1	1	1	3	3	1	39
Exeirus	1	3	1	1	1	1	1	2	3	1	1	1	3	1	1	1	1	1	1	1	3	3	3	1	2	1	1	2	1	1	46
Dienoplus	1	1	1	1	1	1	1	2	3	1	1	1	3	2	1	1	1	3	1	3	3	3	3	2	2	1	1	3	3	1	44
Harpactostigma	1	2	1	1	1	1	1	2	3	1	1	2	3	2	1	1	1	3	1	3	2	1	3	2	2	1	1	3	2	1	48
Gorytes	1	2	1	1	1	1	1	3	3	1	1	1	2	2	1	1	1	3	1	3	3	1	3	2	1	1	1	3	3	1	47
Hoplisoides	1	1	1	1	1	1	1	3	3	1	1	1	3	2	1	1	1	3	1	3	2	3	3	2	1	1	1	3	3	1	48
Ammatomus	3	1	3	1	1	1	1	3	3	1	1	1-3	3	1	1	1	1	2	1	3	3	1	3	3	1	1	1	3	2	1	51-53
Sphecius	3	2	1	1	1	1	1	3	3	1	1	1	3	2	1	1	1	3	1	3	3	1	3	3	1	1	1	3	3	1	53
Kohlia	2	3	2	1	2	1	1	3	3	1	2	2	2	2	1	1	1	3	1	2	3	1	3	3	1	2	1	3	2	1	54
NYSSONINI	1-2	1	3	1	1	1	1	3	3	1	3	1-3	3	3	3	1-3	1-3	1	1	3	2	3	1-3 1-2	1-3	1	1-3	3	3	1	0	50-62

* In this table, and also in tables 41B and 42, a value of 1 for a given character means that in that genus the character is primitive (i.e., most like other wasps), while a value of 3 means that the character is specialized. A value of 2 is awarded to intermediate states. In some genera more than one condition prevails, as indicated by figures such as 1-2 or 1-3. For further explanation see text.

§C. Comparative morphology

TABLE 41B. THIRTY CHARACTERS OF ADULT STIZINI AND BEMBICINI

Genus	1. Ocelli	2. Antennae	3. Labrum	4. Mandibles	5. Proboscis	6. Max. palpi	7. Labial palpi	8. Mesosoma	9. Mesosc. laminae	10. Metanotum	11. Propodeum	12. Epicnem. ridge	13. Oblique groove	14. Scrobal groove	15. Metasternum	16. Tibial spurs	17. Hind femora	18. Pecten	19. Basal vein	20. Discoidal vein	21. Pterostigma	22. 2nd SM cell	23. Recurrent veins	24. TMV hind wing	25. Pygidium ♀	26. Sternite II ♂	27. Tergite VII ♂	28. Sternite VII ♂	29. Sternite VIII ♂	30. Volsellae ♂	TOTAL
STIZINI																															
Stizus	1	3	2	2	1	1	1	3	3	3	1	3	3	1	1	1	1	3	3	3	3	1	3	3	2-3	1	3	1	3	1-2	60-62
Stizoides	1	3	2	3	1	1	1	3	3	3	1	3	3	1	1	1-3	1	3	3	3	3	1-2	3	3	2-3	1	3	1	3	2	63-67
Bembecinus	1	2-3	2	1	1	1	1	3	3	3	3	3	3	2-3	1	1	1	3	3	3	3	2-3	3	2	3	1	3	1	3	2	64-67
BEMBICINI																															
Bicyrtes	3	3	3	1	2	1	3	3	3	3	3	3	1	1	1	3	1	3	3	3	3	1	3	1	1-2	2	1	1	3	2	66-67
Stictiella	2	3	3	1	2	1	3	3	3	3	1	3	3	1	1	3	1	3	3	3	3	1	3	1	3	1-3	3	1	3	3	66-68
Microstictia	2	3	3	1	2	1	1	3	3	3	1	3	3	1	1	3	1	3	3	3	3	1	3	1	3	3	3	1	3	3	68
Glenostictia	2	3	3	1	2	1	1	3	3	3	1	3	3	1	1	3	1	3	3	3	3	1	3	1-2	3	3	3	1	3	3	68-69
Xerostictia	2	3	3	1	3	2	1	3	3	3	1	3	3	1	1	3	1	3	3	3	3	1	3	2	3	3	3	1	3	3	73
Steniolia	2	3	3	1	3	3	3	3	3	3	1	3	3	1	1	3	1	3	3	3	3	1	3	2	3	3	3	1	3	3	74
Stictia	3	3	3	1	2	1	1	3	3	3	1	3	3	1	1	3	1	3	3	3	3	1	3	2	3	2	3	1	3	2	66
Trichostictia	2	3	3	1	2	1	1	3	3	3	1	3	3	1	1	3	1	3	3	3	3	1	3	1	3	1	3	1	3	2	63
Zyzzyx	2	3	3	1	2	2	2	3	3	3	1	3	3	1	1	3	1	3	3	3	3	1	3	1	3	2	3	1	3	2	68
Rubrica	3	3	3	1	2	1	1	3	3	3	1	3	3	1	1	3	1	3	3	3	3	1	3	1	3	3	3	1	3	3	68
Selman	2	3	3	1	2	1	1	3	3	3	1	3	3	1	1	3	1	3	3	3	3	1	3	2	3	1	3	1	3	3	69
Hemidula	2	3	3	3	2	1	1	3	3	3	1	3	3	1	1	3	1	3	3	3	3	1	3	1	3	2	3	1	3	3	68
Editha	2	3	3	1	2	1	1	3	3	3	1	3	3	1	1	3	1	3	3	3	3	1	3	1	3	1	3	1	3	2	67
Carlobembix	2	3	3	3	2	1	1	3	3	3	1	3	3	1	1	3	1	3	3	3	3	1	3	1	3	2	3	1	3	2	67
Bembix	2-3	3	3	1	3	2	3	3	3	3	1	3	3	1	1	3	1	3	3	3	3	1	3	1-2	2-3	1	3	1	3	3	70-72
Microbembex	3	3	3	3	3	2-3	3	3	3	3	1	3	3	1	1	3	1	3	3	3	3	1	3	1	3	3	3	1	3	2	72-73

TABLE 42. TEN CHARACTERS OF LARVAL NYSSONINAE

Character / Genus	31. Body setae	32. Spiracles	33. Spir. opening	34. Front	35. Antennae	36. Mandibles	37. Mand. tooth	38. Labrum	39. Epipharynx	40. Labium	TOTAL	COMBINED TOTALS TABLES 41 + 42
ALYSSONINI												
Alysson	1	1	3	3	3	3	1	1	1	1	18	58
GORYTINI												
Ochleroptera	1	1	3	1	1	3	3	1	1	1	16	51
Gorytes	1	1	1	1	1	3	1	1	1	1	12	59
Hoplisoides	1	1	1	1	1	3	1	1	1	1	12	60
Sphecius	1	1	1	1	1	3	1	1	1	1	12	65
NYSSONINI												
Nysson	1	1	1	2	1	3	1	1	1	1	13	63-75
STIZINI												
Stizus	2	1	1	1	1	2	2	1	1	1	13	73-75
Bembecinus	1	1	1	1	1	2	1	1	1	1	11	75-78
BEMBICINI												
Bicyrtes	1	1	1	1	2	3	3	3	3	3	21	87-88
Stictiella	1	1	1	1	2	1-2	1	3	1	3	15-16	81-84
Glenostictia	1	1	1	1	1	1-2	1	3	1	3	14-15	82-84
Steniolia	1	1	1	1	1	1	1	3	3	3	16	90
Stictia	3	3	1	1	1	1-2	1	3	3	3	20-21	86-87
Bembix	2-3	3	1	1	1	1-2	1	3	3	3	19-21	89-93
Microbembex	3	1	1	1	3	3	3	3	3	3	24	96-97

with no limitations as to time and money. In the present work, which is aimed primarily at an elucidation of the major features of the evolution of behavior, I can hope to do no more than to sketch in lightly a few of the more obvious structural features and the changes they appear to have undergone in this group. My treatment of these characters will often seem arbitrary and superficial, but I believe it will suffice for the purposes at hand. Much of the information is presented in tabular form (Tables 41-43).

§C. Comparative morphology

However, some discussion in the text is necessary, especially since a number of genera of unknown ethology have not been mentioned up to this point and may form important parts of the total picture.

I have made an effort to evaluate each character in terms of whether it is primitive (that is, most like the condition in other families of wasps and in other subfamilies of Sphecidae, especially groups by common consent regarded as primitive) or derived (that

TABLE 43. COMPARISON OF SOME CHARACTERS OF THE GENERA OF BEMBICINI*

Character / Genus	3. Labrum length/max. width	5. Proboscis length/eye height	6. No. segments max. palpi	7. No. segments labial palpi	41. Vertex depressed	42. Ocelli in pits	43. Shortening of wings in ♀	44. Kink in 1st tr. cub. vein	45. Bulla in 1st tr. cub. vein	46. Middle coxae spinose ♂	47. Middle femora notched ♂	48. Middle tarsi modified ♂	49. Processes sternite VI ♂	50. No. prongs sternite VIII ♂	30. Length cuspis/digitus ♂
Bicyrtes	1.2	0.8	6	4	±	-	-	-	-	-	-	-	-	3	0.6
Stictiella	1.3	1.0	6	4	±	+	+	-	+	-	-	+	-	3	0.8
Microstictia	1.2	1.0	6	4	±	+	+	-	+	-	+/-	-	-	4	0.8
Glenostictia	1.2	1.0	6	4	±	+	+	-	+	-	+/-	-	-	4	0.8
Xerostictia	1.9	1.4	4	2	+	+	+	-	+	-	-	+	-	4	1.0
Steniolia	1.2	1.6	3	1-2	+	+	+	-	+	-	-	-	-	4	1.0
Stictia	1.1	0.8	6	4	+	-	-	+	-	-	+	-	±	1	0.7
Trichostictia	1.4	1.1	6	4	+	-	-	+	-	+	-	-	-	1	0.7
Zyzzyx	1.4	1.5	5	3	+	-	-	+	-	+	+	-	-	1	0.5
Rubrica	1.2	1.0	6	4	+	-	-	+	-	+	+	-	-	1	0.8
Selman	1.3	0.9	6	4	±	-	+	±	-	-	+	-	-	1	0.8
Hemidula	1.2	0.8	6	4	±	-	-	±	-	-	-	-	-	3	0.8
Editha	1.2	1.0	6	4	-	-	-	±	-	+	+	-	-	1	0.8
Carlobembix	1.3	1.0	6	4	±	-	-	±	-	+	+	-	-	4	0.6
Bembix	1.5	1.0	4	2	+	-	+/-	+/-	-	-	-	+/-	+/-	1	1.1
Microbembex	1.5	1.3	3-4	1-2	+	-	-	-	+	-	-	-	-	1-3	0.5

* Character numbers are the same as in Table 41, except for the addition of 41-50. Proboscis length (Character 5) is measured on the maxilla, from the base of the palpi. A + means that a given character condition is present, a - that it is absent. A ± means that a character condition is present in some measure but not fully developed, while a +/- means that it is present in some members of the genus and absent in others.

is, unique and exhibiting progressive modification within the Nyssoninae, or paralleling a similar situation in other more advanced groups of wasps). In the tables, the value of 1 is recorded for the primitive condition and the value of 3 for the derived condition; the value of 2 is reserved for intermediate conditions. Of course, I have often had to be arbitrary in deciding whether, for example, in *Sphecius* the proboscis is long enough to be given a value of 1 or 2. Also, I do not pretend to have examined more than a few species of each genus, so some overgeneralization is inevitable. All I hope to do by this relatively crude procedure is (1) to arrive at a preliminary evaluation of the degree of evolutionary advance of each genus with respect to external structure and (2) to point out characters shared in common and therefore possibly indicating common ancestry (for example, the dentate hind femora of the Alyssonini and some Nyssonini, as indicated in Table 41A, col. 17, with a value of 3).

Admittedly there are situations in which one cannot be certain as to which is the primitive condition. A structure may be acquired and then lost, or lost and reacquired (although the ideas epitomized in "Dollo's law" lead us to doubt that a complex structure can be reacquired in its original form). I have discussed some aspects of this problem elsewhere, with respect to larval characters (Evans, 1959a). Not all such difficulties have been eliminated from the tables. I suggest, for example, that even though primitive wasps tend to be relatively elongate and with strong constrictions between some body sections (rather than robust and compact like a *Bembix*), the simple 1-2-3 sequence implied in Tables 41A and B, col. 8, may not be completely correct. The fossil Gorytini have a moderately compact body, and it is possible that the Nyssoninae come from an ancestor of similar form, with both the robust *Bembix*-type and the very elongate *Alysson*-type being derivatives of this. Also, I do not feel completely certain that the ancestral nyssonines had a scrobal groove. Such a groove is present in most Sphecidae, including groups commonly regarded as generalized, but it is not well formed in the Pompilidae, Scoliidae, and some other groups. There are several other characters in which the primitive and derived conditions are somewhat ambiguous. I can only hope that if I have erred, I have erred as often one way as the other. In the vast majority of cases the derived condition is perfectly obvious, so the final tabulation should at least provide a rough indication of evolutionary advance, which is all that is intended. In any event the tables provide a concise summary of much data and permit a ready comparison of the gross features of the various genera.

§D. Important structural features 407

Lists of the 30 adult characters employed in Tables 41A and B and the 10 larval characters employed in Table 42 appear on pages 408–409. Most of the adult characters and their alternatives have been illustrated, and references are given to figures showing the condition implied. The larval characters are not figured here, but all will be found figured by Evans and Lin (1956b) and by Evans (1964a). A complication is introduced by the fact that the larvae of only 15 genera are sufficiently well known for inclusion in the table. Fortunately these are well scattered through the subfamily, and larval characters seem sufficiently constant in each tribe so that one can perhaps assume that no radical departures occur among all or most of the unstudied genera. Thirty-six genera are included in the table of adult characters. Omitted are the several genera of Gorytini also omitted from the table of distributional data (see Chapter XIII:B), as well as some other poorly known forms. The various segregates of the old genus *Nysson* are not analyzed. I have no reason to believe that any of the groups omitted are of special phylogenetic importance; so far as I know all are quite closely related to genera included in the table. Further discussion of various generic complexes follows.

D. Further discussion of important structural features

The following notes pertain especially to genera of unknown or poorly known ethology that were not discussed in earlier chapters, and to certain characters not covered adequately in the tables.

(1) *Entomosericus* Dahlbom. This is a small Palaearctic genus of unknown ethology. *E. kaufmanni* Radoszkowsky, the only species I have studied, is of much the same size and color as the species of *Didineis*, although somewhat hairier. This genus is commonly assigned to the Alyssonini, although the second submarginal cell of the fore wing is four sided and receives the two recurrent veins near its middle (Fig. 192). Sternite VIII of the male is a simple, tongue-shaped structure (Fig. 200), but the genitalia are most unusual, both the digitus and the cuspis being very short, the parameres large and strongly biramous. All in all, this genus provides an interesting mixture of *Alysson*-like, *Gorytes*-like, and rather specialized features. Presumably it is an offshoot of the common ancestor of the Alyssonini and Gorytini.

(2) *Clitemnestra* Spinola. The observations of Janvier on the nesting behavior of two Chilean members of this genus were reviewed

FORTY MAJOR STRUCTURAL CHARACTERS OF NYSSONINAE

Structure	Primitive condition	Derived condition
	Adults	
1. Ocelli	Unmodified (Fig. 1)	Greatly reduced (Fig. 140)
2. Antennae	Inserted low, the clypeus transverse (Fig. 9)	Inserted near middle of face (Fig. 1)
3. Labrum	Small, mostly concealed by clypeus (Fig. 9)	Large, wholly exserted (Fig. 140)
4. Mandibles	Dentate (Fig. 1)	Simple (Fig. 46)
5. Proboscis	Short (Fig. 9)	Very long (Fig. 110)
6. Maxillary palpi	With 6 segments (Fig. 1)	With 3 segments (Fig. 110)
7. Labial palpi	With 4 segments (Fig. 1)	With 1 or 2 segments (Fig. 177)
8. Mesosoma	Elongate, constricted between pro- and mesonota (Fig. 11)	Robust, compact, sclerites smoothly confluent (Fig. 3)
9. Mesoscutal laminae	Narrow, strongly defined, not broken by a carina	Broader, less sharply defined, broken by a carina (Fig. 2)
10. Metanotum	Complete (Fig. 2)	Incomplete on sides (Fig. 73)
11. Propodeum	Simple (Fig. 2)	Produced posterolaterally (Fig. 51)
12. Epicnemial ridge	Present (Fig. 3)	Absent (Fig. 97)
13. Oblique groove	Present (Fig. 40)	Absent (Fig. 3)
14. Scrobal groove	Present (Fig. 3)	Absent (Fig. 11)
15. Metasternum	Simple	Forming a broad plate underlying the hind coxae
16. Mid-tibial spurs	Two (Fig. 53)	One
17. Hind femora	Simple	With an apical dentiform process (Fig. 14)
18. Pecten of fore tarsi ♀	Weak or absent (Fig. 15)	Strongly developed (Fig. 181)
19. Basal vein of fore wing	Reaches subcosta near stigma (Fig. 4)	Reaches subcosta far basad of stigma (Fig. 179)
20. Discoidal vein of fore wing	Continuous with median vein (Fig. 10)	Arises on transverse median vein (Fig. 4)

FORTY MAJOR STRUCTURAL CHARACTERS (*cont.*)

Structure	Primitive condition	Derived condition
21. Pterostigma	Large, rounded below (Fig. 10)	Small, slender (Fig. 4)
22. Second submarginal cell	Four sided (Fig. 4)	Petiolate (Fig. 10)
23. Recurrent veins	Received near corners of 2nd submarginal (Fig. 10)	Received near middle of 2nd submarginal (Fig. 4)
24. Transverse median vein of hind wing	Well basad of origin of cubital vein (Fig. 195)	Well beyond origin of cubital vein (Fig. 193)
25. Pygidium ♀	Defined by carinae	Not defined
26. Processes on sternite II of male metasoma	Absent	Present
27. Tergite VII ♂	Simple (Fig. 5)	With lateral spiracular lobes (Fig. 100)
28. Sternite VII ♂	More or less fully exposed and sclerotized	Covered by sternite VI, weakly sclerotized
29. Sternite VIII ♂	Relatively broad, simple (Fig. 200)	Forming a slender apical pseudosting (Fig. 6)
30. Volsellae ♂	Cuspis very short (Fig. 12)	Cuspis long and slender (Fig. 112)
Larvae		
31. Body setae	Absent or virtually so	Present and fairly conspicuous
32. Spiracles	All circular or nearly so	Elliptical (higher than wide)
33. Spiracular opening	Armed with a circlet of spines	Unarmed
34. Front	Simple	With mammilliform processes
35. Antennal papillae	Short, slightly longer than thick	Very long and slender
36. Mandibles	With 4 teeth	Only 2 teeth
37. Inner mandibular tooth	Simple	Broad and truncate
38. Margin of labrum	Not bristly	With strongly protruding bristles
39. Epipharynx	More or less wholly spinulose	With a bare transverse streak
40. Oral surface of labium	Papillose	Spinulose

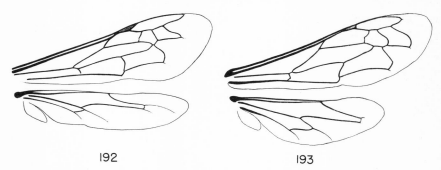

Figs. 192 and 193. Wings: 192, of *Entomosericus kaufmanni* Radoszkowski; 193, of *Kohlia cephalotes* Handlirsch.

briefly in Chapter III:L. Further consideration of the structure of the genus seems appropriate here, since the distribution pattern (Chile and Australia) suggests that this may be an ancient group, and as a matter of fact the genus has the lowest score of any in Table 41. Indeed, the only important departure from the supposed prototype of the Nyssoninae is the reduction of the last two sternites of the male, which are weakly sclerotized and entirely internal. The wing venation is especially generalized, more so even than the Eocene *Hoplisus archoryctes*. The more important structural features are shown in Figs. 194–199.

Ochleroptera differs only slightly from *Clitemnestra*; the male terminalia of these two genera are almost identical. To the minor differences indicated in Table 41 should be added the fact that in *Ochleroptera* the first metasomal segment is small, forming a tapering stalk for the remainder of the metasoma. The small Palaearctic genus *Olgia* Radoszkowsky has a very similar wing venation and enough other characters in common with the above two genera that Beaumont (1953) believes that the three form a single complex.

In these three genera, as in *Alysson* and *Didineis*, the recurrent veins of the fore wing are received at the lower outer corners of the second submarginal cell, and may even be received on the first and third submarginal cells, a single species sometimes being variable in this regard. Apparently the insertion of the recurrent veins, so characteristic of the nyssonine wasps, remains primitive and somewhat unfixed in these genera. One also notes a commonly recurring color pattern in these same genera: the presence of round, whitish spots on the sides of the second metasomal tergite (and sometimes additional tergites). These stand in contrast to the apical tergal bands characteristic of nearly all other Gorytini.

§D. Important structural features

(3) *Remarks on more "typical" Gorytini.* Argogorytes and Exeirus are interesting in having simple mesoscutal laminae as in the genera considered above; the features of the mesopleura of these genera also suggest *Clitemnestra*. However, sternite VIII of the male *Argogorytes* forms a pseudosting, as in *Gorytes* and *Sphecius*, although that of *Exeirus* is less modified. *Harpactostigma* is superficially similar to *Gorytes*, but the oblique groove of the mesopleura has been lost, suggesting *Ammatomus* and *Sphecius*, and sternite VIII of the male is more generalized (Fig. 201), being in fact somewhat intermediate between that of *Ammatomus* and that of *Gorytes*. *Gorytes* itself is interesting in that the oblique and scrobal grooves of the mesopleura

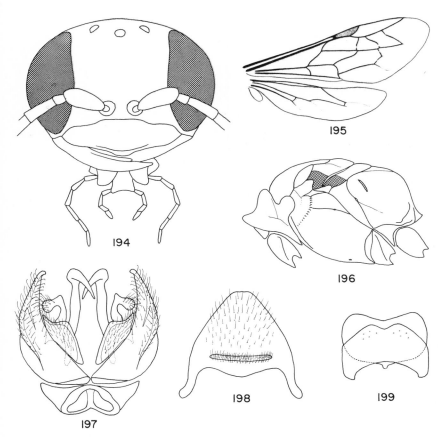

Figs. 194–199. *Clitemnestra gayi* Spinola: 194, head of female, mouthparts extended; 195, wings; 196, mesosoma of female, lateral aspect; 197, male genitalia; 198, sternite VI of male metasoma; 199, sternites VII and VIII of male metasoma.

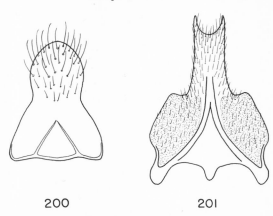

Figs. 200 and 201. Sternite VIII of male metasoma: 200, of *Entomosericus kaufmanni* Radoszkowski; 201, of *Harpactostigma gracilis* (Patton).

form a single horizontal depression, simulating the condition in *Alysson* and *Didineis* (Fig. 23). A precoxal ridge is present in *Gorytes, Hoplisoides,* and *Dienoplus,* while *Hoplisoides* and *Argogorytes* are alike in having the epicnemial ridges connected across the venter. A detailed consideration of the more typical Gorytini would require inclusion of several genera here omitted.

(4) *Kohlia* Handlirsch. This genus is of unusual interest. Although it shares a number of features with *Sphecius* and *Ammatomus,* there are some striking bembicine-like characters: the labrum is large and strongly hinged on the clypeus, the proboscis is somewhat prolonged, and the ocelli are slightly distorted. The tip of the marginal cell of the fore wing is pulled away from the wing margin (Fig. 193), a character found elsewhere only in the genus *Microbembex.* As Beaumont (1954) notes, in *Kohlia coxalis* Morice the parameres of the male genitalia are deeply biramous, as in *Entomosericus* and certain species of *Sphecius.* This genus is indeed a puzzling mixture of characters, and it is unfortunate that its ethology remains unknown.

Stizobembex Gussakovskij (1952) is apparently a synonym of *Kohlia.* In describing *Stizobembex pavlovskii* from a male collected near Mikoyanabad, Tadzhikistan, U.S.S.R. (directly north of Afghanistan), Gussakovskij failed to compare his supposed new genus with *Kohlia,* known up to that time from a species from North Africa and another species from South Africa. Gussakovskij described *Stizobembex*

§D. Important structural features   413

as a "missing link"; its ancestry to *Bembix* "does not seem to allow of any doubt—one can observe in it the beginning of the long proboscis . . . and also some traces of the beginning of change in the ocelli; on the other hand, some primitive features . . . lead directly to *Sphecius*, bypassing the tribe Stizini" (translation from the Russian).

(5) *Nyssonini.* The members of this complex have a characteristic facies, but the group as a whole exhibits a fairly broad spectrum of characters. I found no cuspides in the genitalia of *Nysson lateralis*, and Snodgrass (1941) found none in *Zanysson texanus;* for this reason I have placed a zero in column 30 of Table 41. Characters held in common with *Alysson* include the processes on the front of the larva, several details of wing venation, and the dentate hind femora, the last feature occurring in some but not all Nyssonini.

(6) *Stizini.* The three genera of this tribe are very similar, and as compared with the Gorytini exhibit several important structural advances. Both the Stizini and the Bembicini have sternite VII of the male exposed and sclerotized, a puzzling feature since this sternite is covered by sternite VI in the otherwise more primitive Gorytini and Nyssonini (the Alyssonini are variable in this regard). The larvae of the Stizini are remarkably generalized if I have properly evaluated the characters considered.

(7) *Bembicini.* The 16 genera that make up this tribe exhibit much less structural diversity than do the somewhat larger number of genera making up the Gorytini: all Bembicini are of robust form, the mesopleurum is without ridges or grooves except for the arching scrobal groove, the wing venation is subject to little variation, and so on. For this reason it has seemed desirable to tabulate some of the generic differences in greater detail and to add several additional characters for consideration (Table 43). In this table I have made no attempt to work out the full range of variation for the ratios used, but have merely measured and averaged a few typical species of each genus. As elsewhere, I am concerned with clues to relationships and degree of evolutionary advance rather than taxonomic discrimination of the genera. Further details will be found in the papers of Parker (1917, 1929), Lohrmann (1948), Willink (1947, 1958), and Gillaspy (1963b, 1964).

Even the additional table provided cannot do full justice to all details of structure, such as the modifications of the antennae of the

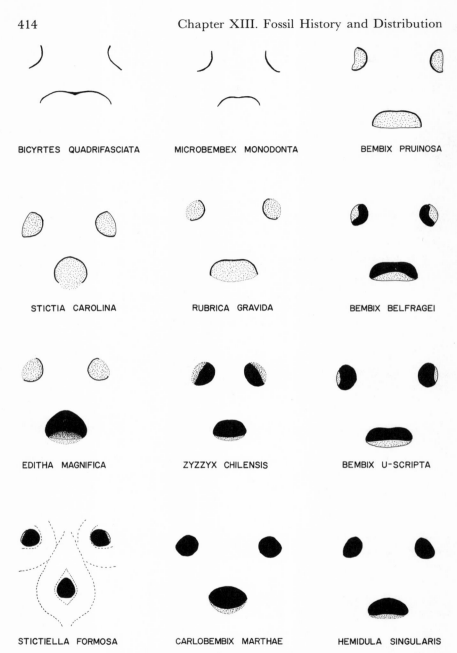

Fig. 202. Ocelli of selected examples of Bembicini. Clear, translucent lenses are here colored black, while smooth surfaces, actually blackish in color but without setigerous punctures (presumably lenses suffused with pigment), are stippled; the white background represents normal, setigerous integument.

male, the protuberant clypeus of *Microbembex* and some other groups, the pyriform labrum of *Steniolia,* and so on. Ocellar structure is not treated in Table 43, but is reviewed in Fig. 202. The ocellar lenses of all Bembicini, where preserved, are nearly or quite flat (except for the slightly convex and nearly circular lenses of *Carlobembex*). In some cases lenses are present but partially suffused with dark pigment, and in *Stictia* the anterior ocellus is wholly covered with a smooth, dark film. It is difficult to judge to what extent the ocelli are still functional in the various genera.

E. Major trends in structural modification

The data summarized in this chapter obviously form a complex and sometimes confusing picture. A more complete fossil record would doubtless solve some of the many riddles posed. For example, how does one explain the presence of an exposed sternite VII in the metasoma of male Stizini and Bembicini when this sternite is reduced and internal in the generally more primitive Gorytini? Even further, how does one explain the fact that both sternites VII and VIII are reduced and internal in what appear to be the most primitive genera of all, *Clitemnestra* and *Ochleroptera?* How does it happen that *Microbembex,* in many ways the most highly evolved genus of the subfamily, has tergite VII of the male simple and without lateral lobes, in contrast to nearly all other Bembicini and Stizini, some of which must have been near the ancestral stock of *Microbembex?* Is the presence of a petiolate third submarginal cell in *Alysson* and the Nyssonini to be considered significant, when the same condition occurs in *Exeirus* and *Bembecinus,* which are surely unrelated to *Alysson* and to each other? If the Gorytini, Stizini, and Bembicini form a linear series, as often supposed, how does one account for the presence in the gorytine genus *Kohlia* of certain features shared with the Bembicini but not the Stizini?

A quick scanning of the tables and of Fig. 203 makes it clear that the same structures have sometimes been lost in otherwise quite dissimilar genera. Consider, for example, the list of genera having mandibles that have lost the tooth on the inner margin: *Olgia, Ammatomus, Nysson, Stizoides, Hemidula, Carlobembix,* and *Microbembex.* Or the genera that exhibit loss of a mid-tibial spur: *Entomosericus,* some Nyssonini, some *Stizoides,* and all the Bembicini. Or the genera of Gorytini in which the female pygidium is reduced; or the genera of Bembicini in which the palpal segmentation is reduced. Also, it seems evident that certain structures may

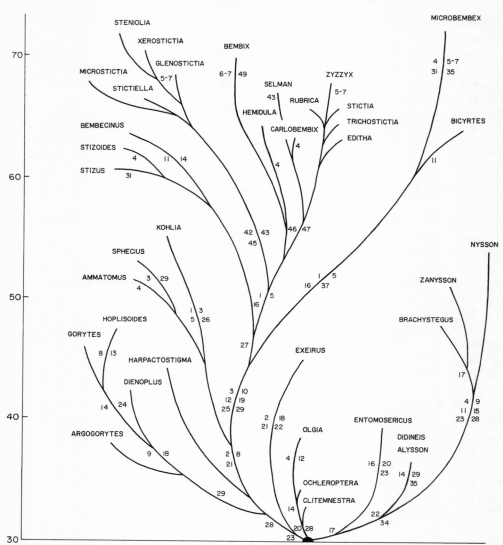

Fig. 203. A tentative phylogenetic arrangement of the major genera of Nyssoninae, based on morphological evidence. The numbers on the lines of ascent refer to features characteristic of that line (as opposed to a divergent line); see Tables 41–43 for characters indicated by the numbers. Numbers on the lefthand margin (30–70) indicate the level of advance as deduced from adult characters (extreme right-hand column in Table 41).

§E. Trends in structural modification

have been enlarged or acquired in different genera independently: for example, the bifid parameres of the male genitalia of *Entomosericus* and some *Kohlia* and *Sphecius;* the produced propodeal angles in *Bembecinus* and *Bicyrtes;* the elongated proboscis of such dissimilar genera as *Zyzzyx* and *Steniolia*. In these and several other cases, if we assume that these similarities are *not* the result of parallel evolution, we must at the same time devaluate the many other characters that suggest that the genera in question are not closely related. Convergences and parallelisms, which are probably no more abundant in the Nyssoninae than in many other large groups of animals, can only be detected by considering the total spectrum of characters held in common and the nature of the structures involved (including whether or not these structures appear to be involved in convergences and parallelisms in other groups). The same sort of evidence must be used in deciding whether a lost structure is likely to have been reacquired (for instance, a tooth on the mandibles, an exposed and sclerotized sternite VII). In no case will our answers to particular questions be in the realm of facts, but in the realm of hypotheses, some of them highly probable, some of them quite uncertain. The attempt to reconstruct the evolution of the group is a master hypothesis which attempts to include all the lesser ones, some of which invariably contradict one another, and some of which are of far greater probability than others.

Be that as it may—and the subject will come up again in the final chapter—there is no question as to the direction of evolution of many characters of the Nyssoninae. Both the fossil record and geographic data point to the Gorytini as including the most primitive Nyssoninae, and our analysis of structure points not only to the Gorytini but to certain specific attributes of this tribe, best exemplified by genera such as *Clitemnestra, Ochleroptera,* and *Argogorytes*. At the other end of the spectrum are the various features characteristic of the Bembicini, particularly as seen in such genera as *Bembix, Microbembex,* and *Steniolia*. These trends may be summarized briefly as follows.

(1) There is a fairly regular increase in body size, from small wasps to among the largest wasps known (with occasional reversals, as in *Bembecinus* and *Microbembex*).

(2) From generally black wasps, often with whitish spots on the second metasomal tergite, have evolved wasps with extensive maculations of white or yellow, involving markings on the head and mesosoma and apical or subapical tergal bands on the metasoma.

(3) There is a tendency for the development of a strong vestiture of pale pubescence and erect hairs, possibly forming reflective surfaces and thereby reducing heat absorption.

(4) There is a tendency for the eyes to become enlarged at the expense of the front, and for the vertex to become depressed in the higher forms. The ocelli become distorted and reduced in size, then gradually covered with dark pigment, finally reduced to mere slits (Fig. 202).

(5) The clypeus becomes more elongate and located between the bottoms of the eyes, the antennae then being inserted higher, near the middle of the front. The clypeus may be protuberant, the base of the labrum swollen, the two forming beneath them a chamber into which the proboscis can be withdrawn.

(6) The labrum becomes fully exserted and enlarged, hinged to the margin of the clypeus. The proboscis becomes gradually more elongate, the galeae, glossae, and paraglossae all becoming very long and slender, but the palpi tending to become shorter and with fewer segments. The proboscis develops a hinge mechanism so that it can be withdrawn at least partially beneath the clypeus and labrum.

(7) The mesosoma becomes more robust and compact, without strong constrictions between the sclerites. The pronotum and propodeum become relatively short, the mesothorax large, the metanotum reduced dorso-laterally. On the mesopleura, the epicnemial ridge and oblique groove are lost, and the scrobal groove comes to form a smooth arc rather than being angulate as it is when the oblique groove is present (Fig. 196).

(8) The pecten of the forelegs becomes increasingly well developed, even in the males of many Bembicini. There is a tendency toward the loss of the smaller midtibial spur and for the development of various spines and concavities on the middle coxae, femora, or tibiae of the males.

(9) In the fore wing, the transverse median, basal, and discoidal veins are at first nearly straight and form a simple trifurcation; later their intersection becomes more complex. The basal vein tends to reach subcosta at a point increasingly far from the stigma, and the latter tends to become smaller and more slender. The recurrent veins tend to become more convergent above, meeting the cubitus near the middle of the second submarginal cell.

(10) In the hind wing, the submedian cell is at first short, the transverse median vein erect and meeting the median vein far basad

§E. Trends in structural modification

of the origin of cubitus; the submedian cell tends to become more elongate, the transverse median vein longer and more arcuate and situated farther toward the wing margin.

(11) The metasoma tends to become more robust; the pygidium of the female tends to lose its lateral carinae; the abdomen of the male developes various ventral and apical processes apparently associated with copulation. Tergite VII of the male developes large lateral lobes bearing the spiracles, these being reflexed ventrally, and may also develop lateral apical processes.

(12) In most Gorytini, Alyssonini, and Nyssonini there are six visible sternites in the male plus a variably developed sternite VIII, sternite VII being largely or wholly covered by sternite VI (but in *Clitemnestra, Ochleroptera,* and *Olgia* sternite VI covers both VII and VIII fully); thus tergite VII is opposed to sternite VI and VIII externally. In the Stizini and Bembicini sternite VII is broadly exposed and forms the apical sternite, opposed to tergite VII.

(13) Sternite VIII of the male is at first a relatively broad and short plate, but it becomes modified in various lines to form one, two, or three apical prongs, and in one line a fourth, ventral prong is added (Fig. 104).

(14) The male genitalia are at first relatively broad and short, in higher forms much more elongate. The digitus and cuspis, in particular, come to form slender processes, the cuspis tending to become relatively longer with respect to the digitus, eventually exceeding it slightly.

These 14 sets of characters—and probably others that are less obvious—appear to have played important roles in the phylogeny of the Nyssoninae. Parallelisms have occurred in the case of nearly every character involved, suggesting that strong selection pressures have been operating to mold *Bembix*-type wasps from the protean Gorytini. In addition, each genus has certain characters of its own, some of them often shared with other genera, these characters giving the genus some of its characteristic facies but bearing little or no obvious relationship to major trends in the subfamily (for example, the precoxal ridges of *Gorytes* and a few related genera, the clavate antennae of *Ammatomus,* the bispinose propodeum of *Nysson,* and others). This general impression of the Nyssoninae is not dissimilar to that which one obtains from many other large groups of animals. Indeed, these wasps seem to demonstrate unusually well a statement of Simpson (1959): "In the origin and evolution of many higher categories there is an interplay of more general adaptive improvements, often affecting many lineages in parallel, and of more specific adaptations of subgroups to particular ecological zones and niches."

For the moment we shall merely assume that the various trends in the Nyssoninae represent adaptive improvements. This is a matter that can be discussed more fruitfully following a summation of what is known of the ethology of the group.

Chapter XIV. Comparative Ethology of the Nyssoninae

In this chapter I shall attempt to summarize the data on ecology and behavior presented in Chapters II through XII. One can foresee that this summary will be far from satisfactory: there are too many gaps in our knowledge, too many uncertainties, too many sources of error. While it is possible to bring together objective data on the external structure of all genera of Nyssoninae, information on behavior is entirely wanting for some genera and distressingly incomplete for others. However, those similarities and differences in behavior that are now apparent need to be defined and juxtaposed. After direct comparison of various elements in the behavior patterns, it should be possible to trace the modifications that have occurred and to correlate these with structural changes and possible selection pressures. In this way gaps, uncertainties, and contradictions should be made to stand out, and future research can be designed to correct them.

This chapter follows much the same outline as the accounts of individual species, although broken down into a greater number of topics. Ecology is treated in an initial section dealing with habitat and physical factors, and in a final section dealing with natural enemies. Needless to say, behavior cannot be discussed apart from its environmental backdrop. As Nielsen (1958) has pointed out, autecology is essentially inseparable from ethology. Such a separation, like that between ethology and psychology, tells us something about the investigator but nothing about the animals under study.

A. Ecology and general features of behavior

(1) *Habitat.* Sand wasps are all closely associated with the soil, typically with soil that has some sand content, vegetation at least

sparse enough to afford them space in which to dig, and a substrate reasonably free of roots. There has been no known radiation into other types of habitats: no Nyssoninae nest in pre-existing cavities of any sort;[1] none of them burrow in soft wood, utilize mud or plant materials in nesting, or nest elsewhere than in the ground. In this respect the Nyssoninae stand in contrast to most other large groups of wasps, where such radiation has occurred (for example, Pompilidae, Vespidae, and several other subfamilies of Sphecidae such as the Sphecinae, Pemphredoninae, and Larrinae). In some groups of wasps the switch from hypogeic to epigeic nests has occurred several times independently. The Nyssoninae and Philanthinae represent the only large groups of wasps that to my knowledge have remained wholly tied to the soil. This is an important preliminary consideration, for it helps explain the lack of diversification in body form and gross behavior which is apparent in groups such as the Larrinae and Crabroninae. In the Nyssoninae, the trend has evidently been toward the development of larger and more effective diggers.

In Chapter II, I characterized *Alysson* and *Didineis* as hygophiles, citing numerous records of these wasps occurring in damp and shady situations. The Raus state that they never found *A. melleus* nesting "elsewhere than in a cool, damp bank of mud or sandy clay, near to a body of water." I have often found this species active on cloudy days, and the greatest concentration of nests I found was in almost continuous shade in damp sand beneath a willow tree. Some of the other species of *Alysson* may nest amid vegetation. I have taken *Didineis texana* in marshes, and Strandtmann found this species with prey along the edge of a turnip field. The European *D. lunicornis* is said to nest in compact, argillaceous soil, especially in hoofprints.

The Chilean *Clitemnestra gayi* is said by Janvier to nest in clay banks in forested areas, the nest entrances often being more or less concealed among mosses and lichens. Apparently *C. chilensis* occurs in more open country, although also nesting in argillaceous soil (Chapter III:L).

Most other Gorytini appear to prefer open tracts of sandy soil and to be active during periods of bright sunshine. However, in genera such as *Gorytes* and *Hoplisoides* these tracts may be quite small and may be surrounded by woodland. Species of *Gorytes* may nest along watercourses, even in situations that are sometimes

---

[1]DeGaulle's record (1908) of the Palaearctic *Sphecius nigricornis* Dufour nesting in stems of brambles has been treated by all subsequent authors as probably erroneous.

§A. Ecology, features of behavior

flooded, as may some Stizini (*Bembecinus*) and some Bembicini (*Stictia signata* and others).

More commonly, gorytine wasps are found in open situations where the soil is well drained, its water content low to moderate. Most species tend to nest in flat or slightly sloping soil, but *Ochleroptera bipunctata* and *Ammatomus moneduloides* nest principally in vertical sandbanks (Chapter III:N, P). The long burrows of *Sphecius speciosus* are dug in various types of fairly compact sandy earth, even in cinders or gravel fill, sometimes among the grass of lawns (Chapter V:B, 1).

The Bembicini as a whole show considerable radiation into various types of soil. However, many species are relatively unspecialized and capable of utilizing small to large tracts of soil of quite variable sand and moisture content (for example, several species of *Bicyrtes* and *Bembix*). Certain Bembicini are restricted to relatively pure, fine-grained sand (*Microbembex monodonta, Bembix pruinosa*, several species of *Stictia*); others tend to nest in relatively hard-packed soil (*Rubrica surinamensis, Bembix nubilipennis*); some species of *Bembix* appear to be restricted to saline soil or other unusual niches (see Evans, 1957b:197–199). The genera with recessed ocelli, in particular, seem adapted for nesting in exceedingly dry soil. Wasps of the genus *Steniolia* I characterized as nesting "not in pure sand, but in various types of powdery, arenaceous earth, often with a high content of loam, manure, pulverized rock, or alluvial pebbles" (Chapter VIII:I, 1).

Nielsen (1945) summarized data on nesting sites of various Bembicini, especially species of the Eastern Hemisphere. He also discussed at some length the relationships of *Bembix rostrata* to physical factors in the environment (see also my discussion under *Microbembex monodonta* in Chapter XII:A, 1).

The type of situation utilized for nesting appears to be of primary importance in determining the distribution of these wasps. So far as is known, all Nyssoninae tend to form localized nesting aggregations which persist year after year unless there is a change in physical conditions. Several authors have described fluctuations in populations or movements of colonies as a result of excessive moisture, floods, or (most commonly) an increasing growth of plants over a formerly bare area. The tendency of wasps to nest in the place where they emerged may be genetically determined or it may be the result of locality imprinting or of a certain amount of trial-and-error searching, the old nesting site being the most suitable area readily available; the fact that some colonies do, in fact, move more or less *in toto* suggests that the latter may be important (see

Nielsen, 1945:56; Evans, 1957b:198). On the other hand, I found a large colony of *Stictia signata* nesting in dense grass, and assumed that this colony was present in such an unusual habitat because it had failed to move when the area was developed as a grass-covered landing strip for planes (Chapter IX:C, 1).

One sometimes observes that late in the nesting season females radiate out from the center of the colony and nest in the periphery or at some distance away; I have noted this, for example, in two different colonies of *Stictia carolina* (Chapter IX:B, 3). On the whole, however, sand wasps are exceedingly localized in their distribution. I have found that marked females of species as diverse as *Hoplisoides nebulosus* and *Bembix sayi* may prepare a series of successive nests in very close proximity.

Sand wasps do, of course, leave the nesting site when taking nectar, hunting for prey, and sometimes for sleeping. The distance they fly is partially dependent upon the availability of prey and suitable vegetation in the vicinity. Smaller Nyssoninae (*Alysson, Bembecinus, Microbembex*) are seldom encountered more than a few meters from their nesting sites, while the larger forms (*Stictia, Sphecius*) may range 1 km or more from their nests.

It is unusual to find a mixed colony containing more than one species of a genus, but it is common to find members of several genera nesting together. This is a result of the fact that within each genus the species tend to be allopatric or to have slightly different ecological requirements; members of different genera will usually differ in prey preferences but not necessarily in choice of nesting site (the competitive exclusion principle). I reported a more or less mixed colony of *Bembix troglodytes, B. u-scripta,* and *B. spinolae similans* in New Mexico (Chapter XI:B), but in a situation where the populations of all three were low (and the hunting behavior not identical). Much more common are mixed nesting aggregations of, for example: *Hoplisoides nebulosus* and *Bicyrtes ventralis; Bembix pruinosa* and *Microbembex monodonta; Stictia signata* and *Rubrica surinamensis; Bicyrtes quadrifasciata, Bembix sayi,* and *Stictiella serrata*. In many cases wasps of other groups (Pompilidae, Philanthinae, Larrinae) also nest in the same area, their nests interspersed with those of Nyssoninae (see also Llano, 1959, for associates of Argentinian sand wasps). The localization of many wasps in restricted areas of suitable soil is important from another point of view: most of their parasites are not host specific and are therefore able to maintain high populations in these areas in spite of population fluctuations in some of their hosts. This may place a premium on behavioral adaptations for reducing parasitism.

§A. Ecology, features of behavior   425

(2) *Size and density of colonies.* Most reports of very large, dense colonies pertain to Bembicini. I have never found a greater concentration of any wasp than the colony of *Bembix cinerea* I studied in southern Florida in 1955 (Evans, 1957b:78–94). Since primitive wasps such as Tiphiidae and Pompilidae are highly solitary, one might anticipate a trend within the Nyssoninae from strictly solitary to highly gregarious nesters. Such a trend is not really discernible. I reported finding a nesting site of *Alysson melleus* containing 300 nests in an area measuring only 0.75 × 1.5 m (Chapter II:A, 3). With respect to *Clitemnestra chilensis,* Janvier (1928) reported a colony of "about a hundred individuals, and more during some summers. The nests are very close to one another, so that it is difficult to ascertain which cells correspond to each nest."

Typical Gorytini such as *Gorytes canaliculatus* and *Hoplisoides nebulosus* commonly occur in small nesting aggregations of 1 to 12 or so females. However, I found a colony of 40–50 females of *H. tricolor,* and Powell and Chemsak found a colony of 60–80 females of *H. adornatus* (Chapter III:B, C, E, H). In the case of *Sphecius speciosus,* various workers have reported nesting aggregations varying in size from 1 to nearly 1,000 individuals. In this species there is evidence of great fluctuations in colony size from year to year (Chapter V:B, 1). It is probably safe to say that in all Gorytini, Alyssonini, and Stizini the year-to-year and colony-to-colony variation in population size is considerable, largely masking any differences that might occur with respect to innate tendencies for females to nest in close proximity. That species such as *Hoplisoides nebulosus* have not been found to form large colonies may reflect in part a tendency for such species to nest in small, circumscribed areas rather than the broad expanses occupied by *Sphecius* and most bembicine wasps. It is possible that such species are genuinely less gregarious and that selection has favored small, impermanent colonies close to sources of prey and in areas with low concentrations of miltogrammine flies and other natural enemies.

In any event, it is an overgeneralization to say that the Bembicini are "more gregarious" than Gorytini and Alyssonini. As I have noted in earlier chapters, the species of *Bicyrtes* typically occur in small aggregations, and not more than three nests of any one species of *Stictiella* have as yet been found in any one locality. Even some species of *Bembix,* most particularly *sayi,* form extremely diffuse colonies, the nest entrances generally separated by 1 m or more. On the other hand, many bembicine wasps are known to form colonies of many hundreds of individuals, often with the nest entrances in the center of the colony only a few centimeters apart. In large colonies of *Bembix amoena* and *B. occidentalis* one sometimes finds two or more nests diverging from a common entrance. Large and populous colonies are also reported in the genera *Microbembex, Rubrica, Stictia,* and *Glenostictia*. It appears that in *Steniolia* the females may nest in scattered, localized sites even though forming large common sleeping clusters at night.

Unfortunately this entire subject, though of much intrinsic interest, rests at present mainly on collections of subjective impressions. Actual year-to-year counts of population size of colonies of sand wasps and their parasites are very much needed. Efforts should be made to correlate fluctuations in colony size with the abundance of parasites and with physical factors such as rainfall, winter temperatures, and the amount of sunlight during the previous year's nesting season. At the northern periphery of the range of some species (for instance, *Bembix pruinosa*), it seems to me that the last two factors may be especially important in controlling the size of colonies—but again that is only a subjective impression. Lin (1964) was able to keep accurate yearly records of population size in *Sphecius speciosus* by counting emergence holes. He found population size positively correlated with rainfall during the breeding season; apparently parasites were not an important factor in the colonies with which he worked.

(3) *Aggression among females.* Although one often speaks of "colonies" of these wasps, these colonies are of course mere aggregations of solitary wasps and not comparable to the colonies of social wasps, bees, and ants. In most Nyssoninae one notes essentially no reactions between females nesting together. Now and then, if two females happen to be working outside nests that are only a few centimeters apart, one may fly toward the other and drive it away temporarily. However, even in populous colonies of some large species, such as *Stictia carolina,* one rarely observes such behavior. Lin (1936b) has observed only a few instances of "density dependent fighting" among female *Sphecius speciosus.*

An exception is provided by certain species of *Bembix,* in which one does note considerable aggression among the females. There is no evidence that females defend a definite territory around their nests, but they do frequently dash after other wasps near their nests. The females of several species also steal prey from one another frequently. As I reported for *B. nubilipennis,* females do not bite or sting one another, but merely pounce upon one another and try to obtain a secure grip on the prey and fly away with it (Chapter X:C, 1). Sometimes a fly will change hands several times

§A. Ecology, features of behavior 427

before finally being carried into a nest. Other relatively "aggressive species" are *B. texana* and *B. cinerea*. I have seen *texana* females try to take flies away from the much larger species *Stictia carolina*, but without success. These species of *Bembix* all form dense colonies, but some other species, which also nest in dense colonies, seem to indulge in prey stealing only infrequently (for example, *B. amoena*, Chapter X:A, 2). Nielsen (1945) discusses variation in aggressiveness of various colonies of the European *B. rostrata*. Tsuneki (1956) regards the Japanese species *B. niponica* as truly "subsocial," since he observed no "hostility" among members of any of the colonies he studied; however, Baba observed prey stealing in one colony of this same species.

Microbembex monodonta is also reported to exhibit much fighting among females working at nests in close proximity. In this species females not only steal prey from one another but will take it from other digger wasps and even from worker ants (Chapter XII:A, 6).

One might have expected that in these gregariously nesting wasps there would have been evolution toward less intraspecific strife, but as a matter of fact it is the higher Bembicini in which most aggression has been reported.

(4) *Feeding*. Adult sand wasps require a certain amount of water and food and probably feed at least once a day except during periods of prolonged inactivity. Sources of food include honeydew secreted by aphids and scale insects, nectar from flowers or extrafloral nectaries, sap exuding from wounds in trees, and the body fluids of captured insects.

Certain Nyssoninae are only rarely encountered on flowers, particularly the species of *Alysson, Ochleroptera, Hoplisoides,* and *Bembecinus*. Members of the first two genera are often abundant at honeydew, and it is probable that many of the smaller Nyssoninae make use of honeydew. In the genus *Gorytes, canaliculatus* has only rarely been taken on flowers, but *atricornis* is sometimes common there; in these cases the flowers visited are chiefly Umbelliferae, which have very shallow corollas and are utilized by many short-tongued wasps. The cicada killers, both *Sphecius* and *Exeirus*, are reported as feeding at exuding plant sap.

Sphecius, Stizus, and *Stizoides* have sufficiently long proboscises to utilize a wide variety of flowers as sources of nectar; *Stizoides unicinctus* is recorded from some 25 genera of 13 families (Chapter VI:F). All genera of Bembicini appear to utilize many different plants, including, commonly, various legumes and composites. There are many published records of various higher Nyssoninae visiting flowers, and it would serve no purpose to list these here. Nielsen (1945, Table I) presented records for several Stizini and *Bembix*, chiefly Palaearctic, and I have listed others (1957b, and in the present work). Proximity to nesting sites and abundance of nectar appear to be more important than the taxonomic affinities of the plants.

The case of *Steniolia* deserves special mention. These wasps have the longest tongues of any Nyssoninae, and seem especially adapted to utilization of Compositae as sources of nectar; nearly two thirds of all records for the genus pertain to this family of plants (Chapter VIII:I, 1). Several species of this genus are strongly deserticolous, and it is probable that under the conditions in which this genus evolved there was severe competition with bees and other wasps for the limited numbers of flowers, many of which were Compositae. Gillaspy (1964) believes that proboscis elongation in *Steniolia* "probably originated in xeric habitats in response primarily to critical moisture requirements although competition for carbohydrate energy may have been a secondary influence." In a more limited sense this may have been true of other genera of Bembicini, all of which are in some measure xerophiles and all of which exhibit some lengthening of the proboscis.

It is a curious fact that although there are relatively few observations of adult Nyssoninae feeding directly upon their prey, these records include a wide diversity of species. I reported a female of the gorytine wasp *Hoplisoides nebulosus* landing on a blade of grass and hanging there by one leg while inserting her proboscis into the cervical membrane of her treehopper prey and appearing to suck its blood (Chapter III:B, 3). Williams (1914) also reported *Dienoplus gyponae* malaxating its leafhopper prey after stinging (that is, chewing it and feeding on the exuding blood), then carrying it to the nest.

These observations are very similar to those of Hartman (1905) on *Microbembex monodonta*, a female of which was seen hanging from a weed malaxating a syrphid fly. Janvier (1928) described and figured *Bembix brullei* feeding upon a fly by inserting her mouthparts in the neck region of the fly; in this case the wasp was in a vertical position, standing on a twig with her hind legs and holding the fly with her front and middle legs. Much the same behavior was described for *Zyzzyx chilensis,* Janvier remarking upon the long bladelike maxillae of this species, which are said to be introduced into the viscera of the prey. Both *Zyzzyx* and *Bembix brullei* are reported to reject the prey after the body fluids have been sucked out. The males of *B. brullei* are said to exhibit this behavior as

§A. Ecology, features of behavior 429

well as the females. Ferton (1897, 1902) observed the European species *B. rostrata* and *B. oculata* feeding upon flies and then rejecting them. He believed that this occurred chiefly during periods of scarcity of flowers. One of the most interesting records is that of Howard, Dyar, and Knab (1912), who reported *Stictia signata* capturing mosquitoes and devouring them on the wing (Chapter IX:C).

Tsuneki (1956:154) on two occasions saw female *Bembix niponica* land on the soil with their fly, push the fly forward, then insert the proboscis through its cervical membrane. Apparently these wasps were not feeding on the blood of the fly but were taking nectar from the foregut. "The action of drinking honey was three or four times repeated," Tsuneki says of one case, "then she pulled out the organs from the wound and recaptured the fly as usual under her body and went flying to her nest." His studies of behavior inside the nest of this species showed that such behavior occurs rather frequently inside the burrow. In the present study (Chapter XI:D), I reported seeing several female *B. pruinosa* stop on the sand for 10–20 sec and use their mouthparts to probe the oral opening of the fly. It appeared that they were imbibing fluids from the foregut of the flies. They then proceeded to their nest with the flies in the usual manner. Similar behavior was reported for *B. spinolae* (Chapter X:F, 2).

Apparently many different Gorytini and Bembicini feed upon their prey under certain conditions, but there is no evidence that any species does this with regularity, with the possible exception of the two Chilean species studied by Janvier. Malaxation of prey occurs widely among wasps. "Honey-robbing" (that is, extraction of the crop contents) is reported especially from wasps that prey upon bees (*Philanthus* and some *Cerceris*). The limited records for the Nyssoninae suggest that some species merely malaxate the prey or drain its crop contents on the way to the nest, while some of the larger species actually capture insects that are consumed directly (*Stictia, Zyzzyx*). Much remains to be learned about the frequency of this behavior and the factors that release it.

(5) *Sleeping.* All wasps and many other insects spend the night and periods of inclement weather in a state of torpor, conveniently termed "sleep," without implying a close similarity to the sleep of higher vertebrates. Many sleeping Hymenoptera assume characteristic, stereotyped poses, some of which have been discussed and figured by Rau and Rau (1916), Rau (1938), Evans and Linsley (1960), Linsley (1962), and others. Sleep is apparently induced by lowered temperatures and reduced light.

The vast majority of wasps, probably including most Nyssoninae, sleep on vegetation not far removed from the nesting sites. Unfortunately almost nothing is known regarding the sleeping situations and postures of most Gorytini, Alyssonini, and Nyssonini. There is evidence that some wasps of the genera *Alysson, Gorytes,* and *Hoplisoides* do not spend the night in their nests, and of course the species of *Nysson* do not have nests in which to sleep. Brues (1903) reported *Nysson plagiatus* Cresson in a mixed sleeping aggregation of wasps and bees, but did not indicate how many were found or what position they assumed. Otherwise, nothing is known as to where these wasps sleep; presumably the smaller forms sleep more or less solitarily in bushes or herbaceous vegetation, the larger forms possibly in trees.

In the Stizini, *Stizoides unicinctus* has been found sleeping on several occasions, on two occasions side-by-side with its host, *Prionyx*. In each case six to eight *Stizoides* occurred together on or near the top of herbs, with no bodily contact between the wasps (Chapter VI:F). In the related genus *Bembecinus* the situation is very different. These wasps form large balls either close to the ground or near the top of various herbs and shrubs. One such ball, reported by Brauns (1911) from South Africa, was said to be as large as a baby's head and to contain more than 1,000 wasps. Some sleeping clusters have been found containing all or mostly one sex, while others contain both males and females. In all, six species of *Bembecinus* have been reported forming ball-like clusters on three continents (Chapter VI:H, 1).

Some Bembicini form small, loose sleeping aggregations like those of *Stizoides*. *Bicyrtes capnoptera* has been found sleeping in the center or at the top of tall herbaceous plants, either solitarily or more often in numbers of from two to five. These wasps have a characteristic posture and often line up in series on horizontal branches (Chapter VII:D; Fig. 94). *Stictiella pulchella* is known to form loose clusters on indigobush, the clusters often of fairly large size and with a small amount of bodily contact among the wasps (Chapter VIII:B, 1). *Glenostictia pulla* is known to cluster (Chapter VIII:E), as are four species of the genus *Steniolia* (Chapter VIII:I, 2). *S. obliqua* is known to cluster on pine trees as well as on herbs and to form clusters of several hundred individuals in close bodily contact (Figs. 113 and 115).

The dense clusters of *Zyzzyx chilensis* may also occur on pines as well as on broadleafed trees and may contain several hundred individuals (Chapter IX:H). *Rubrica surinamensis* is reported to cluster on low vegetation; in this case Vesey-Fitzgerald reports

§A. Ecology, features of behavior

that the males and females segregate, although in all the previous examples the clusters are bisexual (Chapter IX:J). It is known that the species of *Stictia* do not sleep in the nest, and it is probable that they, too, form clusters, possibly in trees.

An anomaly is provided by *Glenostictia scitula,* in which the males form loose clusters on vegetation but the females sleep in the nest, in this case about half way down the burrow (Chapter VIII:G, 1). In the genus *Bembix,* so far as is known, the females always sleep in their nests, generally at the bottom of the burrow, just outside the inner closure (Chapter XI:E, 2). Male *Bembix* dig their own sleeping burrows or occupy available empty burrows. The genus *Microbembex* is apparently unique in that both males and females dig short sleeping burrows, often in large groups somewhat to one side of the main nesting area (Chapter XII:A, 1).

There are a few reports of Gorytini sleeping in their nests. Janvier (1928) speaks of males and females of *Clitemnestra gayi* passing the night together in the nest galleries. I reported finding a female *Hoplisoides tricolor* in a nest in the evening (Chapter III:E), although other species of this genus are known to sleep away from the nest. N. Lin (personal communication) has evidence that female *Sphecius speciosus* sleep in the nests, but that the males do not. The situation in the bembicine genus *Bicyrtes* is puzzling, since *B. parata* is reported to spend the night in special sleeping burrows (like *Microbembex*), although *B. capnoptera* has been found sleeping on plants (Chapter VII:D). Obviously many more data are needed on this subject. Numerous genera are completely unstudied with respect to their sleeping behavior.

The following outline may serve to summarize present knowledge of sleeping in the Nyssoninae:

A. Sleep on vegetation (both sexes)
   1. Solitarily or in small groups with no bodily contact: *Stizoides,* some *Bicyrtes,* probably many others
   2. In dense clusters with considerable body contact: *Bembecinus, Stictiella,* some *Glenostictia, Steniolia, Zyzzyx, Rubrica*

B. Males sleep in loose clusters on vegetation, females in their nests: *Glenostictia scitula,* probably *Sphecius speciosus.*

C. Sleep in the ground
   1. Both females and males together in nest burrows: *Clitemnestra*
   2. Females in their nests, males in short sleeping burrows: *Bembix*
   3. Both males and females in short sleeping burrows: *Microbembex,* some *Bicyrtes*

(6) *Reproductive behavior*. As in nearly all solitary wasps and bees, male Nyssoninae typically emerge several days before the females. The males tend to remain in the area where they emerged; much of their behavior is directed toward finding freshly emerged females and mating with them. The population of males typically declines in 2–3 weeks, and males may be entirely absent during the latter half of the nesting season, at least in univoltine species. An apparent exception is provided by the genus *Steniolia*, in which males have been found clustering with the females until the nesting season is nearly over (Chapter VIII:I, 1).

Virtually nothing is known of the precopulatory behavior of male Alyssonini, Nyssonini, and Gorytini other than *Sphecius*. It is possible that this behavior has simply not been observed because of the small size of these wasps. However, I have spent many hours in places where *Alysson* and *Hoplisoides* nested in considerable numbers, and I question whether the males of these wasps do in fact have sun dances of the type so often described for *Bembix* and other larger Nyssoninae.

In the case of *Alysson melleus*, I observed a male land behind a female, take flight and circle about her, than land behind her again. After repeating this several times, he landed on top of her, straddling her body and thrusting his wings well forward while extending his abdomen and copulating with her for about 15 sec (Chapter II:A, 2).

In the stizine wasp *Bembecinus neglectus*, I reported several males walking in circles on the sand and occasionally taking short, hopping flights. Now and then a male would pounce upon a female and attempt to mate with her, and sometimes males would pounce on other males; on one occasion I observed a ball of four males rolling over and over on the sand (Chapter VI:H, 1).

Virtually all Bembicini that have been observed form sun dances, a term first used by the Raus to describe the more or less continuous, sinuous flight of the males over the nesting area. The number of males participating in this flight may vary from only a few to many hundreds. Most species fly only a few centimeters above the ground, but in large species such as *Stictia carolina* the flight may be 0.5 to 1 m high. Males pause on the sand surface or on vegetation from time to time, usually assuming a characteristic pose with the body close to the ground and the legs extended rather rigidly (Fig. 160). Two species of *Bembix* are known to have an unusual type of precopulatory flight which I have called a "hopping dance" (Evans, 1957b).

A few of the larger Nyssoninae are known to be territorial,

§B. Nesting behavior

that is, the males patrol a specific area day after day, defending this area against intrusion by other males or even other insects or thrown pebbles. This has been best studied in *Sphecius speciosus;* in this species Lin (1963a) found that most males select an emergence hole as a perch from which they fly off to pursue and sometimes grapple with intruding males (Chapter V:B, 2). I have noted that males of *Stictia carolina* behave somewhat similarly, usually selecting low plants or lumps of earth as perches (Chapter IX:B, 2). Male *Stictia vivida* appear to spend long periods of time in the air, hovering and flying about in a restricted area, darting after any moving objects within their territory (Chapter IX:D).

It is curious that territoriality has been observed only in very large species. Presumably such behavior results in a better spacing of the males, insuring that all females will be fecundated and perhaps reducing attempts by males to pair with nesting females, as so commonly occurs in *Bembix*. Why this behavior should have evolved only in the very largest Nyssoninae is unclear.

Although many early authors reported sand wasps mating in flight, it is doubtful if copulation usually takes place in the air. The male and female typically come together in the air, but they fly off and land on a plant or on the ground. I have described the mating posture in *Bembix belfragei* and other species (1957b). Nielsen (1945) described mating in *B. rostrata* and presented several photographs of mating pairs on various plants. The mating posture of *Sphecius* appears to differ from that of other sand wasps (Chapter V: B, 2).

Mating in *Steniolia* (and perhaps other wasps that cluster) occurs at the clusters, at the time these are forming in the late afternoon. In this genus, copulation is preceded by a characteristic undulating flight, described in Chapter VIII:I, 3. The males appear relatively long-lived, clustering with the females throughout much of the season; one wonders if the females require frequent fecundation. In most wasps the females are receptive for only a few hours or at most a very few days after they emerge from their cocoons.

B. Nesting behavior

(1) *Digging the nest.* Many persons have described digging behavior in various Nyssoninae, but for the most part these descriptions are insufficiently detailed to be of much value. Doubtless all species are capable of varying their digging behavior to some extent depending upon the texture of the soil, size and abundance of pebbles, and so on. These factors make it somewhat difficult to

generalize regarding the movements of digging. However, I believe that some important differences in digging can be detected.

The mandibles and front legs provide the major digging organs of nyssonine wasps. In some of the small, slender-bodied species that lack a pecten on the front tarsus, it appears that the mandibles play a very important role and that all of the legs are used in pushing the soil behind the wasp as she drills into the soil. In *Alysson melleus* I noted that after a bit of soil is loosened by the mandibles, it is formed into a small lump which is pushed backward beneath the body by the front legs, then passed out behind the body either from the initial thrust of the front legs or with assistance from the middle and hind legs. When the burrow is a few millimeters deep, the wasp assumes a vertical position in it, head down, rotating in a clockwise direction while drilling into the earth and pushing the soil upward (Chapter II:A, 3). Digging in *Clitemnestra* may not be very different. Janvier (1928) remarks that in *C. chilensis* the females "bite the earth with their mandibles, which are flattened into curved blades, toothed at their tip and furnished with bristles for almost their entire length. These organs work in the manner of pincers, cutting and progressing rather quickly through the loose soil. The anterior legs scrape from time to time, and by backing up periodically the wasp throws the earth behind."

The mandibles retain an important role in loosening soil and dragging out pebbles in most Nyssoninae. They are used least in species that nest in fine-grained, very friable sand, and in such species (for example, *Bembix pruinosa, Microbembex* spp.) one often observes reduction in the tooth on the inner margin of the mandibles. The practice of rotating the body, more especially the head, occurs throughout the subfamily and serves to insure that the burrow is more or less round in cross section. The removal of the soil in lumps, as noted in *Alysson,* has been noted in other species that nest in relatively moist soil (for instance, *Bicyrtes fodiens, Bembecinus* spp.).

The majority of Nyssoninae have a more or less well-developed pecten on the front tarsi, and in these wasps the front tarsi assume the major role in moving the soil. Olberg (1959) has provided an excellent series of figures of *Gorytes laticinctus* digging. He notes that in this species the front legs do not work in perfect unison, one leg typically being slightly in advance of the other (the front legs are used alternately in the Pompilidae and in some Sphecidae, more especially Sphecinae). In *Hoplisoides nebulosus* I remarked (Chapter III:B,1) that the front legs are used synchronously, the abdomen moving up and down rhythmically and allowing the soil

§B. Nesting behavior

to shoot out behind the body. The wasp backs out frequently and sweeps the soil away from the entrance. Motion pictures of *H. spilographus* taken by the writer in Jackson Hole, Wyoming, reveal that the front legs work in perfect synchrony in this species. In this same species, Powell and Chemsak (1959) reported use of the hind legs for kicking the soil behind the body, but I did not confirm this at Jackson Hole.

Use of the hind legs seems especially pronounced in *Sphecius speciosus,* in which I reported (Chapter V:B, 3) use of the forelegs for raking back the earth but much more slowly than in *Bembix.* As the wasp backed out of the burrow periodically, she pushed soil behind her with her hind legs, which opened and closed in a somewhat scissors-like motion. As a result of her manner of backing out the burrow and "bulldozing" the earth behind her, a large trough was maintained through the growing heap of soil at the entrance. I have recorded this interesting and apparently unique behavior on motion picture film.

As compared to Gorytini, Bembicini frequently appear to dig much more rapidly, the forelegs working vigorously and in perfect unison, the body moving up and down rapidly in such a way as to permit the sand to be ejected in spurts beneath the body. So far as I am aware, each elevation of the posterior part of the body is synchronized with a single thrust of the forelegs except in two genera: *Bicyrtes* (Chapter VII:A, 2) and *Microbembex* (Chapter XII:A, 3). In these two genera the body undergoes prominent "bobbing" movements in the nest entrance; that is, the head goes down and the abdomen up for a moment, then the head up and the abdomen down. While the head is down, not one but several strokes of the front legs are made (although some very slight body movements are perceptible between each stroke). These strokes of the front legs occur in quick succession, and the sand is thrown far behind the strongly elevated abdomen so that there is no accumulation of soil at the nest entrance as there is in other genera (see also Chapter XV:C, 1.)

It is noteworthy that the males of some Bembicini (*Bembix, Microbembex;* Figs. 145 and 182) have a fairly strong pecten and are able to dig effectively in a manner similar to the females. It should also be noted that the females of cleptoparasitic genera (*Stizoides, Nysson*) are reasonably effective diggers, somewhat in the manner of *Hoplisoides* and *Gorytes.*

(2) *Leveling.* In the majority of Nyssoninae the mound of earth that accumulates at the nest entrance is left more or less intact.

In *Alysson* the soil accumulates all around the entrance, but in most others it forms a pile only on one side of the burrow. In *Bicyrtes* and *Microbembex,* as noted above, the soil is widely dispersed initially so that no mound is formed.

Weak and irregular leveling movements have been described in a number of Gorytini, for example in *Hoplisoides tricolor* and *Gorytes canaliculatus.* In the latter species I reported that following the initial closure of the nest, the female backs across the mound 5-7 cm and works toward the entrance in almost a straight line, kicking sand. This is repeated several times, over slightly different paths, so that the mound is partially leveled (Chapter III:H, 1). Although leveling movements have not been reported for *Stizus* or for some *Bembecinus* (Stizini), I reported leveling in *B. neglectus* very similar to that in *Gorytes canaliculatus* but somewhat more prolonged and complete (Chapter VI:H, 2).

Very similar leveling behavior, involving more or less straight-line movements across the mound, performed after the initial closure, has been reported for a number of Bembicini, including species of *Stictia* (Chapter IX:B, 3) and *Bembix* (Chapter X:H). In some species of *Bembix* (*spinolae,* and especially *nubilipennis,* Chapter X:C, 2) leveling is especially elaborate and complete, with prominent side-to-side movements as the wasp crosses the mound. In *B. troglodytes* and *B. multipicta* (Chapter XI:C, 2) leveling is similar but is performed before nest closure. Rotational leveling of an unusual type occurs in *B. pruinosa* (for further discussion of *Bembix,* see also Evans, 1957b).

Many Nyssoninae exhibit brief and incomplete leveling movements intermittently during digging; such movements are often little more than clearing movements which are performed at a distance of 1 or 2 cm from the entrance. In the genus *Glenostictia,* stereotyped rotational leveling movements are interspersed with digging in such a way that no mound is present when the nest is finally completed (Chapter VIII:E; G, 3; see also Chapter XV:C, 1).

(3) *False burrows.* False burrows are best known in the genus *Bembix* and were discussed at some length in Chapter XI:E, 6. I suggested there that false burrows evolved from closing behavior in which most of the soil was taken from one place, and that since the resulting pits proved to have survival value as "decoys" for parasites, they became ritualized and divorced from their original function in some species. Initial false burrows of more or less regular occurrence are known to occur in three species of *Bembix* and also in *Stizus pulcherrimus* (Chapter VI:B, 1) and *Rubrica gravida*

(Chapter IX:L). Irregular initial and terminal false burrows are known to occur in *Bembix amoena*, ritualized terminal false burrows in *B. sayi*.

In the Gorytini I noted one female *Hoplisoides nebulosus* making a final closure by taking soil mostly from one spot, making a shallow pit or rudimentary terminal false burrow (Chapter III:B, 5). In the stizine wasp *Bembecinus neglectus*, I found that most females dug the soil for closure from certain spots around the entrance so that each nest entrance tended to have a ring of small depressions around the entrance; but some wasps took most of the soil from a single point, forming a short false burrow up to 1 cm deep (Chapter VI:H, 2).

Thus there is evidence, among at least a few Gorytini and Stizini, of the initial steps I have postulated as leading to the stereotyped false burrows of some species of *Stizus* and *Bembix*. There are records of quite a number of wasps of other groups (Pompilidae, Sphecinae) taking most of their soil for closure from one or a few pits around the nest entrance, so this behavior may be widespread, although evidenced only in widely diverse species. Highly stereotyped false burrows like those of some Nyssoninae are also known in one species of *Sphex* (Sphecinae) (Tsuneki, 1963) and two species of *Philanthus* (Philanthinae) (Tsuneki, 1943b; Evans, 1965).

Evidence that false burrows do divert the attention of parasites (Bombyliidae, Miltogramminae, Mutillidae), at least momentarily, was cited under *Bembix amoena* (Chapter X:A, 7, 11), *B. sayi* (Chapter X:B, 4), and *B. texana* (Chapter XI:A, 4); I have also found this to occur in *Philanthus lepidus* (Evans, 1965). In *Microbembex monodonta* the open sleeping burrows sometimes divert parasites (Chapter XII:A, 11) and thus serve somewhat the same function as false burrows. That bombyliid flies deposit eggs in almost any open hole is well established, and if all of these parasites "waste" a certain percentage of their active hours at false burrows, then their effectiveness as parasites will be reduced.

(4) *Outer closure and concealment*. Present evidence indicates that all species of the following genera of Gorytini and Stizini maintain a temporary outer closure of the nest: *Gorytes, Hoplisoides, Hapalomellinus, Dienoplus,* and *Bembecinus;* also, cleptoparasites of the genera *Stizoides* and *Nysson* are reported to close the nests of their hosts when they leave. *Stizus pulcherrimus* is reported to make a closure, but *S. fasciatus* and *S. distinguendus* apparently do not. So far as is known, outer closure of the nest is always omitted in *Alysson, Didineis, Clitemnestra, Argogorytes, Exeirus,* and *Sphecius*. N. Lin (per-

sonal communication) has found that when nests of *Sphecius specious* are closed artificially, the females are well able to find their nests and to scrape open the entrances, in some cases while still holding their cicadas. It is possible that nest closure has been secondarily lost in *Sphecius* and some other genera, although the ability to enter while holding prey has been retained. In the tribe Bembicini, the following genera maintain a temporary outer closure: *Bicyrtes, Stictiella, Glenostictia, Steniolia, Stictia, Zyzzyx,* and *Microbembex*. In most cases there are records for several species, and no exceptions (or rare exceptions) have been found. *Rubrica surinamensis* is said to make an outer closure when the nest contains an egg or small larva, but to omit it later; *R. gravida* is said to maintain a closure at all times; but *R. denticornis* is said to omit the closure (Chapter IX:J, K, L). *Bembix* exhibits much variation with respect to outer closure (see Chapter XI:E, 4).

The initial closure of a new nest is often more prolonged than later closures and may be followed by movements of concealment. The initial closure is likely to involve some packing of the soil with the deflected tip of the abdomen, as occurs commonly during final closure of the nest (infrequently during temporary closures other than the initial one). Packing with rapid blows of the tip of the abdomen has been reported in *Hoplisoides, Bicyrtes, Stictia, Rubrica, Bembix,* and *Steniolia* (it is noteworthy that only the first of these genera has a well-defined pygidial plate).

Movements of concealment in Gorytini and most Bembicini are very simple. In *Gorytes canaliculatus* I remarked merely that sand is scraped in various directions over the entrance, concealing it completely (Chapter III:H, 1), and my remarks concerning *Bicyrtes quadrifasciata* were very similar (Chapter VII:A, 2). In *Stictiella serrata* I noted that following closure the female spends several minutes going out in various directions from the entrance kicking sand toward the entrance (Chapter VIII:B, 2). A similar pattern of radiating lines has been reported for some species of *Stictia* and *Bembix* (for example, Fig. 157). In *Microbembex monodonta* and *Glenostictia scitula* these radiating lines are quite long and numerous, and all are made on the side in the direction of the burrow (that is, the side of the entrance opposite that on which leveling is performed). I have described this behavior in some detail for the latter species (Chapter VIII:G, 3); in this species the female hovers over the concealed nest entrance in a characteristic manner before flying off (see also Chapter XV:C, 2.)

(5) *Inner closure.* In only two genera, *Stictia* and *Bembix,* is it known that the female maintains a closure separating the burrow from

§B. Nesting behavior 439

an active cell. In these same two genera one finds a tendency for the burrow to level off before reaching the cell, and these are the only two genera in which one often finds a spur in the nest: both of these features are of course associated with the inner closure (see Figs. 131, 151, 174; cf. Figs. 76, 91, 188). In *Stictia* and some species of *Bembix* the inner closure may be several centimeters long, especially in nests containing an egg. This closure may be weaker and may even be omitted entirely during periods of active provisioning. Two species of *Bembix* are not known to make an inner closure at any time (Chapter XI:E, 4).

(6) *Wing elevation and metasomal pumping.* In discussing *Hoplisoides nebulosus* (Chapter III:B), I remarked that when these wasps are walking about on the sand, working outside the nest entrance, or arriving at the entrance with prey, the wings are elevated at about a 35–45° angle with the abdomen. As the female enters the burrow, either when digging or provisioning, she depresses her wings flat against her body. Much the same remarks were made with respect to *H. tricolor* (Chapter III:E, 1) and *Gorytes canaliculatus* (Chapter III:H, 1), and my impression is that wing elevation of this nature is characteristic of all of the more typical Gorytini. In the Bembicini, I have seen female *Bicyrtes quadrifasciata* and *Microbembex monodonta* raise their wings when working outside the nest entrance, but only occasionally, possibly resulting from the presence of a human observer. In *Stictia carolina* males resting on their perches characteristically elevate their wings in a similar manner, from time to time flying off swiftly (Chapter IX:B, 2). Probably these elevations of the wings represent intention movements of flight; they appear to be ritualized in certain Gorytini, but the function they serve is unclear.

Another movement commonly observed in *Alysson* (Chapter II:A, 6), *Hoplisoides* (Chapter III:E), and related genera, may be termed "metasomal pumping." This consists of slow, rhythmic up-and-down movements of the metasoma, performed when the wasp is walking over the ground with prey or standing in front of the burrow holding the prey, just prior to entering. The function of these movements is also unknown. I have never observed such movements in Bembicini, but in this tribe one is often impressed with the rhythmic contraction and extension of the metasoma, involving movement of the segments against one another. This is most often noticed when a wasp lands on the soil after a period of flight, and it is doubtless associated with respiration.

(7) *Orientation.* All sand wasps appear to learn the location of the nest during short flights either while digging the nest or just after

its completion (or both). It is probable that orientation is entirely by means of visual cues (Tsuneki, 1956; see also Chapter XII:A, 5). Few comparative data are available concerning this aspect of the behavior, but some brief remarks may be worthwhile.

After completing a nest, *Gorytes canaliculatus* flies in a series of loops and figure eights only about 5 cm above the nest entrance, then, after five or more such loops, increases her height slightly and takes several more loops at a gradually increasing height from the nest entrance (Chapter III:H, 1). Similar orientation patterns have been described for a number of other Nyssoninae, for example for *Bicyrtes ventralis* (Chapter VII:B, 2), although the species of this genus often take short flights during digging and closure as well as a longer one before leaving the nest for the first time. *Sphecius speciosus* is reported by Savin (1923) to fly about the entrance in circles of increasing size, "the outer one having a diameter of about thirty feet."

Orientation flights in *Hoplisoides nebulosus* (Chapter III:B, 1) and *Bicyrtes quadrifasciata* (Chapter VII:A, 2) occur during the later stages of digging, during closure, and again just before leaving the nest; in these species the flights are unusually high, the wasp rising high in the air above her nest and later returning at an altitude of one half to several meters and descending obliquely to the nest. Several authors have described this behavior at length in the latter species.

In *Glenostictia scitula* I reported a great many short flights interspersed with digging and leveling movements, but no distinct orientation flight following completion of closure and leveling. In this species the flights are low and circling (Chapter VIII:G, 3).

These are only a few of the major variants which have been noted with respect to the orientation flight. It is curious that, in general, the orientation flights of sand wasps are not conspicuous or prolonged. Evidently these wasps are able to memorize a great many visual cues in a short time, as they rarely fail to find their nests. I found that it was easy to misguide *Hoplisoides nebulosus* merely by displacing a leaf or stone a few centimeters. However, bembicine wasps seem to rely upon a great many different landmarks, and even great disturbances to the nesting area (for example, in *Microbembex,* Chapter XII:A, 5) rarely prevent the wasps from finding their nests. The studies of Van Iersel (1952, 1965), Tsuneki (1956), and Chmurzynski (1964), all dealing with species of *Bembix,* doubtless apply in a general way to most higher Nyssoninae.

(8) *Number and arrangement of nest cells.* It is apparent that the majority of Nyssoninae prepare a series of cells from the same burrow.

§B. Nesting behavior

When one cell is completely provisioned, it is closed off, a new cell burrow and cell being constructed at the same time. In certain species the female may appear outside the nest at this time, but more often all of the work occurs within the nest, the soil from the new parts of the nest being used for closing the completed cell. Typically (for example, in various species of *Alysson, Gorytes, Hoplisoides,* and *Bicyrtes;* Figs. 18, 31, and 38) the initial cell is deepest, the following cells being built from short side burrows back toward the entrance, so that the final cell is closer to the surface than the first cell. However, Janvier reported certain Chilean species as proceeding in the opposite way; that is, the burrow is gradually lengthened, later cells being deeper than the initial cell (*Clitemnestra, Zyzzyx*). Some species prepare a very large section of new burrow prior to preparing a new cell (for example, *Bembix amoena, B. belfragei;* Fig. 151) or a series of new cells (*Sphecius speciosus,* Fig. 56). In these two genera, and most other higher Nyssoninae, the cells of one nest tend to be at roughly the same depth rather than progressively shallower.

In *Sphecius* it appears that the female continues to prepare additional cells from a single nest for the duration of her life, or at least for a considerable period of time. This may also be true of other species that prepare many cells per nest (*Exeirus, Stizus, Argogorytes, Clitemnestra*). Most of the typical Gorytini seem to close the nest after only a few cells and then to prepare a new nest nearby. In some cases the nest may occupy 1 day, the number of cells perhaps being determined by success in hunting, as appeared to be the case in *Hoplisoides nebulosus* (Chapter III:B, 2). Much more often, the nest persists over 2 or more days. No information is available as to the factors inducing the female to stop preparing cells in one nest and prepare an entirely new nest. Probably there are innate factors relating to the spacing of cells, and these are influenced by the amount of suitably friable soil surrounding the cell burrow. That is, the female tends to fill up the space around the main burrow with cells, but these must be well spaced and not too deep or too shallow. Experimentation with a typical species might clarify this matter.

While most of the unicellular nests reported for various Gorytini are probably the result of nests being dug out for study before final closure, it is apparent that in some species of the bembicine genus *Bicyrtes* (Chapter VII:A, 3; B, 3), also in some species of *Bembix* (*cinerea, amoena;* Chapter XI:E, 3), either unicellular or bicellular nests may occur in single colonies of one species. Tsuneki (1956) found occasional multicellular nests in aggregations of otherwise unicellular nests of *Bembix niponica,* as well as one colony of this species in which nests of up to six cells were the rule. Such species may represent transitional steps between multicellular and unicellular nests.

Unicellular nests are characteristic of the majority of Bembicini. In the case of certain species (for example, *Glenostictia scitula, Steniolia obliqua, Microbembex monodonta,* and several species of *Bembix;* Figs. 109, 118, 188) extensive studies employing marked females have demonstrated conclusively that these wasps prepare a series of unicellular nests. Such nests may be deep and complex and involve much labor in construction, but they nevertheless contain only one cell (for instance, *Bembix pruinosa*). In the genus *Bembix* multicellular nests seem more characteristic of species nesting in compact soil, while most species nesting in friable sand make unicellular nests. Some of the relatively complex nests of certain species of *Bembix* I discussed in 1957b (see also Chapter XI:E, 3). Complex nests such as those occurring in *B. pruinosa, occidentalis,* and *belfragei* are unknown in other genera of Nyssoninae.

Data on the maximum number of nest cells presently reported for various genera of Nyssoninae are summarized in Table 44. In several cases, some species of a genus are reported to make several cells per nest and others only one. Further study may reveal that some species of the latter group do in fact always make unicellular nests, as in some species of *Bembix* and *Microbembex*.

C. Hunting and provisioning

(1) *Type of prey*. In general terms, it may be said that Alyssonini, Gorytini, and Stizini prey upon adult or immature insects exhibiting simple metamorphosis, chiefly Homoptera (*Stizus* upon Orthoptera). On the other hand, Bembicini prey upon adults of insects exhibiting complete metamorphosis (except *Bicyrtes*, which preys upon Heteroptera). Diptera provide the major prey of Bembicini, but two genera (*Stictiella* and *Editha*) utilize diurnal Lepidoptera, and Lepidoptera have been reported as occasional prey of *Zyzzyx* and certain species of *Bembix*. Small Hymenoptera are known to be used as occasional prey by *Bembix citripes* and *Glenostictia scitula,* and the latter species also utilizes Hemiptera of both the suborders Homoptera and Heteroptera. The breakdown of host specificity observed in *Glenostictia scitula* is carried still further in the species of *Microbembex*, which utilize a wide assortment of arthropods—in the case of *monodonta*, at least, chiefly dead or disabled individuals (Table 45).

The majority of Nyssoninae may be thought of as rather broadly

§C. Hunting and provisioning

TABLE 44. NUMBER OF NEST CELLS IN VARIOUS GENERA OF NYSSONINAE

Genus	Maximum no. cells reported	No. species studied
Alysson	5	4
Clitemnestra	10	2
Ochleroptera	3 (?7)	1
Argogorytes	9	2
Gorytes	4	7
Hoplisoides	4	4
Dienoplus	15	5
Hapalomellinus	2	1
Exeirus	several	1
Sphecius	15.8 (mean)	1
Stizus	9	4
Bembecinus	2	7
Bicyrtes	5	5
Stictiella	17	4
Glenostictia	1	3
Steniolia	1	4
Stictia	1	4
Zyzzyx	6	1
Rubrica	1	2
Bembix *	6	25
Microbembex **	8	3

* But the majority of species make unicellular nests (see summary, Chapter XI:E,3).

** But the well-studied M. monodonta makes unicellular nests; only M. ciliata is reported to make multicellular nests (Chapter XII:A,4;B).

oligophagous, taking insects of several genera and sometimes several related families. A relatively high degree of specificity is exhibited by the species of *Sphecius* and *Exeirus,* which are specialists on cicadas (Chapter V:B, C, E) and apparently by the species of *Argogorytes,* which capture spittle insects (Cercopidae) from their masses of froth (Chapter III:K). Within genera such as *Hoplisoides* and *Gorytes,* some species appear to restrict their predation to Membracidae, others to Cicadellidae or other groups. Among the fly-

TABLE 45. PREY OF GENERA OF NYSSONINAE *

Genus	No. spp. studied	ORTHOPTERA	HOMOPTERA						HETEROPTERA	LEPIDOPTERA	DIPTERA	HYMENOPTERA	OTHERS
			Cicadidae	Cicadellidae	Cercopidae	Membracidae	Fulgoroidea	Psyllidae					
Alysson	6		X	x			X						
Didineis	2		X				X						
Clitemnestra	2					X	X						
Ochleroptera	1		X	x	x	x	x						
Argogorytes	2			X									
Gorytes	9		X	X	x	X							
Hoplisoides	8		X		X	X						x	
Psammaecius	1		x										
Lestiphorus	1			x									
Harpactostigma	1						x						
Psammaletes	1						x						
Dienoplus	9		X	x			x						
Ammatomus	1						X						
Exeirus	1		X										
Sphecius	6		X										
Stizus	9	X											
Bembecinus	11			X	X	X	X	x					
Bicyrtes	8								X				
Stictiella	6									X			
Glenostictia	3			x			x	x			X	x	
Steniolia	6										X		
Stictia	6										X		
Zyzzyx	1									x	X		
Rubrica	3										X		
Editha	1									X			
Bembix	32									x	X	x	Odonata (1 Australian sp.)
Microbembex	3	X		X	X	X	X	X	X	X	X	X	Arachnida, Ephemerida, Psocoptera, Neuroptera, Trichoptera, Coleoptera

* Common use of prey of one order is indicated by X, while isolated records are indicated by x.

catchers, certain species (especially in the genera *Glenostictia* and *Steniolia*) appear to use large numbers of Bombyliidae, while others (especially some species of *Stictia* and *Bembix*) do much hunting about animals and use large numbers of Tabanidae. Some genera appear relatively uniform in choice of prey but have one or more aberrant species. For example, many hundreds of records for North American *Bembix* demonstrate a restriction to Diptera, but an Australian species uses damselflies (Odonata). In *Glenostictia* two

§C. Hunting and provisioning 445

species prey only on flower-inhabiting flies, but *scitula* employs insects of three orders (Chapter VIII:E, F, G).

None of this is surprising when we recall that primitive wasps (for instance, Tiphiidae and many Sphecinae) tend to be relatively strongly host specific, while most examples of lack of specificity occur in higher wasps (for example, many Crabronini, Vespinae). Also, the shift from insects with simple metamorphosis to holometabolous insects may be observed in other groups of wasps (see Evans, 1963b). In these respects the Nyssoninae are fairly typical sphecoid wasps. The occurrence of anomalous records for certain species and of "aberrant" species in some genera is exactly what would be expected in a group that is undergoing evolution in the behavior patterns associated with prey selection. The only real surprise relates to the genus *Microbembex,* for these are the only sphecoid wasps that have become general scavengers (see also Chapter XV:C, 3.)

(2) *Hunting behavior.* Relatively little is known regarding the details of hunting behavior. Such behavior is initiated only after the completion of the first nest and is doubtless influenced by the development of eggs in the ovaries as well as by stimuli received by the female from the nest cell. Presumably the female flies into the habitat of the prey and responds to the image of the prey in somewhat the manner described by Tinbergen (1935) for *Philanthus triangulum.* That these innate behavior patterns are influenced by learning is suggested by the fact that different individuals of one species often fill up their nests with different species of prey, indicating that they have learned a good source of prey and have returned to it again and again. Paralysis of the prey may also improve with experience, but old females may fail to immobilize their prey adequately (Chapter VIII:B, 4). It is believed that Nyssoninae usually sting their prey only once, in the venter of the thorax (Chapter V:B, 5), and it is probable that Rathmayer's (1962) studies on *Philanthus triangulum* apply in a general way to Nyssoninae.

Alysson melleus hunts for its leafhopper prey by flying from branch to branch and running over twigs and leaves much like an ant (Chapter II:A, 6). Certain Gorytini have also been observed walking about on vegetation seeking their prey. *Sphecius speciosus* is said to circle about the trunks and branches of trees and to dart after cicadas, thrusting its sting into the venter of the thorax and persisting until the cicada ceases to struggle; the cicada is held to the tree and then carried over on to the top of a branch before being carried off to the nest (Chapter V:B, 5). Hunting in the stizine genus *Bembecinus*, the species of which prey upon Homoptera, and in the bembicine genus *Bicyrtes*, the species of which utilize Heteroptera, probably resembles in a general way that of the Gorytini.

The majority of Bembicini utilize adult Diptera or Lepidoptera and commonly take their prey when it is in flight or resting upon flowers. Some species of *Stictia* often hunt about animals, but there is evidence that they do not normally grasp horseflies that are actually feeding on the animal, but rather seize them as they fly about (Chapter IX:B, 6). I have seen *Bembix* females seize flies in the air over carcasses and dung and proceed directly to their nests, apparently stinging and adjusting the position of their prey during flight. However, some Bembicini fall to the ground with their prey, perhaps especially when capturing large insects (Chapter IX:J). In my summary of the genera with recessed ocelli, I reported that these wasps most commonly attack flower-visiting insects (*Stictiella* attacks diurnal Lepidoptera, *Glenostictia* and *Steniolia* chiefly bombyliid and syrphid flies); the females maneuver rapidly on their short wings and strike suddenly at their prey, which is then carried away directly in flight (Chapter VIII:I, 6). The several European species of *Bembix* that have short wings are reputed to behave similarly. The short-winged species typically produce a high-pitched whining sound when hunting as well as during some of their activities around the nest.

The scavenger wasp *Microbembex monodonta* flies close above the sand and picks up dead or disabled arthropods from the surface. These wasps will also steal such arthropods from one another and will sometimes pick up paralyzed insects from the nest entrances of other digger wasps (Chapter XII:A, 6). Stealing of prey also occurs commonly within large colonies of certain species of *Bembix* (Chapter XIV:A, 3).

(3) *Paralysis of the prey.* The majority of Nyssoninae paralyze their prey in such a way that it remains alive and fresh for several days. For the first few hours, up to 1 or 2 days, the prey may exhibit weak movements of the legs, especially when prodded. Tsuneki reports that the orthopterous prey of *Stizus pulcherrimus* may defecate in the cell (Chapter VI:B, 3). From time to time, in the nests of many different species, one finds an insect that is imperfectly paralyzed or one that has died and begun to deteriorate. There may also be variation among individuals and local populations in the degree of paralysis of the prey. For example, leafhoppers in the cells of *Alysson melleus* often appear dead, but in some nests they respond to stimuli up to 48 hr after being stung (Chapter II:A, 6).

§C. Hunting and provisioning

I have found nests of *Bicyrtes* in which the bugs appeared to deteriorate within 1 or 2 days after being captured; but Krombein (1955) found that bugs from the nests of *B. quadrifasciata* would remain alive in rearing tins up to 2.5 weeks.

Among the Nyssoninae that exhibit progressive provisioning, the prey may be deeply paralyzed and may fail to respond to prodding within a short time after being stung. This is generally true, for example, in the stizine genus *Bembecinus* and in bembicine genera such as *Stictiella, Rubrica,* and *Zyzzyx*. In the last-named genus, Janvier (1928) found that smaller flies were often killed while large flies were only imperfectly paralyzed. I have found this to be true to some extent in certain species of *Bembix* (Chapter XI:E, 8).

Workers on Bembicini are often vague as to whether the prey is dead or merely deeply paralyzed. Some species, even though they provision progressively, evidently only paralyze their prey, not always deeply (*Stictia carolina, Bembix niponica*). Those species of *Bembix* that oviposit on the initial fly almost invariably kill that fly, which appears dessicated within a few days, even though other flies are merely paralyzed. On the other hand, some species of *Bembix* appear to kill all the flies (especially *B. texana,* in which cell cleaning is well developed). Species in which provisioning is continued until the larva is ready to spin its cocoon (that is, in which truncated progressive provisioning does not commonly occur) also seem to kill their prey. Here I would include the species of *Steniolia, Glenostictia scitula,* and *Bembix u-scripta*.

(4) *Prey carriage*. The sphecoid wasps as a whole exhibit three major types of prey carriage, which I have termed mandibular, pedal, and abdominal (Evans, 1963a). In the great majority of sphecoids and in all other wasps, various mandibular mechanisms are employed, and there is every reason to regard these as primitive. Pedal mechanisms are employed by the majority of Nyssoninae but by relatively few other Sphecidae (chiefly Crabronini, Psenini, and Philanthinae). Abdominal mechanisms are employed by only a few wasps (*Oxybelus, Clypeadon*), none of them Nyssoninae.

Within the Nyssoninae, carriage of the prey with the mandibles is reported for only two genera: *Alysson* and *Clitemnestra*. *Alysson ratzeburgii* is said by Ferton to make use of the legs as well as the mandibles when carrying the leafhopper prey, and it is possible that other species of this genus embrace the prey with the legs when flying. *A. melleus* and several other species are known to employ only the mandibles when walking on the ground, as they often do for considerable distances before entering the nest (see

Fig. 18, and also figures in Olberg, 1959:336–338). Prey-laden females enter the nest directly, the entrance being left open at all times (Chapter II:A, 6).

Janvier reports that in *Clitemnestra gayi* the prey is carried with the mandibles, while in *C. chilensis* both the mandibles and legs are employed. The latter species is said to fly directly to the nest entrance, in contrast to *Alysson* but like most Gorytini, Stizini, and Bembicini. In both species of *Clitemnestra* the nest entrance is left open (Chapter III:L).

All other Nyssoninae that have been studied carry the prey well back beneath their body, the middle legs supplying the major grasp. During flight, the hind legs may also embrace the prey loosely, at least in the case of large prey, and in *Sphecius* the hind legs appear to be especially important in carrying the very large prey (Chapter V:B, 5). When entering the nest, most of these wasps scrape open the entrance with the front legs, which are free and unimpeded; upon entering, the prey slips backward slightly, being released by the middle legs and grasped by the hind legs, so that it follows the wasp down the narrow bore of the burrow (at least in many cases). In various species it has been noted that large prey sometimes become stuck in the entrance, in which case it is drawn in from the inside with the mandibles. These generalizations apply whether the prey is Homoptera, Heteroptera, Lepidoptera, or Diptera, and even apply to nonspecific predators and scavengers such as *Glenostictia scitula* and *Microbembex monodonta*. In the latter species, the "prey" may vary greatly in size and shape, but it is nevertheless carried chiefly with the middle legs.

Something of an exception is provided by the genus *Stizus*, the species of which employ very elongate prey of the order Orthoptera. Although pedal carriage still prevails, Tsuneki reports that in *S. pulcherrimus* the wasp also grasps the antenna of the prey with her mandibles; the female may land a few centimeters in front of the nest entrance and walk to the entrance while still holding the antennae of the grasshopper. However, the antennae are released when the wasp scrapes open the nest entrance, as they must be in order to free the front legs. Sometimes, Tsuneki reports, the wasp leaves the prey at the entrance, then enters the nest, turns around, and draws in the prey by grasping its antenna with the mandibles (Chapter VI:B, 3).

(5) *Oviposition*. Ten different types of oviposition occur in the Nyssoninae, and it is convenient to consider these in the following way. (For further discussion, see Chapter XV:C, 5; see also Fig. 214.)

§C. Hunting and provisioning

GROUP A

Egg fastened to homopterous prey for its full length. Egg is laid on last (or one of last) prey placed in cell (except in the cleptoparasitic *Nysson*).

Type 1. Egg laid longitudinally on venter, beside leg bases (Figs. 20, 33, 39, and 58). Examples: *Alysson, Ochleroptera, Gorytes, Hoplisoides, Dienoplus, Exeirus, Sphecius* (Chapters II, III, and V).

Type 2. Egg laid in various positions of concealment (Figs. 35 and 36). Example: *Nysson*. In this case the egg is rather small and has a strong surface sculpturing; it hatches rather promptly, and the larva destroys the egg or larva of the host (Chapter IV).

GROUP B

Egg fastened by its caudal end, the cephalic end extending free. Egg is laid on first prey placed in cell.

Type 3. Egg laid on venter of homopterous prey, near middle leg bases, either longitudinally or transversely. Example: *Clitemnestra* (Chapter III:L).

Type 4. Egg laid obliquely on side of thorax of orthopterous prey (Figs. 70 and 71). Examples: *Stizus, Stizoides* (but *Stizus imperialis* is said to lay the egg between the front legs, and may thus belong to the following group; Chapter VI:B, C, F).

Type 5. Egg laid on or near ventral midline of heteropterous prey, its cephalic end directly obliquely forward (Figs. 88–90). Example: *Bicyrtes* (Chapter VII).

Type 6. Egg attached to side of thorax of prey (usually Diptera or Lepidoptera), its cephalic end extending more or less vertically upward (Fig. 162). Examples: *Stictiella* (all known spp.), *Glenostictia pulla, G. gilva, Steniolia* (all known spp.), *Stictia* (most spp.), *Rubrica surinamensis, Bembix* (most spp.). In *Bembix* and *Stictia* one wing of the fly is extended and the egg glued to the wing base; in *Bembix*, at least, one leg is dislocated on the side bearing the egg (see especially Tsuneki, 1956). In some species of *Stictia* the prey is often placed dorsum up, in contrast to the usual venter-up position in all other Nyssoninae (Fig. 137).

GROUP C

Egg laid in the cell before any prey is brought in.

Type 7. Egg glued erect to the substrate near the center of the cell or toward its apex (Figs. 175 and 190). Examples: *Zyzzyx chilensis, Microbembex monodonta, M. sulfurea, Bembix* (five known spp.; see Chapter XI:E, 7).

Type 8. Egg glued to top of a small pedestal made of sand grains, inclined in a direction away from the cell entrance (Figs. 77 and 78). Examples: *Bembecinus* (several species; Chapter VI:H, 3).

Type 9. Egg placed against wall at apex of cell, more or less vertically. Example: *Bembix pruinosa* (see Evans, 1957b: 155-157).

Type 10. Egg laid flat in bottom of cell, near center or toward apex. Examples: *Glenostictia scitula* (Chapter VIII:G, 5), *Stictia carolina* (Chapter IX:B, 5), *Bembix occidentalis* (Evans, 1957b:176-177).

(6) *Type of provisioning.* Sand wasps may be thought of as exhibiting either mass or progressive provisioning. This is a less hard-and-fast distinction than once supposed, but these terms remain useful. The following grouping recognizes four types of provisioning under two major headings. (For further discussion, see Chapter XV:C, 4.)

GROUP A

Wasp fills up cell without waiting for egg to hatch.

Type 1. Typical mass provisioning: prey is brought in more or less steadily so that cell is filled within a few hours (or at least before the egg hatches). Examples: all known Alyssonini and Gorytini (Chapters II, III, V); *Bicyrtes fodiens* (Chapter VII:C); *Bembix hinei* (Evans, 1957b:94-104). In *Clitemnestra,* Janvier reports that provisioning may extend over 2 or 3 days, but in these wasps it is said that the egg hatches after 5 or 6 days; Chapter III:L.

Type 2. Delayed provisioning: same as above, except that upon occasion (either following inclement weather, or perhaps innately in some species) the last portion of the prey is brought in after the egg has hatched. Examples: *Stizus pulcherrimus* (Chapter VI:B, 3), *Bicyrtes quadrifasciata* (Chapter VII:A, 5), *B. ventralis* (Chapter VII:B, 4), *Stictiella serrata, S. evansi* (Chapter VIII:B, 4).

GROUP B

Wasp brings in no prey (at least other than egg pedestal) until egg is ready to hatch, then brings in an increasing amount each day as the larva grows.

Type 3. Truncated progressive provisioning: cell filled up rapidly when prey is plentiful and the weather favorable, so that the cell is closed well before the larva is fully grown. Examples: *Bembecinus neglectus, B. hungaricus* (Chapter VI:H, 4), *Microbembex monodonta* (Chapter XII:A, 9), *Bembix* spp. (especially *pruinosa, sayi;* Chapter X:B, 3).

Type 4. Fully progressive provisioning: cell provisioned more slowly and until the larva is ready or nearly ready to spin its cocoon. Examples: *Glenostictia scitula* (Chapter VIII:G, 6), *Steniolia*

§C. Hunting and provisioning

obliqua, S. duplicata (Chapter VIII:I, 6), *Bembix u-scripta* (Evans, 1961).

Clearly there is no sharp distinction between types 1 and 2 or between types 3 and 4. The distinction between types 2 and 3 (groups A and B) is more real, being based upon a fundamental behavioral difference: whether the female continues to provision immediately after oviposition, her activities presumably being directed by stimuli relating to the fullness of the cell, or whether she waits until the hatching of the egg, after which time her activities are apparently directed in large part by stimuli received from the larva each day. Tsuneki (1957) has explored the nature of these stimuli in *Bembix niponica,* and probably his studies apply in a general way to other Bembicini that provision progressively.

Unfortunately, the distinction between types 2 and 3 (delayed and truncated progressive provisioning), although presumably fundamental, is often blurred by lack of knowledge. The genus *Stictiella,* in particular, appears to bridge the gap between these two types, and much more detail is needed regarding this genus. It is essential that we learn to what extent type of provisioning is determined by environmental factors and to what extent it is innate. Wheeler (1928) and others have hypothesized that progressive provisioning arose from delayed provisioning. Presumably this is an example of genetic assimilation of a phenotypic behavioral potentiality, as discussed further in the following chapter.

Tsuneki (1956) objects to the term progressive provisioning as applied to *Bembix* and related wasps. He points out that the behavior of the mother wasp is determined each day by the state of the larva, and that the cell is essentially mass provisioned on a day-to-day basis, the prey being piled up in the cell. This stands in contrast to the social wasps, where the prey is macerated and actually presented to the larva periodically during the day, in response to stimuli received upon intimate and frequent contact between adult and larva. Tsuneki would term the behavior of *Bembix* "progressive mass-provisioning." I concede that Tsuneki has a point, but suggest that the term progressive provisioning be used from *Bembix* and other solitary wasps, as it has been used so frequently in the past. Social wasps may be said, in contrast, to exhibit "direct feeding" of the larva.

(7) *Cell cleaning.* This highly specialized behavior is known to occur only in *Rubrica surinamensis* (Chapter IX:J) and in five species of *Bembix* (Chapter XI:E, 10). I shall not discuss it further here other than to point out that this behavior, too, is characteristic of social

wasps, but that social wasps clean out and reuse cells as required, in contrast to the day-by-day cell cleaning found in some Bembicini (which of course do not reuse the cells). Again, there is evidence that some species clean the cells only in certain situations (*B. niponica*) and that others clean the cells regularly (*B. texana*), thus reintroducing the problem considered above. An experimental study of cell cleaning in one or more species of Bembicini might be especially rewarding.

(8) *Final closure.* I have discussed final closure of the nest in various species of Nyssoninae in Chapters III through XII, but the data are too fragmentary to permit much fruitful comparison. It appears that some Gorytini which prepare a large number of cells from a single burrow may never prepare a definitive outer closure of the nest (this occurs at least at times in *Sphecius speciosus,* perhaps in others). In most Gorytini (for example, *Hoplisoides nebulosus,* Chapter III:B, 5) the final closure is relatively simple, little more than a prolonged temporary closure. The bottom part of the burrow is filled with earth reamed from the walls of the burrow, the upper part of the burrow, with soil scraped in from the periphery of the entrance. In virtually all Nyssoninae (even those without a distinct pygidial plate), the soil is packed in the burrow with rapid, light blows of the deflected tip of the abdomen. Some Nyssoninae make false burrows either occasionally or regularly at the time of final closure, a matter explored in an earlier section of this chapter (B, 3).

In *Hoplisoides nebulosus* I noted that when the burrow is full, the wasp goes out in several directions a distance of 2-4 cm scraping sand toward the entrance (up to 7 cm in *H. tricolor*). Similar movements of concealment occur in many Nyssoninae, for example in *Bicyrtes quadrifasciata,* a species in which the females commonly hover over the covered nest entrance periodically, now and then scraping a little more sand over it. In the higher Bembicini these movements of concealment may become prolonged and elaborate. Some species make many radiating lines away from the filled nest entrance, scraping sand as they proceed, these lines being in some cases as much as 35 cm long (*Bembix sayi,* Chapter X:B, 4). This "radiating line" behavior occurs in species of several genera (*Bembix, Stictia, Microbembex, Glenostictia*). With rare exceptions the lines are made on the side of the burrow away from the threshold; that is, on the side opposite the mound (Fig. 157). Hovering over the covered entrance, with occasional further scraping of sand, has been reported for several Bembicini, and is especially conspicuous in *Glenostictia scitula* (Chapter VIII:G, 7).

§D. Cleptoparasitism   453

Certain species of *Bembix* that exhibit elaborate leveling movements following completion of a new nest also do some leveling of the accumulated soil prior to beginning the final closure (for example, *B. troglodytes;* Evans, 1957b:132-133). On the other hand, those species that do not level often leave a conspicuous mound of sand even after fairly lengthy movements of concealment following final closure (for instance, *B. sayi,* Chapter X:B, 4, and Fig. 155).

In any given species, there are many similarities between the behavior during temporary closures (especially the initial one) and behavior during final closure. Generally speaking, movements of filling, packing, and concealing tend to be accentuated at final closure although differing little qualitatively from similar movements during temporary closures. On the other hand, leveling movements may be reduced, and orientation flights are absent or are replaced by low, hovering flights apparently serving a different function (see also Chapter XV:C, 2.)

D. Cleptoparasitism

The Nyssonini and the stizine genus *Stizoides* have been little mentioned in this chapter up to this point, since these two groups do not construct or provision their own nests, but rather lay their eggs in the nests of other wasps, their larvae developing on the prey provided by the host wasp. Some elements in their behavior have therefore disappeared, while others have become greatly modified. The behavior of several species of *Nysson* and of one species of *Brachystegus* was summarized in Chapter IV; it is assumed that the other genera of this tribe behave similarly, since they are structurally similar. The behavior of *Stizoides* was summarized in Chapter VI:G. These summaries need not be repeated here, but a comparison of the two groups seems in order.

The genera *Nysson* and *Stizoides,* although dissimilar in structure, have a number of behavioral features in common. The females spend most of their time in and around the nesting sites of their hosts. They make short, low flights and stop frequently on the soil, walking about circuitously and inspecting various holes. Available evidence suggests that members of both genera use olfactory as well as visual cues in locating nests of their hosts. Having located a host nest, they are able to dig into it effectively, their digging behavior not differing notably from that of their hosts. When the host wasp is digging or provisioning the nest, the parasite may remain in the vicinity and enter at a later time. However, members of both groups are known at times to enter nests while the host is still inside. In this case the parasite may be driven away by the host (reported for *Stizoides unicinctus* and its host *Prionyx* spp., for *S. tridentatus* and its host *Sphex maxillosus,* for *Brachystegus scalaris* and its host *Tachytes europaeus,* and for *Nysson dimidiatus* and its host *Dienoplus elegans*). However, in the case of *Nysson,* the host very often seems to show little if any reaction to the close presence of the parasite, and may even close the nest while the parasite is inside. Both *Nysson* and *Stizoides* typically close the nest of the host when they leave, making a brief outer closure not dissimilar to that of their host's.

Some of the differences between these genera reflect their taxonomic affinities. *Stizoides* sleeps on vegetation somewhat gregariously (often with its host!), while *Nysson* is assumed to sleep solitarily (at least no aggregations have been found). Being an apparent derivative of gorytine wasps, *Nysson* restricts its attention to nests provisioned with Homoptera (at least so far as has been proved). On the other hand, *Stizoides* is closely related to *Stizus,* and like that genus utilizes Orthoptera. In the case of *Stizoides* the known hosts belong to another subfamily, the Sphecinae, but I have postulated that *Stizus* was probably the original host, and perhaps still one of the hosts. Apparently the genus *Brachystegus* is exceptional in that it oviposits on prey (grasshoppers) different from that utilized by its presumed ancestral stock. This genus is considered a highly evolved member of the Nyssonini; presumably it has undergone a transfer from Homoptera to Orthoptera and from Gorytini to Larrinae as hosts. There is some suggestion that other Nyssonini have also transferred to other hosts and prey, but much remains to be learned of this phenomenon (Chapter IV:B).

An important difference concerns the manner of oviposition and behavior of the first-instar larva. *Nysson* lays its egg in a concealed position on the prey before the cell is fully provisioned and the egg of the host laid. The parasite egg hatches before that of the host, and the first-instar parasite larva destroys the egg or larva of the host. Members of the genus *Stizoides* attack wasps that oviposit on the first prey in the cell or that use only one prey. Thus the host egg is already *in situ* when the parasite enters; this egg is destroyed by the *Stizoides,* which then lays its egg on the side of the grasshopper in much the manner of *Stizus.* Apparently *Brachystegus* resembles *Stizoides* in this respect. The oviposition behavior of *Stizoides* and *Brachystegus* appears less specialized than that of *Nysson,* as does the egg itself (which in *Nysson* has strong surface sculpturing as well as a shorter incubation period). The behavior of the newly hatched larva of *Nysson* is of course also highly specialized. It is worth noting that *Nysson* has many more specializations of adult structure than does *Stizoides*, indicating fairly clearly that it evolved from typical Nyssoninae much earlier than did *Stizoides*.

The cleptoparasitic Nyssoninae parallel very closely the situation in the spider wasps (Pompilidae). In this family, *Evagetes* shows but slight structural divergence from its hosts, and its manner of attacking the host is almost identical to that of *Stizoides*. On the other hand, *Ceropales* is placed in a different subfamily from its hosts; in this genus, the egg is laid in a concealed position on the spider and the first-instar larva destroys that of the host, much as in *Nysson*. However, *Ceropales* differs from *Nysson* in that it is not a digger but inserts the egg into the book lungs of the spider while the latter is being taken to the nest.

Throughout the higher Hymenoptera, it appears that cleptoparasites have evolved from the same stock as their hosts (Wheeler, 1919). Some have apparently diverged recently and still resemble their hosts closely and exhibit relatively simple parasitic behavior (*Stizoides, Evagetes*), while others apparently split off from their hosts long ago and have become much more specialized structurally and ethologically (*Nysson, Evagetes*). Further examples of these and other stages may be found in the ants and in the bees. The early stages in the evolution of cleptoparasitism are unknown; some authors have assumed prey stealing to have been an initial step.

E. Cocoon spinning

Although this study has concerned itself primarily with adult behavior, it would be unwise to omit mention of the one manifestation of complex behavior on the part of the larva: the spinning of the cocoon. This behavior is especially complex in the Nyssoninae, all of which utilize sand grains in the walls of the cocoon and many of which construct a series of pores in the walls.

There are many brief and fragmentary published descriptions of cocoon spinning in various Nyssoninae, chiefly the larger species, but a definitive study of this behavior has yet to be made. To date, the best descriptions are those of Janvier (1928) on *Bembix brullei* and of Tsuneki on *Stizus pulcherrimus* (1943b) and *Bembix niponica* (1956). Tsuneki's discussion of *B. niponica* is based on more than 20 larvae observed in vials or petri dishes containing sand similar to that in the nesting site. He recognized six stages in cocoon construction and presented sketches of several of these. In the summer of 1956 I studied two larvae of *B. pruinosa* and one of *B. spinolae* which spun their cocoons in glasstopped rearing tins. My observations largely confirm those of Tsuneki, but I suggest recognizing a seventh stage. These stages are outlined briefly below. For further details the reader is referred to the accounts of Janvier and of Tsuneki. The length of time taken for each stage is given, mostly following Tsuneki; these are approximate times under ideal conditions.

Stage 1: preparing the cell. Shortly after it stops feeding, or even slightly before, the larva moves its mouthparts over the cell walls, clearing the walls of loose sand and fragments of prey and packing these beneath it (5–10 hr).

Stage 2: spinning the hammock. The larva moves about and stretches its head into various parts of the cell, attaching silken threads to many places on the walls. In the center of this meshwork a sheet or "hammock" is formed, on which the larva remains for the following stages (4–7 hr).

Stage 3: spinning the shroud. The larva surrounds itself with an elongate ellipsoid of silk, its posterior end slightly lower and attached to the floor with a bundle of silk, its anterior end (toward the entrance to the cell) forming a somewhat funnel-shaped opening (4–5 hr).

Stage 4: coating the walls with sand grains. The larva reaches out the opening, scoops a number of sand grains into a pile under its head, making much use of the mandibles, then gradually withdraws into the shroud, moving the lump of sand along with it. The sand grains are then applied to the walls of the cocoon, presumably with the aid of oral secretions. This is repeated many times, the larva reversing its direction to complete the posterior end of the cocoon (2–4 hr).

Stage 5: making the cap. The larva "gnaws the brim of the opening with the mandibles and pulls it inwards," then spins a silken disc over the opening and lines it with sand grains previously stored inside (30 min to 1 hr).

Stage 6: making the pores. The larva works over the inner walls of the cocoon for several hours, in the course of which it forces its mandibles out through several places near the widest part of the cocoon, opening and closing the mandibles and rotating them to some extent. With the aid of oral secretions, these holes are formed into small open cones. The pores are by no means evenly spaced. They tend to be relatively constant in number within most species, as shown in Table 46 (6–12 hr).

Stage 7: lining the inside of the cocoon. The larva then lines the entire inner surface of the cocoon with a layer of silk, which is especially thick beneath the pores, forming slight swellings beneath them. Actually there are three layers of silk beneath each pore, the middle layer forming a solid disc, the outer layer a disc of lighter silk which is generally perforated by a single hole (Fig. 204). During this stage the walls lose their elasticity and dry to form a very hard cocoon (about 1 day).

Thus at least two full days are required for the performance of these elaborate behavior patterns, the details of which might well provide the substance of a lengthy treatise in themselves.

§E. Cocoon spinning 457

Fig. 204. Cross section of a typical "pore" in the wall of the cocoon of a *Bembix* (somewhat diagrammatic). The rim of the pore, shown in black, is actually dark brown, resembling heavy parchment; the stippled portion is the middle of the three layers of silk.

There is no evidence that *Bembix niponica, B. pruinosa,* or *B. spinolae* differ materially in their cocoon-spinning behavior; *Stizus pulcherrimus* also exhibits essentially the same behavior (Chapter VI:B, 5; but see Fabre's rather different observations on *S. ruficornis,* Chapter VI:D). Tsuneki remarks that certain species (*S. pulcherrimus, B. rostrata*) apparently select sand grains of a certain size and push away those which are too large. The species of *Bembix* studied by Tsuneki and myself nest in sand of rather uniform texture and do not exhibit selection of grain size. Since the cocoons of all Nyssoninae tend to show only a limited amount of variation in the size of the sand grains in the walls, it is probable that many species are capable of exercising some selection of particle size. It goes without saying that the cocoons of all sand wasps reflect the type of soil in which they nest. That is, the cocoons of species that nest in pure, fine-grain sand tend to be lined with translucent quartz grains of rather uniform size (for example, *Microbembex monodonta, Bembix pruinosa*), while those of species that nest in heavy clay-loam tend to be lined with much smaller particles of dark color, rendering the walls completely opaque (for example, *Bembix cinerea, Sphecius speciosus*). Cocoons from two different colonies of one species may present a somewhat different appearance if the substrate is different.

No detailed descriptions of cocoon spinning have been made on any species other than those cited above. However, comparative study of the cocoons, plus what is known of *Bembix* and *Stizus,* as well as fragmentary notes on several other species, permit a number of deductions as to comparative spinning behavior. I have brought together, in Tables 46A–46C, data on all the cocoons I have been able to collect, as well as some published data. Photographs of selected cocoons are presented in Figs. 205–208. On the basis of available data, the following generalizations seem justified.

(1) Most Nyssoninae spin ovoid cocoons, more tapered toward the posterior end, which is directed toward the apex of the cell (the excrement being voided into the tapered portion); the anterior

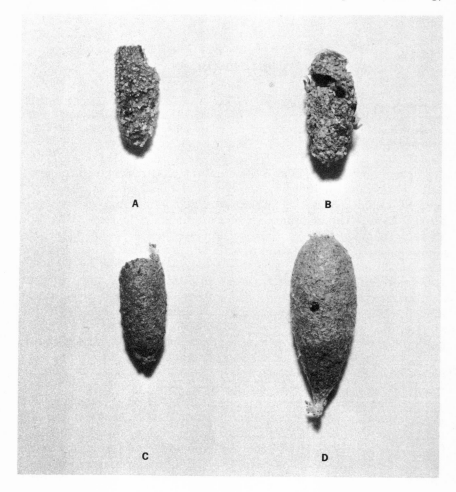

A B C D

end is broader and blunter and has ill-defined cap. In the smaller Gorytini and in *Alysson* there is a tendency for the cocoon to be more cylindrical and less tapered posteriorly. A small terminal nipple-like protuberance has been noted in *Gorytes*[2] and in *Stizus* (Fig. 207A).

(2) In the genera *Glenostictia* and *Steniolia* the initial shroud of stage 3 is apparently double walled, the outer shroud persisting around the final product (Fig. 208); in all other Nyssoninae the

---

[2]Maneval (1939) shows the cocoon of *Gorytes planifrons* as being shaped essentially as in *Bembix*, that is, tapered posteriorly and without a terminal nipple.

§E. Cocoon spinning

Fig. 205. Cocoons of various Alyssonini and Gorytini, all shown on same scale of magnification and with anterior end uppermost: (A) *Alysson melleus* (cap missing); (B) *Ochleroptera bipunctata* (slightly damaged); (C) *Gorytes canaliculatus* (cap missing); (D) *Hoplisoides costalis* (the dark spot is not a pore but a hole made by an insect pin on which this cocoon had been mounted); (E) *Sphecius speciosus,* lower surface (left) and upper surface (right).

Fig. 206. Cocoons of various Stizini and Bembicini, all of about the same scale of magnification and with anterior end uppermost: (A) *Bembecinus neglectus* (cap missing); (B) *Glenostictia pulla*; (C) *Microbembex monodonta* (the dark spot is the single pore); (D) *Bicyrtes quadrifasciata*; (E) *Bembix cinerea*; (F) *Stictia vivida*.

§E. Cocoon spinning

Fig. 207. Cocoon: (A) of *Stizus pulcherrimus,* anterior end (cap) uppermost; (B) of *Bembix texana,* anterior end uppermost.

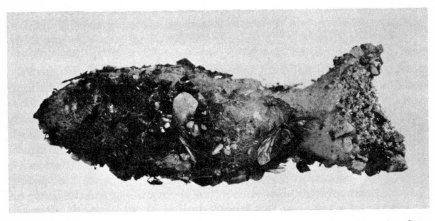

Fig. 208. Cocoon of *Steniolia nigripes* surrounded by its outer shroud bearing fragments of the prey. (Photograph supplied by P. D. Hurd, Jr., from material collected in San Diego County, California.)

Chapter XIV. Comparative Ethology

TABLE 46A. DATA ON COCOONS OF ALYSSONINI, NYSSONINI, GORYTINI, AND STIZINI

Species	No. of cocoons examined	Mean length cocoons (mm)	No. of pores	General form	Sources of material; references
1. Alysson melleus	2	8	0	Narrowly elliptical Sand grains adhere somewhat loosely to walls (Fig. 205A)	Palmdale, Fla. (no. 1968) Pottawatomie Co., Ks. (no. 47)
2. Ochleroptera bipunctata	8	8	0	Narrowly elliptical Sand grains more firmly built into walls (Fig. 205B)	Ithaca, N.Y. (no. 1571) Huntington, L.I., N.Y., March 1924 (J.C. Bridwell) [U.S. Nat. Mus.]
3. Gorytes canaliculatus	3	9	0	Narrowly elliptical Small terminal nipple (Fig. 205C)	Ithaca, N.Y. (nos. 1087, 1385) Pottawatomie Co., Ks. (no.111)
4. Hoplisoides costalis	2	12	0	Ovoid, somewhat pointed posteriorly (Fig. 205D)	Woodstock, Md. (E.J. Reinhard) [U.S. Nat. Mus.]
5. Sphecius speciosus	3	29	2-13	Narrowly ovoid-elliptical Pores crowded together on one side (Fig. 205E)	Pittsford, N.Y. (no. 1122) Reinhard, 1929
6. Nysson dimidiatus	1	7	0	Ovoid Shaped much as in Bembix	Maneval, 1939
7. Stizus pulcherrimus	1	23	5	Ovoid-elliptical Terminal nipple Pores not all on one side (Fig. 207A)	Tsuneki, 1943b; cocoon supplied by Tsuneki
8. Bembecinus neglectus	5	13	4-6	Narrowly ovoid No terminal nipple (Fig. 206A)	Pottawatomie Co., Ks. (nos. 375, 377, 381)
9. Bembecinus mexicanus	1	16	5	As above	Muna, Yucatan, Mexico (no. 1907C)
10. Bembecinus argentifrons	1	11	3	As above	Capetown, So. Africa (J.C. Bridwell) [U.S. Nat. Mus.]

§E. Cocoon spinning

TABLE 46B. DATA ON COCOONS OF BEMBICINI (OTHER THAN BEMBIX)

Species	No. of cocoons examined	Mean length cocoons (mm)	No. of pores	General form	Sources of material; references
11. Bicyrtes quadrifasciata	5	19	5-7	Ovoid (Fig. 206D)	Pottawatomie Co., Ks. (no. CY39) Ithaca, N.Y. (no. 948) Kill Devil Hills, N.C. (Krombein)
12. Bicyrtes ventralis	1	17	5	Ovoid (as above)	Pottawatomie Co., Ks. (no. 106)
13. Bicyrtes fodiens	4	16	5-7	Ovoid (as above)	Pottawatomie Co., Ks. (nos. 146, 200, 201)
14. Bicyrtes capnoptera	1	16	7	Ovoid (as above)	Austin, Texas, July 23 (1913) [U.S. Nat. Mus.]
15. Glenostictia pulla	1	13	3	Narrowly ovoid (Fig. 206B) Pores all with double apertures	Jerome Co., Idaho (Gillaspy)
16. Glenostictia scitula	3	15	4-5	Ovoid; shroud persists Pores lack elevated rims in available material	Brewster Co., Texas (nos. 1854, 1857)
17. Steniolia nigripes	9	25	4-5	Ovoid; shroud persists (Fig. 208) Pores lack elevated rims	San Diego Co., Calif. (P.D. Hurd)
18. Steniolia obliqua	4	23	4-5	As above, but pores have elevated rims and nearly all have double apertures	Jackson Hole, Wyo. (no. 1821)
19. Stictia carolina	2	32	10	Ovoid	Pottawatomie Co., Ks. (no. 136) Kill Devil Hills, N.C. (Krombein)
20. Stictia vivida	1	26	10	Ovoid (Fig. 206F)	Progreso, Yucatan, Mex. (no. 1915A)
21. Rubrica surinamensis	several	30	2	Ovoid	Brèthes, 1902
22. Rubrica gravida	several	28	2-3	Ovoid	Llano, 1959
23. Microbembex monodonta	4	13	1	Narrowly ovoid (Fig. 206C) Pore tends to be toward posterior end of cocoon	Pottawatomie Co., Ks. (no. 18) Andover, Mass. (nos. 1956, 1959)

TABLE 46C. DATA ON COCOONS OF 15 SPECIES OF THE GENUS BEMBIX

Species	No. of cocoons examined	Mean length cocoons (mm)	No. of pores (mean + range of variation)	Remarks	Sources of material; references
24. Bembix amoena	20	23	4.5 (3-7)	Made of very fine-grained soil	Yellowstone Pk., Wyo. (nos. 1786-1799) Smithfield, Utah (no. 1825)
25. B. belfragei	35	24	5.5 (3-8)		Pottawatomie Co., Ks. (nos. 87-614) Highlands Co., Fla. (no. 1732)
26. B. cinerea	2	20	28.5 (27-30)	Made of very fine particles of dark silt (Fig. 206E)	Flamingo, Fla. (no. 1027)
27. B. comata	10	18	3.7 (3-5)		Alameda, Calif. (nos. 783-795)
28. B. multipicta	1	22	9.0		Progreso, Yucatan, Mex. (no. 1912)
29. B. niponica	101	22	10.5 (6-17)		Tsuneki, 1956
30. B. nubilipennis	15	22	5.0 (3-7)	Made of fine-grained clay-sand	Versailles, Ind. (nos. 998, 1455, 1460, 1469)
31. B. occidentalis beutenmulleri	5	25	7.0 (5-10)		Antioch, Calif. (no. 1005)
32. B. pruinosa	11	22	6.7 (5-8)	Nearly all pores have double apertures	Oswego Co., N.Y. (nos. 958, 1109, 1205)
33. B. rostrata	several	25	9.0 (8-10)		Wesenberg-Lund, 1891
34. B. sayi	1	21	5.0		Highlands Co., Fla. (no. 1030)
35. B. spinolae	3	20	6.3 (6-7)		Reno Co., Ks. (no. 563) Ithaca, N.Y. (no. 944)
36. B. texana	1	23	6.0	Fig. 207B	Highlands Co., Fla. (no. 1059)
37. B. troglodytes	4	19	5.0 (3-7)		Pottawatomie Co., Ks. (nos. 23, 48, 138)
38. B. u-scripta	14	21	16.0 (9-32)	Surface unusually rough	Cameron Co., Texas (nos. 1171, 1172, 1173, 1180)

§E. Cocoon spinning

finished cocoon rests more or less free in the cell except for some fine silk threads left over from stage 2.

(3) All Nyssoninae incorporate particles of soil into the walls of the cocoon, probably in much the manner described for *Bembix*.[3] It appears that in *Alysson* the particles are incorporated more loosely and incompletely than in other genera; I am uncertain as to whether sand grains are actually drawn into the cocoon and applied from the inside as in other Nyssoninae.

(4) In *Alysson, Nysson*,[4] and the smaller Gorytini, no pores are prepared in the walls of the cocoon, the sixth of the seven stages described for *Bembix* apparently being omitted. The pores of *Sphecius* and of the Stizini and Bembicini are strikingly similar in structure aside from minor differences noted in the tables and figures. It is difficult to believe that the unique, complex behavior involved in pore construction could have arisen more than once.

(5) The number of pores is roughly correlated with the size of the cocoon, as shown in Fig. 209, with a few notable exceptions which stand out clearly in the figure. A number of species of diverse genera have between three and six pores, but in *Stictia* and a few of the larger species of *Bembix* the number is close to ten. The amount of interspecific variation in pore number is not great, again with a few notable exceptions which stand out in the figure.

The species of *Rubrica* form one of the major exceptions to the general trend. Brèthes (1902) states that cocoons of *R. surinamensis* measure 30 mm long, while Llano (1959) gives 22–33 mm as the length of the cocoons of *R. gravida* (although this is a larger species). Brèthes states that there are only two pores in the cocoons of *surinamensis*, Bodkin (1917) indicates that there are four in *denticornis*, while Llano figures only two in the cocoons of *gravida* but suggests there may be three or more at times. Thus these three species will all fall in the extreme lower right-hand corner of the graph (Fig. 209, nos. 21 and 22; I have omitted *denticornis* for want of data on cocoon size). It would appear that the pores of these species are unusually large and strongly elevated; possibly the species of this

---

[3]I reported in 1957b that *Bembix hinei* prepares a cocoon "totally different from that of other Bembicini. No sand grains are incorporated into its walls, which are brown and parchmentlike [also without pores]. Possibly this cocoon is adapted for survival in soil saturated with salt water." I based these remarks on a single cocoon spun in a rearing tin. I have since seen a few other cocoons of diverse Bembicini, all spun in abnormal situations, which were only partially coated with sand grains. Also, Tsuneki mentions that 4 of 20 cocoons of *B. niponica* spun in vials were abnormal—all of them much like my *B. hinei* cocoon! I now feel almost certain that this cocoon was abnormal and that my remarks on its adaptive significance were unjustified. I have omitted this species from Table 46C.

[4]Maneval (1939) presented an excellent figure of the cocoon of *Nysson dimidiatus*, which is shaped much like a *Bembix* cocoon but lacks pores in the walls.

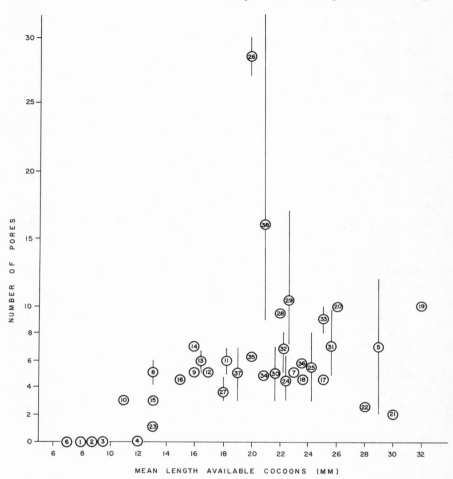

Fig. 209. Relationship of size of cocoons to the number of pores. Each circle bears the number of the species (as taken from Tables 46A–46C) and is located at the mean for that species (or the single specimen when only one is known); the vertical lines above and below each circle indicate the known range of variation.

genus compensate for a reduction in number of pores by an increase in their size. However, further studies are in order.

I am unable to provide an explanation for the great variation in the number of pores in the cocoons of *Bembix niponica* (Fig. 209, no. 29) and *B. u-scripta* (no. 38). The *niponica* sample was large (101); but 105 cocoons of ten species of North American *Bembix*, when pooled, do not show anywhere near as much variation in pore number. In the case of *u-scripta* I had only 14 cocoons, all from one locality, but the number of pores varied from 9 to 32. Actually, only 2 cocoons had more than 17 pores, so much of the variation might be the result of behavioral abnormalities on the part of a few individuals. The fact that this colony was located in the heavy, somewhat moist and saline soil of a beach (near Port Isabel, Texas), apparently an unusual habitat for this species, is suggestive. One notes that *B. cinerea*, which typically nests in the heavy soil of salt marshes, where there is a considerable content of salt water, makes 27-30 pores per cocoon (Fig. 206E). It would be interesting to know how many pores *u-scripta* makes in more inland localities. It is possible that selection favors a greater number of pores in dense, saline soil. The Port Isabel colony of *u-scripta* may not have been established there long enough for the number of pores to have become stabilized at the higher figures; or there may be enough ingress of individuals from other localities to maintain a large amount of variation.

Unfortunately there is no good experimental evidence as to the true function of these pores. The fact that there are more pores in certain species nesting in moister soil suggests that they do not function to let in moisture, and there is nothing in their structure that suggests that they form a device for drawing water out. Most authors have assumed that they are respiratory in function. Riley (1892b) made this suggestion, and in the discussion following his paper it was mentioned that Trouvelot had tried shellacking the cocoons of Bembicini but had failed to prevent them from developing normally. Reinhard (1929) found that air would pass through the pores in the cocoons of *Sphecius*, but not readily through the walls. He coated several cocoons with wax and then opened them after 2 weeks, but could notice no ill effects on the larva. He postulated that the resting larva needed little oxygen, but that during pupation and emergence of the adult much more might be needed. As yet no one has attempted to verify his hypothesis.

Two facts lead me to believe that the pores may well be respiratory in function: (1) pores are most numerous in cocoons found in damp, heavy soil, where there are surely fewer and smaller air spaces than in sand; (2) larger species, which have greater oxygen requirements, tend in general to have more pores in the walls of their cocoons.

F. Natural enemies

Finally, an attempt must be made to summarize the available data on the various types of parasites and nest scavengers that attack Nyssoninae. In Table 47 this information is brought together with respect to the families and genera involved; the reader is referred to Chapters II-XII for the names of the species. Little is known about the behavior and incidence of most of these parasites. Any generalizations made here are likely to require revision in the relatively near future if these insects receive anywhere near the amount of attention they deserve.

(1) MILTOGRAMMINAE. It will be noticed immediately that the most prevalent and non-host specific enemies are the miltogrammine flies (Sarcophagidae). Members of the genus *Senotainia* are known to attack nine genera of Nyssoninae; they may well attack all the genera, as certain species of *Senotainia* are known to attack wasps of several subfamilies of Sphecidae as well as wasps of the family Pompilidae. Adult miltogrammine flies deposit live larvae on the prey of wasps, and these larvae develop primarily at the expense of the prey, although they very often destroy the wasp egg or small larva. In the case of wasps that provision progressively, it is known that miltogrammine maggots that gain access to the cell after the wasp larva is fairly large often do not destroy the host; if the female wasp brings in sufficient prey, both host and parasite may reach maturity (for example Chapter X:G). However, in nearly all other cases, the host perishes either because the prey is consumed first by the maggots or (apparently much more commonly) because the host egg or larva is actually destroyed by the maggots.

Certain miltogrammine flies enter the burrow and larviposit in the burrow or cell. Ristich (1956) speaks of certain of these as "hole crawlers" (*Metopia* spp.) and others as "hole searchers" (*Opsidia gonioides, Phrosinella fulvicornis*). In *Phrosinella* the female fly has flattened fore tarsi and is able to dig in the soil, but the species of *Metopia* apparently enter only burrows that are already open. These flies apparently respond to the burrow rather than to the wasp. On the other hand, flies of the genus *Senotainia* appear to respond to the image of the wasp. These flies trail the wasps to their nests and larviposit on the prey as it is being taken into the nest; or they may follow the wasp into the nest, larvipositing and emerging almost immediately. In this genus the eyes of the female are larger than usual and the facets on the front part of the eyes are enlarged. Allen (1926) considers these as probably "specializations permitting the female to keep herself oriented more readily with the rapidly moving form of the wasp she follows, while in flight."

Miltogrammine flies also often perch near burrows, flying up when the female arrives and following her in (Fig. 210). These

§F. Natural enemies   469

TABLE 47. NATURAL ENEMIES OF THE NYSSONINAE*

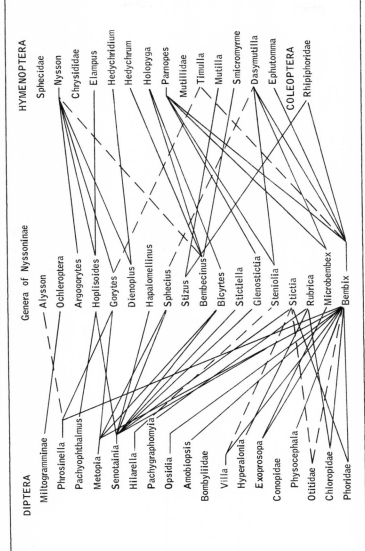

* Dashed lines indicate an element of doubt either as to the identification of the parasite or that the parasite was successfully attacking this host. Other still more doubtful records are mentioned in the text but not included here.

470 Chapter XIV. Comparative Ethology

Fig. 210. A miltogrammine fly perching upon a twig overlooking the nesting site of a sand wasp. (From Evans, 1963d.)

flies are also sometimes seen hovering behind wasps that are digging or otherwise engaged, or even trailing Mutillidae, but in general they seem to respond most fully to the image of a wasp laden with prey. I have noted that *Senotainia rufiventris* is most often seen following female *Microbembex monodonta* carrying large prey rather than small (Chapter XII:A, 11). Ristich (1956) found *S. trilineata* to be more successful in attacking *Aphilanthops frigidus,* which carries large queen ants rather loosely, than in attacking species of *Philanthus,* which carry small bees tightly beneath their bodies.

(2) BOMBYLIIDAE. Female Bombyliidae (bee flies) lay large numbers of eggs, which are ejected into various open holes in the sand. The small larvae are adapted for finding their way into the cell, although a great many of them undoubtedly perish without finding a host. In the cell, they wait until the wasp larva enters the prepupal stage, after which time they develop as external parasitoids

§F. Natural enemies 471

of the wasp larva. Bohart and MacSwain (1939) described parasitism of *Bembix occidentalis* by *Exoprosopa eremita*, and Bohart, Stephen, and Eppley (1960) described parasitism of the alkali bee by *Heterostylum robustum*. In the Nyssoninae, Bombyliidae are known to attack only certain Bembicini.

(3) CONOPIDAE. The conopid fly *Physocephala texana* is now known to attack several species of the genus *Bembix* (*P. affinis*, discussed by Bohart and MacSwain, 1939, is a synonym of *texana*). These flies attack the adult wasps, especially the males, laying their egg between the abdominal tergites. The fly larva develops within the abdomen of the wasp and eventually forms its puparium there (see Chapter X:A, 11).

(4) OTITIDAE, CHLOROPIDAE, PHORIDAE. Certain flies of these three families have been reared from the cells of Bembicini or have been found entering the burrows. The maggots are probably best regarded as cell scavengers. In several cases they have been found living on the fly debris in the cell without preventing the wasp larva from developing normally (see Chapters IX:B, 8; X:A, 11; X:E).

(5) CHRYSIDIDAE. Cuckoo wasps are known to attack many different wasps and bees; their heavy integument and ability to roll into a ball apparently serve to protect them against the stings of their hosts. These wasps enter the nests of their host and lay their eggs on the larva. Some species dig effectively and are able to enter nests even though they are closed. The larva remains small until the host spins its cocoon, after which time the larva develops rapidly and eventually spins its cocoon inside that of the host (see especially Chapter VIII:I, 8).

(6) MUTILLIDAE. Mutillid wasps (or "velvet ants") are also known to attack many different bees and wasps, developing as parasitoids on the host larva after the cocoon has been spun and spinning their own cocoon inside that of the host. Mutillids also dig into nests, but they do not normally lay their egg until after the cocoon has been spun. The egg is laid through a hole in the wall of the cocoon, which is later sealed up by the mutillid (for details, see Bohart and MacSwain, 1939).

Mutillidae also sometimes attack adult *Bembix* and feed upon their body fluids. Mellor (1927) found that female *Ephutomma continua* seek out adult *Bembix olivacea* in their burrows at night and, without stinging their victims, feed at their throats and consume their body contents. Other mutillids are known to attack adult honeybees in their hives (Clausen, 1940).

(7) RHIPIPHORIDAE. Rhipiphorid beetles are known to attack, as

parasitoids, the larvae of a considerable variety of bees and wasps. *Macrosiagon flavipenne* is known to attack *Bembix spinolae* and probably *B. nubilipennis* (Chapter X:C, 6), while an African species is said to attack *Stizus imperialis*. Rhipiphoridae typically lay their eggs on the soil or on plants, and the larvae attach themselves to the hosts and are carried to the nest (Clausen, 1940:548–557).

(8) DISCUSSION. Cleptoparasitism by wasps of the genus *Nysson* was discussed in an earlier section. *Nysson* attacks Gorytini, while most of the other natural enemies (except Miltogramminae) appear to attack principally the Stizini and Bembicini. The large number of parasites recorded for *Bembix* (Table 47) is a reflection of the fact that this genus is much better studied than any other; it does not necessarily mean that this genus is heavily parasitized as compared with others.

In the next chapter we shall need to consider natural enemies as possible selective agents in molding the behavior patterns of these wasps. It may therefore be useful to conclude this chapter by attempting to categorize the known enemies with respect to the way in which they attack their hosts:

Predators on adult wasps: certain Mutillidae, certain ant lions (Tsuneki, 1956:125), kingbirds (Chapter X:C, 6)
Parasitoids of adult wasps: Conopidae
Parasitoids of wasp larvae: Bombyliidae, Chrysididae, Mutillidae, Rhipiphoridae
Cleptoparasites: Nysson, Miltogramminae
Scavengers: Phoridae, Chloropidae, Otitidae

It may also be found useful to group these natural enemies according to their behavior leading to oviposition or larviposition:

Enter cell via closed or open burrows: Nysson, Chrysididae, Mutillidae, some Miltogramminae
Oviposit or larviposit in open burrows: Bombyliidae, Otitidae, Phoridae, Chloropidae, some Miltogramminae
Trail adults and larviposit on prey: some Miltogramminae, especially Senotainia
Oviposit away from host, the larva attaching itself to the host and being carried to the nest: Rhipiphoridae
Oviposit on adult wasps: Conopidae

Chapter XV. The Evolution of the Behavior of Sand Wasps

E. T. Nielsen's "Moeurs des *Bembex*" (1954) concluded on a pessimistic note. Certain behavioral features, Nielsen felt, "traverse the classification of the bembicine wasps in an extremely troubling manner. Sometimes we find a difference in different colonies of the same species, sometimes we find wholly different lines of evolution terminating in almost identical biological types . . . The biology of *Bembex* teaches us the instability of the fundamental postulate that upholds the congruence of biological and morphological variation."

The term *"Bembex"* was used by Nielsen to denote the tribes Bembicini and Stizini plus the gorytine genus *Sphecius*. I believe that his conclusions as to the lack of concordance of structure and behavior were in part the result of the unrealistic classification that he accepted. There is, in fact, no sound way of defining *"Bembex"* in his sense on the basis of structure. Also, Nielsen did not accept the generic status of *Bembecinus, Stizoides, Stictiella, Rubrica, Zyzzyx,* and several other groups. Furthermore, his own field experience involved only one species, *Bembix rostrata* (L.), and he lacked some of the perspective necessary to evaluate the reported behavior patterns of other sand wasps.

Lest this seem a harsh criticism, I hasten to add that "Moeurs des *Bembex*" provided an important summation of knowledge of the group and did much to stimulate further work. A great deal has been learned since 1945. Perhaps the strongest compliment that can be paid to any work of science is to say that its publication engendered its own obsolescence.

A. Behavior and the classification of the Nyssoninae

Although I cannot accept Nielsen's conclusions, I do not pretend to find perfect concordance between behavioral features and the classification based upon structure. My survey of the ethology of the Gorytini (Chapter III) showed, with few exceptions, a monotonous uniformity, even though the genera involved are separated by greater morphological gaps than separate the genera of Bembicini. Conversely, the Stizini and Bembicini display an array of behavioral differences much more impressive than the minor structural features that separate the genera in this section of the Nyssoninae. For example, present knowledge indicates that members of the genera *Gorytes, Hoplisoides,* and *Dienoplus* make similar nests, utilize much the same types of prey, lay the egg in the same manner, and so forth, and so far as we know highly distinctive genera such as *Ochleroptera, Argogorytes,* and *Ammatomus* do not differ radically in their behavior. On the other hand, one finds many differences in behavior between such genera as *Stictiella* and *Stictia,* which Nielsen was content to leave as subgenera of the old blanket genus *Monedula*. Indeed, even within certain genera (*Stictia, Glenostictia, Bembix*), one finds important differences in manner of oviposition which contrast strongly to the uniform manner of oviposition of Gorytini (including *Sphecius*).

It appears that in the Gorytini there has been much structural evolution unaccompanied by important changes in behavior, and that in the Stizini and Bembicini there has been rapid evolution in behavior unaccompanied by pronounced changes in structure. A person impressed with the importance of behavior in evolution might insist that most of the Gorytini be lumped into one genus and that *Bembix* be split into two or more genera. I do not so insist, for two reasons: (1) our knowledge of the Gorytini is not so far advanced that we can be sure there are not subtle but important ethological differences, and (2) palaeontological and comparative morphological evidence suggests that the Stizini and Bembicini have undergone their radiation very recently: in a sense, their structure has not had time to "catch up" with their behavior. While ethological differences on the species level must influence our classification, especially when they involve isolating mechanisms (for instance, in the species of the cricket genus *Gryllus;* Alexander, 1962), and while ethological differences have sometimes been used for discrimination of genera (for example, in the case of *Glenostictia;* Gillaspy, Evans, and Lin, 1962), we cannot entirely afford to sacrifice the utilitarian aspects of classification. If we were to split *Bembix* into several genera based on manner of oviposition, we would have to leave a great many species unclassified until such time as their eggs were discovered.[1] Furthermore, how can we decide which

---

[1]Minkiewicz did, in fact, split *Bembix* on this basis, but he has not been followed by subsequent workers (see Evans, 1957b:210 for details).

§A. Behavior and classification 475

characters are "important"? Or, if we evade that question by weighing all characters equally, how can we do full justice to all aspects of behavior and physiology, which are surely as essential to the organism and to the classifier as structural details? These are questions that, for the moment, we must learn to live with. Even two different sets of more than 70 morphological characters each (one set from the larvae and one from the adults), when treated by sophisticated statistical methods, yielded classifications far from completely congruent (Rohlf, 1963).

To me, it is reassuring to find so much concordance between structure and behavior in the nyssonine wasps. In behavior, as in structure, the Nyssoninae appear to form a coherent and probably monophyletic group (after removal of *Mellinus, Ampulex,* and certain other elements discordant both in structure and behavior). Many of the genera are as readily definable on the basis of behavior as on structural criteria: *Stizus, Bembecinus, Bicyrtes, Steniolia,* and others. So far as we know, *Bembix* is unique in at least one feature of behavior (Chapter XI:E, 2) in spite of much variation within the genus. A number of complexes of genera hold together as well on the basis of behavior as on the basis of structure: for example, *Gorytes, Hoplisoides,* and *Dienoplus;* the genera of Bembicini with recessed ocelli. *Bicyrtes* and *Microbembex,* which are not commonly placed together in classifications, have at least one important structural feature of the adults and one of the larvae in common (Tables 41B and 42, Fig. 203); these two genera also have strikingly similar digging behavior, in which the female throws out several jets of sand each time she dips her body, the sand being scattered so widely that no mound ever accumulates.

On the other hand, there are genera that appear very similar on the basis of behavior but are so different structurally that we cannot believe them to be closely related. One example is provided by the two genera of cicada killers, *Sphecius* and *Exeirus* (Chapter V). These wasps dig similar nests, hunt their prey in a similar manner, and so forth. Their structural modifications for carrying their very large prey are, however, very different: unusually long legs in *Exeirus,* flattened and curved tibial spurs in *Sphecius.* Another striking example is provided by *Glenostictia scitula,* a species that lays its egg in the empty cell and provisions with a wide variety of different insects, strongly suggesting *Microbembex,* although there are many reasons for believing that these represent widely separate phyletic lines. Nielsen is perfectly correct that different evolutionary lines have sometimes resulted in animals of very similar behavior. There are as many parallelisms and convergences in behavior in the Nyssoninae as there are in structure, perhaps more. I do not find this a cause for pessimism, but rather a major source of suggestions regarding selection pressures and directions of evolution.

In general, ethology suggests that the classification of the Nyssoninae requires no major revisions. The various genera created by the splitting up of the old genus *Monedula* appear sound enough so far as the sketchy behavioral evidence goes (*Stictia, Zyzzyx, Rubrica, Editha, Stictiella*). In the case of *Stictiella*, recent studies of behavior suggested that the genus still contained disparate elements, and we erected the genus *Glenostictia* for those species employing progressive provisioning and utilizing Diptera rather than Lepidoptera (Gillaspy, Evans, and Lin, 1962). This step suggested still further splitting up of this complex (Gillaspy, 1963b); whether these later steps will appear justified on the basis of ethology remains to be seen. It is probable that as our knowledge of the behavior of sand wasps grows, further refinements in the classification will suggest themselves.

There is one realm in which behavioral characters fail to confirm the usual classification of these wasps fully. This is with respect to the tribes. In Chapters II and III we found nothing to justify strong separation of *Alysson* from *Gorytes* and its relatives (especially *Clitemnestra*). If the phylogeny expressed in Fig. 203 is anywhere near correct, recognition of the tribe Alyssonini on the basis of structure is also scarcely justifiable. One also looks in vain for any hard-and-fast behavioral features separating the Stizini and Bembicini. Structurally, these tribes are distinguished on the basis of the distorted ocelli of the Bembicini (but the ocelli are little modified in *Carlobembix,* and somewhat distorted in the gorytine genus *Kohlia*) and the loss of a mid-tibial spur in the Bembicini (but a mid-tibial spur has also been lost in *Entomosericus* and in some *Nysson* and *Stizoides*).

The truth would seem to be not that ethology and morphology are here incongruent, but that the tribes of Nyssoninae are weakly supported on any grounds. Larval structure is relatively uniform throughout the subfamily, although the tribes can be separated (after a fashion) by selection of certain characters (Evans, 1959a). Differences in cocoon structure do not fall at the usual breaks between the tribes (Table 46A). The Gorytini (plus the Alyssonini) might be characterized ethologically on the basis of their common manner of oviposition (Figs. 20, 33, and 58); but selection of certain other behavioral features (for instance, type of prey) would alter the tribal limits. Only *Nysson* and its segregates are truly distinctive in ethology and in adult structure (although not in structure of the larvae or cocoons!); if Figure 203 is reasonably correct, the Nyssonini do in fact form a discrete (but short) phyletic line. Although I have found the five conventional tribal names useful in the present study and have used them freely, I doubt if most of them represent entities that are any more real than several other groupings of genera which might be proposed. A more realistic classification would recognize at most two tribes, the Nyssonini and the Bembicini. Inclusion of the Alyssonini, Gorytini, and Stizini in the Bembicini might help to indicate that these wasps appear to form a single phyletic complex, as is strongly suggested by both structure and behavior

B. Phyletic trends in the behavior of sand wasps

That the behavior of *Bembix* is more complex and contains more unique elements than that of *Gorytes* is obvious enough. However, before proceeding further it is desirable to list some of the behavior patterns that differ and to evaluate them in a preliminary way. The criteria for deciding which attributes should be considered primitive and which advanced have been discussed briefly elsewhere (XIII:C; Evans, 1957b:216–217). Applying these criteria, we may have some measure of confidence that

Primitive Nyssoninae	More advanced Nyssoninae
1a. Tend to be hygrophiles or mesophiles;	b. are often xerophiles;
2a. Nest in small, unspecialized niches;	b. nest in broader expanses of bare soil, or in unusual habitats;
3a. Feed at honeydew or at flowers with shallow corollas;	b. feed at flowers with deep corollas;
4a. Sleep on vegetation, not in dense clusters;	b. sleep in dense clusters; *or* c. sleep in special burrows; *or* d. sleep in the brood nests;
5a. Exhibit simple prenuptial behavior;	b. are territorial ($\delta\delta$) *or* c. exhibit sun dances or hopping dances ($\delta\delta$);
6a. Dig with the mandibles, with some assistance from the legs;	b. dig with strong thrusts of front legs synchronized with tilting of body; *or* c. dig with several thrusts of front legs at each tilting of body;

7a. Make simple nests with an indeterminate number of cells (usually several);
　b. make unicellular nests (sometimes of unusual form);

8a. Do not level the mound at the nest entrance;
　b. level the mound with elaborate, stereotyped scraping movements;

9a. Do not maintain an outer closure of the nest;
　b. maintain an outer closure at all times;

10a. Do not make an inner closure of the nest;
　b. maintain an inner closure at least when the egg is present or the larva small;

11a. Do not make false burrows;
　b. make false burrows;

12a. Hunt for prey close to nesting area, chiefly on vegetation;
　b. range widely for prey;

13a. Prey upon insects with gradual metamorphosis (Hemiptera, Orthoptera);
　b. prey upon adult insects with complete metamorphosis (chiefly Diptera, Lepidoptera);

14a. Sting the prey into permanent paralysis;
　b. kill the prey by stinging;

15a. Carry the prey to the nest with the mandibles;
　b. carry the prey to the nest with the middle legs (often assisted by the hind legs in flight);

16a. Fasten the egg to the prey;
　b. fasten the egg to the floor of the empty cell *or*
　c. deposit the egg loosely in the empty cell;

17a. Employ mass provisioning;
　b. provision progressively;

18a. Leave the prey remains in the cell;
　b. clean the cell periodically;

19a. Make a simple final closure (or none at all);
　b. make an elaborate final closure, including stereotyped movements of concealment;

20a. Make a solid-walled cocoon;
　b. make a series of pores in the walls of the cocoon.

　　This list includes a good many oversimplifications, as reference to the previous chapter will reveal. It is also not exhaustive, although most other characters which might be added are more nebulous: for example, patterns of orientation (Chapter XIV:B, 7) and wing elevation (Chapter XIV:B, 6). One seemingly important character

§B. Phyletic trends

has had to be omitted: whether the egg is laid on the last prey placed in the cell or on the first. Perusal of the Nyssoninae suggests that the former is the primitive condition, but Janvier has reported oviposition on the first prey in two species of *Clitemnestra*. The development of progressive provisioning is contingent upon the egg being laid upon the first prey in the cell.

Although the primitive versus advanced alternatives stated for the above 20 features provide useful hypotheses, we cannot rule out the possibility of reversals in specific cases. For example, members of the genus *Bembix* are advanced in nearly every feature of behavior listed; yet some species of this genus omit the outer closure of the nest (Chapter XI:E, 4), in contrast to the great majority of higher Nyssoninae. Clearly nest closure has been lost secondarily in a few species of *Bembix*. Even the type of provisioning has apparently undergone a reversal within the genus *Bembix* (Evans, 1957b:217-218). There is no obvious reason why mound-leveling movements, false burrows, and other specializations may not be lost in particular cases. We can detect secondary simplifications in behavior only by applying our same criteria at a lower level. That is, *within* the genus *Bembix* 9b is evidently primitive (that is, like nearly all other higher Nyssoninae) while 9a is specialized (that is, unique in a portion of the genus).

Still further complications are provided by behavioral convergences. We cannot assume, for example, that all wasps that lay the egg in the empty cell (16b,c) do so because they are derived from a common ancestor which did so. As a matter of fact it appears that this seemingly important behavioral advance, which in fact involves a considerable reorganization in the sequence of behavioral acts, arose at least eight times independently in the Nyssoninae (see below, C, 5).

It is disconcerting to find that most, if not all, of the behavioral features we have tabulated after so much deliberation may be subject to multiple convergence and to reversals. However, by proceeding cautiously it should be possible to detect most of these. Furthermore, there are certain types of behavioral modifications of a less quixotic nature. For example, while it is relatively easy for complex behavior to become simplified, once complex behavioral elements have been lost we can be fairly sure that they will not be reacquired in the same form (Dollo's law, more commonly applied to structures). Thus, once the fly pedestal was lost (16b), it is difficult to conceive of it being regained, and there is no evidence that any reversal has occurred here. Similarly, it is difficult to imagine *Nysson* once again becoming a hunting wasp, for in becoming a cleptoparasite it has lost many complex behavioral features concerned with the capture and immobilization of the prey.

A few of the major behavioral modifications in the Nyssoninae have been superimposed upon the dendrogram developed in Chapter XIII, permitting an easier visualization of some of the major trends (Fig. 211). This arrangement should have a certain amount of predictive value if our synthesis has been reasonably sound up to this point. Because of the great gaps in knowledge of the behavior of numerous genera, detailed quantification such as we attempted with structure in Chapter XIII is impractical. It is, however, of interest to test a few of the better-known genera, using the system outlined in Chapter XIII:C but employing the 20 behavioral features outlined above. It is reassuring to find a close parallel between structural and behavioral advance.

Selected genera	Degree of evolutionary advance	
	On basis of behavior: 20 characters (Chap. XV:B)	On basis of structure: 40 characters (Chap. XIII:C)
Alysson	20	58
Hoplisoides	27	60
Sphecius	30	65
Bicyrtes	38–43	87–88
Bembix	47–60	89–93

C. Examination of specific behavior patterns

We are now in a position to turn our attention to specific aspects of behavior and to attempt to trace their modifications in detail. Hopefully, we shall be able to detect the interplay of structural and behavioral change and to draw some conclusions regarding probable selection pressures. The following behavior patterns have been selected for discussion: digging and leveling, closure, prey selection and prey carriage, provisioning, and oviposition.

(1) *Digging and leveling.* There is no greater antithesis among wasps than that between *Alysson* corkscrewing its way through the moist soil of a shady streamside and *Stictia* scraping out a tunnel through a sand dune like a great digging machine. *Alysson*, a small, slender wasp, drills a vertical tunnel by breaking the soil with its mandibles and pushing the soil along behind it with its legs. In contrast,

§C. Specific behavior patterns   481

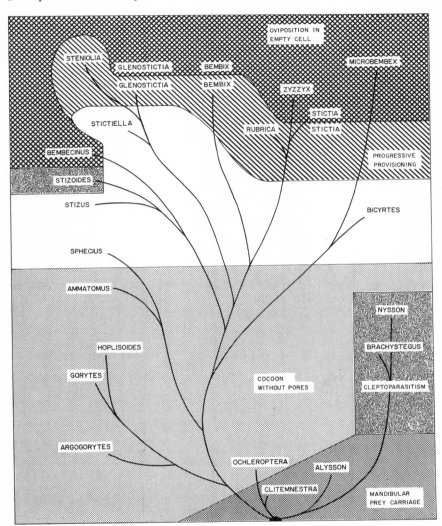

Fig. 211. A somewhat simplified version of the dendrogram presented in Fig. 203 (genera of unknown or poorly known ethology being omitted), with five major ethological characters superimposed upon it. It is to be assumed that genera outside a particular labeled area employ contrasting behavior; for example, the genera outside the "mandibular prey carriage" compartment exhibit pedal prey carriage or (in the case of the Nyssonini) none at all.

Stictia is robust and among the largest of digger wasps (the related genus *Editha* contains the largest of all); its oblique burrow is excavated rapidly with vigorous thrusts of the front legs which are perfectly synchronized with one another and with elevations of the abdomen which permit the sand to shoot out in spurts behind the body. However, an almost perfect series of intermediates connects these two extremes of the Nyssoninae, a series graded with respect to speed and efficiency of digging and with respect to the robust body form and better development of the pecten of the front tarsus (cf. Figs. 15, 26, and 85; Chapter XIV:B, 1).

This increase in the speed and effectiveness of digging appears correlated with a trend toward the occupation of broader expanses of bare soil, eventually including sand dunes and deserts. Doubtless the more rapid digging movements of a *Stictia* or *Microbembex* as compared with an *Alysson* or *Hoplisoides* reflect to some extent the higher body temperatures of these wasps, nesting as they often do in extremely hot situations. It is also probable that bembicine wasps cannot afford to dig slowly because of the exceedingly high temperatures of the sand surface: these wasps must be able to fly and to orient rapidly, to hover effectively, and to dig quickly through the surface layers during the heat of the day. Chapman *et al.* (1926) showed that *Bembix* and *Microbembex* die within a few minutes when forceably exposed to the surface heat of a sand dune.

In their ability to throw the sand far behind them, the species of *Bicyrtes* and *Microbembex* have surpassed even *Stictia*, *Bembix*, and other higher sand wasps. Members of these two genera actually throw several spurts of sand each time the abdomen is elevated. That is, a unit of digging behavior, which in most sand wasps is:

Head: down   Abdomen: up   Forelegs: backward (scraping sand)
      up   down   forward

repeated many times, has become modified somewhat as follows:

Head: down   Abdomen: up   Forelegs: backward (scraping sand)
                                                     forward
                                                   backward (scraping sand)
                                                   forward
                                                 backward (scraping sand)
     up   down   forward

followed by a brief pause before repetition. Such a change in behavior may be the result of a very simple alteration in nervous triggering. The stronger and more prolonged tilting of the body in *Bicyrtes* and *Microbembex* permits the sand to shoot out behind at a

§C. Specific behavior patterns

higher angle than in other genera and therefore to fall at a greater distance behind. Thus no mound of earth accumulates as it does in other sand wasps, many of which have developed special behavior patterns serving to remove this mound.

In the vertical burrow of *Alysson,* the soil is merely pushed up and out the entrance, forming a large mound surrounding the entrance, as in the nests of many ants. In nearly all other Nyssoninae the burrow is oblique, permitting much more manipulation of the soil because of the reduced effect of gravity. In these wasps the loosened soil is moved up the burrow in stages and then cleared from the entrance before the wasp plunges back into the burrow to repeat the operation. When clearing, the wasps remain in the entrance or within 1 or 2 cm of it, scraping sand vigorously. When the wasp backs out farther and then proceeds toward the entrance kicking sand, we speak of these as simple leveling movements interspersed with digging and clearing, although they are barely distinguishable from the latter. These very simple leveling movements may be subject to three further developments: (1) they may increase in number, (2) they may become more stereotyped, and (3) they may be wholly or largely postponed until after the completion of the nest, even until after the initial closure (see Chapters XI:E, 5; XIV:B, 2). The more elaborate leveling movements result in a more complete dispersal of the mound, which in fact is entirely removed in many species. Presumably this also involves refinements in their powers of orientation, since these wasps must locate a nest completely unmarked by a mound or depression, often in a broad expanse of bare sand.

Movements of concealment of the closed nest entrance differ little from those of leveling except that they are often transposed forward, in such a way that the sand is thrown *toward* the nest entrance rather than away from it. Such movements are almost always performed on the side of the entrance away from the mound. This suggests that in the short flights interspersed with movements of concealment, the wasp still orients to the nest entrance in the usual manner. That is, it is a simpler step for the wasp to obtain the soil for concealment on the side opposite the mound than it is for the wasp to turn around and use some of the sand dispersed from the mound (since this would involve an orientation in the opposite direction).

The elaborate "radiating lines" behavior of some of the higher sand wasps is doubtless no more than a prolonged and stereotyped development of transposed leveling movements. These radiating lines have also often been postponed to the time of the final closure, when leveling is mostly or wholly absent. It is interesting to find "radiating lines" behavior at the final closure in species in which mound leveling has been lost altogether (*Bembix sayi*), greatly modified (*Bembix pruinosa, Glenostictia scitula*), or even rendered obsolete by the development of highly efficient clearing behavior (*Microbembex monodonta*). Thus we may speak of this behavior as being "ritualized" and "displaced" in the sense that it has become more stereotyped as well as divorced from its original function and from its original position in the behavioral sequence. I have already employed the concept of ritualization in discussing the false burrows of certain species (Chapter XI:E, 6).

(2) *Closure.* Outer closure developed relatively early in the evolution of the Nyssoninae and may, in fact, have occurred in the ancestral nyssonines (Sphecidae of several different subfamilies make outer closures). Within the Nyssoninae, this behavior has apparently been lost many times independently. Some species, or even some colonies of one species, show considerable variation with respect to outer closure (for instance, *Bembix amoena*, Chapter X:A, 6). However, in most species, and even in many genera (for example, *Gorytes, Sphecius, Steniolia*), closing behavior is quite fixed (that is, completely absent or characteristic of all individuals at all times).

Closure may be thought of as "digging in reverse." That is, the movements are those of digging, but the wasp faces away from the nest. In the final stages of closure, the wasp may come out the entrance scraping sand behind it, then take a short flight and land in the entrance facing the nest, then rotate quickly 180° and resume closure. This suggests the importance of orientation toward the nest entrance; no sand wasps are known to land in the entrance facing away from the nest. This rapid 180° rotation is especially conspicuous in the final closure of *Microbembex monodonta* (Chapter XII:A, 10) and some other higher Nyssoninae. During closure the wasp may move several centimeters from the entrance scraping sand toward the nest. I suggest that such movements of concealment as occur on the threshold of the burrow are actually modified movements of closure rather than being derived from leveling behavior as described in the preceding section.

In contrast to outer closure, which occurs in a wide variety of Nyssoninae, and in more elaborate form in nearly all Nyssoninae at the time of final closure, inner closure occurs only in the species of *Stictia* and in most species of *Bembix*. Since inner closure is more characteristic of nests containing eggs or small larvae, and is often omitted when the larva is large, and since it occurs in two genera

§C. Specific behavior patterns

that often nest in hot, dry situations, one might assume that it represents a device for protecting the egg and small larva from excessive heat and dessication. However, the inner closure is wanting in xerophiles such as *Microbembex* and *Steniolia*. Doubtless it is principally a mechanism for deterring parasites, perhaps especially hole-searching Miltogramminae (Chapter XIV:F, 1). It is apparent that miltogrammine maggots do not usually destroy large larvae, but they usually cause the death of the egg or small larva. Hence it seems a logical assumption that the inner closure functions as a protection against these flies rather than chrysidid or mutillid wasps, which dig effectively and are more particularly attracted to larger larvae or cocoons.

Outer closure probably functions as a deterrant to all of those Diptera and Hymenoptera that enter or oviposit in open burrows (Chapter XIV:F, 8). Its effects are doubtless enhanced by more complete leveling and concealment, which may result in complete obliteration of all visual evidence of the nest, and by the presence of false burrows, which may actually deceive the parasites. It should be pointed out that the presence of a closure imposes a penalty upon a provisioning wasp: she must pause in the entrance long enough to remove the closure, and this may provide an opportunity for another female to steal the prey. Also, such a delay at the entrance is actually advantageous for miltogrammine flies of the genus *Senotainia*, which commonly succeed in larvipositing while the wasp is entering the nest (Evans, 1957b:233–234).

That the distribution and nature of closure, leveling, concealment, and false burrows are so variable within the Nyssoninae (and yet quite fixed in most species and in some genera) is doubtless a reflection of the relative importance of various natural enemies at different times and places in the past. That is, in situations in which Bombyliidae or hole-searching Miltogramminae were important causes of mortality, selection has favored the development of more complete closure and concealment. But when flies of the genus *Senotainia* caused reduction of the populations year after year, closure tended to be host. The development of cell cleaning in certain species may reflect a period in the past history of these species when miltogrammines or certain types of cell scavengers were important causes of mortality. The intraspecific variation in closure, leveling, false burrows, and cell cleaning observed in a few species may reflect inconstancies in selection pressures. I have, for example, noted that in *Bembix pruinosa* and *B. troglodytes* false burrows tend to occur more regularly at the periphery of the range, where other selective agents (especially physical factors) are presumably at their maxima (Chapter XI:B, D).

It does not follow from these considerations that more advanced sand wasps are freer from natural enemies than the less advanced forms. While *Bembix* and its allies undoubtedly suffer less mortality from Miltogramminae than *Bicyrtes, Gorytes,* and other genera, parasites of several other groups have apparently taken advantage of the success of the Bembicini and have come to attack them successfully (*Parnopes, Dasymutilla, Macrosiagon*). Furthermore, we would expect that as sand wasps evolved more effective means of escaping parasitism, the parasites would themselves develop keener sense organs, more effective digging behavior, and other devices for "keeping up" with their hosts. It is important to remember that most sand wasps of more advanced development occur in large local aggregations, in some cases in unusual ecological niches. As I pointed out in 1957, ecological restriction very often means

a lack of opportunity for a colony to radiate out into new nesting sites in the immediate vicinity. Such a colony may be annihilated by parasites or other natural enemies. Since most of the natural enemies of *Bembix* [and other sand wasps] are not host specific, they need not decline in numbers with their hosts. Thus many of the more specialized species . . . possess behavioral modifications apparently functioning to reduce the incidence of parasitism (complex nests, mound leveling, cell cleaning, false-burrow construction, and the like).

(3) *Prey selection and prey carriage.* Obviously, one of the most conservative elements in the behavior of sand wasps is prey selection. A very large portion of the more generalized Nyssoninae prey upon leafhoppers and treehoppers. Relatively few stocks have broken away from their primitive prey preferences, and each of these stocks has (with a few exceptions) remained with its newly acquired type of prey (Table 45). This suggests that a change in type of prey represents a more profound nervous reorganization than do the modifications of the behavior patterns considered in the previous two sections. This reorganization must result in adoption of a different situation in which hunting is performed, a change in the wasp's response to visual, olfactory, and tactile stimuli occurring in that situation, and modifications in the way the prey is handled. There have been no detailed studies of hunting behavior in the Nyssoninae (see Chapter XIV:C, 1, 2 for a brief review of this subject). However, the studies of Tinbergen (1935) on *Philanthus triangulum* doubtless apply in a general way to most higher Sphecidae.

Adoption of a new type of prey represents a major shift in adaptive zone, such that the adaptive grid of Simpson (1953) is useful

§C. Specific behavior patterns 487

(Fig. 212). The higher Nyssoninae have shifted from an ancient (although still successful) group of insects, the Hemiptera, to groups of insects with complete metamorphosis that have themselves undergone much of their radiation in the middle and late Cenozoic (Lepidoptera and Diptera Brachycera). Thus these wasps have availed themselves of an expanding and almost inexhaustible source of food. That the Lepidoptera appear to have supplied something of a "stepping stone" from the Hemiptera to the Diptera may reflect the fact that the Lepidoptera are, on the whole, relatively weaker fliers and more closely associated with plants than the Diptera; thus the transition from Hemiptera to Lepidoptera may have involved less profound changes in hunting behavior than a direct transition to Diptera. The Lepidoptera provided a less ample field for exploitation than the Diptera, since the vast majority of Lepidoptera are nocturnal.

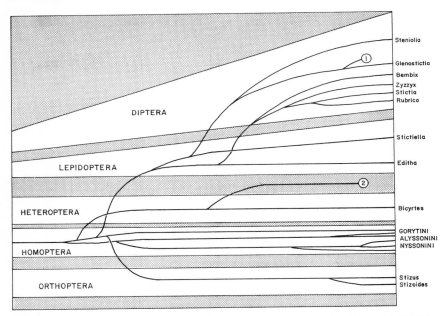

Fig. 212. Prey selection in the Nyssoninae shown on an adaptive grid (highly diagrammatic and oversimplified). The time scale is presumed to be from the beginning of the Tertiary (left margin) to the present (right margin), although of course the time of origin of various stocks and of entries into new adaptive zones is hypothetical. Two genera which have lost their host specificity may be said to have entered still another adaptive zone, which may be visualized as projecting into a third dimension; these are (1) *Glenostictia* and (2) *Microbembex*. (For further details and examples of exceptional prey records see p. 442 and Table 45.)

Prey specificity is apparently a characteristic that solitary wasps inherited from their parasitoid and sawfly ancestors (Evans, 1963b). It is interesting to find that those Nyssoninae that exhibit a breakdown in prey specificity all belong to more advanced groups. Thus, one finds that many species of *Bembix* take an exceedingly broad spectrum of flies and that a few species take insects other than Diptera either occasionally or regularly; these species include the Australian *atrifrons*, which is reported to use damselflies (the only wasps known to utilize Odonata as prey), the South American *citripes*, which uses flies but takes an occasional bee, and an unknown African species which was taken with a skipper (see Chapter XI:E, 9). *Zyzzyx chilensis* (Chapter IX:H) preys upon a wide variety of Diptera and rarely upon skippers (Lepidoptera). It is noteworthy that skippers provide the usual prey of *Editha adonis* and of certain species of *Stictiella*, so that the use of an occasional skipper by *Zyzzyx* and *Bembix* might be considered an atavism. The use of an occasional bee by *Bembix citripes* is also interesting, as *Glenostictia scitula* uses bees in considerable numbers, in addition to flies and bugs. One wonders if the Bembicini may be in the process of entering still another adaptive zone: the use of Hymenoptera as prey (a zone perhaps already too filled by the Philanthinae to provide a broad field for expansion by the Nyssoninae).

The breakdown of prey specificity in *Glenostictia scitula* is especially interesting, since the other species of this genus, so far as is known, utilize Diptera. The use of Hemiptera by this species might be considered another atavism, though one would have predicted the use of Lepidoptera by this species also (perhaps it does use moths or butterflies in certain situations). *G. scitula* nests in the spring, before most other digger wasps are active. It is possible that the species was able to occupy this niche only by losing its specificity for flies, which may not be sufficiently abundant at this season to fully supply its large nesting aggregations. *G. scitula* appears to hunt and handle its prey in much the same manner as *Steniolia* and *Bembix*, but its responses to potential prey have been "loosened" to the extent that it accepts a wide variety of small insects (see Table 18).

The case of *Microbembex* is still more interesting, for here there is no semblance of specificity whatever. Even the hunting behavior has been greatly modified, the females merely cruising about low over their nesting area and picking up arthropods from the soil. The species are collectively known to utilize arachnids of 3 orders and insects of 11 orders. Experimentally, *M. monodonta* will even accept millepedes, centipedes, and terrestrial Crustacea. When pre-

§C. Specific behavior patterns 489

sented with pieces of earthworms and vertebrate tissue mixed in with dead arthropods, female *monodonta* picked out the arthropods and left the others; however, I found pieces of plant tissue in two nests (Chapter XII:A, 6–8). Although *monodonta* is definitely a scavenger, two South American species are presumed to attack living arthropods. One of them, *uruguayensis*, is reported to specialize on carabid beetles, but this matter deserves further study. No wasps other than *Microbembex monodonta* are known to have become scavengers. That this genus has not undergone radiation in this adaptive zone is probably a consequence of the fact that it entered the zone quite recently, so that there has been insufficient time for radiation. Furthermore, this adaptive zone was amply filled long ago by a large (and ecologically much more versatile) group, the ants.

The manner of prey carriage is remarkably fixed throughout most Nyssoninae; even *Microbembex* carries its strange assortment of corpses with its middle legs. Only two genera of sand wasps, *Alysson* and *Clitemnestra*, employ the mandibles rather than the middle legs as the major device for grasping the prey (Chapter XIV:C, 4; Figs. 19, 211). I have outlined the evolution of prey-carrying mechanisms in wasps elsewhere (Evans, 1963a). At least four separate stocks of Sphecidae underwent a shift from mandibular to pedal prey carriage. This suggests that there are major selective advantages in pedal carriage. I believe the major one to be the fact that wasps returning to the nest with prey held well back beneath them with the middle legs (for example, Fig. 167) are able to open the nest entrance without depositing the prey on the ground, where it will be subject to the attacks of miltogrammine flies, tiger beetles, ants, and so on. Two other selective advantages also suggest themselves: (1) when the prey is held close beneath the center of the body it is well protected from larviposition by *Senotainia* while the wasp is in flight, and (2) when the prey is held close to the center of gravity of the wasp, the latter is able to fly more rapidly and to maneuver more effectively.

(4) *Provisioning*. The four types of provisioning were outlined earlier (Chapter XIV:C, 6). Briefly, the vast majority of Nyssoninae employ mass provisioning, which may be "delayed" in a few cases; in a number of genera of higher Nyssoninae, the cells are provisioned progressively, the female waiting 24 hr or more until the egg is ready to hatch, then bringing an amount of prey each day suitable to the size of the larva. It should be pointed out that wasps that provision progressively lay fewer eggs (one every 4–8 days,

while mass provisioners may lay 1–3 per day). It is well known that the ovaries of wasps that provision progressively contain only one large oocyte at a time, thus limiting these wasps to oviposition once every several days (see especially Iwata, 1955, 1964b, also Fig. 213).

The selective advantage in progressive provisioning is commonly considered to be the greater protection from natural enemies afforded the egg and larva. There is little question that there is far less mortality to eggs and small larvae in progressive provisioners than in mass provisioners. It is probably safe to say that in species that oviposit in the empty cell, provision progressively, remain in the nest at all times when not feeding or hunting, and clean the cells periodically (*Bembix texana,* for example), it is virtually impossible for any parasite, predator, or scavenger to develop in the cell. The available evidence suggests that the only really important enemies of the higher Nyssoninae are those that develop on the wasp larva only after it has been abandoned by the mother (Bombyliidae, Mutillidae, Chrysididae; see Chapter XIV:F). By contrast, in the lower Nyssoninae, as in a great many other solitary wasps, Miltogramminae often cause high mortality of eggs and small larvae. There are few quantitative data on this point, but I have found that in the nests of *Gorytes canaliculatus* that I have

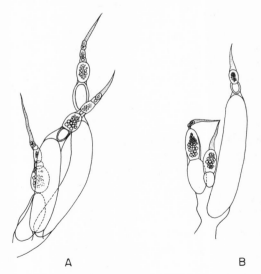

Fig. 213. Ovary of (A) a species of *Gorytes* as compared with that of (B) a species of *Bembecinus,* the latter being a progressive provisioner. (From Iwata, 1955.)

§C. Specific behavior patterns 491

studied about one third of the larvae were destroyed by miltogrammine maggots (Chapter III:H, 7), while in various colonies of *Bicyrtes quadrifasciata* from 20 to 40 percent were destroyed (Chapter VII:A, 8). In *Sphecius speciosus*, Reinhard reported about 50 percent mortality from *Senotainia trilineata*, and my own limited data on this species are in close agreement (Chapter V:A, 7).

It is true, of course, that if these wasps produce eggs 3-4 times as rapidly as *Bembix* and its allies, as seems to be the case, they are able to suffer 50 percent mortality and still produce more offspring per unit of time. It seems probable that there is sufficient increase in the length of life of the females to permit them to indulge in the lavish expenditure of time required by progressive provisioning without undergoing a reduction in reproductive potential. Again, there are few quantitative data to back this up, but I would venture to guess that a female *Bembix* lives 2-4 times as long as a *Gorytes*. It should also be remembered that wasps adapted for xeric conditions lose very little time as a result of cloudy, rainy, or cool weather; and the higher sand wasps are for the most part xerophiles.

Nevertheless, it would seem that progressive provisioning must have had strong survival value of and by itself at one time, since it appears to have arisen at least four times independently in the Nyssoninae (Fig. 211). Increased length of adult life and utilization of xeric habitats may well have been preadaptations rather than postadaptations. It is probable that progressive provisioning arose at times and places when these wasps were suffering exceedingly high losses from miltogrammine flies or some other source of mortality to the eggs or small larvae.

It is often assumed that progressive provisioning arose from delayed provisioning (for example, by Wheeler, 1928). The circumstances that might have led to this may be visualized somewhat as follows. During a prolonged period of drought, possibly in circumscribed areas, prey became exceedingly scarce, and the wasps were consistently unable to stock their cells fully on the initial day; or there may have been a prolonged period of above-normal rainfall, severely limiting the hunting opportunities of the wasps. Under these conditions, pressure of eggs developing in the ovaries may have caused lowering of the threshold for oviposition to the extent that the egg was laid on the first prey in the cell rather than the last (an important prerequisite for progressive provisioning). Prey was brought in over a period of days, some of it after the larva was fairly large (as sometimes occurs in *Bicyrtes* at the present time).[2] As a result, eggs and small larvae were less exposed to attacks by miltogrammine flies—each entry with a new prey representing a new exposure. Maggots introduced into a nest with a larger larva did not usually cause the death of the larva, as is generally true in *Bembix* and its relatives today (see, for instance, Chapters X:B, 5; X:G). Under these unfavorable conditions of climate, there would be exceedingly strong selection to reduce mortality from biotic factors, and the effect of genes facilitating this behavior would be enhanced. I do not consider this a Lamarckian interpretation, since it assumes a considerable initial behavioral flexibility which is, under unusually strong selection pressure, shifted to one extreme and then ritualized at that extreme. There need be no implication that an individual wasp acquired a new behavior and passed this on to its offspring (see below, Chapter XV:D).

(5) *Oviposition.* The ten types of oviposition behavior occurring in the Nyssoninae were outlined in Chapter XIV:C, 5, and references were given there to examples of each type and to figures illustrating most of the types. The phylogeny of the ten types seems reasonably clear (Figs. 211 and 214), and presents the by-now-familiar pattern of multiple parallelisms. In this case it seems probable that oviposition in the empty cell (group C) has arisen at least five, and perhaps as many as eight times independently (types 8 and 9 are represented by one known genus or species each, but types 7 and 10 are each represented by three genera that are unlikely to have derived this behavior from a common ancestor).

The omission of the pedestal on which the egg is laid involves a considerable reorganization in behavior, as I discussed in 1957 (b:226–227). As I pointed out at that time, the omission of the pedestal may, in itself, result in a heightened reproductive potential. The search for prey is always attendant upon a number of uncertainties:

the weather, the time of day, the abundance of suitable [prey] in the vicinity, and the like . . . If conditions are not favorable (if it is late in the afternoon, for example, and the next day is inclement), the female may have to wait a considerable time before laying her egg. In the more advanced species of *Bembix,* the egg can be laid at any time. Although I have no actual data on this point, I feel certain that such a species as

---

[2]Baerends and Baerends van Roon (1950) found that in *Ammophila pubescens* the development of the brood was "about nil at an air temperature that forces the wasp to inertia." However, my experience has been that in the much deeper nests of Nyssoninae temperature of the cells is more constant and development of the brood more or less continuous even when cool or cloudy weather prevents the adult from carrying out her activities.

§C. Specific behavior patterns 493

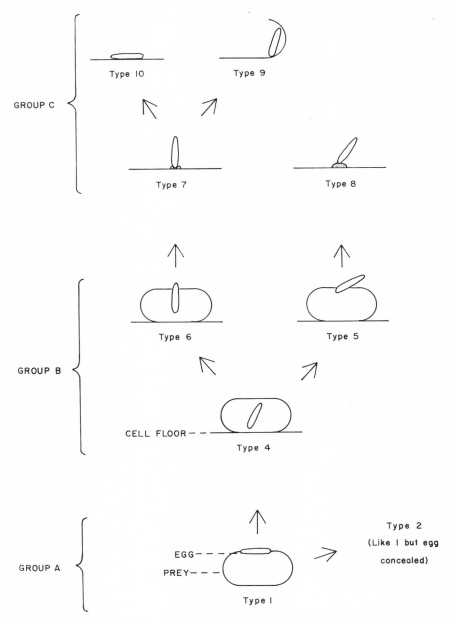

Fig. 214. Diagram showing probable evolution of major types of oviposition behavior in the Nyssoninae. Type 3 is omitted from the figure since it is not well documented; apparently it is somewhat intermediate between types 1 and 4. (For explanation, see Chapters XIV:C, 5 and XV:C, 5.)

pruinosa is often able to rear more progeny per unit of time than, for example, *spinolae*.

I now feel that a more important consideration is the fact that oviposition in the empty cell completely eliminates the possibility of introducing miltogrammine maggots with the initial prey. Since in many wasps these maggots are known to destroy the egg directly, the omission of the prey pedestal may have had a very considerable selective advantage. It is perhaps worth noting that miltogrammine maggots have never been found in the cells of any of those species of *Bembix* that oviposit in the empty cell (although the nests studied number in the hundreds). In several other species of *Bembix* such maggots have been found; even though they apparently do not often cause the death of the larva, they may force the wasp to bring in more flies than usual and thus delay completion of the nesting cycle.

Omission of the fly pedestal may have originated from a simple lowering of the oviposition threshold under severe climatic conditions such as those visualized in the preceding section. That is, wasps were frequently unable to obtain prey at the time the eggs were ready to be laid, and the eggs were deposited *"in vacuo"* on the floor of the cell, at first glued to the substrate in a manner similar to that in which they would have been fastened to the prey. Individuals that possessed a genetic background favorable to oviposition under these conditions would be favored by selection (that is, the more readily its constitution permitted it to lay the egg without waiting for prey the more likely it would be to leave more progeny). One can then visualize the development of a local population that omitted the egg pedestal and the spreading of this behavioral improvement throughout the range of the species. It is probable that in the periods of climatic stress hypothesized, population size would fall quite low. With the removal of the stress there would be rapid expansion in population size, with all that this implies with respect to the spreading of evolutionary changes.

The belief that oviposition may still occur even under conditions not allowing deposition of the egg in the usual position is supported by Steiner's (1962) findings with respect to *Liris nigra* Van der Linden, a wasp of the subfamily Larrinae. This wasp typically lays the egg near the base of the front legs. However, when the anterior half of the prey is excised under certain conditions, the wasp oviposits at the base of the hind legs, but only sometimes and only after considerable delay, during which time a gradual lowering of threshold presumably occurs. That oviposition on the substrate in the absence of prey may occur in such cases has yet to be shown, but the possibility that oviposition in the empty cell began as a vacuum reaction at least provides an appealing hypothesis not entirely without experimental support.

D. Some general considerations regarding the evolution of behavior

The above hypotheses are far from completely satisfying, but they may at least serve as food for further thought. As C. O. Whitman said in his classic essay on animal behavior (1899): "We may seldom, if ever, be able to trace the whole development of an instinct . . . The main reliance in getting at the phyletic history must be comparative study." The reconstruction of events in the evolution of behavior which occurred at an unknown time in the distant past and under unknown physical and biotic conditions would seem futile in the extreme, and it is not surprising that there are those who reject comparative ethology almost *in toto*. For example, W. R. Thompson (1964) has remarked that this field "has relied simply on the procedure of lining up degrees in adaptive behaviour, putting the series on end and projecting it into the past. In fact, of all unverifiable speculations, those on the origin of instincts have the least title to serious consideration."

To accept this statement fully is to dismiss this entire approach as not worth pursuing. Thompson, like Sokal and Sneath (1963) and others of their school, seems to believe that in the absence of a fossil record, phylogeny is pure speculation. It is not. Phylogeny includes the gathering of a great many facts and the processing of these facts in the light of widely accepted criteria (Maslin, 1942). The resulting phylogenetic schemes are no more than rough approximations, but as such they have much heuristic value and often permit the successful prediction of unknown behaviors. I know of no other approach to the question of how behavior patterns originate, and human curiosity will simply not let this matter be laid aside with a shrug.

It is axiomatic in comparative ethology that behavior patterns can be treated in much the same ways as structures and that the modern synthetic theory of evolution applies to behavior as well as to structure. Indeed, it is foolish to attempt to separate structure and behavior; as I said in the opening pages of this book, structures are what animals use to behave, behavior, the use to which animals put their structures. It seems safe to say that most adaptations have both structural and behavioral components.

Nevertheless, the study of the evolution of behavior poses unusual problems and suggests answers that may differ slightly from answers based strictly upon morphology. For one thing, an animal's structure is always with it, so to speak, but its various behavior patterns unfold only under specific circumstances. For another, even the most stereotyped of motor patterns show infinitely more variation in expression than one ever finds in the case of structures. The female *Bembix pruinosa*, for example, has seven pecten spines on the basal segment of the front tarsus; small individuals may have six, but I have never seen a female with five or eight. However, the burrow that results from the use of these spines is known to vary from 23 to 84 cm in length! In this case it is probable that soil moisture or friability influence the discontinuance of digging behavior. Even fixed motor patterns such as the stinging of the prey are far from immune to environmentally induced variation (Steiner, 1962). We must also remember that the structure of an adult insect is fixed (although subject to wear), while some aspects of its behavior are clearly influenced by learning (nest and hunting sites, at least), and probably much behavior is improved by performance (although we know little of this).

One often notes that behavior that is highly variable in one species is more or less fixed in a related species: for example, false burrows in *Bembix amoena* and *B. texana*. In birds, behavior that appears as an intention movement or displacement activity in one species may be incorporated into the courtship repertory of another species. Considerations such as these lead us to ask whether behavioral changes initiated by learning or habituation are capable of becoming genetically controlled. Mayr (1958) believes that "this is one of the few evolutionary phenomena where the 'Baldwin effect' might have played a role," defining the Baldwin effect as "the hypothesis that a nongenetic plasticity of the phenotype facilitates reconstruction of the genotype."[3]

As Waddington (1960) points out, the plasticity of the phenotype cannot truly be considered nongenetic, since "it must be an expression of genetically transmitted potentialities." In developing

---

[3] In social insects, much intraspecific variation in structure and behavior is induced by trophic factors (polyphasy). Caste determination in some of the higher stingless bees is, however, genetically controlled. Emerson (1958) cites this as an example of the Baldwin effect, since "it may be assumed that a genetic substitution that triggers the growth processes—formerly triggered by biochemical and physiological processes only—has occurred . . . Of course, the capacity to react to any environmental, physiological, or genetic stimulus is almost invariably based upon a complex genetic pattern that is inherited and itself evolves. But it is clear that an evolutionary feedback occurs through natural selection from the function to the development and genetic mechanisms."

§D. Some general considerations

his concept of the "genetic assimilation of acquired characters," Waddington has performed a number of experiments which, although concerned with structural and physiological features, may be pertinent to the present discussion. For example, he reared *Drosophila* for 21 generations on a medium containing a concentration of sodium chloride sufficient to prevent 70-80 percent of each successive generation from reaching maturity. The resulting stock had slightly larger anal papillae in the larval stage, this feature being associated with a heightened adaptation to high salt content of the medium. These adaptations were not reversible by one or two generations in the normal medium. Waddington speaks of these features as being "canalized," that is, caused by strong selection pressure to evolve irreversibly along a restricted portion of their former adaptability.

Mayr (1963) points out that "an extreme environment may bring out developmental potencies that are not expressed under more normal conditions; it permits genetic factors to manifest themselves that do not normally reach the threshold of phenotypic expression." Mayr prefers to speak of "threshold selection," feeling that Waddington's term "genetic assimilation" fails to bring out that "the treatment merely reveals which among a number of individuals already carry polygenes or modifiers of the desired phenotype."

Virtually all behavior patterns have a certain latitude of execution, that is, they are capable of following various pathways depending upon the immediate environmental situation. When certain types of stimuli recur repeatedly, and when there is a premium upon the smooth performance of a certain response, we may very well expect a measure of canalization. Possibly a great many behavioral changes have followed the simple model shown in Fig. 215A. Certain major changes in behavior (for example, progressive provisioning and oviposition in the empty cell) may be the result of threshold selection as outlined in the preceding section and as diagrammed in Fig. 215B. One assumes that during the periods of stress visualized, populations would be low and greatly fragmented, thus facilitating genetic change. In both models, as applied to sand wasps, I have postulated that behavioral shifts relating to physical factors came to have selective value in reducing the incidence of parasitism; this explains their persistence when the physical stresses have disappeared (Chapters XI:E, 6 and XV:C, 4-5).

Bateson (1963) has recently presented a lucid discussion of the role of somatic change in evolution. He points out that "any change

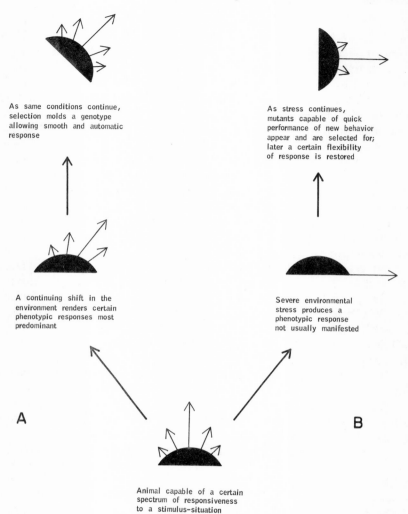

Fig. 215. Model suggesting two possible courses of behavioral change. (For explanation, see pp. 496–499.)

of environment which requires adaptive change in the species will be lethal unless, by somatic change, the organisms (or some of them) are able to weather out a period of unpredictable duration, until either appropriate genotypic change occurs (whether by mutation or by redistribution of genes already available in the population), or because the environment returns to the previous normal." However, "when the internal organization of the organisms of a

§D. Some general considerations

species has been limited by environmental or mutational pressure to some narrow subset of the total range of living states, further evolutionary progress will require some sort of genotypic change which will compensate for this limitation."

Bateson's remarks would seem especially pertinent to behavior patterns, all of which must make some allowance for variation in the precise environmental situation. Bateson speaks of the importance of genotypic changes that "simulate Lamarckian inheritance." The prevalence of such changes in the evolution of behavior doubtless accounts for the strong flavor of Lamarckianism that pervades much of the literature on insect behavior. It is little wonder that entomologists have so often thought of instincts as "inherited habits"[4] or "lapsed intelligence," and even Darwin can be forgiven for surmising that "some intelligent actions, after being performed during several generations, become converted into instincts and are inherited."

Wecker (1964) has found that habitat selection in field mice is largely innate, but after 12 to 20 generations in the laboratory, certain stocks underwent a loss of hereditary control of habitat selection, although retaining a capacity to learn from early field experience associated with their normal habitat (though not other habitats). He feels that the innate pattern has been superimposed on the originally learned pattern, citing this as an example of genetic assimilation of acquired features. He points out that increased hereditary control is advantageous in a stable environment since it limits the ways in which the organism can respond to a particular stimulus; but in a fluctuating environment the retention of a capacity to modify its behavior may be all important.

It is difficult to know to what extent, if any, the innate behavior patterns of wasps and other insects arose from learned patterns. Since we know that many invertebrates exhibit simple forms of learning (Thorpe, 1963), and since we know that wasps do exhibit learning with respect to their nesting sites and their hunting arenas, it does seem possible that this phenomenon occurs. Steiner (1962) was able to produce some conditioning of *Liris* females to displaced antennae on the prey. It seems to me probable that most of the behavior patterns of solitary wasps evolved by a process of canalization of somatic plasticity in response, with or without learned elements, such canalization usually being followed by a

---

[4]Entomologists seem forever addicted to the use of the word "habits" when they mean behavior. Behavior means simply what an animal does, and is analogous to the word "structure." But habits are modes of behavior acquired by repetition, and to title a paper "Habits of . . ." implies a postulate as to the origin of the behavior described.

restoration of latitude of performance of the "new" behavior. Such matters as "tool-using" in *Ammophila* (Evans, 1959c) and "hiding the egg" in *Nysson* (Chapter IV:A) provide especially interesting examples of plastic innate behavior simulating intelligence, but probably having evolved in a manner similar to that outlined above.

E. A final look at the sand wasps

The sand wasps are first known from the Eocene, when they are represented by small, *Gorytes*-like forms which probably dug simple nests in the ground and preyed upon leafhoppers in the manner of contemporary species of *Gorytes*. The genus *Gorytes* (in the broad sense) persisted through the Miocene and down to the present, but no other groups of Nyssoninae are represented in the fossil record. The higher Nyssoninae probably evolved quite recently (since the Miocene). *Stizus* and *Stizoides* apparently underwent most of their evolution in the Old World, while the Bembicini are exclusively a New World group except for *Bembix* itself, which has become cosmopolitan. The more generalized sand wasps exhibit fairly pronounced generic differences in structure, especially of the mesosoma, but relatively minor differences in behavior. In contrast, the more advanced sand wasps are relatively homogeneous in body form but quite diverse in behavior. This may reflect the fact that the latter group has evolved so recently that structural change has not had time to "catch up" with behavior. More probably, it reflects the fact that these wasps have achieved a body form capable of little further improvement with respect to the business of digging, flying, maneuvering with prey, and so on.

The reduction of the ocelli in the more advanced sand wasps is difficult to explain. It may represent an adaptation for life in the brilliant light of broad expanses of light sand, a conclusion suggested by the fact that nocturnal wasps tend to have large ocelli and one species of *Bembix* with relatively well-preserved ocelli is crepuscular. Little is known regarding the function of the ocelli of insects. Lindauer and Schricker (1963) blinded the ocelli of honeybees and found no effect upon the frequency of collecting flights during the middle of the day. However, such bees began to collect food later in the morning and stopped collecting earlier in the afternoon than normal bees. They concluded that the ocelli "provide information concerning either the absolute brightness or the speed of change in brightness of the light at dusk." The reduction and loss of ocelli in sand wasps restricted in their activities

§E. A final look

to periods of sunshine in areas of great light intensity may be comparable to the loss of eyes in cave animals.

It should be noted that certain wasps of quite another subfamily, the Larrinae, show convergence toward *Bembix* and its allies with respect to several important features. These include an increase in body size, similar configuration of the mesosoma, development of pale pubescence and hairs on the body, and a similar reduction in the ocelli. Possibly these are all "desirable attributes" of effective xerophilic predatory wasps.

The development of larger body size in the Nyssoninae permitted exploitation of larger prey, but it is doubtful if this in itself was of major importance (small insects on the whole being much more abundant than large ones, and the numbers of cicadas and large flies, and so forth, in any one area being fairly limited). Larger size probably resulted in more effective control over and independence from the environment (a major criterion of evolutionary progress as defined by Huxley, 1943). I refer to the fact that larger wasps are better able to move soil particles *en masse* because of the greater size of their legs and pecten and the greater thrust of their forelegs. Furthermore, larger and stronger-flying wasps are able to carry their prey greater distances, and thus they may hunt at some distance from their nests and still return swiftly to them. The fact that the more highly developed sand wasps have been released from a close bondage to their nesting areas may have been of considerable evolutionary significance.

Increase in body size may have had other selective advantages. Rensch (1960) points out several advantages of larger size. For example, larger insects tend to have more ommatidia in the compound eyes, thus improving their visual acuity and possibly their powers of orientation. Larger insects also tend to have larger corpora pedunculata in the brain. According to Rensch (after Goossen, 1949) the small vespid wasp *Ancistrocerus parietinus* (7.9 mm long) has about 500 globuli cells in the corpora pedunculata, while the much larger *Vespa crabro* (22.2 mm long) has about 1,000 globuli cells. Rensch claims that as a general rule smaller bees, wasps, and beetles exhibit simpler types of behavior. Hence, he states, the degree of differentiation in the corpora pedunculata "is more than a mere index of the phylogenetic level of a certain insect group." That the corpora pedunculata play important roles in complex behavior appears to be well established (for example, Vowles, 1961; Roeder, 1963), though it would be a rash assumption at this stage of our knowledge to claim that *Bembix* outperforms *Gorytes* because it has more globuli cells.

502 Chapter XV. Evolution of Behavior

The major trends in structural modification in sand wasps were summarized in Chapter XIII:E, the major trends in behavior in Chapter XV:B. There are several obvious correlates in the two lists, as indicated in the tabulation below. However, several of the structural modifications pertain to secondary sexual characters of the male. These may be associated with differences in the details of courtship and mating, but we know little about this except that some of the leg modifications assist in holding the female during copulation. Many of the behavior patterns (closure, cell cleaning, and so forth) have no obvious morphological components except insofar as these relate to general adaptive improvements.

It is of some interest to attempt to tabulate some of the adaptive features of sand wasps with reference to their presumed selective advantages. Most of these features have been discussed in earlier sections of this chapter. I do not pretend that this is anything more than a very preliminary classification of behavioral and structural features, or that any of the supposed relationships have been proved. Skeptics are invited to design and carry out appropriate experiments.

ADAPTIVE SIGNIFICANCE OF MAJOR CHARACTERS OF NYSSONINAE

A. General adaptive improvements
  1. Increase in size; more robust form
       Greater strength and maneuverability in flight
          Predation at greater distances from nest
       Increased size of eyes and number of ommatidia
          Greater visual acuity in hunting and orientation
       Increase in size and differentiation of corpora pedunculata
          Capacity for more complex behavior
       Construction of pores in walls of cocoon
  2. Greater longevity of adults
       Capacity for progressive provisioning
  3. Shift in type of prey (chiefly Hemiptera to Diptera)
       Shortened wings for predation on certain types of flies
  4. Utilization of more numerous sources of nectar
       Elongation of proboscis
  5. Greater success in mating
       Various secondary sexual characters
       Greater density of aggregations
       Sun dances and other precopulatory behavior
       Clustering
       Territoriality

§E. A final look

B. Adaptations for more xeric habitats
 1. More rapid flight
     Enlarged mesosoma and flight musculature
     Refinements in wing venation
 2. More rapid digging
     Stronger pecten and leg musculature
 3. Paler coloration and pale vestiture
 4. Reduction in ocelli
 5. Deeper nests

C. Mechanisms for reducing attacks of natural enemies
 1. Deeper and more complex nests
 2. Closure: outer and inner, temporary and final
 3. Mound leveling
 4. Nest concealment
 5. False burrows
 6. Progressive provisioning
 7. Oviposition in empty cell
 8. Spending of inactive periods in nest
 9. Prey carriage with middle legs
 10. Cell cleaning

We may say that the history of the sand wasps has been one of adaptive improvement with respect to the physical environment combined with the development of complex behavioral devices for avoiding the attacks of parasites. It is interesting to attempt to tabulate some of the behavioral changes that have occurred with respect to the apparent nature of the change, bringing together various hypotheses proposed earlier. This list is in every sense tentative, and it makes no attempt to be exhaustive.

TYPE OF BEHAVIORAL CHANGE	POSSIBLE EXAMPLE
1. Omission of a behavioral element	Loss of egg pedestal
2. Change in sequence of behavioral elements	Oviposition on first prey rather than last
3. Interpolation of a period of inactivity	Progressive provisioning
4. Repetition of behavioral elements	Elaborate leveling behavior
5. Change in intensity of movements	Digging behavior of *Bicyrtes*
6. Greater speed of performance	Digging behavior of higher sand wasps

7. Greater stereotypy	False burrows
8. Performance of behavior in a different place	"Radiating lines" behavior
9. Alteration in threshold of response	Oviposition in empty cell
10. Narrowing of responsiveness to a spectrum of stimuli	Prey specificity (for example, *Argogorytes*)
11. Broadening of responsiveness to stimuli	Loss of prey specificity (for example, *Glenostictia scitula*)

There are several instances in the sand wasps of behavior in the apparent process of change: for example, delayed provisioning, leveling movements, and false burrows in various stages of elaboration and ritualization. One of the best examples is provided by species undergoing a breakdown in prey specificity. Thus we find *Bembix citripes*, for example, preying upon flies but occasionally taking a small bee; *Glenostictia scitula* taking flies, bees, male ants, small wasps, and small bugs; and the species of *Microbembex* taking virtually any type of arthropod available in their habitat. In each case we are dealing with an "alteration of threshold of response" and a "broadening of responsiveness to stimuli," which may or may not be the same thing. It would be instructive to know if *Bembix spinolae* will use bees if flies are not available, and if *Glenostictia pulla* will accept bugs in some circumstances, when its congener *scitula* uses them regularly.

Many of these problems are amenable to experimentation, but unfortunately these wasps are difficult to rear in the laboratory and not always easy to subject to controls in the field. We can only hope that much more comparative data will someday become available and that a few well-planned experiments will suggest answers to some of the critical problems. There is an urgent need for more workers in this field; only by the concerted effort of many workers in various parts of the world, who will criticize one another's efforts and gradually develop better techniques and conceptual schemes, will the ethology of digger wasps be able to reach a level that will enable it to contribute importantly to the main stream of biology.

One thing seems certain. Sand wasps often thrive in man-made excavations and fills: indeed, such places often provide the best laboratories for wasp ethologists. The more advanced sand wasps are a young, successful, evolving group. As man continues to denude the earth and to spread about his fly-attracting refuse, he is likely to render the earth more suitable for sand wasps and less suitable

§E. A final look   505

for himself. We may visualize some post-historic human survivor standing in a vast barren, contemplating a digger wasp and envying its incurious eye and its simple attachment to the earth. It is, of course, just barely possible that man will some day learn enough about the fundamentals of behavior so that he will never find himself in that predicament.

Bibliography

Adlerz, G.
1903   Lefnadsförhallenden och instinkter inom familjerna Pompilidae och Sphegidae. I. *Handl. K. Svenska Vetensk. Akad.* 37(5):1-181.
1906   Lefnadsförhallenden och instinkter inom familjerna Pompilidae och Sphegidae. II. *Handl. K. Svenska Vetensk. Akad.* 42(1):1-48.
1910   Lefnadsförhallenden och instinkter inom familjerna Pompilidae och Sphegidae. III. *Handl. K. Svenska Vetensk. Akad.* 45(12):1-75.

Alexander, R. D.
1962   The role of behavioral study in cricket classification. *Syst. Zool.* 11:53-72.

Allen, H. W.
1926   North American species of two-winged flies belonging to the tribe Miltogrammini. *Proc. U.S. Nat. Mus.* 68(9):1-106.

Arens, L. E., and E. L. Arens
1953   The behavior of the wasp *Stizoides tridentatus* F. (Hymenoptera, Sphecidae). *Ent. Obozr.* (Moscow) 33:190-193. [In Russian.]

Arnold, G.
1929   The Sphegidae of South Africa. Part XII. *Ann. Transvaal Mus.* 13:217-418.

Ashmead, W. H.
1894   The habits of the aculeate Hymenoptera. III. *Psyche* 7:59-66.
1899   Classification of the entomophilous wasps, or the superfamily Sphegoidea. *Canad. Ent.* 31:145-155, 161-174, 212-225, 238-251, 291-300, 322-330, 345-357.

Baba, K.
1937   On some ecological observations on *Bembix niponica* Sm. *Kontyu* 11:21-28. [In Japanese, with English summary.]

Baerends, G. P., and J. M. Baerends-Van Roon
1950   Embryological and ecological investigations on the development of the eggs of *Ammophila campestris* Jur. *Tijdschr. Ent.* 92:53-112.

Balduf, W. V.
1941   Take offs by prey-laden wasps (Pompilidae and Sphecidae). *Ent. News* 52:91-92.

Barber, H. S.
1915   *Macrosiagon flavipennis* in cocoon of *Bembix spinolae* (Coleoptera, Rhipiphoridae). *Proc. Ent. Soc. Washington* 17:187-188.

Barth, G. P.
1907   Observations on the nesting habits of *Gorytes canaliculatus* Pack. *Bull. Wisconsin Nat. Hist. Soc.* 5:141-149.

Bates, H. W.
 1863 *The naturalist on the river Amazons.* 2 vols. John Murray, London.
Bateson, G.
 1963 The role of somatic change in evolution. *Evolution* 17:529–539.
Beaumont, J. de
 1953 Le genre *Olgia* Radoszk. (Hym. Sphecid.). *Rev. Suisse Zool.* 60:205–223.
 1954 Remarques sur la systématique des Nyssoninae paléarctiques (Hym. Sphecid.). *Rev. Suisse Zool.* 61:283–322.
Belt, T.
 1874 *The naturalist in Nicaragua.* John Murray, London. 403 pp.
Berland, L.
 1925 *Faune de France.* Vol. 10, Hyménoptères Vespiformes. I. 364 pp.
 1941 Nôte sur les *Sphecius* chasseurs de cigales (Hym., Sphegidae). *Rev. Franc. Ent.* 8:1–2.
Bernard, F.
 1934 Observations sur les proies de quelques Hyménoptères. *Bull. Soc. Ent. France* 39:247–250.
Bodkin, G. E.
 1917 "Cowfly tigers," an account of the hymenopterous family Bembecidae in British Guiana. *J. Bd. Agri. Brit. Guiana* 10:119–125.
 1918 Notes on some British Guiana Hymenoptera (exclusive of the Formicidae). *Trans. Ent. Soc. London, 1917,* pp. 297–321.
Bohart, G. E., and J. W. MacSwain
 1939 The life history of the sand wasp, *Bembix occidentalis beutenmuelleri* Fox and its parasites. *Bull. So. Calif. Acad. Sci.* 38:84–97.
 1940 Notes on two chrysidids parasitic on western bembicid wasps. *Pan-Pac. Ent.* 16:92–93.
Bohart, G. E., W. P. Stephen, and R. K. Eppley
 1960 The biology of *Heterostylum robustum* (Diptera: Bombyliidae), a parasite of the alkali bee. *Ann. Ent. Soc. Amer.* 53:425–435.
Bondar, G.
 1930 Vespas que protegem os animaes domesticos contra as muscas. *Correio Agricola* 8:179–181.
Bradley, J. C.
 1920 [Report on *Sphecius grandis,* in Minutes of Ent. Sect., Acad. Nat. Sci. Philadelphia.] *Ent. News* 31:112–113.
Brauns, H.
 1911 Biologisches über südafrikanishe Hymenopteren. *Z. f. wissen. Insektenbiologie* 7:90–92.
Brèthes, J.
 1902 Notes biologiques sur trois Hyménoptères de Buenos Aires. *Rev. Mus. La Plata* 10:193–205.
 1918 Un bembécido cazador de hemípteros. *Physis* (Buenos Aires) 4:348–349.
Bridwell, J. C.
 1937 [Report on *Bembecinus moneduloides* (Smith), in Minutes of 478th Meeting of Ent. Soc. Washington.] *Proc. Ent. Soc. Washington* 39:14–15.
Brischke, D.
 1886 Die Hymenopteren des Bernsteins. *Schrift. Naturfor. Ges. Danzig* 6:278–279.
Bristowe, W. S.
 1948 Notes on the habits and prey of twenty species of British hunting wasps. *Proc. Linn. Soc. London* 160:12–37.

Brues, C. T.
  1903 On the sleeping habits of some aculeate Hymenoptera. *J. N.Y. Ent. Soc.* 11:228–230.
Brues, C. T., and A. L. Melander
  1932 Classification of insects. *Bull. Mus. Comp. Zool.* (Harvard) 73:1-669.
Bryant, A. H. R.
  1870 The handsome digger wasp as a horse guard. *Amer. Ent.* 2:87.
Burroughs, J.
  1881 *Pepacton.* Houghton Mifflin Company, Boston. 260 pp.
Callan, E. McC.
  1945 A wasp preying on house-flies and stable-flies. *Nature* 155:146.
  1954 Observations on Vespoidea and Sphecoidea from the Paria Peninsula and Patos Island, Venezuela. *Bol. Ent. Venez.* 9:13–27.
Cazier, M. A., and M. A. Mortenson
  1965 Studies on the bionomics of sphecoid wasps. III. *Hapalomellinus albitomentosus* (Bradley) (Hymenoptera: Sphecidae). *Wasmann J. Biol.* 22:261–276.
Chapman, R. N., C. E. Mickel, J. R. Parker, G. E. Miller, and E. G. Kelly
  1926 Studies in the ecology of sand dune insects. *Ecology* 7:416–426.
Chmurzynski, J. A.
  1964 Studies on the stages of spatial orientation in female *Bembex rostrata* (Linné, 1758) returning to their nests (Hymenoptera, Sphegidae). *Acta Biol. Exper.* (Warsaw) 24:103–132.
Clausen, C. P.
  1940 *Entomophagous Insects.* McGraw-Hill, New York. 688 pp.
Cockerell, T. D. A.
  1906 Fossil Hymenoptera from Florissant, Colorado. *Bull. Mus. Comp. Zool.* (Harvard) 50:33–58.
  1922 An ancient wasp. *Nature* 110:313.
  1924 Fossil insects in the United States National Museum. *Proc. U.S. Nat. Mus.* 64(13):1–15.
Copello, A.
  1933 Biologia de *Hyperalonia morio* (Dip. Bomb.). *Revista Soc. Ent. Argentina* 5:117–120.
Curran, C. H.
  1951 *Insects in your life.* Sheridan House, New York. 316 pp.
Dambach, C. A., and E. Good
  1943 Life history and habits of the cicada killer (*Sphecius speciosus*) in Ohio. *Ohio J. Sci.* 43:32–41.
Darwin, C.
  1859 *On the origin of species by means of natural selection.* John Murray, London. 490 pp.
Davis, W. T.
  1920 Mating habits of *Sphecius speciosus*, the cicada-killing wasp. *Bull. Brooklyn Ent. Soc.* 15:128–129.
  1924 Cicada-killing wasps and flies. *J. N.Y. Ent. Soc.* 32:113.
  1926 The wasp *Bembidula quadrifasciata*. *J. N.Y. Ent. Soc.* 34:89–90.
De Gaulle, J.
  1908 Catalogue systématique et biologique des Hyménoptères de France. *Feuille des Jeunes Naturalistes,* 1906–1908. 171 pp.
Deleurance, E. P.
  1941 Contribution à l'étude biologique de la Camargue. Ethologie I. Observations entomologiques. *Bull. Mus. Hist. Nat. Marseille* 1:275–289.

1943	Notes sur la biologie de quelques prédateurs de la région de Montignac (Dordogne). *Bull. Mus. Hist. Nat. Marseille* 3:56–73.
1944	Contributions à l'étude biologique de la Camargue. *Bull. Mus. Hist. Nat. Marseille* 4:56–61.
1946	Etudes sur quelques éléments de la faune entomologique du bois des Rieges (Camargue). *Ann. Soc. Ent. France* 113:31–70.

Dow, R.
1935	The prey of the sphecoid wasps. Unpub. diss. Harvard University. 246 pp.
1942	The relation of the prey of *Sphecius speciosus* to the size and sex of the adult wasp (Hym.: Sphecidae). *Ann. Ent. Soc. Amer.* 34:310–317.

Dubois, E.
1921	Sphegidae de Belgique. II. *Bull. Soc. Ent. Belgique* 3:190–216.

Emerson, A. E.
1958	The evolution of behavior among social insects. [In Roe, A., and G. G. Simpson. *Behavior and Evolution*. Yale University Press, New Haven, Chap. 15, pp. 311–335.]

Evans, H. E.
1955	An ethological study of the digger wasp *Bembecinus neglectus*, with a review of the ethology of the genus. *Behaviour* 7:287–303.
1957a	Notes on a *Stictia* new to the United States (Hymenoptera: Sphecidae: Bembicini). *Ent. News* 68:76–77.
1957b	*Studies on the comparative ethology of digger wasps of the genus Bembix.* Comstock Publ. Assoc., Cornell Univ. Press, Ithaca, N.Y. 248 pp.
1958a	Studies on the nesting behavior of digger wasps of the tribe Sphecini. Part I: Genus *Prionoyx* Dahlbom. *Ann. Ent. Soc. Amer.* 51:177–186.
1958b	Ethological studies on digger wasps of the genus *Astata* (Hymenoptera, Sphecidae). *J. N.Y. Ent. Soc.* 65:159–185.
1959a	Studies on the larvae of digger wasps (Hymenoptera, Sphecidae). Part V: Conclusion. *Trans. Amer. Ent. Soc.* 85:137–191.
1959b	Adaptation of a sand wasp. *Natural History* 68:380–383.
1959c	Observations on the nesting behavior of digger wasps of the genus *Ammophila*. *Amer. Midl. Nat.* 62:449–473.
1961	A study of *Bembix u-scripta*, a crepuscular digger wasp. *Psyche* 67:45–61.
1962	The geyser-loving digger wasps of Yellowstone. *The Explorer* 4:6–11.
1963a	The evolution of prey-carrying mechanisms in wasps. *Evolution* 16:468–483.
1963b	Predatory wasps. *Scientific Amer.* 208:145–154.
1963c	The clustering wasps—and why they cluster. *Audubon Mag.* 65:236–237.
1963d	*Wasp Farm.* Natural History Press, New York. 178 pp.
1964a	Further studies on the larvae of digger wasps (Hymenoptera: Sphecidae). *Trans. Amer. Ent. Soc.* 90:235–299.
1964b	Notes on the nesting behavior of *Philanthus lepidus* Cresson (Hymenoptera, Sphecidae). *Psyche* 71:142–149.

Evans, H. E., and J. E. Gillaspy
1964	Observations on the ethology of digger wasps of the genus *Steniolia* (Hymenoptera: Sphecidae: Bembicini). *Amer. Midl. Nat.* 72:257–280.

Evans, H. E., and C. S. Lin
1956a	Studies on the larvae of digger wasps (Hymenoptera, Sphecidae). Part I: Sphecinae. *Trans. Amer. Ent. Soc.* 81:131–153.
1956b	Studies on the larvae of digger wasps (Hymenoptera, Sphecidae). Part II: Nyssoninae. *Trans. Amer. Ent. Soc.* 82:35–66.

Evans, H. E., C. S. Lin, and C. M. Yoshimoto

1954   Biological notes on *Psammaecius tricolor* (Cresson) (Hymenoptera: Sphecidae: Gorytini). *Ent. News* 65:6–11.
Evans, H. E., and E. G. Linsley.
1960   Notes on a sleeping aggregation of solitary bees and wasps. *Bull. So. Calif. Acad. Sci.* 59:30–37.
Fabre, J. H.
1886   *Souvenirs Entomologiques.* 3 ser. Ch. Delagrave, Paris. 433 pp.
Ferton, C.
1899   Observations sur l'instinct des *Bembex* Fabr. *Actes Soc. Linn. Bordeaux* 54:331–345.
1901   Notes detachées sur l'instinct des hyménoptères mellifères et ravisseurs. *Ann. Soc. Ent. France* 70:83–148.
1902a   Sur les moeurs de *Stizus fasciatus* Fabr. *Comptes Rendus Assoc. Franc. Av. Sci.* Congrès d'Ajaccio, 1901, pp. 680–683.
1902b   Notes detachées sur l'instinct des hyménoptères mellifères et ravisseurs. 2e ser. *Ann. Soc. Ent. France* 71:499–530.
1905   Notes sur l'instinct des hyménoptères mellifères et ravisseurs. 3e ser. *Ann. Soc. Ent. France* 74:56–101.
1908   Notes detachées sur l'instinct des hyménoptères mellifères et ravisseurs. 4e ser. *Ann. Soc. Ent. France* 77:535–584.
1909   Notes detachées sur l'instinct des hyménoptères mellifères et ravisseurs. 5e ser. *Ann. Soc. Ent. France* 78:401–422.
1910   Notes detachées sur l'instinct des hyménoptères mellifères et ravisseurs. 6e ser. *Ann. Soc. Ent. France* 79:145–178.
1911   Notes detachées sur l'instinct des hyménoptères mellifères et ravisseurs. 7e ser. *Ann. Soc. Ent. France* 80:351–412.
Froggatt, W. W.
1903   Cicadas ("locusts") and their habits. *Agri. Gaz. New So. Wales* 14:334–341, 418–425.
1907   *Australian Insects.* Wm. Brooks & Co., Sydney. 449 pp.
Gillaspy, J. E.
1963a   The genus *Stizoides* (Hymenoptera: Sphecidae: Stizini) in North America, with notes on the Old World fauna. *Bull. Mus. Comp. Zool.* (Harvard) 128:371–391.
1963b   Two new genera and a new species of Bembicini (Sphecidae) from North America, with a key to genera having recessed ocelli. *Ent. News* 74:187–199.
1963c   The identity of *Stictiella corniculata* Mickel (Sphecidae: Bembicini), with a note on synonymy in *Stictiella*. *Ent. News* 74:251–252.
1964   A revisionary study of the genus *Steniolia* (Hymenoptera: Sphecidae: Bembicini). *Trans. Amer. Ent. Soc.* 89:1–117.
Gillaspy, J. E., H. E. Evans, and C. S. Lin
1962   Observations on the behavior of digger wasps of the genus *Stictiella* (Hymenoptera: Sphecidae) with a partition of the genus. *Ann. Ent. Soc. Amer.* 55:559–566.
Gittins, A. R.
1958   Nesting habit and prey record of *Harpactostigma* (*Arcesilas*) *laminiferum* (Fox) (Hymenoptera: Sphecidae). *Pan-Pac. Ent.* 34:142.
Goossen, H.
1949   Untersuchungen an Gehirnen verschieden grosser, jeweils verwandter Coleopteren-und Hymenopteren-Arten. *Zool. Jahrb., Abt. Allg. Zool.* 62:1–64.

Grandi, G.
　1961　　Studi di un entomologo sugli imenotteri superiori. *Boll. Ist. Ent. Univ. Bologna* 25:1-659.

Gussakovskij, V. V.
　1952　　New and little-known species of Psammocharidae and Sphecidae (Hymenoptera) of W. Tadzhikistan. *Trud. Zool. Inst. Akad. Nauk SSSR* 10:199-288. [In Russian.]

Hamm, A. H., and O. W. Richards
　1930　　The biology of the British fossorial wasps of the families Mellinidae, Gorytidae, Philanthidae, Oxybelidae, and Trypoxylonidae. *Trans. Ent. Soc. London* 78:95-131.

Handlirsch, A.
　1887-1895　Monographie der mit *Nysson* und *Bembex* verwandten Grabwespen. *Sitzber. Akad. Wissen. Wien* 95:246-421, 96:219-309, 97:316-565, 98:440-517, 101:25-181, 102:657-942, 104:801-1079.

Hartman, C.
　1905　　Observations on the habits of some solitary wasps of Texas. *Bull. Univ. Texas,* No. 65. 72 pp.

Hine, J. S.
　1906　　A preliminary report on the horseflies of Louisiana, with a discussion of remedies and natural enemies. *Louisiana State Crop Pest Commission.* Circular No. 6. 43 pp.
　1907　　Second report upon the horseflies of Louisiana. *Louisiana Agricultural Experiment Station Bull.* No. 93. 59 pp.

Howard, L. O.
　1901　　*The Insect Book.* Doubleday, Page, and Co., New York. 429 pp.

Howard, L. O., H. G. Dyar, and F. Knab
　1912　　*The mosquitoes of North and Central America and the West Indies.* Vol. I. Carnegie Inst. Washington Publ., No. 159. 520 pp.

Howes, P. G.
　1919　　*Insect behavior.* Gorham Press, Boston. 176 pp.

Hudson, W. H.
　1892　　*The naturalist in La Plata.* London, Chapman and Hall, Ltd. 388 pp.

Huxley, J.
　1943　　*Evolution: the modern synthesis.* Harper & Brothers, New York and London. 645 pp.

Iersel, J. J. A. Van
　1952　　On the orientation of *Bembex rostrata* L. *Trans. 9th Internat. Congr. Ent., Amsterdam, 1951* 1:384-393.

Iersel, J. J. A. Van, and J. Van den Assem
　1965　　Aspects of orientation in the digger wasp *Bembix rostrata. Animal Behaviour.* Suppl. No. 1: 145-162.

Iwata, K.
　1936　　On the habits of *Stizus* and *Bembix* which occur in Japan. *Kontyu* 10:233-250. [In Japanese.]
　1937　　On the habits of *Harpactus laevis* Latr. *Kontyu* 11:404-409. [In Japanese.]
　1939　　Habits of some solitary wasps in Formosa (IV). *Trans. Nat. Hist. Soc. Formosa* 29:161-178. [In Japanese.]
　1955　　The comparative anatomy of the ovary in Hymenoptera. Part I. Aculeata. *Mushi* 29:17-34.
　1964a　Bionomics of non-social wasps in Thailand. *Nature and Life SE Asia* 3:323-383.

1964b   Egg giantism in subsocial Hymenoptera, with ethological discussion on tropical bamboo carpenter bees. *Nature and Life SE Asia* 3:399-434.

James, M. T.
- 1957   The genus *Eulalia* in Florida and the West Indies. *Fla. Ent.* 40:15-18.

Janvier, H. (M. F. Claude-Joseph)
- 1925   Le sommeil et l'orientation chez les *Monedula*. *Rev. Chilena Hist. Nat.* 29:214-216.
- 1928   Recherches biologiques sur les prédateurs du Chili. *Ann. Sci. Nat., Zool.* (10)11:67-207.

Jensen-Haarup, A. C.
- 1924   Hemipterological notes and descriptions III. *Ent. Meddel.* 14:323-338.

Katayama, H.
- 1933   On the prey of *Stizus pulcherrimus* Smith. *Trans. Kansai Ent. Soc.* 4:86-87. [In Japanese.]

Klopfer, P. H.
- 1962   *Behavioral aspects of ecology*. Prentice-Hall, Englewood Cliffs, New Jersey. 173 pp.

Kohl, F. F.
- 1880   Die Raubwespen Tirol's nach ihrer horizontalen und verticalen Verbreitung mit einem Anhange biologischen und Kritisher Notizen. *Z. Ferdinandeums Innsbruck* (3)24:97-242.
- 1896   Die Gattungen der Sphegiden. *Ann. Naturhist. Hofmus. Wien* 11:233-516.

Krombein, K. V.
- 1936   Biological notes on some solitary wasps (Hymenoptera: Sphecidae). *Ent. News* 47:93-99.
- 1951   Tribe Gorytini [In Muesebeck, C. F. W., K. V. Krombein, and H. K. Townes. *Hymenoptera of America North of Mexico: Synoptic Catalog*. U.S. Dept. Agri. Monogr. 2, pp. 986-993.]
- 1952   Preliminary annotated list of the wasps of Westmoreland State Park, Virginia, with notes on the genus *Thaumatodryinus* (Hymenoptera: Aculeata). *Trans. Amer. Ent. Soc.* 78:89-100.
- 1953   Biological and taxonomic observations on the wasps in a coastal area of North Carolina (Hymenoptera: Aculeata). *Wasmann J. Biol.* 10:257-340.
- 1955   Some notes on the wasps of Kill Devil Hills, North Carolina, 1954 (Hymenoptera, Aculeata). *Proc. Ent. Soc. Washington* 57:145-160.
- 1958a   Additions during 1956 and 1957 to the wasp fauna of Lost River State Park, West Virginia, with biological notes and descriptions of new species. *Proc. Ent. Soc. Washington* 60:49-64.
- 1958b   Biological notes on some wasps of Kill Devil Hills, North Carolina, and additions to the faunal list (Hymenoptera, Aculeata). *Proc. Ent. Soc. Washington* 60:97-110.
- 1958c   Biology and taxonomy of the cuckoo-wasps of coastal North Carolina (Hymenoptera, Chrysididae). *Trans. Amer. Ent. Soc.* 84:141-168.
- 1958d   *Hymenoptera of America north of Mexico: Synoptic Catalog*. U.S. Dept. Agri. Monogr. 2. First Supplement. 305 pp.
- 1959   Biological notes on some ground-nesting wasps at Kill Devil Hills, North Carolina, 1958, and additions to the faunal list (Hymenoptera, Aculeata). *Proc. Ent. Soc. Washington* 61:193-199.
- 1961   Some insect visitors of mat Euphorbia in southeastern Arizona (Hymenoptera, Diptera). *Ent. News* 72:80-83.
- 1964a   Miscellaneous prey records of solitary wasps. V. (Hymenoptera: Aculeata). *Bull. Brooklyn Ent. Soc.* 58:118-120.

1964b   Results of the Archbold Expeditions. No. 87. Biological notes on some Floridian wasps (Hymenoptera, Aculeata). *Amer. Mus. Novitates,* No. 2201. 27 pp.

Krombein, K. V., and A. Willink
1950   The North American species of *Bembecinus* (Hymenoptera, Sphecidae, Stizini). *Amer. Midl. Nat.* 44:699-713.

Kurczewski, F. E., and E. J. Kurczewski
1963   An annotated list of digger wasps from Presque Isle State Park, Pennsylvania (Hymenoptera: Aculeata). *Proc. Ent. Soc. Washington* 65:141-149.

Lafler, H. A.
1896   A new parasite. *Ent. News* 7:62-63.

La Rivers, I.
1942   Notes on the bembicid *Stictiella pulla* (Handlirsch). *Pan-Pac. Ent.* 18:4-8.

Lin, N.
1963a   Territorial behavior in the cicada killer wasp, *Sphecius speciosus* (Drury) (Hymenoptera: Sphecidae). I. *Behaviour* 20:115-133.
1963b   Observations of suspected density dependent fighting between females of the cicada killer wasp *Sphecius speciosus. Bull. Brooklyn Ent. Soc.* 58:121-123.
1964   Weather and the natural regulation of three populations of the cicada killer wasp. Unpub. diss. University of Kansas, 1964.

Lindauer, M.
1962   Ethology. *Ann. Rev. Psychology* 13:35-70.

Lindauer, M., and B. Schricker
1963   Über die Funktion der Ocellen bei den Dämmerungsflügen der Honigbiene. *Biol. Zentralbl.* 82:721-725.

Linsley, E. G.
1962   Sleeping aggregations of aculeate Hymenoptera—II. *Ann. Ent. Soc. Amer.* 55:148-164.

Llano, R. J.
1959   *Observaciones biologicas de insectos bonaerenses.* Suplemento Revista Educacion, Ministerio Educacion Provincia Buenos Aires, La Plata. 136 pp.

Lohrmann, E.
1948   Die Grabwespengruppe der Bembicinen. Überschau und Stammesgeschichte. *Mitt. Münchner Ent. Ges.* 34:420-471.

McCulloch, A. R.
1923   War in the garden. *Australian Mus. Mag.* 1:209-212.

Maidl, F., and A. Klima
1939   Sphecidae I (*Astatinae-Nyssoninae*). [*In* Hedicke, H. *Hymenopterorum catalogus.* W. Junk, Gravenhage. Part 8. 150 pp.]

Maillard, F.
1847   Note sur le nid d'un Hyménoptère ovither zoophage, le Gorytes à large ceinture, découvert le 10 juillet 1847. *Mem. Soc. Acad. Archeol. Sci. Arts Oise* 1:92-94.

Malyshev, S. I.
1959   *The Hymenoptera, their origin and evolution.* Sovetskaya Nauka, Moscow. 291 pp. [In Russian; reviewed by O. L. Kryzhanovskij in *Ent. Obozr.* in translation as *Ent. Reviews* 42(1963):377-378.]

Maneval, H.
1928   Notes sur quelques Hyménoptères fouisseurs. *Bull. Soc. Ent. France* 1928, pp. 29-32.
1932   Notes recueillies sur les Hyménoptères. *Ann. Soc. Ent. France* 101:85-110.

1936   Nouvelles notes sur divers Hyménoptères et leurs larves. *Rev. Franc. Ent.* 3:18–32.
1937   Notes sur les Hyménoptères. *Rev. Franc. Ent.* 4:162–181.
1939   Notes sur les Hyménoptères. 6e ser. *Ann. Soc. Ent. France* 108:49–108.

Maslin, T. P.
1942   Morphological criteria of phyletic relationships. *Syst. Zool.* 1:49–70.

Mayr, E.
1958   Behavior and systematics. [*In* Roe, A., and G. G. Simpson. *Behavior and Evolution.* Yale University Press, New Haven. chap. 16, pp. 341–362.]
1963   *Animal species and evolution.* Harvard University Press, Cambridge, Mass. 797 pp.

Mellor, J. E.
1927   A note on the mutillid *Ephutomma continua* Fabr. and of *Bembix mediterranea* Hdl. in Egypt with a summary of the distribution and of some previously recorded habits of the Mutillidae. *Bull. Soc. R. Ent. Egypt* 1927, pp. 69–79.

Meunier, F.
1915   Über einige fossile Insekten aus den Braunkohlenschichten (Aquitanien) von Rott (Siebengebirge). *Z. Deutsch. Geol. Ges.* 67:205–217.

Michener, C. D.
1944   Comparative external morphology, phylogeny, and a classification of the bees. *Bull. Amer. Mus. Nat. Hist.* 82:157–326.

Mickel, C. E.
1924   An analysis of a bimodal variation in size of the parasite *Dasymutilla bioculata* Cresson (Hymen: Mutillidae). *Ent. News* 35:236–242.

Musgrave, A.
1925   The sand wasp's burrow. *Australian Mus. Mag.* 2:243.

Nielsen, E. T.
1945   Moeurs des Bembex. *Spolia Zool. Mus. Hauniensis* 7:1–174.
1958   The method of ethology. *Proc. 10th Internat. Congr. Ent.* (Montreal) 2:563–565.

Olberg, G.
1959   *Das Verhalten der solitären Wespen Mitteleuropas.* Deutscher Verlag Wissenschaften, Berlin. 401 pp.

Oldroyd, H.
1954   *The horse-flies (Diptera: Tabanidae) of the Ethiopian region.* Vol. II. British Museum, London. 341 pp.

Parker, J. B.
1910   Notes on the nesting habits of *Bembex nubilipennis. Ohio Nat.* 10:163–165.
1917   A revision of the bembicine wasps of America north of Mexico. *Proc. U.S. Nat. Mus.* 52:1–155.
1925   Notes on the nesting habits of *Bembix comata* Parker. *Proc. Ent. Soc. Washington* 27:189–195.
1929   A generic revision of the fossorial wasps of the tribes Stizini and Bembicini with notes and descriptions of new species. *Proc. U.S. Nat. Mus.* 75(5):1–203.

Pate, V. S. L.
1936   Studies in the nyssonine wasps. II. The subgenera of *Sphecius* (Hymenoptera: Sphecidae: Gorytini). *Bull. Brooklyn Ent. Soc.* 31:198–200.
1937   The generic names of the sphecoid wasps and their type species (Hymenoptera: Aculeata). *Mem. Amer. Ent. Soc.* No. 9. 103 pp.

1938	Studies in the nyssonine wasps (Hymenoptera: Sphecidae). IV. New or redefined genera of the tribe Nyssonini, with descriptions of new species. *Trans. Amer. Ent. Soc.* 64:117-190.
1941	Two new species of sphecoid wasps from Trinidad (Hymenoptera: Aculeata). *Notulae Nat., Acad. Nat. Sci. Phila.* No. 91. 8 pp.
1946	Prey records of gorytine wasps (Hymenoptera, Sphecidae). *Bull. Brooklyn Ent. Soc.* 41:99.
1947	On the genus *Ochleroptera* Holmberg (Hymenoptera, Sphecidae, Gorytini). *Bull. Brooklyn Ent. Soc.* 42:65-70.

Poulton, E. B.
1917 Predaceous reduviid bugs and fossors, with their prey, from the S. Paulo district of southeast Brazil. *Proc. Ent. Soc. London* 1917, pp. xxiv-xli.

Powell, J. A., and J. A. Chemsak
1959 Biological observations on *Psammaecius adornatus* (Bradley). *Pan-Pac. Ent.* 35:195-201.

Rathmayer, W.
1962 Paralysis caused by the digger wasp *Philanthus*. *Nature* 196:1148-1151.

Rau, P.
1922	Ecological and behavior notes on Missouri insects. *Trans. Acad. Sci. St. Louis* 24:1-41.
1934	Behavior notes on certain solitary wasps. *Canad. Ent.* 66:259-261.
1938	Additional observations on the sleep of insects. *Ann. Ent. Soc. Amer.* 31:540-556.

Rau, P., and N. Rau
1916	The sleep of insects; an ecological study. *Ann. Ent. Soc. Amer.* 9:227-274.
1918	*Wasp studies afield.* Princeton University Press, Princeton, New Jersey. 372 pp.

Reinhard, E. G.
1925a	The wasp *Hoplisus costalis*, a hunter of tree-hoppers. *J. Washington Acad. Sci.* 15:107-110.
1925b	The wasp *Nysson hoplisivora*, a parasitic relative of *Hoplisus costalis*. *J. Washington Acad. Sci.* 15:172-177.
1929	*The Witchery of Wasps.* Century Co., New York. 291 pp.

Rensch, B.
1960 *Evolution above the species level.* Columbia University Press, New York. 419 pp.

Richards, O. W.
1937	Results of the Oxford University Expedition to British Guiana, 1929. Hymenoptera, Sphecidae and Bembicidae. *Trans. R. Ent. Soc. London* 86:101-118.
1956	*Hymenoptera: Introduction and keys to families.* Handbook for Ident. British Insects, Vol. 6, pt. 1. Royal Entomological Society, London. 94 pp.

Riley, C. V.
1892a	The larger digger wasp. *Insect Life* 4:248-252.
1892b	On the larva and some peculiarities of the cocoon of *Sphecius speciosus*. *Proc. Ent. Soc. Washington* 2:170-172.

Ristich, S. S.
1953	A study of the prey, enemies, and habits of the great golden digger wasp, *Chlorion ichneumoneum* (L.). *Canad. Ent.* 85:374-386.
1956	The host relationships of a miltogrammid fly *Senotainia trilineata* (VDW). *Ohio J. Sci.* 56:271-274.

Rodeck, H. G.
  1931 Nesting habits of *Bembicinus godmani* (Cameron). *J. Colo. Wyo. Acad. Sci.* 1:61.
Roeder, K. D.
  1963 *Nerve cells and insect behavior.* Harvard University Press, Cambridge, Mass. 188 pp.
Rohlf, F. J.
  1963 Congruence of larval and adult classifications in *Aedes* (Diptera: Culicidae). *Syst. Zool.* 12:97–117.
Rohwer, S. A.
  1909 Three new fossil insects from Florissant, Colorado. *Amer. J. Sci.* (4) 28:533–536.
  1911 Descriptions of new species of wasps with notes on described species. *Proc. U.S. Nat. Mus.* 40:551–587.
  1916 Sphecoidea [*In* Viereck, H. L. *The Hymenoptera, or wasp-like insects, of Connecticut.* Connecticut Geological and Natural History Survey Bull. No. 22, pp. 645–697.]
Sabrosky, C. W.
  1956 *Musca autumnalis* in upstate New York. *Proc. Ent. Soc. Washington* 58:347.
Savin, W. M.
  1923 A wasp that hunts cicadas. *Nat. Hist.* 23:569–575.
Say, T.
  1837 Descriptions of new North American Hymenoptera, and observations on some already described. *Boston J. Nat. Hist.* 1:361–416.
Scudder, S. H.
  1890 *The tertiary insects of North America.* Report United States Geological Survey. Vol. 13. 734 pp.
Scullen, H. A.
  1965 Review of the genus *Cerceris* in America North of Mexico (Hymenoptera: Sphecidae). *Proc. U.S. Nat. Mus.* 116:333–548.
Shappirio, D. G.
  1946 Notes on District of Columbia wasps (Hym.: Sphecidae). *Ent. News* 57:229–230.
Shuckard, W. E.
  1837 *Essay on the indigenous fossorial Hymenoptera.* C. Roworth & Sons, London. 259 pp.
Simpson, G. G.
  1953 *The major features of evolution.* Columbia University Press, New York. 434 pp.
  1959 The nature and origin of supraspecific taxa. [*In Genetics and Twentieth Century Darwinism,* Cold Spring Harbor Symp. Quant. Biol. 24:255–272.]
Smith, H. E.
  1915 The grasshopper outbreak in New Mexico during the summer of 1913. *U.S. Dept. Agri. Bull.* No. 293. 12 pp.
Smith, M. R.
  1923 The life history and habits of *Bicyrtes quadrifasciata* Say. *Ann. Ent. Soc. Amer.* 16:238–246.
Snodgrass, R. E.
  1910 The thorax of the Hymenoptera. *Proc. U.S. Nat. Mus.* 39:37–91.
  1941 The male genitalia of Hymenoptera. *Smithsonian Misc. Coll.* Vol. 99. No. 14. 86 pp.

1956   *Anatomy of the honeybee.* Comstock Publ. Assoc., Cornell Univ. Press, Ithaca, N.Y. 334 pp.

Sokal, R. R., and P. H. A. Sneath
1963   *Principles of numerical taxonomy.* W. H. Freeman, San Francisco. 359 pp.

Steiner, A.
1962   Etude du comportement prédateur d'un Hyménoptère sphégien: *Liris nigra* V.D.L. (= *Notogonia pompiliformis* Pz.). *Ann. Sci. Nat., Zool.* (12)4: 1–125.

Stoehr, L. M.
1917   *Microbembex monodonta. Nat. Canad.* 43:113–119.

Strandtmann, R. W.
1945   Observations on the nesting habits of some digger wasps (Sphecidae). *Ann. Ent. Soc. Amer.* 38:305–313.

Telford, A. D.
1964   The Nearctic *Parnopes* with an analysis of the male genitalia in the genus (Hymenoptera: Chrysididae). *Univ. Calif. Publ. Ent.* 36(1):1–42.

Thompson, W. R.
1964   The work of Jean Henri Fabre. *Canad. Ent.* 96:62–70.

Thorpe, W. H.
1963   *Learning and instinct in animals.* 2nd ed. Methuen & Co., London. 558 pp.

Tinbergen, N.
1935   Über die Orientierung des Bienenwolfes. II. Die Bienenjagd. *Zeitschr. Vergl. Physiol.* 21:699–716.

Tsuneki, K.
1943a   A naturalist at the front. *Osaka.* Pp. 289–296. [In Japanese.]
1943b   On the habit of *Philanthus coronatus* Fabricius (Hymenoptera, Philanthidae). *Mushi* 15:33–36. [In Japanese.]
1943c   On the habits of *Stizus pulcherrimus* Smith (Hym. Stizidae). *Mushi* 15:37–47. [In Japanese.]
1956   Ethological studies on *Bembix niponica* Smith, with emphasis on the psychobiological analysis of behaviour inside the nest (Hymenoptera, Sphecidae). I. Biological Part. *Mem. Fac. Lib. Arts, Fukui Univ.* Ser. II. No. 6, pp. 77–172.
1957   Ethological studies on *Bembix niponica* Smith, with emphasis on the psychobiological analysis of behaviour inside the nest (Hymenoptera, Sphecidae). II. Experimental Part. *Mem. Fac. Lib. Arts, Fukui Univ.* Ser. II. No. 7, pp. 1–116.
1958   Ethological studies on *Bembix niponica* Smith, with emphasis on the psychobiological analysis of behaviour inside the nest (Hymenoptera, Sphecidae). III. Conclusive Part. *Mem. Fac. Lib. Arts, Fukui Univ.* Ser. II. No. 8, pp. 1–78.
1963a   Comparative studies on the nesting biology of the genus *Sphex* (s.l.) in East Asia (Hymenoptera, Sphecidae). *Mem. Fac. Lib. Arts, Fukui Univ.* Ser. II, No. 13, pp. 13–78.
1963b   The tribe Gorytini of Japan and Korea (Hymenoptera, Sphecidae). *Etizenia, Occ. Pap. Biol. Lab. Fukui Univ.,* No. 1. 20 pp.
1965a   Variation in characters of *Bembecinus hungaricus* Frivaldzky occurring in East Asia, with taxonomic notes on hitherto known species (Hymenoptera, Sphecidae). *Etizenia, Occ. Publ. Biol. Lab. Fukui Univ.,* No. 8:1–17.
1965b   The nesting biology of *Stizus pulcherrimus* F. Smith (Hym., Sphecidae) with special reference to the geographic variation. *Etizenia, Occ. Publ. Biol. Lab. Fukui Univ.,* No. 10:1–21.

Bibliography

Turner, R. E.
1912 Studies in the fossorial wasps of the family Scoliidae, subfamilies Elidinae and Anthoboscinae. *Proc. Zool. Soc. London* 1912, pp. 696-754.

Vesey-Fitzgerald, D.
1940 Notes on Bembicidae and allied wasps from Trinidad (Hym.: Bembicidae and Stizidae). *Proc. R. Ent. Soc. London* 15(A):37-39.
1956 Notes on Sphecidae (Hym.) and their prey from Trinidad and British Guiana. *Ent. Month. Mag.* 92:286-287.

Vowles, D. M.
1961 Neural mechanisms in insect behaviour [*In* Thorpe, W. H., and O. L. Zangwill. *Current Problems in Animal Behavior.* Cambridge University Press, Cambridge, England. chap. i, pp. 5-29.]

Waddington, C. H.
1960 Evolutionary adaptation. [In Tax, S. *The evolution of life.* University of Chicago Press, Chicago. Pp. 381-402.]

Walsh, B. D., and C. V. Riley
1868 Wasps and their habits. *Amer. Ent.* 1:122-143.

Wecker, S. C.
1964 Habitat selection. *Sci. Amer.* 211:109-116.

Wesenberg-Lund, C.
1891 *Bembex rostrata,* dens Liv og Instinkter. *Ent. Meddel.* 3:19-44.

Wheeler, W. M.
1919 The parasitic Aculeata, a study in evolution. *Proc. Amer. Phil. Soc.* 58:1-40.
1928 *The social insects: their origin and evolution.* Harcourt, Brace, & Co., New York. 378 pp.

Wheeler, W. M., and R. Dow
1933 Unusual prey of *Bembix. Psyche* 40:57-59.

Whitman, C. O.
1899 Animal behavior. *Biol. Lectures Marine Biol. Lab. Wood's Hole, No. 16,* pp. 285-338. Ginn & Co., Boston.

Williams, F. X.
1914 Notes on the habits of some wasps that occur in Kansas, with the description of a new species. *Kansas Univ. Sci. Bull.* 8:223-230.
1928 Studies in tropical wasps—their hosts and associates. *Bull. Experiment Station Hawaiian Sugar Planters' Association, Ent. Ser.,* no. 19. 179 pp.
1954 *Xenosphex xerophila,* an apparently new genus and species of wasp from southern California (Hymenoptera, Sphecidae, Nyssoninae, Gorytini). *Wasmann J. Biol.* 12:97-103.

Willink, A.
1947 Las especies argentinas de "Bembicini" (Hymenoptera: Sphecidae: Nissoninae). *Acta Zool. Lilloana* 4:509-651.
1949 Las especies neotropicales de "Bembecinus" (Hymenoptera, Sphecidae, Nyssoninae, Stizini), *Acta Zool. Lilloana* 7:81-112.
1958 Descripcion de un nuevo genero y especie de Bembicini y observaciones sobre otra (Hymenoptera: Sphecidae). *Acta Zool. Lilloana* 16:47-54.

Wolcott, G. N.
1923 Insectae portoricensis. *J. Dept. Agri. Porto Rico* 7(1):1-313.

Wolf, H.
 1951 Die parasitische Lebenweise der Grabwespengattung *Nysson* Latr. (Hym. Crabronidae). *Nachr. Samm. Schmarotzerbestim.* (Aschaftenburg), No. 33, pp. 77–80.

Yasumatsu, K.
 1939 An observation on *Bembix niponica*. *Mushi* 12:67. [In Japanese.]

Yasumatsu, K., and H. Masuda
 1932 On a new hunting wasp from Japan. *Ann. Soc. Hist. Nat. Fukuokensis* 1:53–65. [In Japanese, with English summary.]

Index

Accessory burrows, 355
Adaptive improvements, 420, 494, 500–504
Adaptive radiation, 474, 486–487
Adlerz, G., 69, 70–71
Aggression, 426
Allen, H. W., 159, 167, 468
Alysson: distribution, 399, 400; ethology, summary of, 29–30, 443, 444; natural enemies, 469; phylogeny, 416, 480, 481; structure, 16–17, 402, 404, 410
  cameroni, 29; *fuscatus*, 28; *guignardi*, 28; *melleus*, 17–28, 458, 459, 462; *oppositus*, 28; *ratzeburgii*, 28; *tricolor*, 29
Alyssonini, 4, 16, 476–477
Ammatomus, 78–80, 402, 444; distribution, 399, 400; phylogeny, 416, 481
  moneduloides, 78–80
Amobia aurifrons, 52
Amobiopsis, 469
Ampulicini, 4, 475
Ant lions, as predators, 472
Ants. See Formicidae
Arens, L. E., 131
Argogorytes, 68–69; distribution, 399, 400; ethology, summary of, 80–82, 443, 444; natural enemies, 469; phylogeny, 416, 481; structure, 402, 411
  campestris, 69, 81; *mystaceus*, 69
Arigorytes, 399
Arnold, G., 85, 111, 267
Ashmead, W. H., 3, 104, 225

Back burrows, 13, 354–355; in *Bembix amoena*, 279–281; in *Bembix sayi*, 296–297
Baldwin effect, 496
Barth, G. P., 35, 58, 60
Bates, H. W., 243–246
Bateson, G., 497–499
Beaumont, J. de, 7, 8, 410, 412
Behavior: advanced features of, 477–479; changes in, 503–504; evolution of, 477–500, 502–504; primitive features of, 477–479; variation in, 496–499. See also Digging behavior, Hunting behavior, Leveling behavior, Mating behavior, Sleeping behavior
Bembecinus, 116, 132–143, 443, 444; distribution, 399–401; natural enemies, 469; phylogeny, 416, 481; structure, 403, 404, 490
  agilis, 137–139, 142; *argentifrons*, 142, 462; *bicinctus*, 135; *cinguliger*, 135; *consobrinus*, 135; *errans*, 137, 139, 142; *fertoni*, 137, 138, 142; *godmani bolivari*, 138, 141; *godmani godmani*, 134–142; *hungaricus*, 137–142; *japonicus*, 137, 142; *mexicanus*, 134–142, 462; *nanus*, 139, 141; *neglectus*, 132–143, 460, 462; *oxydorcus*, 135; *prismaticus*, 137, 138, 139, 142; *rhopalocerus*, 135; *tridens*, 87, 132, 137–143
Bembicini, 4, 144, 405, 413, 474–477
Bembix, 1, 3; cocoons, 455–464; distribution, 399–401; ethology, summary of, 349–359, 443, 444; larvae, 269, 404; natural enemies, 469; phylogeny, 416, 480, 481; structure, 267–269, 403, 405
  amoena, 269–288; *atrifrons*, 357, 488; *belfragei*, 315–317, 414; *brullei*, 14, 455; *cameroni*, 310–311; *cinerea*, 42, 460; *citripes*, 356, 488, 504; *comata*, 351, 352; *hinei*, 350, 465; *integra*, 357; *multipicta*, 322, 336–344; *niponica*, 1, 14, 374, 429, 441, 451; *nubilipennis*, 298–308; *occidentalis*, 349, 356; *olivacea*, 352, 356; *pruinosa*, 322, 345–349, 414; *rostrata*, 1, 358, 374, 423; *sayi*, 288–298, 353–355; *spinolae*, 268, 311–315, 361; *texana*, 289, 322–335, 461; *troglodytes*, 322, 335–337; *truncata*, 308–309; *u-scripta*, 267, 289, 317–321, 414
Berland, L., 111, 126
Bernard, F., 69
Bicyrtes: cocoons, 463; digging behavior, contrasted to other sand wasps, 482; distribution, 399, 401; ethology, summary

Bicyrtes: (continued)
of, 174–175, 435, 443, 444; larvae, 146, 404; natural enemies, 469; phylogeny, 416, 480, 481; structure, 144–145, 403–405
burmeisteri, 173, 175; *capnoptera*, 171–172, 174–175; *discisa*, 173–175; *fodiens*, 167–170, 174–175; *parata*, 170–171; *quadrifasciata*, 145–160, 414, 460; *variegata*, 173–175; *ventralis*, 158, 160–167, 175
Birds, as predators, 308, 472
Bodkin, G. E., 244, 246, 261–262, 465
Bohart, G. E., 387, 390, 471
Bombyliidae, as parasites, 469–471, 472; of *Bembix*, 288, 315, 336, 349; of *Microbembex*, 387; of *Rubrica*, 261; of *Stictia*, 246
Bondar, G., 258–259, 262–263, 265
Bothynostethus, 3, 16, 399
Brachystegus, 83, 89, 90, 416, 481
scalaris, 89
Bradley, J. C., 110
Brauns, H., 125, 135
Brèthes, J., 173, 258–261, 465
Bristowe, W. S., 70–71
Burroughs, J., 93, 101, 105
Burrows. *See* Accessory burrows, Back burrows, Cell burrows, False burrows, Nest structure

Callan, E. M., 142, 173, 258–261, 364
Canalization, 497
Carlobembix, 399, 403, 405, 414, 416
Cazier, M. A., 71–72, 89–90
Cell burrows, 13–14, 353
Cell cleaning, 451–452, 485; in *Bembix*, 329–333, 358; in *Rubrica*, 260
Cells, number and arrangement of, 440–443
Centipedes, as prey, 383
Cerceris, 90
Cercopidae, as prey, 444; of *Alysson*, 29; of *Argogorytes*, 69; of *Bembecinus*, 142; of *Dienoplus*, 70–71; of *Gorytes*, 67, 68; of *Microbembex*, 380; of *Ochleroptera*, 77
Chapman, R. N., 364, 482
Chemsak, J., 46–49
Chloropidae, as cell scavengers, 261, 311, 469, 471, 472
Chrysididae, as parasites, 53, 71, 143, 469, 471, 472. See also *Holopyga*, *Parnopes*
Cicadellidae, as prey, 443–444; of *Alysson*, 25–29; of *Bembecinus*, 139–142; of *Didineis*, 30; of *Dienoplus*, 70–71; of *Editha*, 265; of *Glenostictia*, 201; of *Gorytes*, 62–63, 66; of *Hapalomellinus*, 72; of *Hoplisoides*, 54, 56; of *Microbembex*, 380, 382; of *Ochleroptera*, 76–77; of *Psammaecius*, 70
Cicadidae, as prey, 91, 444; of *Exeirus*, 114; of *Sphecius*, 103–106, 110–111
Classification of Nyssoninae, 3–5, 473–477
Cleptoparasitism, 132, 453–455, 481

Clitemnestra: distribution, 399, 400; ethology, 73–74, 80–82, 443, 444; phylogeny, 416, 481; structure, 402, 407, 410–411
chilensis, 73–74, 81; *gayi*, 73
Closure: final, 13, 452–453; inner, 438–439, 484–485; temporary, 13, 352–353, 437–438, 484–486
Clustering, 430–431, 433; in *Bembecinus*, 135; in *Glenostictia*, 191–192; in *Steniolia*, 209–213; in *Zyzzyx*, 255–256
Cockerell, T. D. A., 394–395, 396–397
Cocoon spinning, 455–467, 481
Coleoptera: as parasites, *see* Rhipiphoridae; as prey, of *Microbembex ciliata*, 389, of *M. monodonta*, 377, 380, 382–383, of *M. uruguayensis*, 390, of *Rubrica*, 263
Colony localization, 423–424
Competition for nesting sites, 350
Concealment of nest, 438, 453, 483–484
Conopidae, as parasites, 287, 315, 349, 469, 471, 472
Convergence, 417, 475, 479, 501
Copello, A., 259, 261
Corpora pedunculata, 501
Cuckoo wasps. *See* Chrysididae
Curran, C. H., 103, 104, 105, 109

Dambach, C. A., 93–110
Dasymutilla, 469; *asopus cassandra*, 297; *bioculata*, 387; *californica*, 288; *creusa bellona*, 288; *klugii*, 110; *nigripes*, 387; *pyrrhus*, 317, 326, 333–334; *ursula*, 315; *vesta*, 42, 219
Davis, W. T.: on *Bicyrtes quadrifasciata*, 146, 155–156; on *Sphecius speciosus*, 93–97, 104–105
Deleurance, E. P., 71, 89, 126–127, 131
Delphacidae, as prey, 26, 30
Density of colonies, 425–426
Didineis: distribution, 399, 400; ethology, 30, 444; phylogeny, 416; structure, 16–17, 402, 410
lunicornis, 30; *solidescens*, 393; *texana*, 30
Dienoplus: distribution, 399; ethology, 70–71, 88, 443, 444; natural enemies, 469; phylogeny, 416; structure, 68, 402
affinis, 71; *concinnus*, 71, 81; *elegans*, 71; *fertoni*, 71; *gyponae*, 70; *laevis*, 70; *leucrurus*, 71; *lunatus*, 71; *tumidus*, 70
Digging behavior, 433–434, 480–483, 496
Diptera: as parasites, *see* Bombyliidae, Conopidae, Miltogramminae; as prey, 356–358, 442, 444, 487, of *Bembix amoena*, 281–284, of *B. belfragei*, 317, of *B. cameroni*, 311, of *B. multipicta*, 344, of *B. nubilipennis*, 304–306, of *B. pruinosa*, 347–348, of *B. sayi*, 292–294, of *B. spinolae*, 313–314, of *B. texana*, 329–330, of *B. troglodytes*, 336–337, of *B. truncata*, 309, of *B. u-scripta*, 319–320, of *Glenostictia*, 188, 189, 200–202, of *Micro-*

Index

Diptera: (continued)
  bembex, 365, 380–383, 389, of *Rubrica*, 259, 260, 262, 263, of *Steniolia*, 217–218, of *Stictia*, 236–240, 243–246, 250–254, of *Zyzzyx*, 257
Distribution patterns, 398–401, 423
Dohrniphora cornuta, 241
Dollo's law, 406, 479
Dow, R., 101, 104, 108, 126, 173

Editha, 222; distribution, 399; ethology, 265–266, 444, 487; phylogeny, 416; structure, 255, 264, 403, 405, 414
  adonis, 265; *magnifica*, 255, 264, 265
Elampus viridicyaneus, 53
Emerson, A. E., 496
Entomosericus, 16; distribution, 399; phylogeny, 416; structure, 401, 402, 407, 410, 412
Ephemerida, as prey, 377, 379, 382
Ephutomma continua, 471
Epicnemial ridge, 5–7, 408
Evolution: advances in, 480, 501–504; reversals in trends, 479; trends in behavioral, 477–481; trends in structural, 415–419. *See also* Behavior, Convergence, Parallelism
Exeirus, 91; distribution, 399; ethology, 112–115, 443, 444; phylogeny, 416; structure, 111–113, 402, 411
  lateritius, 111–115
Exoprosopa, 469; *arenicola*, 349; *dorcadion*, 288, 315; *fascipennis*, 336, 387

Fabre, J. H., 126
False burrows, 13, 436–437, 485; in *Bembix*, 353–355; in *B. amoena*, 279–281; in *B. pruinosa*, 346; in *B. sayi*, 295–297; in *B. texana*, 325–328; in *Rubrica*, 263; in *Stizus*, 119
Feeding by adults, 427–429
Ferton, C., on *Alysson*, 28–30; on *Bembecinus*, 137, 142; on *Bembix*, 429; on *Hoplisoides*, 56; on *Stizus*, 124–125
Flies. *See* Diptera
Formicidae: as natural enemies, 241; relationships with *Microbembex*, 375, 377, 378, 381, 383, 389
Fossil record, 392–398
Froggatt, W. W., 112–114
Fulgoroidea, as prey, 444; of *Alysson*, 29; of *Ammatomus*, 78–79; of *Bembecinus*, 141–142; of *Clitemnestra*, 73; of *Dideineis*, 30; of *Dienoplus*, 71; of *Gorytes*, 62–63, 67, 68; of *Harpactostigma*, 69; of *Hoplisoides*, 56; of *Microbembex*, 380; of *Ochleroptera*, 77; of *Psammaletes*, 69

Genitalia, male, 11
Gillaspy, J. E.: on *Glenostictia*, 186–192, 198, 200; on *Steniolia*, 207, 428; on *Stictiella*, 179–184; on *Stizoides*, 127, 128, 131
Gittins, A., 69
Glenostictia: cocoons, 205, 460, 463; distribution, 399; ethology, 186–205, 220–221, 443, 444, 487; natural enemies, 469; phylogeny, 416, 481; structure, 185, 403, 404, 405
  clypeata, 186; *gilva*, 188; *pictifrons*, 186; *pulla*, 186–188, 460, 463; *scitula*, 185, 189–205, 463, 488
Gorytes: distribution, 399, 400; ethology, 58–68, 443, 444; fossils, 394–398; natural enemies, 469, 490; phylogeny, 416, 481; structure, 31–32, 57, 402, 404, 411
  asperatus, 58; *atricornis*, 67; *brasiliensis*, 68; *canaliculatus*, 34, 58–66, 458, 462; *deceptor*, 67; *laticinctus*, 67; *phaleratus*, 67; *planifrons*, 67; *simillimus*, 32, 66, 80
Gorytini, 4, 31, 400, 474–477
Grandi, G., 68, 142, 143
Grasshoppers. *See* Orthoptera
Gussakovskij, V. V., 412–413

Habitat of sand wasps, 421–424, 482
Hamm, A. H., 70, 85, 88
Handlirsch, A., 3, 31, 176
Handlirschia, 399
Hapalomellinus, 68, 399, 443, 469
  albitomentosus, 71–72, 80
Harpactostigma, 68, 444; distribution, 399; structure, 402, 411, 412
  laminiferum, 69
Harpactus, 70
Hartman, C.: on *Alysson*, 17–25; on *Bicyrtes*, 146–159, 170–171; on *Microbembex*, 363–364, 372–388; on *Stictia*, 225–240
Hedychridium integrum, 71; *roseum*, 71
Hedychrum chalybaeum, 143
Hemidula, 399, 403, 405, 414, 416
Heteroptera, as prey, 442, 444, 487; of *Bicyrtes*, 158, 170–174; of *B. fodiens*, 169; of *B. quadrifasciata*, 154–156; of *B. ventralis*, 164–166; of *Glenostictia*, 201; of *Microbembex*, 380, 382, 389
Hilarella hilarella, 219, 287, 469
Hine, J. S., 225–241
Holopyga, 469; *chrysonota*, 143; *ventralis*, 160, 169
Honey-robbing, 429
Hoplisidia kohliana, 397
Hoplisoides: distribution, 399, 400; ethology, 34–57, 443, 444; natural enemies, 469; phylogeny, 416, 480, 481; structure, 33, 402, 404
  costalis, 49–53, 458, 459, 462; *denticulatus*, 56; *latifrons*, 56; *nebulosus*, 33–46, 80; *punctatus*, 56; *punctatus manjikuli*, 56; *spilographus*, 46–49, 80; *spilopterus*, 55; *tricolor*, 53–55, 80; *umbonicida*, 56

Hoplisus, 395; *archoryctes*, 394–395, 398; *sepultus*, 396–398
Howard, L. O., 246
Howes, P. G., 93, 107
Hudson, W. H., 253, 262–263
Hunting behavior, 442–446, 486–487
Hymenoptera: as parasites, *see* Chrysididae, Mutillidae, Nyssonini, *Stizoides;* as prey, 444, 488, of *Bembix,* 357, of *Glenostictia,* 186, 199–203, of *Hoplisoides,* 56, of *Microbembex,* 381–383, 389
Hyperalonia morio, 261

Iersel, J. van, 349, 374, 440
Intelligence, 500
Iwata, K., 70, 137, 142, 490

Janvier, H.: on *Bembix,* 14, 428, 455; on *Bicyrtes,* 173–175; on *Clitemnestra,* 73–74, 425, 434; on *Microbembex,* 388–390; on *Zyzzyx,* 254–257, 428

Kohl, F. F., 3, 28
Kohlia, 399, 402, 410, 412, 416
Krombein, K. V.: on *Ammatomus,* 78–79; on *Bembix,* 323–324; on *Bicyrtes,* 146–160; on *Gorytes,* 58, 66, 67; on *Hoplisoides,* 35–42, 49–53, 56; on *Microbembex,* 363, 377, 379, 388; on *Stictia,* 225–241; on *Stictiella,* 179, 181–184
Kurczewski, F. E.: on *Alysson,* 18, 28; on *Bembix,* 311; on *Bicyrtes,* 166; on *Gorytes,* 58, 63–65, 67; on *Microbembex,* 363, 365, 378, 379

Lamarckism, 492, 499
LaRivers, I., 186–188
Larvae, 11–12, 404, 407, 409
Lasiopleura grisea, 311
Leafhoppers. *See* Cicadellidae
Learning, 496, 499–500
Lepidoptera, as prey, 442, 444, 487, 488; of *Bembix,* 357; of *Editha,* 265; of *Microbembex,* 375, 380, 383; of *Stictiella,* 182–183; of *Zyzzyx,* 257
Lestiphorus, 33, 68, 399, 444
  bicinctus, 69
Leveling behavior, 13, 353, 435–436, 453, 483
Lin, C. S.: on *Bembix,* 299, 302–308, 336; on *Bicyrtes,* 146, 171; on *Stictiella,* 181, 182, 183
Lin, N., 93–101, 426, 431, 437
Linsley, E. G., 171, 186, 429
Liohippelates pusio, 261
Llano, R. J.: on *Bembix,* 357; on *Microbembex,* 390; on *Rubrica,* 258, 260, 262–264, 465

McCulloch, A. R., 112–115
Macrosiagon flavipenne, 308
Malaxation of prey, 428–429
Malyshev, S. I., 374
Maneval, H.: on *Dienoplus,* 70–71; on *Gorytes,* 67, 458; on *Hoplisoides,* 56; on *Nysson,* 57, 87, 462, 465
Mantidae, as prey, 126
Mating behavior, 432–433
Mayr, E., 496, 497
Megaselia, 287
Megistommum, 399
Mellinus, 3, 4, 398, 475
Membracidae, as prey, 444; of *Bembecinus,* 141–142; of *Clitemnestra,* 74; of *Gorytes,* 67; of *Hoplisoides,* 38–41, 46, 48–51, 55–56; of *Microbembex,* 380, 389; of *Ochleroptera,* 77
Metanysson, 83, 87, 90
  arivaipa, 90; *coahuila,* 90
Metasomal pumping, 439
Metopia, 468, 469
  argyrocephala, 64, 109, 170, 287
Meunier, F., 396
Mickel, C. E., 387
Microbembex: digging behavior, contrasted to other sand wasps, 435, 482; distribution, 399; ethology, summary of, 390–391, 443, 444, 488–489; larvae, 361, 404; natural enemies, 469; phylogeny, 416, 481; structure, 360–362, 403, 405
  argentina, 390; *aurata,* 388, 390; *ciliata,* 388–390, 391; *monodonta,* 361–388, 414, 460, 463; *sulfurea,* 388; *uruguayensis,* 390, 391
Microstictia, 177, 399, 403, 405, 416
Millepedes, as prey, 383
Miltogramma, 123
Miltogramminae, as parasites, 468–472, 485, 491–492, 494; of *Alysson,* 28; of *Bembix,* 287; of *Gorytes,* 64; of *Hoplisoides,* 52; of *Rubrica,* 261; of *Steniolia,* 219. See also *Metopia, Senotainia*
Monedula, 176, 476
Mortenson, M. A., 71–72
Musgrave, A., 112–114
Mutilla merope, 126
Mutillidae, as parasites, 469, 471, 472. See also *Dasymutilla, Mutilla, Smicromyrme, Timulla*

Natural enemies, 467–472, 486, 490, 503
Nest structure, 13–14, 351, 440–443
Neuroptera, as prey, 380, 383
Nielsen, E. T., 1, 350, 351, 421, 473, 475
Nysson: distribution, 399–400; ethology, 85–89, 469, 472; larvae, 84–85, 404; phylogeny, 416, 481; structure, 83–84
  bellus, 55, 86; *daeckei,* 44–46, 64–65, 86; *dimidiatus,* 57, 71, 85–87, 462; *fidelis,* 65, 86; *hoplisivora,* 52–53, 86; *interruptus,* 86;

Index

Nysson: (continued)
  lateralis, 65, 84, 86; *maculatus,* 71, 86; *moestus,* 49, 86; *niger,* 86; *opulentus,* 42; *plagiatus,* 430; *pumilus,* 49, 86; *rottensis,* 396; *rusticus,* 49, 86; *spinosus,* 69, 86; *tridens,* 86; *trimaculatus,* 71, 86; *tuberculatus,* 42–44, 86
Nyssonini, 4, 476–477; distribution, 401; ethology, 83–90, 453–455; structure, 402, 413

Oblique groove, 8, 408
Ocelli, 6, 144, 414, 500–501
Ochleroptera: distribution, 399, 400; ethology, 80–82, 443, 444; natural enemies, 469; phylogeny, 416, 481; structure, 74, 402, 404, 410
  bipunctata, 75–77, 458, 459, 462
Odonata, as prey, 357, 383, 389, 444, 488
Olberg, G., 28, 68, 71, 143, 434
Olgia, 399, 400, 402, 410, 416
Opsidia, 468, 469
Orientation, 373–374, 439–440
Orthoptera, as prey, 444, 487; of *Microbembex,* 378, 379; of *Stizoides,* 128–131; of *Stizus,* 120–122, 124–125
Oscinella columbiana, 247
Otitidae, as cell scavengers, 247, 469, 471, 472
Oviposition, 448–450, 481, 492–495; in *Bembix,* 356

Pachygraphomyia spinosa, 261, 469
Pachyophthalmus, 52, 469
Parallelism, 417, 419–420, 475
Paralysis of prey, 356, 446–447
Parasites. *See* Natural enemies
Parker, J. B.: on *Bembix,* 298–308; on *Bicyrtes,* 146–156, 161–167; on classification of sand wasps, 144, 176; on *Microbembex,* 363–368, 372–388
Parnopes, 469; *concinnus,* 205, 344; *chrysoprasinus,* 307–308, 314–315; *edwardsii,* 219, 288, 315, 349; *fulvicornis,* 387–388, 390
Pate, V. S. L.: on classification of sand wasps, 4, 31, 83, 90, 92; on *Gorytes,* 67; on *Hoplisoides,* 40, 55; on *Ochleroptera,* 74, 75; on *Psammaletes,* 69
Pecten, 9, 434–435, 483, 496
Phalangida, as prey, 379
Phoridae, as cell scavengers, 241, 287, 469, 471, 472
Phrosinella, 468, 469
  fulvicornis, 28, 64, 468
Phylogeny of Nyssoninae, 416, 481
Physocephala texana, 287, 315, 349
Poulton, E. B., 173, 265
Powell, J. A., 46–49
Precoxal ridge, 6
Prey carriage, 447–448, 481, 489

Prey selection, 15, 442–445, 486–489, 504
Prey stealing, 426–427, 446
Primitive features: of behavior, 477–479; of structure, 405–406, 408–409
Prionyx, 128
Provisioning, 14, 450; delayed, 489, 491; mass, 489; progressive, 481, 489–492
Psammaecius, 33, 68, 70, 399, 444
  punctulatus, 70
Psammaletes, 33, 68, 399, 444
  pechumani, 69
Pseudosting, 11
Psocoptera, as prey, 380, 382
Psyllidae, as prey, 444; of *Bembecinus,* 141–142; of *Glenostictia,* 201; of *Ochleroptera,* 76–77; of *Microbembex,* 380, 382
Pygidium, 10, 409

Rau, P.: on *Alysson,* 18, 21–26; on *Bembix,* 298–308; on *Bicyrtes,* 146, 149–155, 167; on *Hoplisoides,* 49; on *Microbembex,* 363, 372, 384; on *Sphecius,* 94–95; on *Stizoides,* 128–130
Reinhard, E. G., 49–53, 93–107, 467
Rensch, B., 501
Rhipiphoridae, as parasites, 126, 308, 469, 471–472
Richards, O. W.: on *Bembecinus,* 137, 141; on *Bicyrtes,* 173; on *Microbembex,* 388–390; on *Stictia,* 243–247
Riley, C. V., 93, 95, 100–110, 467
Ristich, S. S., 85
Ritualization, 354, 484
Rodeck, H., 137, 141
Rohwer, S. A., 3, 396
Rubrica: cocoons, 463, 465; distribution, 399; ethology, 258–264, 443, 444; natural enemies, 469; phylogeny, 416, 481; structure, 257–258, 403, 405
  denticornis, 261–262, 266; *gravida,* 262–264, 414, 463; *surinamensis,* 258–261, 265–266, 358, 463

Savin, W. M., 98, 100, 105
Scavenger wasps, 374–383, 391
Scrobal groove, 6–7, 406, 408
Scudder, S. H., 393
Selman, 399, 403, 405, 416
Senotainia, 205, 337, 468–470, 485
  inyoensis, 349; *opiparis,* 297; *rubriventris,* 159, 184, 287, 297; *rufiventris,* 387, 470; *trilineata* as predator on *Bembix,* 287, 297, 308, 317, 334, on *Bicyrtes,* 159–160, 167, on *Hapalomellinus,* 72, on *Hoplisoides,* 42, 49, on *Sphecius,* 109, on *Stictia,* 243; *vigilans,* 159–160, 167, 315
Shuckard, W. E., 85
Signum, 6–7
Simpson, G. G., 419, 486
Size, of body, 417, 501; of colonies, 425–426

Sleeping behavior, 350, 429–431
Smicromyrme viduata, 143
Smith, H. E., 128–131
Smith, M. R., 146–150, 153–159
Solpugida, as prey, 389
Sowbugs, as prey, 383
Sphecius: distribution, 399; ethology, 91–111, 443, 444; natural enemies, 469; phylogeny, 416, 480, 481; structure, 91–93, 402, 404
  convallis, 110; grandidieri, 111; grandis, 110; milleri, 111; nigricornis, 111; speciosus, 5–11, 91–110, 459, 462; spectabilis, 110
Sphex, 131, 354, 437
Spiders, as prey, 375, 379, 382
Spur: of nest, 13–14, 351; of tibia, 9, 92, 93, 144
Steiner, A., 494, 496, 499
Steniolia, 176; cocoons, 461, 463; distribution, 399, 401; ethology, 207–221, 428, 443, 444; larvae, 205–206, 404; natural enemies, 469; phylogeny, 416, 481, 487; structure, 205–207, 403, 405
  duplicata, 207, 209, 216–218; elegans, 207, 209, 217; eremica, 207, 217; longirostra, 207, 216–217; nigripes, 207, 216–217, 461; obliqua, 206, 219, 463; scolopacea albicantia, 207, 209; tibialis, 207, 217
Stictia: cocoons, 460, 463; distribution, 399, 401; ethology, 223–253, 443, 444; larvae, 223, 404; natural enemies, 469; phylogeny, 416, 481, 487; structure, 223, 403, 405
  carolina, 223–243, 414, 463; heros, 233, 252–253, 266; pantherina, 253; punctata, 253; signata, 233, 243–247; vivida, 247–252, 339, 460, 463
Stictiella, 176; distribution, 399; ethology, 179–184, 443, 444; larvae, 177, 404; natural enemies, 469; phylogeny, 416, 481; structure, 177–178, 403, 405
  callista, 179, 183; emarginata, 179, 183; evansi, 179–184; formosa, 178–184, 414; pulchella, 179–184; serrata, 179–184
Stinkbugs. See Heteroptera
Stizini, 4, 116, 401, 413, 474–476
Stizobembex, 412–413
Stizoides: distribution, 399, 401; ethology, 127–132, 453–455; phylogeny, 416, 481, 487; structure, 127, 403
  crassicornis, 131; tridentatus, 131; unicinctus, 127–131
Stizus: distribution, 399, 401; ethology, 118–127, 443, 444; natural enemies, 469; phylogeny, 416, 481, 487; structure, 117–118, 403, 404
  atrox, 126; brevipennis, 117, 125; chrysorrhoeus, 126; dewitzi, 126; distinguendus, 126; fasciatus, 124–125; imperialis, 125; marshalli, 126; pulcherrimus, 118–123, 455, 461, 462; ruficornis, 126; texanus, 125
Stoehr, L. M., 363–369, 372–382
Strandtmann, R. W., 30, 75, 136–137, 141
Structure, 5–12, 401–420; derived, 405–406, 408–409; primitive, 405–406, 408–409; trends in evolution of, 415–419

Tachytes, 89, 90
Taxigramma heteroneura, 219
Telford, A. D., 388
Territoriality, 432–433; in Sphecius, 95–97; in Stictia, 228, 247–249
Thompson, W. R., 495
Thorpe, W. H., 350, 374, 499
Threshold selection, 491, 494, 497, 504
Timulla leona, 65
Treehoppers. See Membracidae
Trichogorytes, 399
Trichoptera, as prey, 380–381
Trichostictia, 222, 399, 403, 405, 416
Tsuneki, K.: on classification of sand wasps, 70, 137; on cocoon spinning, 455–457; on ethology of Bembix, 1, 14, 349–358, 427, 429, 441; on ethology of Stizus, 118–123

Vacuum reaction, 495
Variation in behavior, 496–499
Velvet ants. See Mutillidae
Vesey-Fitzgerald, D.: on Bembecinus, 141; on Bicyrtes, 173–174; on Microbembex, 363, 379–381; on Rubrica, 258–259, 261
Villa atrata, 349; melasoma, 288

Waddington, C. H., 496–497
Wecker, S. C., 499
Wheeler, W. M., 85, 132, 451, 455, 491
Whitman, C. O., 495
Williams, F. X., 68, 70, 125, 127–131
Willink, A.: on classification of sand wasps, 132, 144, 258, 267; on ethology, 135, 260, 390
Wing elevation, 439
Wing venation, 9
Wolcott, G., 243, 246
Wolf, H., 85, 86

Xenosphex, 399
Xerostictia, 205, 399, 403, 405, 416

Yasumatsu, K., 29, 354
Yoshimoto, C. M., 225

Zanysson, 83, 89–90, 416; tonto, 89
Zyzzyx: distribution, 399, 400; ethology, 254–257, 443, 444; phylogeny, 416, 481, 487; structure, 222, 254–255, 403, 405
  chilensis, 14, 254–257, 414

DATE DUE	
NOV 17 1981	
DEC 03 1981	
NOV 26 1986	
DEC 7 1987	
OCT 08 1992	
NOV 11 1992	
APR 19 1994	
APR 09 1995	
APR 06	
APR 11 2000	
APR 19 2000	
MAR 12 2005	
MAR 30 2005	